Leishmania

Current Biology and Control

Edited by

Subrata Adak

Structural Biology & Bio-Informatics Division
CSIR - Indian Institute of Chemical Biology
Kolkata
India

and

Rupak Datta

Department of Biological Sciences
Indian Institute of Science Education and Research Kolkata
Mohanpur Campus
Mohanpur
India

Caister Academic Press

Caister Academic Press
Norfolk, UK

www.caister.com

British Library Cataloguing-in-Publication Data
A catalogue record for this book is available from the British Library

ISBN: 978-1-908230-52-2 (hardback)
ISBN: 978-1-908230-53-9 (ebook)

Cover design adapted from Figure 2.2

Contents

Contributors

Subrata Adak
Division of Structural Biology and Bio-informatics
CSIR-Indian Institute of Chemical Biology
Kolkata
India

adaks@iicb.res.in

Nahid Ali
Infectious Diseases and Immunology Division
Indian Institute of Chemical Biology
Jadavpur
Kolkata
West Bengal
India

nali@iicb.res.in

Saswat Kumar Bal
Special Centre for Molecular Medicine
Jawahralal Nehru University
New Delhi
India

saswat255@gmail.com

Sudarshana Basu
RNA Biology Research Laboratory
Molecular and Human Genetics Division
CSIR-Indian Institute of Chemical Biology
Kolkata
India

sudarshana_basu@yahoo.com

Kavita Bharati
Special Centre for Molecular Medicine
Jawahralal Nehru University
New Delhi
India

kavitabharati@gmail.com

Vikash Bhardwaj
Special Centre for Molecular Medicine
Jawahralal Nehru University
New Delhi
India

vikashbhardwaj@gmail.com

Suvendra N. Bhattacharyya
RNA Biology Research Laboratory
Molecular and Human Genetics Division
CSIR-Indian Institute of Chemical Biology
Kolkata
India

suvendra@iicb.res.in

Jaya Chakravarty
Department of Medicine
Institute of Medical Sciences
Banaras Hindu University
Varanasi
India

tapadar@gmail.com

Sayan Chowdhury
Infectious Diseases and Immunology Division
CSIR-Indian Institute of Chemical Biology
Jadavpur
Kolkata
India

mail2sayanchowdhury@gmail.com

Joachim Clos
Bernhard Nocht Institute for Tropical Medicine
Bernhard-Nocht-Strasse 74
Hamburg
Germany

clos@bnitm.de

Pijush K. Das
Infectious Diseases and Immunology Division
CSIR-Indian Institute of Chemical Biology
Kolkata
India

pijushdas@iicb.res.in

June Ghosh
RNA Biology Research Laboratory
Molecular and Human Genetics Division
CSIR-Indian Institute of Chemical Biology
Kolkata
India

june.ghosh@gmail.com

Soumya kanti Ghosh
National Centre for Cell Science
Ganeshkhind
Pune
India

soumyakanti21ghosh@gmail.com

Antje Hombach
Bernhard Nocht Institute for Tropical Medicine
Bernhard-Nocht-Strasse 74
Hamburg
Germany

hombach@bnitm.de

Kalavati M. Lalsare
National Centre for Cell Science
Ganeshkhind
Pune
India

swatilalsare98@gmail.com

Hemanta K. Majumder
Infectious Diseases and Immunology Division
CSIR-Indian Institute of Chemical Biology
Jadavpur
Kolkata
India

hkmajumder@iicb.res.in

Sunil Martin
Cancer Immunology Programme
Dana-Farber/Harvard Cancer Centre
Harvard Medical School
Boston, MA
USA

sunil_martin@dfci.harvard.edu

Radhika Mathur
Cell Death and Differentiation Research
National Institute of Immunology
New Delhi
India

radhikamathur@nii.ac.in

Chinmay K. Mukhopadhyay
Special Centre for Molecular Medicine
Jawahralal Nehru University
New Delhi
India

ckm2300@mail.jnu.ac.in

Gayathri Natarajan
Department of Microbiology
Ohio State University
Columbus, OH
USA

natarajan.40@osu.edu

Steve Oghumu
Department of Oral Biology
College of Dentistry
Ohio State University; and
Department of Pathology
Ohio State University Medical Center
Columbus, OH
USA

oghumu.1@osu.edu

Chiranjib Pal
Cellular Immunology and Experimental Therapeutics Laboratory
Department of Zoology
West Bengal State University
Barasat
West Bengal
India

cpcu.immunology@gmail.com

Swati Pal
Developmental and Stem Cell Biology Program
The Hospital for Sick Children
Toronto Medical Discovery Tower
Toronto, ON
Canada

swatipal_49@yahoo.co.in

Abdus Sabur
Infectious Diseases and Immunology Division
Indian Institute of Chemical Biology
Jadavpur
Kolkata
West Bengal
India

absab2007@gmail.com

Bhaskar Saha
National Centre for Cell Science
Ganeshkhind
Pune
India

sahab@nccs.res.in

Shalini Saini
Special Centre for Molecular Medicine
Jawahralal Nehru University
New Delhi
India

shalini_msbio@yahoo.co.in

Abhay R. Satoskar
Department of Pathology
Ohio State University Medical Center; and
Department of Microbiology
Ohio State University
Columbus, OH
USA

abhay.satoskar@osumc.edu

Chandrima Shaha
Cell Death and Differentiation Research
National Institute of Immunology
New Delhi
India

cshaha@nii.ac.in

Supriya Srivastav
Infectious Diseases and Immunology Division
CSIR-Indian Institute of Chemical Biology
Kolkata
India

supriya114@gmail.com

James Stock
Department of Microbiology
Ohio State University
Columbus, OH
USA

stock.75@osu.edu

Shyam Sundar
Department of Medicine
Institute of Medical Sciences
Banaras Hindu University
Varanasi
India

drshyamsundar@hotmail.com

Cesar Terrazzas
Department of Pathology
Ohio State University Medical Center
Columbus, OH
USA

ces_terra@hotmail.com

Anindita Ukil
Department of Biochemistry
Calcutta University
Kolkata
India

aninditau@yahoo.com

Sanjay Varikuti
Department of Pathology
Ohio State University Medical Center
Columbus, OH
USA

sanjay.varikuti@osumc.edu

Current Books of Interest

Preface

With an estimated 1.3 million new cases and causing more than 20,000 deaths every year, leishmaniasis continues to be a menace in countries across the globe. Unavailability of a vaccine and limited chemotherapeutic options coupled with rapid emergence of drug-resistant *Leishmania* strains challenges researchers to come up with novel strategies to combat the disease. With this common goal various research groups are actively investigating different aspects of the parasite biology, its interaction with the host immune system, the disease pathogenesis, drug development and molecular mechanism of drug resistance. This book aims to cover these broad areas of modern *Leishmania* research and highlight insightful discoveries in the field. We hope that this book will not only be widely read and appreciated by the *Leishmania* community but will also benefit researchers working in related fields.

Subrata Adak and Rupak Datta

Modulation of Host Cell miRNA Expression During *Leishmania* Infection and Emergence of miRNA as a New Therapeutic Molecule

1

Suvendra N. Bhattacharyya, June Ghosh and Sudarshana Basu

Abstract

The kinetoplastid protozoan *Leishmania donovani* infects liver and spleen of the mammalian hosts and resides inside the tissue macrophages. The pathogen resides within the parasitophorous vacuoles, the specialized subcellular structures in the infected macrophage cells, and affects gene expression in infected and also the non-parasitized neighbouring cells. Expression of miRNAs, the tiny regulatory RNA of eukaryotic cells, is also altered in *Leishmania*-infected macrophages and also in the hepatocytes of infected mouse liver. Restoration of miR-122, a hepatic miRNA that regulates serum cholesterol and is reduced upon *Leishmania* infection in liver, is associated with an increase in serum cholesterol level and lowering of liver parasite load. In this chapter we discuss the importance of host cell miRNA alteration and its mechanism in the context of visceral leishmaniasis.

miRNA-mediated gene regulation in metazoan animals

miRNAs constitute a family of small 21–25 nucleotide non-coding RNAs that regulate gene expression in plants and animals by base pairing to their target genes. They function as antisense regulators of target mRNAs and thereby negatively regulate gene expression post-transcriptionally. Originally discovered in *Caenorhabditis elegans*, miRNAs show tissue-specific expression pattern during different phases of animal development. In the early 1990s, temporal expression of the heterochronic gene *lin-14* was reported to be post-transcriptionally regulated by *lin-4* transcripts (Lee *et al.*, 1993; Wightman *et al.*, 1993). This regulation was found to be mediated by the *lin-14* 3′UTR (Wightman *et al.*, 1993) that has complementary binding sites to *lin-4* transcripts (Lee *et al.*, 1993). Later it was found that the transcript of *let-7*, another gene in the *C. elegans* heterochronic pathway, acts to promote the transition from late larval to adult cell fates by base pairing to the 3′UTR of *lin-14*, *lin-28*, *lin-41*, *lin-42* and *daf-12* (Reinhart *et al.*, 2000; Slack *et al.*, 2000). Later homologues of *let-7* were also identified in human and fly genomes and also in 11 other bilateral animals (Pasquinelli *et al.*, 2000). Since then there have been many reports on discovery and functional analysis of thousands of miRNAs present in different organism.

Certain miRNA genes are located within introns of other protein-coding and non-coding genes; others are clustered in polycistronic transcripts. The location of the miRNA gene dictates the transcriptional regulation mechanism of that miRNA as these genes are expressed at specific organs and at specific points of time (Lagos-Quintana *et al.*, 2003). Nascent miRNA transcripts (pri-miRNA) are first processed by the nuclear RNase Drosha into ~70 nucleotide precursors (pre-miRNA) which are then exported from the nucleus into the cytoplasm by Exportin 5. Pre-miRNAs are further processed by the RNase Dicer to generate ~21–25 nucleotide mature miRNAs (Lee *et al.*, 2003; Lee *et al.*, 2002; Lund *et al.*, 2004) which are incorporated into ribonucleoprotein complexes called miRISCs. Argonaute (AGO) and GW182 family proteins are the predominant protein components of miRISCs.

The predominant mechanism of regulation by miRNAs in animals is translational repression of target genes. miRNAs bind to the 3′UTR of their target mRNAs and negatively regulate their translation. Lin-4 causes inhibition of translation of lin-14 without reduction in mRNA levels or a shift in polysomes, thereby leading to the conclusion that miRNAs inhibit mRNA translation at the elongation step (Lee *et al.*, 1993; Olsen and Ambros, 1999; Wightman *et al.*, 1993). Other results in HeLa (Humphreys *et al.*, 2005; Pillai *et al.*, 2005), Huh7 (Bhattacharyya *et al.*, 2006), HEK293T (Huang *et al.*, 2007), and *C. elegans* (Ding and Grosshans, 2009) point to defects in the control of translational initiation and mRNA stability. These studies raised the possibility that miRNAs interfered with recruitment of translation initiation factors eIF4E to the 5′ cap structure of miRNA targeted mRNAs. The conclusions are supported by other studies using cell-free systems (Mathonnet *et al.*, 2007; Thermann and Hentze, 2007; Wakiyama *et al.*, 2007). miRNAs were also found to inhibit ribosome initiation complex formation either by inhibiting 40S ribosomal subunit recruitment (Thermann and Hentze, 2007) or the joining of the 60S subunit (Chendrimada *et al.*, 2007; Wang *et al.*, 2008). Kiriakidou in 2007 observed that AGO2 binds to the cap structure and suggested that this binding competes with eIF4E and results in inhibition of translation initiation (Kiriakidou *et al.*, 2007).

In plants, most miRNAs studied until now regulate gene expression by destruction of their target mRNAs (Hake, 2003). These miRNAs show perfect base pairing to their corresponding targets, and their complementary sites are located throughout the transcribed regions of the mRNA, not just the 3′UTR. miRNAs also cause mRNA target degradation in human cells, *C. elegans*, *Drosophila* S2 cells and zebra fish (Bagga *et al.*, 2005; Behm-Ansmant *et al.*, 2006; Giraldez *et al.*, 2006; Schmitter *et al.*, 2006). miRNAs destabilize target mRNAs by removal of the poly(A) tail using 3′–5′ exoribonucleases (CCR4–NOT1 complex, poly(A)-specific ribonuclease (PARN) and poly(A) nuclease (PAN); Meyer *et al.*, 2004; Yamashita *et al.*, 2005). This is followed by decapping by the DCP1–DCP2 complex and 5′–3′ exonucleolytic digestion by Xrn 1 (Coller and Parker, 2004).

In mammals, miRNAs are predicted to control the activity of more than 60% of all protein-coding genes (Friedman *et al.*, 2009) and participate in the regulation of almost every cellular process investigated to date. These findings corroborate bioinformatic predictions which indicate that mammalian miRNAs can regulate ~30% of all protein-coding genes. miRNA targeted pathways in animals include pathways in stem cells, neurons, homeostasis in various metabolic organs, lipid metabolism, glucose homeostasis, during ageing and in immune function (O'Connell *et al.*, 2010; Sun and Lai, 2013). Changes in their expression are observed in, and might underlie, various human pathologies including cancer (Esquela-Kerscher and Slack, 2006; Lujambio and Lowe, 2012). Analysis of miRNA expression in cancer samples and their corresponding normal tissues have revealed that miRNA 'signature' profiles are different in tumoral and normal cells, and are associated with the prognosis and progression of cancer. miRNA genes can act as either tumour suppressors or oncogenes depending on the fact that whether their loss of function or overexpression contribute to the malignant transformation of a normal cell. miRNA genes have been found to be frequently located in cancer-associated genomic regions, such as fragile sites, minimal regions of loss of heterozygosity and minimal regions of amplification (Calin *et al.*, 2004). Genomic deletion, mutation, epigenetic silencing and/or miRNA processing alterations may lead to loss of function of a miRNA, whereas amplification of miRNA gene containing genomic regions may contribute to increased cellular proliferation, inhibition of apoptosis and tumour progression. Factors that are required for the biogenesis of miRNAs have also been associated with various cancers and might themselves function as tumour suppressors and oncogenes.

miRNA as regulator of cholesterol metabolism in mammals

miRNA-mediated regulation of metabolic processes in higher animals

The ability to control the rates of metabolic processes in response to changes in the internal or external environment is indispensable for all living cells. Mechanisms that are essential for metabolic control and maintenance of homeostasis are complex processes that involve transcriptional, translational, post-translational, and allosteric regulation of gene expression primarily happening in all higher metazoan animals.

Doubling of total body triacylglycerides upon loss of miR-14 expression was discovered as the first evidence of participation of miRNAs in metabolic control by Xu *et al.* (2003) by using a forward genetic screen in the fruit fly *Drosophila melanogaster*. Teleman and Cohen later identified a role for miR-278 in energy homeostasis in Drosophila. Flies with homozygous mutations for miR-278 that prominently expressed in fat body of flies have a smaller fat body and reduced ratio of total body triglycerides to total protein. Importantly, the phenotype could be rescued by miR-278 expression in flies (Teleman *et al.*, 2006). Sokol and Ambros (2005) demonstrated that miR-1, a highly conserved miRNAs in the animals, is also essential for proper skeletal muscle function in the fruit fly and loss-of-function of muscle-specific miR-1 in *Drosophila* results in small immobilized larvae with severely deformed musculature. In another invertebrate organism, *C. elegans*, the miRNA lin-4 is critical in development and was shown to play important role regulating life span of this organism (Boehm and Slack, 2005).

miRNAs plays a role in hormone secretion control in vertebrates. An unbiased cloning approach of small RNAs from the pancreatic β-cell line MIN6 (Poy *et al.*, 2004) identified 11 novel miRNA sequences including two pancreatic-islet-enriched miRNAs, miR-375 and miR-376. While miR-375 decreases insulin secretion, inhibition of miR-375 function enhances insulin release from pancreatic β-cells. In muscle, miRNAs have been implicated in controlling myoblast differentiation to mature myocytes. MiR-1 and miR-133 are transcribed as a common polycistronic transcript and expressed in a tissue-specific manner during muscle development (Kim *et al.*, 2006). However, they have opposite roles in modulating skeletal muscle proliferation and differentiation. In *in vitro* cultured myoblasts miR-1 promotes myogenesis by targeting histone deacetylase 4 (HDAC4), a transcriptional repressor of muscle specific gene expression. In contrast, miR-133 enhances myoblast proliferation by repressing the serum response factor (SRF). A predominant and tissue specific miRNA, miR-122, has been shown to regulate cholesterol and lipid homeostasis in liver. miR-122 is abundantly expressed in human and rodent liver, with estimates ranging from 50,000 to 80,000 copies of mature miRNA per cell. Using antagomirs, a new class of miRNA inhibitors, Krützfeldt *et al.* (2005) studied the effect of miR-122 on glucose and lipid metabolism in mice. Antagomir-122 reduces miR-122 in the liver and led to a significant decrease in plasma cholesterol levels even after 4 days of antagomir treatment.

miRNA as regulator of lipid metabolism

To date several miRNAs including miR-122, miR-370, miR-378/378*, miR-335, miR-125a-5p, and miR-33 have been identified as the regulator of lipid metabolism. miR-122 regulate liver metabolism and is utilized by hepatitis C virus for its replication and propagation of infection. Down-regulation of miR-122 has been reported in hepatocellular carcinoma (HCC) (Elmen *et al.*, 2008a; Elmen *et al.*, 2008b; Esau *et al.*, 2006; Lanford *et al.*, 2010). Antisense miR-122 treatment resulted in increased expression of hundreds of genes that are normally repressed in hepatocytes, suggesting a function of miR-122 to maintain the liver phenotype. In addition, several genes involved in fatty acid synthesis and oxidation were altered in mice treated with anti-miR-122 including FAS, ACC1, and ACC2 (Elmen *et al.*, 2008a,b; Esau *et al.*, 2006). Furthermore, miR-122 inhibition in the liver of mice fed a high-fat diet reduced hepatic steatosis, which resulted a reduction in cholesterol synthesis and stimulation of fatty acid oxidation (Esau *et al.*, 2006).

Another miRNA, miR-370, was also shown to have effects on lipid metabolism. miR-370 targets carnitine palmitoyltransferase (Cpt1a), a mitochondrial enzyme that mediates the transport of long-chain fatty acids across the membrane by binding them to carnitine to reduce fatty acid oxidation (Iliopoulos *et al.*, 2010). miR-378/378*, an intronic miRNA located within the peroxisome proliferator-activated receptor gamma coactivator-1 alpha (PGC1α) genomic sequence, also regulate lipid metabolism (Gerin *et al.*, 2010). miR-143, miR-2, and miR-335 others miRNAs are found to be associated with lipid metabolism or adipocyte differentiation (Esau *et al.*, 2004). miR-143 regulate ERK5 to control adipocyte differentiation (Esau *et al.*, 2004; Kajimoto *et al.*, 2006; Lin *et al.*, 2009), while, miR-27 inhibits the expression of PPARγ and C/EBPa and is down-regulated during adipocyte differentiation (Lin *et al.*, 2009). miR-335 is up-regulated in response to lipid loading and there is increased expression of this miRNA in the liver and adipose tissue of obese mice (Nakanishi *et al.*, 2009). However, the exact role of miR-335 in regulating lipid metabolism and adipogenesis remains unknown. miR-33a, an intronic miRNA located within intron 16 of human SREBF-2, controls genes involved in cholesterol uptake and synthesis (Marquart *et al.*, 2010; Najafi-Shoushtari *et al.*, 2010; Rayner *et al.*, 2010).

Cholesterol regulates pathogenesis of *Leishmania donovani*

Serum cholesterol level changes during pathogenesis of *Leishmania*

Importance of cholesterol in visceral leishmaniasis (VL) is well established. Serum cholesterol content is found to be compromised in VL patients with progression of disease and, more interestingly, bears an inverse correlation to splenic parasite load (Ghosh *et al.*, 2011). Membrane cholesterol content was monitored and found to be lowered in the infected condition and resulted in membrane raft disruption (Chakraborty *et al.*, 2005). Membrane lipid raft, which acts as an important pathogen portal, is a membrane microdomain consisting of cholesterol and sphingolipids (Rosenberger *et al.*, 2000). Disruption of lipid rafts leads to defective antigen presentation and poor immune response in infected individuals (Chakraborty *et al.*, 2005). This defect can be corrected by liposomal delivery of cholesterol (Chakraborty *et al.*, 2005) that rectifies the IFN-γR1 and R2 dimerization problem on cholesterol-reduced membrane (Sen *et al.*, 2011). In the experimental hamster model of VL, liposomal cholesterol administration rescues the infected animals from organ parasite burden (Banerjee *et al.*, 2009). Membrane cholesterol level was restored with concomitant increase in membrane rigidity. Further studies in hypercholesterolaemic mice showed that cholesterol level is directly related to the severity and progression of the disease (Ghosh *et al.*, 2012).

Importance of cholesterol in pathology and immunity development

The role of membrane cholesterol in the T-cell–antigen-presenting cell (APC) interaction has been well studied in various systems. Cholesterol depletion results in inhibition of phytohaemagglutinin-dependent lymphocyte cytotoxicity against allogeneic target cells, and the opposite effect were observed after cholesterol enrichment (Dabrowski *et al.*, 1980). Cholesterol treatment enhances the antigen-presenting capacity of monocytes by promoting phagocytosis of *Helicobacter pylori* by APCs, and enhances an antigen-specific T-cell response during bacterial challenge, possibly as a result of up-regulation in the expression of major histocompatibility complex (MHC) class II (Hughes and Nei, 1992) leading to a T-cell-dependent reduction in the number of bacteria in the stomach (Wunder *et al.*, 2006). Treatment of APCs with nystatin or methyl cyclodextrin, which are known quenchers of cholesterol from the membrane, reduces the antigen-presenting ability without affecting the cell surface expression of class II molecules (Anderson *et al.*, 2000). Thus, cholesterol may play a decisive role in the cell-mediated immune response of the host. Membrane cholesterol affects membrane embedded receptors in terms of their affinity state, binding capacity, and signal transduction (Burger *et al.*, 2000). Cholesterol is also necessary for raft assembly (Simons and Toomre, 2000). Lipid rafts are intimately involved in controlling the dynamics of immune synapse formation and are essential in T-cell receptor-mediated signal transduction (Razzaq *et al.*, 2004).

Cholesterol as therapeutic agent

Therapy with cholesterol administration is in clinical use either directly or as an adjunct to a primary drug-based treatment. Anti-tubercular drug along with a cholesterol-rich diet together increased the rate of sterilization of sputum in patients with pulmonary tuberculosis (Perez-Guzman *et al.*, 2005). Smith–Lemli–Opitz syndrome patients treated with a pure cholesterol suspension orally showed improvement in symptoms associated with the disease. During its intracellular life cycle *Leishmania donovani* disrupts the membrane rafts of infected macrophages by increasing membrane fluidity. This accelerated the idea that cholesterol-rich liposomes may have a therapeutic role in treatment of leishmaniasis. The increase in fluidity corrected either by exposing the infected macrophage to a temperature lower than the phase transition temperature or by liposomal delivery of excess cholesterol, resulted correction of antigen presentation problem in infected macrophage (Chakraborty *et al.*, 2005). This and other findings that cholesterol could act as a potential therapeutic agent against *Leishmania* infection (Pucadyil and Chattopadhyay, 2007), brought the research focus on the therapeutic efficacy of cholesterol-rich liposomes in *Leishmania donovani*-infected hamsters.

Leishmania-mediated changes in cholesterol production in mammals

Leishmaniasis and cholesterol metabolism

Altered lipid metabolism is a characteristic feature of human visceral leishmaniasis. Various case reports of abnormal lipid profile with severe hypocholesterolaemia, decreased serum lipoprotein concentration and marked hypertriglyceridaemia in clinically diagnosed patients were reported early (Lal *et al.*, 2007; Liberopoulos *et al.*, 2002). According to a recent report an inverse correlation of serum cholesterol to parasite count was found in splenic aspirate in patients from endemic regions of Bihar, India (Ghosh *et al.*, 2011). Determination of the lipid profile in experimental mice model of progressive *Leishmania donovani* infection confirmed a reduction in total cholesterol, LDL, HDL with gradual infection. Serum albumin and glucose remains unaltered, specifically highlighting the defect in lipid metabolism (Ghosh *et al.*, 2013).

Gene expression changes with infection – relation of lipid metabolic genes with level of infection

Leishmania infection initiates alteration in specific gene expression patterns. Whole genome microarray analysis of RNA isolated from liver of the normal and infected animal liver showed differential gene expression in infected condition. Cluster analysis studies of differentially expressed genes and further scrutiny of these clusters revealed that a major group of genes showing down-regulated expression are related to lipid metabolism (477 showed changes among 599 genes analysed). The gene expression data suggested that a major alteration in lipid metabolism occurs in infected mouse liver. This gene group with altered expression includes HMGCR (hydroxyl-3-methylglutaryl coenzyme A reductase); CES1 (carboxyesterase), a hydrolyase which hydrolyses long-chain fatty acid esters and thioesters and is necessary for cholesterol esterification; SCD1 (stearyl-CoA-9-desaturase), an oxidoreductase which participates in PPAR signalling and PUFA synthesis; CPT1B (carnitine-O-palmitoyltransferase 1muscle isoform), a rate-determining enzyme of long-chain fatty acid β-oxidation in muscle mitochondria; ACAT2 (acetyl CoA cholesterol-acyl transferase 2), required for the esterification and storage of cholesterol; CYP7B1(25-hydroxycholesterol-7alpha-hydroxylase), a member of the mono-oxygenase cytochrome P450 superfamily, which catalyses the first step of catabolism of cholesterol to bile acid; and SMGS2 (phosphatidylcholine:ceramide choline phosphotransferase 2), an enzyme catalysing the conversion of phosphatidylcholine and ceramide to sphingomyelin and diacylglycerol and vice versa (Ghosh *et al.*, 2013).

miRNA expression affected by parasite

miRNAs altered in macrophages and other immune cells during infection

During infection, the innate immune response is the primary initiator of inflammatory responses. It involves various cells of the immune system, namely monocyte/macrophages, granulocytes and dendritic cells, and is initiated mainly by the signalling cascades upon Toll-like receptor (TLR) activation. This results in activation of the interferon (IFN) response and NF-κB, resulting in inflammation.

The macrophage inflammatory response to infection involves up-regulation of several miRNAs such as miR-155, miR-146, miR-147, miR-21 and miR-9 (O'Connell *et al.*, 2010, 2012). Various TLR ligands and the proinflammatory cytokines TNF and interferon β induce miR-155 expression in mouse bone marrow-derived macrophages. miR-155 targets and down-regulates the negative regulators SHIP1 and SOCS1, thus leading to increased activation of AKT and IFN response genes. Mice deficient in miR-155 have decreased immune responses, whereas miR-155 overexpression leads to chronic inflammation and haematopoietic cancers. miR-146a, a negative regulator of the immune response, is induced by NF-κB and prevents further production of TLR downstream signal transducer proteins by inhibiting expression of TRAF6 and IRAK1. miR-9 represses NF-κB subunit 1 thereby maintaining a constant level of the protein during TLR4-mediated activation of monocytes and neutrophils. miR-21, also induced by NF-κB in TLR4 induction of macrophages, targets PDCD4 – a tumour suppressor proinflammatory protein which activates NF-κB. This leads to down-regulation of NF-κB signalling and a switch to anti-inflammatory response. Other miRNAs involved in immune responses are miR-125b and let-7. Both are down-regulated by LPS stimulation of macrophages: miR-125b targets TNF-α mRNA, whereas let-7 targets IL-6 mRNA.

miRNAs are also involved in the regulation of T-cell development and function. miR-181a contributes to proper clonal selection of immature T-lymphocytes during thymic development. miRNAs in the miR-17 to miR-92 cluster are involved in managing T-cell survival by repressing pro-apoptotic BIM and PTEN. miR-155 deficiency in T-cells results in a Th2 bias in T-cell differentiation. MiR-155 is involved in producing Th1 and Th17 cells during autoimmune inflammation. miR-326 is also a promoter of Th17 responses. miR-182, up-regulated in T-cells by IL-2, is important in the clonal expansion of T-cells. miR-214 is also up-regulated during this process. miR-146a is also expressed in T-cells during infection and its absence leads to an unchecked Th1 response. miR-155 is also induced during LPS activation of B-cells in mice. Thus, miRNAs are seen to play an important role in various cells of the immune system during infection.

miRNA changes during *Leishmania* infection

Regulation of gene expression is often governed by small RNA molecules. In this respect, the connection between miR-122 and lipid metabolism is very well documented (Girard *et al.*, 2008). Being the most abundant miRNA in liver, miR-122 has an indirect effect on lipid metabolism, although key lipid metabolic genes are not the direct targets of miR-22 but are indirectly regulated by miR-122. A more detailed analysis of the microarray data highlighted a very interesting finding. A cluster of lipid-metabolizing genes showing differential expression was found to be regulated by miR-122. Total RNA were isolated from the liver of normal and infected animals and miR-122 expression was monitored. A significant down-regulation of miR-122 with progressive infection was noted (Ghosh *et al.*, 2013).

Effect of miRNA alternation on pathogenesis of virus and bacteria

Apart from a few known examples of virus-mediated down-regulation of miRNA function in host cells, not much is known about the pathogen-mediated down-regulation of host miRNA machineries in mammalian cells. In the event of virus infection, miRNA inactivation can be achieved by expressing either virus-encoded target RNAs for specific miRNAs or by specific virus-encoded miRNA binding proteins (Cazalla and Steitz, 2010; Cazalla *et al.*, 2010; Libri *et al.*, 2012). *Helicobacter pylori*, a stomach-colonizing pathogenic bacterium, uses CagA protein to interfere with miR-372/373 cluster to arrest cell cycle progress in the AGS cell line (Belair *et al.*, 2011). In a more recent report, it has been shown that upon *Leishmania* infection there is a change in cellular miRNA profile. In their work Kelada *et al.* (2013) identified miR-182 and miR-10a as the key regulators of $CD4^{+}Foxp3^{+}$ regulatory T-cells (T-regs). Using human macrophage, Lemaire *et al.* (2013) identified miRNAs undergoing altered expression in *Leishmania*-infected cells. A comprehensive analysis of miRNA expression profiles in *L. major*-infected human primary macrophages of three healthy donors were assessed at different time points post infection (3–24 hours). These showed a change in expression of 64 out of 365 analysed miRNAs that were consistently deregulated upon infection with *L. major*.

Mechanism of miRNA alteration by *Leishmania*

miRNA biogenesis steps

In an earlier section of this chapter, it has been described how miRNA precursors (pri-miRNAs) are either transcribed from independent miRNA genes or are portions of introns of protein-coding mRNAs and mature miRNAs (miRNAs) are processed from the pri-miRNAs (Lee *et al.*, 2002). A single pri-miRNA often contains sequences for several different miRNAs. Many miRNA genes are transcribed as polycistronic transcripts clustering several miRNAs encoded by a single transcript. According to their genomic location, miRNAs can be grouped into several categories like intronic miRNAs in protein-coding genes; exonic miRNAs in non-coding genes or intronic miRNAs in non-coding genes (Rodriguez *et al.*, 2004). Pri-miRNAs are processed by the RNase III type endonucleases Drosha to pre-miRNAs and subsequently by Dicer to mature miRNA. Both Drosha and Dicer function in complexes with proteins having dsRNA-binding domains (dsRBDs).

Effect of infection on miRNA biogenesis

L. donovani infection reduces expression of miR-122 in mouse liver and alter lipid metabolism. Target RNA repressive ability of miR-122 is also compromised in hepatic cells after interaction with pathogens. Interestingly, pre-miR-122 expression increases after infection. pre-miR-122 accumulation with progressive infection was documented in infected mice that rises to 3-fold above normal level after 60 days post infection (Ghosh *et al.*, 2013). After pre-miRNAs are cleaved by the Dicer, the mature miRNA is loaded to Argonaute 2 to form the minimal RISC complex in mammalian cells (Bhattacharyya *et al.*, 2006; Liu and Uzonna, 2012). After formation of the miRNP, the mRNA targets are recognized through the 3′-UTR seed sequence and subsequently mRNA cleavage or translational repression takes place (Filipowicz *et al.*, 2008). A drastic reduction in Ago 2 bound miR-122 level (almost 6-fold) was observed in *L. donovani*-treated Huh7. This unique finding sheds some light on how *L. donovani* reduces miR-122 activity in hepatic cells after their exposure to the pathogens (Ghosh *et al.*, 2013).

Factors important for host cell miRNA alteration during infection

Leishmanial exosomes and factors present

Exosomes serve as communicators between mammalian cells (Bhatnagar and Schorey, 2007; Bhatnagar *et al.*, 2007). *Leishmania* secrete exosomes that act as modulators of host cell machineries necessary for parasite survival and infection (Silverman *et al.*, 2010). Exosomes, probably being the key player in cellular communication, may answer the most important question of how Küpffer cell-resident parasites could modulate the status of a molecule exclusively expressed in the hepatocytes. It can be hypothesized that exosomes released from infected Küpffer cells are taken up by the surrounding hepatocytes where they deliver various virulence factors necessary for miR-122 down-regulation. In experiments using anti-GP63 antibody or *o*-phenanthroline, a Zn-chelator, it has been shown that GP63 is the factor necessary for miR-122 down-regulation in *Leishmania*-interacting cells. Entry of the exosomes to target hepatocytes is Dynamin 2 dependent and internalization of GP63 in hepatic cells were confirmed both *in vivo* and also *in vitro*. Previous reports suggest that dynamin regulates the entry of HIV1 and cell–cell fusion (Lai *et al.*, 2011). Its role in the entry of a heparan sulfate-binding HRV8 variant is well known (Khan *et al.*, 2011). siRNA-based knockdown of dynamin 2 could relief the anti-miR-122 activity down-regulation in Huh7 cells upon interaction with *Leishmania*. Internalization of this delivery vehicle occurs possibly via an endosome-dependent mechanism evident by the proximal localization of gp63 with early endosomal markers (Ghosh *et al.*, 2013).

Exosomes as intercellular communicators

Discovery of exosomes dates back to the mid-1980s, when an observation was made that, during maturation, reticulocytes released or shed excess transferrin receptors in microvesicles (Pan *et al.*, 1985). This observation sparked two decades of research into released vesicles and led to the characterization of a new organelle, the exosomes. Exosomes are formed within endosomes or secretory lysosomes, by invagination of the limiting membrane, resulting in the formation of a multi-vesicular body (MVB) (Blott and Griffiths, 2002; Thery *et al.*, 2009). Exosomes are known to be released by numerous mammalian cell types, including cells of both the adaptive and innate immune systems, non-immune cells, and various tumours. Exosomes have been found in diverse human body fluids including urine and blood plasma and it has been suggested that they may be useful as disease biomarkers (Admyre *et al.*, 2007; Asea *et al.*, 2008; Pan *et al.*, 2005). Only three non-mammalian cell types have been shown to release bona fide exosomes: *C. elegans* (Liegeois *et al.*, 2006), the pathogenic fungus *Cryptococcus neoformans* (Rodrigues *et al.*, 2008) and *Leishmania* spp. (Silverman *et al.*, 2010).

Mechanism of exosome uptake by host cells

Silverman *et al.* (2010) reported exosome release as a general mechanism for protein secretion by *Leishmania*. These vesicles were seen to deliver cargo and communicate with macrophages. They showed exosome uptake by macrophages as early as two hours after infection. *L. major*-infected macrophages have been seen to have parasite derived GP63 in their cytosolic compartment and the inner leaflet of the plasma membrane (Gomez *et al.*, 2009). Naive macrophages incubated with *Leishmania* cell-free culture supernatants also show GP63 in these locations indicating transfer by secreted vesicles (Gomez *et al.*, 2009; Hassani *et al.*, 2011; Silverman *et al.*, 2010). *Leishmania* exosomes could be released from either recently inoculated promastigotes, by free amastigotes after cell ruptures, or both. The uptake of *Leishmania* exosomes by macrophages may be by surface binding, fusion with the plasma membrane, or by endocytosis. The infected macrophages release *Leishmania* exosomes into the extracellular environment which are again taken up by naive cells (Silverman *et al.*, 2010; Silverman and Reiner, 2011).

Proteins in *Leishmania* exosomes (GP63 and EF-1α) interfere with both the innate and adaptive immune systems in the host (Hosseini *et al.*, 2013; Silverman *et al.*, 2010; Silverman and Reiner, 2011). Wild-type exosomes attenuate interferon-γ induced proinflammatory cytokine production (TNF-α) by *Leishmania*-infected monocytes while conversely enhancing production of the anti-inflammatory cytokine IL-10. This effect is mediated by the activation of multiple host protein-tyrosine phosphatases (PTPs). Activation of these PTPs negatively regulates IFN-γ signalling, and this prevents effective proinflammatory response. These vesicle also dampen the immune response of monocyte-derived dendritic cells and $CD4^+$ T-lymphocytes. Exosomes from WT *L. donovani* failed to prime monocyte-derived dendritic cells to drive the differentiation of naïve CD4 T-cells into IFN-γ-producing Th1 cells. Thus, *Leishmania* exosomes cause immunosuppression in host cells and primes the cells for *Leishmania* invasion. They produce permissive environments for parasite infection *in vivo*. Exposure of resistant C57BL/6 mice to *L. donovani* exosomes resulted in higher parasite loads in the spleen after subsequent infection and a cytokine profile dominated by suppressor T-cells producing both IL-10 and IFN-γ (Silverman *et al.*, 2010)

miRNA as therapeutic agent

miRNA application in treating diseases

For the majority of *Leishmania* species, genetic and bioinformatic studies have not been successful in identification of AGO proteins or Dicer in their genome. This suggests absence of miRNAs and RNAi machinery in *Leishmania* species and thus similar machines in host system could be targeted for therapeutic intervention. Antisense oligonucleotide approaches and siRNA-like technologies for inhibiting miRNA function are being explored as potential therapeutic agents. The current strategy to constitutively synthesize siRNA molecules is based on viral vectors that express short hairpin RNA (shRNA), which share features of miRNAs (Carmona *et al.*, 2006). Fatality in mice due to oversaturation of cellular microRNA/short hairpin RNA pathways (Grimm *et al.*, 2006) is reported earlier while, hepatitis B virus is known to be inhibited by RNA interference *in vivo* (Ying *et al.*, 2007). Therefore, a particular caveat of this approach lies in the dosage. Indeed, intravenous infusion of AAV/shRNA vectors in mice results in strong competition of small RNAs for limiting cellular factors required for their processing (102) although no change was noted in the expression of miR-122. In fact, the exciting first step towards miRNA therapy in the liver was with miR-122 antagomir (Krutzfeldt *et al.*, 2005). In this study, the pharmacological approach proposed by Krützfeldt *et al.* (2005) was remarkably efficient, specific and stable. Injection of the antagomir into the tail veins of the mice selectively degraded miR-122 but not other miRNAs from the liver, even after more than 20 days indicating a durable effect. Thus, in contrast to the challenges raised by gene therapy, the efficacy of intravenous antagomir to target the liver suggests that some issues such as the mode of delivery and the specificity will be more easily alleviated. Taken together, these studies show that while some safety issues still need to be carefully addressed, treatment of hepatic disorders through RNAi is becoming a plausible scenario.

Furthermore, other recent strategies are coming up with better potential. Lipofection of an antisense oligonucleotide based on a locked nucleic acids (LNA)/2′-O-methyl mixmer or electroporation of a peptide nucleic acid (PNA) oligomer is effective at blocking miR-122 activity in human and rat liver cells. These oligonucleotide analogues, evaluated for the first time in microRNA inhibition, are more effective than standard 2′-O-methyl oligonucleotides in binding and inhibiting microRNA action. It is also shown that miRNA inhibition can be achieved without the need for transfection or electroporation of the human or rat cell lines, by conjugation of an antisense PNA to the cell-penetrating peptide R6-Penetratin, or merely by linkage to just four Lys residues, highlighting the potential of PNA for future therapeutic applications as well as for studying miRNA function.

miR-122 and hepatitis C virus (HCV)

miR-122 is a liver-specific miRNA representing 70% of the liver miRNA population (Wang *et al.*, 2010), and it has been characterized for its multiple roles in liver physiology and metabolism, namely regulation of genes involved in fatty acid and cholesterol biosynthesis (Elmen *et al.*, 2008a; Esau *et al.*, 2006). Silencing of miRNAs by LNA-anti-miRs exhibited a suppressive effect on HCV expression in human hepatoma cells harbouring an HCV replicon. Jopling, in 2005, first showed the dependence of HCV replication on hepatic miR-122 (Jopling *et al.*, 2005). miR-122 binds at two sites, S1 and S2, in the 5′UTR of the HCV RNA and is required to maintain viral abundance in the infected cell. In the presence of miR-122, there is an increase in translation of viral RNA and infectious virus production. This occurs due to accelerated association of the small ribosomal subunit in the 48S complex during the initiation phase (Henke *et al.*, 2008). Furthermore, translational enhancement requires uncapped mRNA, the IRES of the HCV RNA and the 3′ region of the miRNA (Roberts *et al.*, 2011). miR-122 also causes increased stability of the viral genome by binding to several internal sites of the HCV RNA (Machlin *et al.*, 2011). The miR-122 molecule binds to the 5′ terminus of the HCV RNA with 3′ overhanging nucleotides masking the 5′ terminal sequences of the HCV genome. It was proposed that the formation of this complex prevents the viral RNA from being recognized by the immune system and protects it from nucleolytic degradation.

Effect of miR-122 on cholesterol and *Leishmania*

After administration of miR-122 expression plasmid in the normal animals via tail miR-122 level in liver were monitored for 1, 3, 5 and 7 days post treatment (Ghosh *et al.*, 2013). miR-122 expression was maximum at 3 days and reduces to the normal level by 7 days. Interestingly, the administration of miR-122 expression plasmid and higher miR-122 level prevent *Leishmania* infection (Ghosh *et al.*, 2013). Parasite burden was monitored, with an unrelated plasmid shRNA under the same promoter showed no alternation in infectivity. Similar experiments were conducted on infected animals and interestingly there was a remarkable reduction in hepatic parasite load in the experimental group whereas the mock group did not show any rescue. The cholesterol level was also found to be almost equal to that of normal animals (Ghosh *et al.*, 2013).

To better highlight the protective role of miR-122, miR-122 expression was restored in liver after the initial establishment of infection. After only three doses of miR-122 expression plasmid injection an increase in miR-122 level with a concomitant elevation in cholesterol level was noted. More importantly, the hepatic parasite load was reduced as previous showing that the protective action of miR122 is observed even after initial infection is established (Ghosh *et al.*, 2013).

Dicer1 restoration and *Leishmania* infection

Leishmanial exosomes contain GP63 metalloprotease, which down-regulates host cell proteins of important function to control the inflammatory response. Dicer-1 is also down-regulated by GP63 in hepatic cells, and restoration of Dicer1 Expression in parasite-infected livers rescues miR-122 expression and reduces liver parasite burden (Ghosh *et al.*, 2013). When expressed exogenously, NHA-DICER1 expression was found to be low in *L. donovani*-infected mouse livers. This observation was consistent with the prediction of Dicer1 as a target that *L. donovani* degrades to down-regulate miR-122 in infected mouse liver. With infection, the level of GP63 was increased in mouse livers and purified gp63 from *L. donovani* entrapped in liposomes when delivered to Huh7 cells, a substantial decrease in miR-122 activity was noted in cells receiving GP63-containing liposomes (Ghosh *et al.*, 2013). During infection, exosomes may be the vehicles for GP63 transfer to hepatocytes. Delivery of purified *L. donovani* exosomes down-regulated Dicer1 but Dicer1 restoration can clear *L. donovani* from infected mouse liver. Excess Dicer1 in liver increased liver miR-122 expression and also restored the serum cholesterol level in infected animals (Fig. 1.1). This was accompanied by a drastic reduction in liver parasite load. Importantly, no apparent change in liver cell morphology, tissue integrity, or production of serum albumin was documented (Ghosh *et al.*, 2013).

Future of miRNA as a therapeutic agent against visceral leishmaniasis

The potential of small regulatory RNAs as therapeutic agents has not yet been explored. miR-122 has been found to be a key molecule, overexpression of which can cause clearance of parasite from infected animal liver. It will be interesting to test the efficacy of this miRNAs against drug resistant parasites induced infection in different mammalian host. It seems to act as anti-leishmanial factor primarily via up-regulated expression of cholesterol in infected tissue but whether miRNA expression could directly affect parasite number is not known. The initial findings are promising and miR-122 may be useful in future as anti-leishmanial gent in *L. donovani*-infected individual.

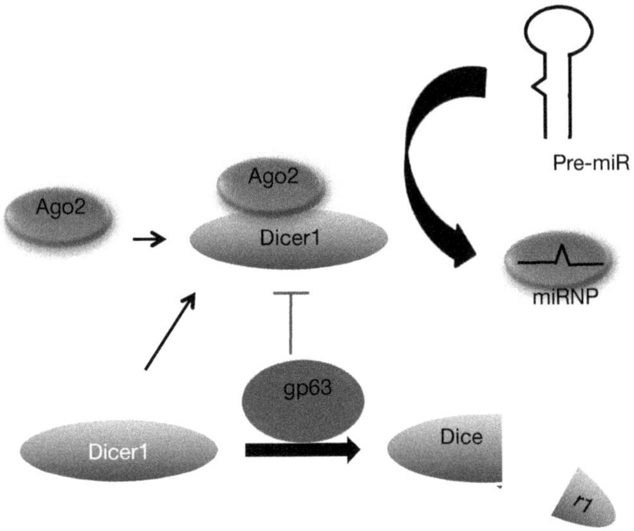

Figure 1.1 *Leishmania* derived exosomes contain GP63 metalloprotease that target Dicer1 in host cells and cleave it. Dicer1 depletion prevents pre-miRNA processing and production of mature miR-122. This subsequently prevents miR-122 activity and resulted lowering of cholesterol production in infected liver. A colour version of this figure is available in the plate section at the back of the book.

References

Admyre, C., Johansson, S.M., Qazi, K.R., Filen, J.J., Lahesmaa, R., Norman, M., Neve, E.P., Scheynius, A., and Gabrielsson, S. (2007). Exosomes with immune modulatory features are present in human breast milk. J. Immunol. *179*, 1969–1978.

Anderson, H.A., Hiltbold, E.M., and Roche, P.A. (2000). Concentration of MHC class II molecules in lipid rafts facilitates antigen presentation. Nat. Immunol. *1*, 156–162.

Asea, A., Jean-Pierre, C., Kaur, P., Rao, P., Linhares, I.M., Skupski, D., and Witkin, S.S. (2008). Heat shock protein-containing exosomes in mid-trimester amniotic fluids. J. Reprod. Immunol. *79*, 12–17.

Bagga, S., Bracht, J., Hunter, S., Massirer, K., Holtz, J., Eachus, R., and Pasquinelli, A.E. (2005). Regulation by let-7 and lin-4 miRNAs results in target mRNA degradation. Cell *122*, 553–563.

Banerjee, S., Ghosh, J., Sen, S., Guha, R., Dhar, R., Ghosh, M., Datta, S., Raychaudhury, B., Naskar, K., Haldar, A.K., *et al.* (2009). Designing therapies against experimental visceral leishmaniasis by modulating the membrane fluidity of antigen-presenting cells. Infect. Immun. *77*, 2330–2342.

Behm-Ansmant, I., Rehwinkel, J., Doerks, T., Stark, A., Bork, P., and Izaurralde, E. (2006). mRNA degradation by miRNAs and GW182 requires both CCR4:NOT deadenylase and DCP1:DCP2 decapping complexes. Gene. Dev. *20*, 1885–1898.

Belair, C., Baud, J., Chabas, S., Sharma, C.M., Vogel, J., Staedel, C., and Darfeuille, F. (2011). Helicobacter pylori interferes with an embryonic stem cell micro RNA cluster to block cell cycle progression. Science 2, 7.

Bhatnagar, S., and Schorey, J.S. (2007). Exosomes released from infected macrophages contain Mycobacterium avium glycopeptidolipids and are proinflammatory. J. Biol. Chem. *282*, 25779–25789.

Bhatnagar, S., Shinagawa, K., Castellino, F.J., and Schorey, J.S. (2007). Exosomes released from macrophages infected with intracellular pathogens stimulate a proinflammatory response *in vitro* and *in vivo*. Blood *110*, 3234–3244.

Bhattacharyya, S.N., Habermacher, R., Martine, U., Closs, E.I., and Filipowicz, W. (2006). Relief of microRNA-mediated translational repression in human cells subjected to stress. Cell *125*, 1111–1124.

Blott, E.J., and Griffiths, G.M. (2002). Secretory lysosomes. Nat. Rev. Mol. Cell Biol. 3, 122–131.

Boehm, M., and Slack, F. (2005). A developmental timing microRNA and its target regulate life span in C. elegans. Science *310*, 1954–1957.

Burger, K., Gimpl, G., and Fahrenholz, F. (2000). Regulation of receptor function by cholesterol. Cell. Mol. Life Sci. *57*, 1577–1592.

Calin, G.A., Liu, C.G., Sevignani, C., Ferracin, M., Felli, N., Dumitru, C.D., Shimizu, M., Cimmino, A., Zupo, S., Dono, M., *et al.* (2004). MicroRNA profiling reveals distinct signatures in B-cell chronic lymphocytic leukemias. Proc. Natl. Acad. Sci. U.S.A. *101*, 11755–11760.

Carmona, S., Ely, A., Crowther, C., Moolla, N., Salazar, F.H., Marion, P.L., Ferry, N., Weinberg, M.S., and Arbuthnot, P. (2006). Effective inhibition of HBV replication *in vivo* by anti-HBx short hairpin RNAs. Mol. Ther. *13*, 411–421.

Cazalla, D., and Steitz, J.A. (2010). Down-regulation of a host microRNA by a viral noncoding RNA. Cold Spring Harbor Symp Quant Biol *75*, 321–324.

Cazalla, D., Yario, T., and Steitz, J.A. (2010). Down-regulation of a host microRNA by a Herpesvirus saimiri noncoding RNA. Science *328*, 1563–1566.

Chakraborty, D., Banerjee, S., Sen, A., Banerjee, K.K., Das, P., and Roy, S. (2005). *Leishmania donovani* affects antigen presentation of macrophage by disrupting lipid rafts. J. Immunol. *175*, 3214–3224.

Chendrimada, T.P., Finn, K.J., Ji, X., Baillat, D., Gregory, R.I., Liebhaber, S.A., Pasquinelli, A.E., and Shiekhattar, R. (2007). MicroRNA silencing through RISC recruitment of eIF6. Nature *447*, 823–828.

Coller, J., and Parker, R. (2004). Eukaryotic mRNA decapping. Annu. Rev. Biochem. 73, 861–890.

Dabrowski, M.P., Peel, W.E., and Thomson, A.E. (1980). Plasma membrane cholesterol regulates human lymphocyte cytotoxic function. Eur. J. Immunol. *10*, 821–827.

Ding, X.C., and Grosshans, H. (2009). Repression of C. elegans microRNA targets at the initiation level of translation requires GW182 proteins. EMBO J. *28*, 213–222.

Elmen, J., Lindow, M., Schutz, S., Lawrence, M., Petri, A., Obad, S., Lindholm, M., Hedtjarn, M., Hansen, H.F., Berger, U., *et al.* (2008a). LNA-mediated microRNA silencing in non-human primates. Nature *452*, 896–899.

Elmen, J., Lindow, M., Silahtaroglu, A., Bak, M., Christensen, M., Lind-Thomsen, A., Hedtjarn, M., Hansen, J.B., Hansen, H.F., Straarup, E.M., *et al.* (2008b). Antagonism of microRNA-122 in mice by systemically administered LNA-antimiR leads to up-regulation of a large set of predicted target mRNAs in the liver. Nucleic Acids Res. *36*, 1153–1162.

Esau, C., Kang, X., Peralta, E., Hanson, E., Marcusson, E.G., Ravichandran, L.V., Sun, Y., Koo, S., Perera, R.J., Jain, R., *et al.* (2004). MicroRNA-143 regulates adipocyte differentiation. J. Biol. Chem. *279*, 52361–52365.

Esau, C., Davis, S., Murray, S.F., Yu, X.X., Pandey, S.K., Pear, M., Watts, L., Booten, S.L., Graham, M., McKay, R., *et al.* (2006). miR-122 regulation of lipid metabolism revealed by *in vivo* antisense targeting. Cell Metab. *3*, 87–98.

Esquela-Kerscher, A., and Slack, F.J. (2006). Oncomirs – microRNAs with a role in cancer. Nat. Rev. Cancer *6*, 259–269.

Filipowicz, W., Bhattacharyya, S.N., and Sonenberg, N. (2008). Mechanisms of post-transcriptional regulation by microRNAs: are the answers in sight? Nat. Rev. Genet. *9*, 102–114.

Friedman, R.C., Farh, K.K., Burge, C.B., and Bartel, D.P. (2009). Most mammalian mRNAs are conserved targets of microRNAs. Genome Res. *19*, 92–105.

Gerin, I., Bommer, G.T., McCoin, C.S., Sousa, K.M., Krishnan, V., and MacDougald, O.A. (2010). Roles for miRNA-378/378* in adipocyte gene expression and lipogenesis. Am. J. Physiol. Endocrinol. Metab. *299*, E198–206.

Ghosh, J., Lal, C.S., Pandey, K., Das, V.N., Das, P., Roychoudhury, K., and Roy, S. (2011). Human visceral leishmaniasis: decrease in serum cholesterol as a function of splenic parasite load. Ann. Trop. Med. Parasitol. *105*, 267–271.

Ghosh, J., Das, S., Guha, R., Ghosh, D., Naskar, K., Das, A., and Roy, S. (2012). Hyperlipidemia offers protection against *Leishmania donovani* infection: role of membrane cholesterol. J. Lipid Res. *53*, 2560–2572.

Ghosh, J., Bose, M., Roy, S., and Bhattacharyya, S.N. (2013). *Leishmania donovani* targets Dicer1 to down-regulate miR-122, lower serum cholesterol, and facilitate murine liver infection. Cell Host Microbe *13*, 277–288.

Giraldez, A.J., Mishima, Y., Rihel, J., Grocock, R.J., Van Dongen, S., Inoue, K., Enright, A.J., and Schier, A.F. (2006). Zebrafish MiR-430 promotes deadenylation and clearance of maternal mRNAs. Science *312*, 75–79.

Girard, M., Jacquemin, E., Munnich, A., Lyonnet, S., and Henrion-Caude, A. (2008). miR-122, a paradigm for the role of microRNAs in the liver. J. Hepatol. *48*, 648–656.

Gomez, M.A., Contreras, I., Halle, M., Tremblay, M.L., McMaster, R.W., and Olivier, M. (2009). *Leishmania* GP63 alters host signaling through cleavage-activated protein tyrosine phosphatases. Science Signal. *2*, ra58.

Grimm, D., Streetz, K.L., Jopling, C.L., Storm, T.A., Pandey, K., Davis, C.R., Marion, P., Salazar, F., and Kay, M.A. (2006). Fatality in mice due to oversaturation of cellular microRNA/short hairpin RNA pathways. Nature *441*, 537–541.

Hake, S. (2003). MicroRNAs: a role in plant development. Curr. Biol. *13*, R851–852.

Hassani, K., Antoniak, E., Jardim, A., and Olivier, M. (2011). Temperature-induced protein secretion by *Leishmania mexicana* modulates macrophage signalling and function. PloS One *6*, e18724.

Henke, J.I., Goergen, D., Zheng, J., Song, Y., Schuttler, C.G., Fehr, C., Junemann, C., and Niepmann, M. (2008). microRNA-122 stimulates translation of hepatitis C virus RNA. EMBO J. *27*, 3300–3310.

Hosseini, H.M., Fooladi, A.A., Nourani, M.R., and Ghanezadeh, F. (2013). The role of exosomes in infectious diseases. Inflamm. Allergy Drug Targets *12*, 29–37.

Huang, J., Liang, Z., Yang, B., Tian, H., Ma, J., and Zhang, H. (2007). Derepression of microRNA-mediated protein translation inhibition by apolipoprotein B mRNA-editing enzyme catalytic polypeptide-like 3G (APOBEC3G) and its family members. J. Biol. Chem. *282*, 33632–33640.

Hughes, A.L., and Nei, M. (1992). Models of host–parasite interaction and MHC polymorphism. Genetics *132*, 863–864.

Humphreys, D.T., Westman, B.J., Martin, D.I., and Preiss, T. (2005). MicroRNAs control translation initiation by inhibiting eukaryotic initiation factor 4E/cap and poly(A) tail function. Proc. Natl. Acad. Sci. U.S.A. *102*, 16961–16966.

Iliopoulos, D., Drosatos, K., Hiyama, Y., Goldberg, I.J., and Zannis, V.I. (2010). MicroRNA-370 controls the expression of microRNA-122 and Cpt1alpha and affects lipid metabolism. J. Lipid Res. *51*, 1513–1523.

Jopling, C.L., Yi, M., Lancaster, A.M., Lemon, S.M., and Sarnow, P. (2005). Modulation of hepatitis C virus RNA abundance by a liver-specific MicroRNA. Science *309*, 1577–1581.

Kajimoto, K., Naraba, H., and Iwai, N. (2006). MicroRNA and 3T3-L1 pre-adipocyte differentiation. RNA *12*, 1626–1632.

Kelada, S., Sethupathy, P., Okoye, I.S., Kistasis, E., Czieso, S., White, S.D., Chou, D., Martens, C., Ricklefs, S.M., Virtaneva, K., *et al.* (2013). miR-182 and miR-10a are key regulators of Treg specialisation and stability during Schistosome and *Leishmania*-associated inflammation. PLoS Pathog. *9*, e1003451.

Khan, A.G., Pickl-Herk, A., Gajdzik, L., Marlovits, T.C., Fuchs, R., and Blaas, D. (2011). Entry of a heparan sulphate-binding HRV8 variant strictly depends on dynamin but not on clathrin, caveolin, and flotillin. Virology *412*, 55–67.

Kim, H.K., Lee, Y.S., Sivaprasad, U., Malhotra, A., and Dutta, A. (2006). Muscle-specific microRNA miR-206 promotes muscle differentiation. J. Cell Biol. *174*, 677–687.

Kiriakidou, M., Tan, G.S., Lamprinaki, S., De Planell-Saguer, M., Nelson, P.T., and Mourelatos, Z. (2007). An mRNA m7G cap binding-like motif within human Ago2 represses translation. Cell *129*, 1141–1151.

Krützfeldt, J., Rajewsky, N., Braich, R., Rajeev, K.G., Tuschl, T., Manoharan, M., and Stoffel, M. (2005). Silencing of microRNAs *in vivo* with 'antagomirs'. Nature *438*, 685–689.

Lagos-Quintana, M., Rauhut, R., Meyer, J., Borkhardt, A., and Tuschl, T. (2003). New microRNAs from mouse and human. RNA *9*, 175–179.

Lai, W., Huang, L., Ho, P., Montefiori, D., and Chen, C.H. (2011). The role of dynamin in HIV type 1 Env-mediated cell–cell fusion. AIDS Res. Hum. Retrov. *27*, 1013–1017.

Lal, C.S., Kumar, A., Kumar, S., Pandey, K., Kumar, N., Bimal, S., Sinha, P.K., and Das, P. (2007). Hypocholesterolemia and increased triglyceride in pediatric visceral leishmaniasis. Clin. Chim. Acta *382*, 151–153.

Lanford, R.E., Hildebrandt-Eriksen, E.S., Petri, A., Persson, R., Lindow, M., Munk, M.E., Kauppinen, S., and Orum, H. (2010). Therapeutic silencing of microRNA-122 in primates with chronic hepatitis C virus infection. Science *327*, 198–201.

Lee, R.C., Feinbaum, R.L., and Ambros, V. (1993). The C. elegans heterochronic gene lin-4 encodes small RNAs with antisense complementarity to lin-14. Cell *75*, 843–854.

Lee, Y., Jeon, K., Lee, J.T., Kim, S., and Kim, V.N. (2002). MicroRNA maturation: stepwise processing and subcellular localization. EMBO J. *21*, 4663–4670.

Lee, Y., Ahn, C., Han, J., Choi, H., Kim, J., Yim, J., Lee, J., Provost, P., Radmark, O., Kim, S., *et al.* (2003). The nuclear RNase III Drosha initiates microRNA processing. Nature *425*, 415–419.

Lemaire, J., Mkannez, G., Guerfali, F.Z., Gustin, C., Attia, H., Sghaier, R.M., Dellagi, K., Laouini, D., and Renard, P. (2013). MicroRNA expression profile in human macrophages in response to *Leishmania major* infection. PLoS Negl. Trop. Dis. *7*, e2478.

Liberopoulos, E., Alexandridis, G., Bairaktari, E., and Elisaf, M. (2002). Severe hypocholesterolemia with reduced serum lipoprotein(a) in a patient with visceral leishmaniasis. Ann. Clin. Lab. Sci. *32*, 305–308.

Libri, V., Helwak, A., Miesen, P., Santhakumar, D., Borger, J.G., Kudla, G., Grey, F., Tollervey, D., and Buck, A.H. (2012). Murine cytomegalovirus encodes a miR-27 inhibitor disguised as a target. Proc. Natl. Acad. Sci. U.S.A. *109*, 279–284.

Liegeois, S., Benedetto, A., Garnier, J.M., Schwab, Y., and Labouesse, M. (2006). The V0-ATPase mediates apical secretion of exosomes containing Hedgehog-related proteins in Caenorhabditis elegans. J. Cell Biol. *173*, 949–961.

Lin, Q., Gao, Z., Alarcon, R.M., Ye, J., and Yun, Z. (2009). A role of miR-27 in the regulation of adipogenesis. FEBS J. *276*, 2348–2358.

Liu, D., and Uzonna, J.E. (2012). The early interaction of *Leishmania* with macrophages and dendritic cells and its influence on the host immune response. Front. Cell. Infect. Microbiol. *2*, 83.

Lujambio, A., and Lowe, S.W. (2012). The microcosmos of cancer. Nature *482*, 347–355.

Lund, E., Guttinger, S., Calado, A., Dahlberg, J.E., and Kutay, U. (2004). Nuclear export of microRNA precursors. Science *303*, 95–98.

Machlin, E.S., Sarnow, P., and Sagan, S.M. (2011). Masking the 5′ terminal nucleotides of the hepatitis C virus genome by an unconventional microRNA-target RNA complex. Proc. Natl. Acad. Sci. U.S.A. *108*, 3193–3198.

Marquart, T.J., Allen, R.M., Ory, D.S., and Baldan, A. (2010). miR-33 links SREBP-2 induction to repression of sterol transporters. Proc. Natl. Acad. Sci. U.S.A. *107*, 12228–12232.

Mathonnet, G., Fabian, M.R., Svitkin, Y.V., Parsyan, A., Huck, L., Murata, T., Biffo, S., Merrick, W.C., Darzynkiewicz, E., Pillai, R.S., *et al.* (2007). MicroRNA inhibition of translation initiation *in vitro* by targeting the cap-binding complex eIF4F. Science *317*, 1764–1767.

Meyer, S., Temme, C., and Wahle, E. (2004). Messenger RNA turnover in eukaryotes: pathways and enzymes. Crit. Rev. Biochem. Mol. Biol. *39*, 197–216.

Najafi-Shoushtari, S.H., Kristo, F., Li, Y., Shioda, T., Cohen, D.E., Gerszten, R.E., and Naar, A.M. (2010). MicroRNA-33 and the SREBP host genes cooperate to control cholesterol homeostasis. Science *328*, 1566–1569.

Nakanishi, N., Nakagawa, Y., Tokushige, N., Aoki, N., Matsuzaka, T., Ishii, K., Yahagi, N., Kobayashi, K., Yatoh, S., Takahashi, A., *et al.* (2009). The up-regulation of microRNA-335 is associated with lipid metabolism in liver and white adipose tissue of genetically obese mice. Biochem. Bioph. Res. Comm. *385*, 492–496.

O'Connell, R.M., Rao, D.S., Chaudhuri, A.A., and Baltimore, D. (2010). Physiological and pathological roles for microRNAs in the immune system. Nat. Rev. Immunol. *10*, 111–122.

O'Connell, R.M., Rao, D.S., and Baltimore, D. (2012). microRNA regulation of inflammatory responses. Annu. Rev. Immunol. *30*, 295–312.

Olsen, P.H., and Ambros, V. (1999). The lin-4 regulatory RNA controls developmental timing in Caenorhabditis elegans by blocking LIN-14 protein synthesis after the initiation of translation. Dev. Biol. *216*, 671–680.

Pan, B.T., Teng, K., Wu, C., Adam, M., and Johnstone, R.M. (1985). Electron microscopic evidence for externalization of the transferrin receptor in vesicular form in sheep reticulocytes. J. Cell Biol. *101*, 942–948.

Pan, Q., Bao, L.W., Kleer, C.G., Sabel, M.S., Griffith, K.A., Teknos, T.N., and Merajver, S.D. (2005). Protein kinase C epsilon is a predictive biomarker of aggressive breast cancer and a validated target for RNA interference anticancer therapy. Cancer Res. *65*, 8366–8371.

Pasquinelli, A.E., Reinhart, B.J., Slack, F., Martindale, M.Q., Kuroda, M.I., Maller, B., Hayward, D.C., Ball, E.E., Degnan, B., Muller, P., *et al.* (2000). Conservation of the sequence and temporal expression of let-7 heterochronic regulatory RNA. Nature *408*, 86–89.

Perez-Guzman, C., Vargas, M.H., Quinonez, F., Bazavilvazo, N., and Aguilar, A. (2005). A cholesterol-rich diet accelerates bacteriologic sterilization in pulmonary tuberculosis. Chest *127*, 643–651.

Pillai, R.S., Bhattacharyya, S.N., Artus, C.G., Zoller, T., Cougot, N., Basyuk, E., Bertrand, E., and Filipowicz, W. (2005). Inhibition of translational initiation by Let-7 MicroRNA in human cells. Science *309*, 1573–1576.

Poy, M.N., Eliasson, L., Krutzfeldt, J., Kuwajima, S., Ma, X., Macdonald, P.E., Pfeffer, S., Tuschl, T., Rajewsky, N., Rorsman, P., *et al.* (2004). A pancreatic islet-specific microRNA regulates insulin secretion. Nature *432*, 226–230.

Pucadyil, T.J., and Chattopadhyay, A. (2007). Cholesterol: a potential therapeutic target in *Leishmania* infection? Trends Parasitol. *23*, 49–53.

Rayner, K.J., Suarez, Y., Davalos, A., Parathath, S., Fitzgerald, M.L., Tamehiro, N., Fisher, E.A., Moore, K.J., and Fernandez-Hernando, C. (2010). MiR-33 contributes to the regulation of cholesterol homeostasis. Science *328*, 1570–1573.

Razzaq, T.M., Ozegbe, P., Jury, E.C., Sembi, P., Blackwell, N.M., and Kabouridis, P.S. (2004). Regulation of T-cell receptor signalling by membrane microdomains. Immunology *113*, 413–426.

Reinhart, B.J., Slack, F.J., Basson, M., Pasquinelli, A.E., Bettinger, J.C., Rougvie, A.E., Horvitz, H.R., and Ruvkun, G. (2000). The 21-nucleotide let-7 RNA regulates developmental timing in *Caenorhabditis elegans*. Nature *403*, 901–906.

Roberts, A.P., Lewis, A.P., and Jopling, C.L. (2011). miR-122 activates hepatitis C virus translation by a specialized mechanism requiring particular RNA components. Nucleic Acids Res. *39*, 7716–7729.

Rodriguez, A., Griffiths-Jones, S., Ashurst, J.L., and Bradley, A. (2004). Identification of mammalian microRNA host genes and transcription units. Genome Res. *14*, 1902–1910.

Rodrigues, M.L., Nakayasu, E.S., Oliveira, D.L., Nimrichter, L., Nosanchuk, J.D., Almeida, I.C., and Casadevall, A. (2008). Extracellular vesicles produced by Cryptococcus neoformans contain protein components associated with virulence. Eukaryot. Cell *7*, 58–67.

Rosenberger, C.M., Brumell, J.H., and Finlay, B.B. (2000). Microbial pathogenesis: lipid rafts as pathogen portals. Curr. Biol. *10*, R823–825.

Schmitter, D., Filkowski, J., Sewer, A., Pillai, R.S., Oakeley, E.J., Zavolan, M., Svoboda, P., and Filipowicz, W. (2006). Effects of Dicer and Argonaute down-regulation on mRNA levels in human HEK293 cells. Nucleic Acids Res. *34*, 4801–4815.

Sen, S., Roy, K., Mukherjee, S., Mukhopadhyay, R., and Roy, S. (2011). Restoration of IFN-γammaR subunit assembly, IFN-γamma signaling and parasite clearance in *Leishmania donovani* infected macrophages: role of membrane cholesterol. PLoS Pathog. *7*, e1002229.

Silverman, J.M., Clos, J., de'Oliveira, C.C., Shirvani, O., Fang, Y., Wang, C., Foster, L.J., and Reiner, N.E. (2010). An exosome-based secretion pathway is responsible for protein export from *Leishmania* and communication with macrophages. J. Cell Sci. *123*, 842–852.

Silverman, J.M., and Reiner, N.E. (2011). *Leishmania* exosomes deliver preemptive strikes to create an environment permissive for early infection. Front. Cell. Infect. Microbiol. *1*, 26.

Simons, K., and Toomre, D. (2000). Lipid rafts and signal transduction. Nat. Rev. Mol. Cell Biol. *1*, 31–39.

Slack, F.J., Basson, M., Liu, Z., Ambros, V., Horvitz, H.R., and Ruvkun, G. (2000). The lin-41 RBCC gene acts in the C. elegans heterochronic pathway between the let-7 regulatory RNA and the LIN-29 transcription factor. Mol. Cell. *5*, 659–669.

Sokol, N.S., and Ambros, V. (2005). Mesodermally expressed Drosophila microRNA-1 is regulated by Twist and is required in muscles during larval growth. Genes Dev. *19*, 2343–2354.

Sun, K., and Lai, E.C. (2013). Adult-specific functions of animal microRNAs. Nat. Rev. Genet. *14*, 535–548.

Teleman, A.A., Maitra, S., and Cohen, S.M. (2006). Drosophila lacking microRNA miR-278 are defective in energy homeostasis. Genes Dev. *20*, 417–422.

Thermann, R., and Hentze, M.W. (2007). Drosophila miR2 induces pseudo-polysomes and inhibits translation initiation. Nature *447*, 875–878.

Thery, C., Ostrowski, M., and Segura, E. (2009). Membrane vesicles as conveyors of immune responses. Nat. Rev. Immunol. *9*, 581–593.

Wakiyama, M., Takimoto, K., Ohara, O., and Yokoyama, S. (2007). Let-7 microRNA-mediated mRNA deadenylation and translational repression in a mammalian cell-free system. Genes Dev. *21*, 1857–1862.

Wang, B., Yanez, A., and Novina, C.D. (2008). MicroRNA-repressed mRNAs contain 40S but not 60S components. Proc. Natl. Acad. Sci. U.S.A. *105*, 5343–5348.

Wang, G.K., Zhu, J.Q., Zhang, J.T., Li, Q., Li, Y., He, J., Qin, Y.W., and Jing, Q. (2010). Circulating microRNA: a novel potential biomarker for early diagnosis of acute myocardial infarction in humans. Eur. Heart J. *31*, 659–666.

Wightman, B., Ha, I., and Ruvkun, G. (1993). Posttranscriptional regulation of the heterochronic gene lin-14 by lin-4 mediates temporal pattern formation in C. elegans. Cell 75, 855–862.

Wunder, C., Churin, Y., Winau, F., Warnecke, D., Vieth, M., Lindner, B., Zahringer, U., Mollenkopf, H.J., Heinz, E., and Meyer, T.F. (2006). Cholesterol glucosylation promotes immune evasion by *Helicobacter pylori*. Nat. Med. *12*, 1030–1038.

Xu, P., Vernooy, S.Y., Guo, M., and Hay, B.A. (2003). The *Drosophila* microRNA Mir-14 suppresses cell death and is required for normal fat metabolism. Curr. Biol. *13*, 790–795.

Yamashita, A., Chang, T.C., Yamashita, Y., Zhu, W., Zhong, Z., Chen, C.Y., and Shyu, A.B. (2005). Concerted action of poly(A) nucleases and decapping enzyme in mammalian mRNA turnover. Nat. Struct. Mol. Biol. *12*, 1054–1063.

Ying, R.S., Zhu, C., Fan, X.G., Li, N., Tian, X.F., Liu, H.B., and Zhang, B.X. (2007). Hepatitis B virus is inhibited by RNA interference in cell culture and in mice. Antivir. Res. 73, 24–30.

Heat Shock Proteins of *Leishmania*: Chaperones in the Driver's Seat

2

Joachim Clos and Antje Hombach

Abstract

The roles of so-called heat shock proteins (HSPs) in *Leishmania* have long been assumed to revolve around protective functions against temperature and adverse milieu inside macrophages. In recent years, however, a picture has emerged showing heat shock proteins and co-chaperone proteins as effectors and regulators of signal transduction and immune evasion pathways, possibly compensating for the lack of regulated gene expression and regulatory factors in *Leishmania* spp.

Introduction: heat shock proteins

Stress induction

Heat shock proteins were first described as a group of proteins that showed elevated synthesis when cells were subjected to heat stress (Lindquist, 1981, 1986). Research soon began to utilize heat shock genes as easily inducible models for eukaryotic transcription regulation. The structures of heat shock promoter regions and the nature and regulation of heat shock transcription factors provided insight into the cells' abilities to switch protective gene expression on and off (Morimoto, 1998; Wu, 1995, 1994; Xiao *et al.*, 1991). In addition to the transcriptional control, the synthesis of heat shock proteins is also regulated at a post-transcriptional level, mostly during translation initiation (Sierra and Zapata, 1994; Yost, 1990). Moreover, most heat shock genes in eukaryotes lack introns, making their expression independent of the highly temperature-sensitive RNA splicing apparatus (Czarnecka *et al.*, 1988; Kay *et al.*, 1987; Muhich and Boothroyd, 1989; Yost and Lindquist, 1986, 1988), whereas elevated HSP synthesis may protect and restore splicing (Vogel *et al.*, 1995; Yost and Lindquist, 1986).

Coping with cell stress

It soon became clear that other physiological stresses induced the same group of proteins, leading to the assumption that their role lay in the protection against many forms of cellular stress (Lindquist, 1986, 1992; Parsell and Lindquist, 1994). Overexpression of selected heat shock proteins and resulting stress tolerance confirmed this view (Dillmann, 1999; Lindquist, 1992; Sanchez and Lindquist, 1990; Sanchez *et al.*, 1992; Solomon, 1991). Even more convincing are the effects of natural variations in stress protein induction displayed in sessile marine animals living in intertidal ranges. Here, variations can be correlated to the stress levels expected in distinct habitats (Lockwood *et al.*, 2010; Lockwood and Somero, 2011; Zhang *et al.*, 2012). Therefore, the concept of stress protection through elevated synthesis of certain heat shock proteins is validated.

Functions in homeostasis – the chaperone concept

The discovery of heat shock cognate (Hsc) genes that showed constitutive expression but coded for *bona fide* heat shock proteins was a first hint at a more universal role played by HSPs when cells are in homeostasis. It was subsequently discovered that heat shock proteins interact with nascent polypeptides and aid them in attaining native structure. Other roles include the unfolding of proteins prior to trans-membrane transport, the targeting of damaged proteins for degradation, and the prevention of protein aggregation. This led to the term 'chaperone' being coined for heat shock proteins. The constitutive roles played by many heat shock proteins are also evident from the fact that null-mutants of major HSPs are not viable under any conditions and from the high degree of primary structure conservation within the major HSP families.

Chaperoning life

An indirect role in regulatory pathways was discovered for the 90 kDa heat shock proteins (Hsp90), as this chaperone family interacts with and is required for the function of a large variety of client proteins that include many protein kinases, transcription factors, oncogenes and cytoskeletal proteins (Buchner, 1999; Rutherford, 1994). It is this role that makes Hsp90 a promising target for anti-cancer and anti-microbial drug leads, as pharmacological inhibition of Hsp90 blocks cell division (Whitesell and Lindquist, 2005). Its dual roles as a highly specific regulatory chaperone and as a basic, stress-alleviating heat shock protein, also implicate Hsp90 as a 'capacitor of evolution' (Queitsch *et al.*, 2002; Rutherford and Lindquist, 1998; Sangster *et al.*, 2004). Hsp90 chaperone complexes can assist proteins in maintaining their function in spite of minor mutations, allowing populations to accumulate a pool of mutations in regulatory proteins. Once the population then faces a stress, e.g. through a changing environment, Hsp90 is recruited to denatured proteins and can no longer keep the mutations silent. Their phenotypes become expressed and can be selected under the new living conditions.

Emerging functions

In recent years, HSPs were also identified as client proteins for post-translational modifications, including protein phosphorylation (Mollapour *et al.*, 2011b; Muller *et al.*, 2013; Sherman and Goldberg, 1993; Soroka and Buchner, 2012; Soroka *et al.*, 2012; Truman *et al.*, 2012; Xu *et al.*, 2012), S-nitrosylation (Retzlaff *et al.*, 2009), and methylation (Cho *et al.*, 2012; Donlin *et al.*, 2012; Jakobsson *et al.*, 2013). These modifications have been shown to modulate chaperone functions, establishing heat shock proteins as effector proteins of signal transduction cascades.

Heat shock proteins in immunology and drug development

In antigen-presenting cells, heat shock proteins play a central role in the presentation of antigens, both for the innate and the adaptive responses (Srivastava, 2002). This has led to immunotherapeutic applications of purified, antigen-laden HSPs (Przepiorka and Srivastava, 1998; Srivastava, 2006) as personalized anti-cancer vaccines. In spite of their high degree of conservation, the major heat shock protein families are also strong antigens by themselves, inducing both the humoral and the cellular arms of the immune response. This is in part due to their high abundance in all cells.

Heat shock proteins in *Leishmania* spp.

Interest in the heat shock proteins of *Leishmania* spp. was piqued by the finding that a temperature increase paired with acidification of growth media sufficed to induce promastigote-to-amastigote conversion for species such as *L. mexicana* and *L. donovani* (Barak *et al.*, 2005; Bates, 1993, 1994; Bates *et al.*, 1992; Zilberstein and Shapira, 1994). The prevalent concept of heat shock proteins at the time was that they protect amastigotes against the harsh environment in macrophage phagolysosomes, adding to the interest and spawning efforts to identify and characterize heat shock genes, their products, and their regulation. In the following paragraphs, we shall describe the basic characteristics of heat shock protein families and of their representatives in *Leishmania* spp.

Hsp70

Hsp70 family members are involved both in stress protection and in basic cellular functions, including protein folding, protein quality control, and transmembrane transport of proteins into organelles. The Hsp70s have ATPase activity, and the hydrolysis of ATP is crucial for the binding and release of client proteins. ATP-bound Hsp70 has poor client protein affinity, but the ADP-bound form will bind non-native proteins. The cyclic binding and release of clients due to ATPase activity is an inherent part of the functionality (Young, 2010). Hsp70, together with Hsp40, is also part of early Hsp90 foldosome complexes (Buchner, 1999), interacting with Hsp90 via the co-chaperone Sti1/HOP.

In the leishmaniae, there are a number of Hsp70 family members (Table 2.1). With the exception of *LinJ.28.2960*, which plays a role as cell death antagonist (Raina and Kaur, 2012), their functions have not been determined by reverse genetics, mostly due to the variety and number of coding genes. Copy numbers for Hsp70 genes appear to vary between species and even laboratory strains (Folgueira and Requena, 2007; Lee *et al.*, 1988; Wallace *et al.*, 1992), possibly reflecting adaptation. Hsp70 is expressed constitutively and amounts to ~2% of the soluble protein in promastigotes (Brandau *et al.*, 1995). Hsp70 interacts with at least two co-chaperones, Sti1 and LdSGT (Hombach *et al.*, 2012; Ommen *et al.*, 2010; Webb *et al.*, 1997).

Hsp40

The Hsp40 chaperones (a.k.a. J proteins) mediate client protein specificity, modulate ATPase activity of Hsp70 and are considered Hsp70-specific co-chaperones. In *Leishmania* spp. there is a multitude of Hsp40 variants (Folgueira and Requena, 2007) that may reflect a larger functional scope for chaperones in kinetoplastid protozoa (Kim *et al.*, 2013). Much like Hsp70, *Leishmania* Hsp40 proteins are still awaiting in-depth genetic and biochemical analysis. Given the large variety of Hsp40 genes, one may assume that they play important regulatory roles in the life cycle stages.

Hsp60/Hsp10

Hsp60 and Hsp10 are the two subunits of a large chaperone complex. Much is known about their orthologues in bacteria, the GroEL/GroES chaperone system. In *E. coli* two heptameric GroEL rings form a barrel-shaped structure with a single heptameric GroES ring as an asymmetrical lid (Kim *et al.*, 2013). The central cavity of the barrel structure houses the client protein preventing its interaction with other unfolded proteins and thus aggregation. In Eukaryotes, the Hsp60 family is split into the cytosolic TRiC chaperones and the mitochondrial chaperonins (CPN) 60 and 10. Only the latter are present in *Leishmania* spp. There are four genes coding for CPN60 in the *Leishmania* genomes. *CPN60.1* is transcribed, but its product could not be detected by Western blot (Schlueter *et al.*, 2000). CPN60.2 is encoded by two gene copies, locates to the mitochondrion, and interacts with CPN10 (Schlueter *et al.*, 2000; Zamora-Veyl *et al.*, 2005). The third chaperonin, CPN60.3, has not been analysed so far. CPN10 is encoded by two identical gene copies. Like CPN60.2, it is confined to the mitochondrion and forms stable complexes with its partner (Zamora-Veyl *et al.*, 2005). CPN60.2 and CPN60.3 are also part of the protein payload of *L. donovani* exosomes, and depend on Hsp100 for selection (Silverman *et al.*, 2010b). Both CPN60.2 and CPN10 show up-regulation under the conditions of axenic amastigote conversion (Rosenzweig *et al.*, 2008) and may thus be in greater demand during the mammalian stage.

Hsp90

In all eukaryotic cells examined to date, the 90 kDa heat shock proteins are essential for proliferation and survival. Hsp90 plays a dual role as a general chaperone of denatured proteins and as highly specialized facilitator of folding for regulatory proteins. This latter function was first discovered when it was identified as the 'inhibitor' of glucocorticoid receptor (Catelli *et al.*, 1985). Since then, its role as a chaperone for signal transduction proteins such as transcription factors, protein kinases, oncogenes, cell cycle regulators, and cytoskeletal proteins was established (Rutherford, 1994), making it a promising target of chemotherapeutic intervention in cancer and infectious diseases (Cowen *et al.*, 2009; Whitesell and Lindquist, 2005; Workman *et al.*, 2007).

Hsp90 is found as a highly interactive homodimer (Fig. 2.1). Each subunit consists of three major domains: (i) the N-terminal 180 amino acids form the ATPase domain of Hsp90 chaperones and serve as an interface for co-chaperones such as P23 and SGT (Bergerat *et al.*, 1997; Li *et al.*, 2012); (ii) the middle domain is responsible for client interaction (Pearl and Prodromou, 2006; Wandinger *et al.*, 2008), and (iii) the C-terminal region mediates dimerization and the binding of

Table 2.1 List of annotated chaperone genes in L. infantum sorted by HSP family. The annotation, gene accession number (http://www.genedb.org/Homepage/Linfantum), theoretical protein mass, functional information and references for the latter are listed

HSP family	Name	Gene(s)	Theoretical mass	Function in *Leishmania* spp.	Reference
CPN10	Chaperonin 10 = CPN10	LinJ.26.0610 LinJ.26.0590	10.6k	Mitochondrial, interacts with CPN60.2	Zamora-Veyl *et al.* (2005)
CPN60	Chaperonin 60.1 = CPN60.1	LinJ.32.1940	64.3k	Not detectable	Schlueter *et al.* (2000)
	Chaperonin 60.2 = CPN60.2	LinJ.36.2130 LinJ.36.2140	59.3k	Mitochondrial	Schlueter *et al.* (2000)
	Chaperonin 60.3	LinJ.30.2830	58.2k	Unknown	
α-Crystallins	Hsp20	LinJ.29.2560	17.5k		
	Hsp23	LinJ.34.0230	23k	Stress tolerance	Unpublished
	P23	LinJ.35.4540	21.6k	Geldanamycin resistance	Unpublished
HSP40	Chaperone protein DNAJ,	LinJ.04.0940	52.1k	Unknown	
HSP70	Heat shock protein 70	LinJ.28.2950 LinJ.28.2960 LinJ.28.3000 LinJ.28.3060	68.3k 71.6k 71.3k 80.4k	Unknown	
	Heat shock protein 70.4	LinJ.26.1220	70.5k	Unknown	
	Glucose-regulated protein 78, putative	LinJ.28.1310	71.9k	Unknown	
	Heat shock 70-related protein 1, mitochondrial precursor, putative	LinJ.30.2470 LinJ.30.2480 LinJ.30.2530 LinJ.30.2540	68.8–71.7k	Unknown	
HSP90	HSP90 = HSP83	LinJ.33.0350 LinJ.33.0360 LinJ.33.0370	78.8k 80.6k 80.6k	Promoting proliferation, life cycle control, phosphoprotein, interaction with co-chaperones Sti1 and LdSGT	Hombach *et al.* (2012), Morales *et al.* (2010), Ommen *et al.* (2010), Webb *et al.* (1997), Wiesgigl and Clos (2001)
	LPG3=GRP94	LinJ.29.0790	87k	Antigen, involvement with LPG expression	Beverley and Turco (1995), Larreta *et al.* (2000)
TPR proteins	Sti1 = stress-inducible protein 1	LinJ.08.1020	62.2k	Interaction with Hsp70 and Hsp90, phosphoprotein, essential protein	Hombach *et al.* (2012), Morales *et al.* (2010), Webb *et al.* (1997)
	LdSGT	LinJ.30.2740	45.8k	Essential protein, interaction with Hsp70, Hsp90, Sti1	Ommen *et al.* (2010)
	HOP 2	LinJ.36.0080	29.0k	Non-essential protein	Ommen *et al.* (2009)
	HIP	LinJ.29.0330	36.9k	Non-essential protein	Ommen *et al.* (2009)
HSP100	HSP100/ClpB	LinJ.29.1360	97k	Non-essential protein for promastigotes and axenic amastigotes, essential for intracellular survival, control of exosome loading	Hubel *et al.* (1997), Krobitsch *et al.* (1998), Krobitsch and Clos (1999), Silverman *et al.* (2010b)
	Heat shock protein 78	LinJ.02.0680	90.8k		

Table 2.1 (Continued)

HSP family	Name	Gene(s)	Theoretical mass	Function in *Leishmania* spp.	Reference
HSL VU	HSL VU complex, proteolytic subunit-like	LinJ.36.4180	24.1k	Essential protein for promastigotes	Chrobak *et al.* (2012)
	HSL VU complex, regulatory subunit-like	LinJ.15.0090	47.5k	Essential protein for promastigotes	Chrobak *et al.* (2012)
	HSL VU complex, regulatory subunit-like	LinJ.09.0380	55.7k	Non-essential protein	Chrobak *et al.* (2012)

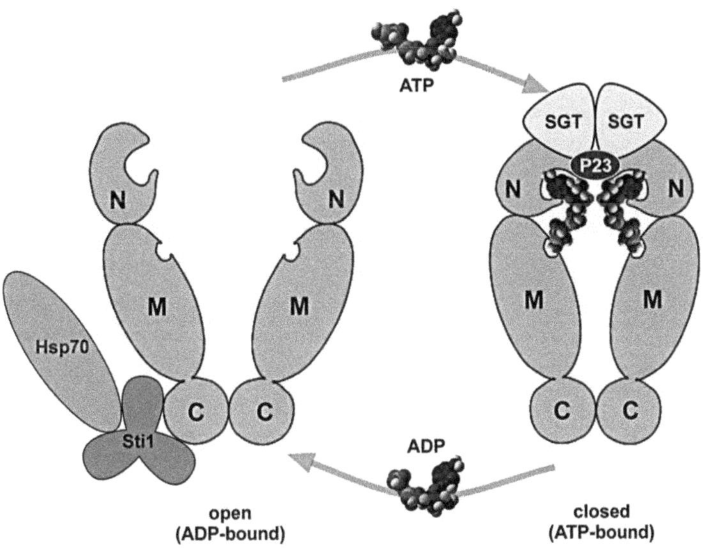

Figure 2.1 Current understanding of *L. donovani* Hsp90 foldosome composition based upon known co–chaperone/chaperone interactions and basic Hsp90 functions and structure. The presumed open conformation is assumed to interact with Hsp70 via Sti1, with the latter known to bind via the Hsp90 C-terminus (Hombach *et al.*, 2012). The closed foldosome is known to include the co-chaperones SGT and P23 bound to the N-terminal domain and client proteins bound to the middle domain (M). We know SGT to interact with Hsp90 as a dimer (Ommen *et al.*, 2010) and P23 to impact on the ATP binding domain (Hombach *et al.*, unpublished).

certain co-chaperones such as Sti1/HOP (Li *et al.*, 2011, 2012). The ATPase domain of Hsp90 is also recognized by two classes of inhibitors, geldanamycin (GA) and radicicol (RAD), and their derivatives. These inhibitors bind only to Hsp90 family members and block ATP hydrolysis, rendering the chaperone non-functional. In the absence of viable Hsp90 null-mutants, inhibition with either compound class has given valuable insight into the scope of functions of Hsp90 in cells (Byrd *et al.*, 1999; Munster *et al.*, 2001; Srethapakdi *et al.*, 2000; Whitesell and Cook, 1996; Whitesell *et al.*, 1994).

In *Leishmania* spp., Hsp90, or Hsp83, is encoded by up to 17 identical, tandemly arranged gene copies on Chromosome 33, spanning up to 70 kb of DNA (Hubel and Clos, 1996; Ivens *et al.*, 2005; Zilka *et al.*, 2001). The high gene copy number is expressed in a constitutively high level of Hsp90 in all *Leishmania* life cycle stages.

Hsp90 accounts for ~3% of the parasite's soluble protein, but is only transiently inducible during promastigote-to-amastigote conversion (Brandau *et al.*, 1995; Rosenzweig *et al.*, 2008). Like Hsp70, its localization is cytoplasmic and it is exported as part of the exosome payload (Brandau *et al.*, 1995; Hombach *et al.*, 2012; Silverman *et al.*, 2010a,b). A recent structural and biochemical analysis of the *L. braziliensis* Hsp90 showed this protein to be structurally compatible with the Hsp90s from eukaryotic model organisms, confirmed its interaction with several co-chaperones, demonstrated its chaperone function, and validated its ATPase activity (Silva *et al.*, 2013).

Hsp90 co-chaperones

It is known that to fulfil its range of functions, Hsp90 depends on a cohort of so-called co-chaperones that can modulate the specificity and turnover rate of the Hsp90 foldosome (Fig. 2.1) complex (Johnson and Brown, 2009). The leishmaniae, too, express an assortment of co-chaperones. Notably absent from the leishmaniae is the cdc37 co-chaperone which is responsible for the recognition of substrate protein kinases (Caplan *et al.*, 2007; Miyata, 2009; Whitesell and Lindquist, 2005).

Other co-chaperone genes could only be annotated tentatively and based on limited sequence conservation. In the following we list the co-chaperones that have been described to some degree in *Leishmania* spp.

Sti1 or HOP links the Hsp70 and Hsp90 chaperone complexes and is essential for client protein transfer into the foldosome *(Li et al., 2012)*. Typical for Sti1 is the presence of three tetratricopeptide repeat (TPR) domains that facilitate interaction with C-terminal consensus sequences found in Hsp70 and Hsp90. Its existence and interaction with both Hsp90 and Hsp70 were shown early for *L. major* (Webb *et al.*, 1997). Its crucial interaction with Hsp90 and its role as a phosphoprotein will be discussed below.

A separate gene, *LinJ.08.1020*, coding for a protein with a single TPR domain is also annotated as a Sti1 homologue in the database (http://tritrypdb.org/tritrypdb/showRecord.do?name=GeneRecordClasses.GeneRecordClass&project_id=TriTrypDB&source_id=LinJ.36.0080). Our results show, however, that this gene, dubbed *HOP2*, can be deleted without phenotypical consequences (Ommen *et al.*, 2009).

SGT, small glutamine-rich TPR protein, was described as a co-chaperone interacting with Hsc70 and modulating the activity of the Hsp70/Hsp40 chaperone complex (Angeletti *et al.*, 2002; Liu *et al.*, 1999). It is also known to be associated with Hsp90 chaperone complexes (Yin *et al.*, 2006). In spite of its TPR domain, SGT interacts not with the C-terminus of Hsp90, but rather with the N-terminus (Li *et al.*, 2012; Pearl and Prodromou, 2006). The *Leishmania* homologues of SGT are atypical because they lack the eponymous glutamine-rich domain. They are found as homodimers associated with Hsp70, Hsp90 and Sti1, and are essential for *L. donovani* viability (Ommen *et al.*, 2010).

Cyp40 (cyclophilin 40) has a peptidyl prolyl isomerase domain typical for cyclophilins and a TPR domain characteristic for co-chaperones. It is known to associate with Hsp90 complexes (Johnson and Brown, 2009). The *Leishmania* homologue was identified as an amastigote-specific phosphoprotein (Morales *et al.*, 2010). Moreover, treatment of *L. donovani* with cyclosporin A, which binds to Cyp40, induced axenic amastigote conversion (Yau *et al.*, 2010).

α-Crystallin family

The α-crystallin domain (ACD) family of chaperones includes small heat shock proteins (small Hsps) and P23 co-chaperones. The defining feature is the ~90 amino acid α-crystallin domain which forms a nine-stranded β-sheet barrel flanked by a highly heterogeneous N-terminus and a short C-terminus. The small HSPs were shown to confer resistance to various cellular stresses, including heat stress (Landry *et al.*, 1989; Lavoie *et al.*, 1993), ischaemic stress (Dillmann, 1999) and drought stress in plants (Sato and Yokoya, 2008), and they appear to be induced in Alzheimer's disease (Renkawek *et al.*, 1994).

There are three genes in *Leishmania* that code for proteins with α-crystallin signature (Table 2.1). Two of these, *LinJ.34.0230* and *LinJ.35.4540*, show an atypical structure with an extremely short N-terminal domain and an extended C-terminal tail, whilst the third, *LinJ.29.2560*, displays a more conventional structure. Ongoing reverse mutagenesis studies of *LinJ.34.0230* and *LinJ.35.4540* in our group (A. Hombach and G. Ommen, unpublished) shows the former to play a role in temperature tolerance, whilst the latter shows phenotypic characteristics of a P23 co-chaperone. In a recent study (Fraga *et al.*, 2013), the putative Hsp20 gene *LinJ.29.2560* was used successfully as a phylogenetic marker for *Leishmania* species differentiation.

Hsp100/ClpB

Chaperones of the Hsp100/ClpB family are characterized by one or two AAA ATPase domains. Some family members, such as ClpA, ClpX, and HslU, are regulatory subunits of bacterial peptidase systems (ClpA/P, ClpX/P, HslVU). The true Hsp100/ClpB chaperones are distinct from those peptidase subunits and were first described as mediators of stress tolerance (Gottesman, 1990, 1996; Parsell *et al.*, 1991; Sanchez and Lindquist, 1990; Sanchez *et al.*, 1992). Rather than preventing protein aggregation, Hsp100 is able to dissolve aggregates and restore function (Parsell and Lindquist, 1994; Vogel *et al.*, 1995) to thermosensitive enzymes. Even more intriguing is the impact of Hsp100 on prion-like protein folding disorders. Here, a balanced expression is required to maintain the aggregation-prone protein structures (Chernoff, 1995; Krobitsch and Lindquist, 2000; Ma and Lindquist, 1999; Newnam *et al.*, 1999; Patino *et al.*, 1996; Schirmer and Lindquist, 1997).

However, the most conserved function of Hsp100/ClpB is its role as mediator of stress tolerance. Stress-induced expression of yeast Hsp104 can protect the cells against a subsequent severe stress (Lee *et al.*, 1994; Parsell and Lindquist, 1994; Sanchez and Lindquist, 1990; Sanchez *et al.*, 1992; Schirmer *et al.*, 1996). It was for a suspected function in thermotolerance that Hsp100 from *L. major* was first identified and analysed (Hubel et al., 1995).

Hsp100 is encoded by a single gene copy on chromosome 29. It is one of the few proteins in *Leishmania* with a significant heat-inducibility, showing a 6- to 8-fold increase in axenic amastigotes compared with cultured promastigotes (Hubel *et al.*, 1995; Krobitsch *et al.*, 1998; Rosenzweig *et al.*, 2008). Its 3′-non-coding region shows sequence similarities to the amastin 3′ UTR. In contrast to the yeast and bacterial homologues which are hexameric (Parsell *et al.*, 1994a), *L. donovani* Hsp100 forms trimers in solution (Krobitsch *et al.*, 1998). It usually locates to foci along the parasite's plasma membrane and along the flagellar pocket (Fig. 2.2B and C), in agreement with it being part of the *Leishmania* exosome payload (Krobitsch *et al.*, 1998; Silverman *et al.*, 2010a).

Leishmania and the cell stress

Heat shock response in *Leishmania* spp.

In keeping with the general lack of individual transcription regulation in Kinetoplastida (Clayton, 2002), the synthesis of heat shock gene mRNAs is unaffected by temperature (Argaman *et al.*, 1994; Brandau *et al.*, 1995; Quijada *et al.*, 1997). While mRNA stability and abundance were reported to increase under heat stress in some studies (Aly *et al.*, 1994; Shapira, 1990; Shapira *et al.*, 2001), nothing more than marginal increases were observed in other reports (Brandau *et al.*, 1995; Hubel *et al.*, 1995; Miller, 1988). It would appear that much like other *Leishmania* genes (Monnerat *et al.*, 2004; Myler *et al.*, 2001), the heat shock genes are transcribed in an unregulated fashion and as part of polycistronic transcription units (Wiesgigl and Clos, 1999), with the possible exception of the Hsp70 gene cluster (Dresel and Clos, 1997). Changes in the RNA stability under heat stress (Aly *et al.*, 1994) cannot account for the rapid increase of HSP synthesis under stress (Brandau *et al.*, 1995). Moreover, induced HSP synthesis is in fact independent of *de novo* nucleic acid synthesis (Clos *et al.*, 1998). The conclusion from this was that HSP synthesis is regulated at the translation level.

More recent studies have confirmed that certain heat shock proteins such as Hsp100 and Hsp60 have an increased abundance during promastigote-to-amastigote conversion (Rosenzweig *et al.*, 2008). Interestingly, the steady state mRNA levels of Hsp60 (CPN60.2 and CPN60.3) and Hsp100 are also increased (Lahav *et al.*, 2011).

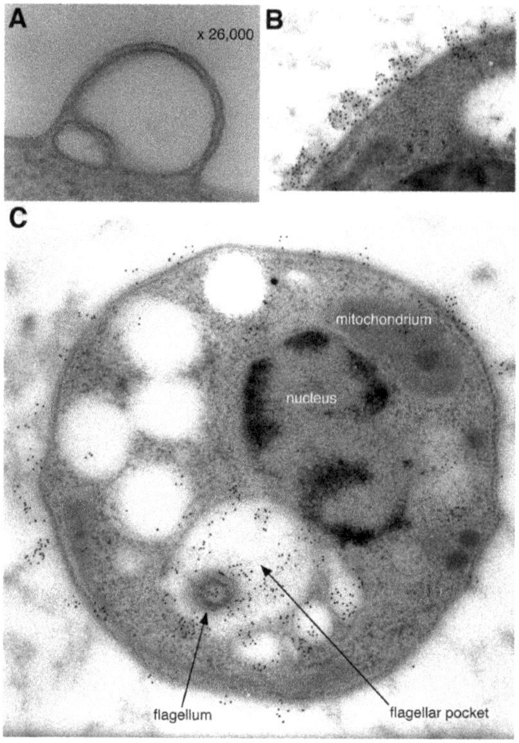

Figure 2.2 (A) Transmission electron microscopic image presumably of budding exosome on the plasma membrane of a *L. mexicana* promastigote. Magnification 26,000. (B) Immune gold electron microscopy of *L. donovani* promastigote kept at 37°C and decorated with anti-Hsp100 antibodies and anti-chicken IgG immunogold particles (10 nm). (C) As before, but additional labelling with anti-α-tubulin mAb and anti-mouse IgG immunogold particles (6 nm). Images produced by M. Krömer.

Other studies found no heat shock mRNAs increased during differentiation (Alcolea *et al.*, 2010a,b). However, correlation between mRNA abundance and corresponding protein abundance is very low in *Leishmania* (Cohen-Freue *et al.*, 2007; Lahav *et al.*, 2011), casting doubt on the significance of quantitative RNA data.

The promastigote view

Much of the early work on the *Leishmania* heat shock response was based on the assumption that the parasites encounter the mammalian tissue temperature and the hostile environment of phagolysosomes upon transmission and require the elevated synthesis of stress proteins for protection (Hubel *et al.*, 1995; Hunter *et al.*, 1984; Lawrence and Robert-Gero, 1985; van der Ploeg *et al.*, 1985). This stemmed from the then prevalent perception of HSPs as effectors of stress protection (Lindquist, 1992; Parsell and Lindquist, 1993; Parsell and Lindquist, 1994; Sanchez and Lindquist, 1990; Solomon, 1991). Also, since *Leishmania* promastigotes were easily cultivatable, the transient insect stage was viewed as the standard. Consequently, the temperatures in the mammalian tissue were seen as stress.

Several heat shock proteins are implicated in stress protection, from the small heat shock proteins (Basha *et al.*, 2012; Dillmann, 1999; Hilton *et al.*, 2013) to the highly abundant Hsp70 (Solomon, 1991; Vogel *et al.*, 1995) and the Hsp100/ClpB-type chaperones. The latter in particular confer stress tolerance to a wide range of organisms, including bacteria, yeast, and plants (Glover and Lindquist, 1998; Parsell and Lindquist, 1994; Schirmer *et al.*, 1996). This is achieved by the ability of Hsp100 family members to dissolve cytotoxic protein aggregates (Chernoff, 1995; Parsell *et al.*, 1994b; Vogel *et al.*, 1995). In *Leishmania*, however, the effects of Hsp100 on temperature tolerance are small (Hubel *et al.*, 1997) to non-existent (Krobitsch *et al.*, 1998). Moreover, functional cloning strategies to identify genes mediating temperature tolerance (Hoyer *et al.*, 2001) did not result in the selection of any known heat shock gene loci.

The amastigote view

When looking at the *Leishmania* life cycle, one cannot ignore the simple fact that the amastigote stage is the standard. Leishmaniae proliferate as amastigotes for months and years, whereas the promastigote is a transient stage of 7 to 10 days. Consequently, one must accept that for *Leishmania*, the conditions inside the mammalian macrophages and other APCs cannot be considered 'stress', but simply an alternative milieu to which the parasites are supremely adapted.

Functions in stress tolerance

The default candidates for mediation of stress tolerance are the Hsp100, Hsp70 and α-crystallin families. As detailed below, the Hsp100 of *Leishmania* spp. do not contribute significantly to the stress tolerance of promastigotes, whilst their crucial role in amastigotes lies in the sorting of exosome payload proteins.

The various Hsp70 genes and their products have not yet been analysed in a functional way. Therefore, their impact on thermotolerance is unknown. In *Drosophila melanogaster*, which lacks Hsp100/ClpB genes, Hsp70 overexpression can induce thermotolerance (Solomon, 1991), but this role is superseded in Hsp100-producing organisms (Lindquist, 1992; Parsell and Lindquist, 1994) to which *Leishmania* belongs. It is also unknown whether and which Hsp70 and Hsp70-related genes are essential for *Leishmania* spp. There are at least 4 different cytosolic Hsp70 family members which await genetic and functional analysis.

Ongoing work in our laboratory shows a 23 kDa small HSP as essential for the survival at mammalian tissue temperatures and, concomitantly, for infectivity (A. Hombach, unpublished). It remains to be seen whether the specific thermotolerance patterns of *Leishmania* species may hinge on the structure and/or expression rates of small HSPs.

Heat shock proteins and drug resistance

So far, two heat shock proteins have been implicated in the basic resistance against anti-parasitic drugs. Overexpression of Hsp70 and Hsp90, respectively, was observed in parasites that were selected under trivalent antimony (Sb^{III}) *in vitro* (Brochu *et al.*, 2004; Vergnes *et al.*, 2007). It was hypothesized that a higher stress protein level may counteract the induction of programmed cell death by antimonials (Vergnes *et al.*, 2007). Another possible role may lie in the protection against oxidative stress inside the phagolysosomes of activated macrophages (Wilson *et al.*, 1994). Recent results also show that an antimony hypersensitivity caused by histone H1 overexpression can be nullified by the simultaneous overexpression of Hsp90 in *L. donovani* (Alexandratos *et al.*, 2013). So far, heat shock proteins have not been linked to resistance against other anti-leishmanials.

The pitfalls of analogy-based assumptions

Many annotations in the TriTryp database are inferred from structural similarities with proteins from eukaryotic and prokaryotic model organisms. In several cases, such structural similarities do not correspond to functional homologies. Therefore, functional assumptions should not be based entirely on amino acid sequence comparisons.

Hsp100 – lack of tolerance

Chaperones of the Hsp100 family are known to confer inducible thermotolerance in organisms as diverse as bacteria, yeast, and higher plants (Lindquist, 1992; Parsell and Lindquist, 1994; Parsell *et al.*, 1991; Sanchez and Lindquist, 1990; Sanchez *et al.*, 1992; Schirmer *et al.*, 1996). It was natural to assume that the *Leishmania* homologues would have a similar function, in particular in the context of arthropod-to-mammal transmission (Hubel *et al.*, 1995). However, phenotype analyses of Hsp100 null mutants showed marginal effects on thermotolerance if at all. In *L. donovani*, lack of Hsp100 reduced the production of amastigote-specific A2 protein during axenic stage conversion. More importantly, Hsp100$^{-/-}$ mutants of both *L. major* and *L. donovani* proved unable to survive and proliferate inside mouse macrophages after *in vitro* infection. The *L. major* Hsp100$^{-/-}$ mutants were also avirulent in BALB/c mice (Hubel *et al.*, 1997; Krobitsch *et al.*, 1998; Krobitsch and Clos, 1999).

In yeast, Hsp70 overexpression can partially overcome the thermosensitive phenotype of Hsp104 null mutants (Sanchez *et al.*, 1993). It was telling that a screen for genes that could overcome the Hsp100 null mutant phenotype in *L. major* produced no heat shock genes, but a general virulence factor unrelated to chaperones (Reiling *et al.*, 2010).

Hsp100 and the exosomes of *Leishmania* spp.

A hint at the function of Hsp100 was obtained from a proteome analysis of *Leishmania* exosomes, export vesicles (Fig. 2.2A) that can transport cytoplasmic proteins to the host cell that lack Golgi-specific signal sequences (Silverman *et al.*, 2008; Silverman *et al.*, 2010a). Among the protein payload of those exosomes were several heat shock proteins, including Hsp100 (Fig. 2.2B), Hsp90 and Hsp70, but also cytoskeletal proteins such as α-tubulin (Fig. 2.2C) and β-tubulin. By including Hsp100$^{-/-}$ mutants in the functional analysis of exosomes, it was found that Hsp100 is critical for the proper sorting of proteins into exosomes, and thus for the immune modulatory function of these vesicles (Lambertz *et al.*, 2012; Silverman and Reiner, 2011a,b). It thus became clear why Hsp100$^{-/-}$ mutants were perfectly viable in axenic culture but failed to survive and proliferate as intracellular amastigotes, both *in vitro* and in mouse infection models (Krobitsch *et al.*, 1998; Krobitsch and Clos, 1999).

HslV – redundant but essential

Model bacteria such as *E. coli* have redundant peptidase systems, including the Lon peptidases, the ClpA/P complexes and lastly, the HslV/U peptidase. The latter was originally identified as a heat-inducible operon in *E. coli*; hence the name *heat shock locus V/U* (Chuang *et al.*, 1993). The HslV product is structurally related to eukaryotic proteasome β subunits and forms the catalytic subunit of the HslV/U complex (Bochtler *et al.*, 1997; Kessel *et al.*, 1996; Rohrwild *et al.*, 1996; Rohrwild *et al.*, 1997). HslU forms the regulatory subunit and is related to ClpB/Hsp100 (Gottesman, 2003). In bacteria, HslV/U forms a complex consisting of two inner, hexameric HslV rings, and two outer, also hexameric, HslU rings. The presence of HslV/U peptidases seemed to be restricted to the prokaryota. However, when several Kinetoplastida and Apicomplexa genomes were sequenced, genes coding for HslV/U homologues were found (Couvreur *et al.*, 2002; Gille *et al.*, 2003).

Whilst HslV/U null mutants in *E. coli* cause no phenotype by themselves (Kanemori *et al.*, 1997), both HslV and one of the HslU genes in *Leishmania* are essential since no viable null mutants could be raised (Chrobak *et al.*, 2012). In addition, no interaction between HslV – which forms a dodecamer by itself – and HslU could be observed, indicating that the *Leishmania* HslV/U homologues have a highly specific and important role that differs from the redundant function in bacteria. The essential role of HslV and its structural similarity to proteasome β-subunits suggests targeting this protein with derivatives of known proteasome inhibitors.

The chaperone machinery and signal transduction in *Leishmania*

Hsp90 and life cycle control

The high gene copy number and intracellular concentration of Hsp90 in *Leishmania* spp. piqued the interest of researchers early on. However, the high copy number and the presumed essential function precluded the use of reverse genetics, i.e. homologous recombination-based gene replacement, to analyse its role. The advent of specific Hsp90 inhibitors, namely the ansamycin antibiotic geldanamycin (GA) and the macrocyclic antimycotic radicicol (RAD) allowed researchers to assess the importance of the protein by chemical knock-down. The effects in the case of *L. donovani* were manifold. Treatment of promastigotes kept at 25°C and at neutral pH resulted in induced heat shock protein synthesis and a growth arrest in the

G2 cell cycle phase. Moreover, the treatment led to the expression of the amastigote-specific A2 marker proteins and the conversion to amastigote-like cell forms (Wiesgigl and Clos, 2001). Several lines of evidence point at Hsp90 as the relevant drug target. Firstly, adaptation of *L. donovani* to intermediate doses of GA coincided with a ~10-fold amplification of Hsp90 genes. Secondly, the targeted overexpression of Hsp90 from episomes increased the IC50 for GA. Thirdly, RAD had identical effects on the morphology and A2 protein expression (Wiesgigl and Clos, 2001). Fourth, expression of a RAD-resistant form of Hsp90 abrogated the effects of the drug on cell growth and morphology (Hombach *et al.*, 2012). These results point at Hsp90 as a pivotal regulator of cell fate.

The role of Hsp90 in protozoan life cycle control may not be unique to *L. donovani*. GA treatment of *Trypanosoma cruzi* epimastigotes resulted in the formation of spheromastigotes, a stage normally associated with starvation (Graefe *et al.*, 2002). In *Plasmodium falciparum*, GA treatment of ring stages blocked the further development towards trophozoites (Banumathy *et al.*, 2003). The signals transposed by Hsp90 are therefore not restricted to temperature.

Hsp90 and the foldosome members

In *Leishmania* spp. the interaction of Hsp90 with several co-chaperones was demonstrated. As mentioned, the highly conserved and ubiquitous co-chaperone Sti1 is found complexes and in co-localization with Hsp70 and Hsp90 (Hombach *et al.*, 2012; Webb *et al.*, 1997). Moreover, it was shown that Sti1 and the Hsp90 – Sti1 interaction are both crucial for *L. donovani* viability and infectivity, and depend on a variant C-terminal pentapeptide recognition sequence in Hsp90 (Hombach *et al.*, 2012; Morales *et al.*, 2010).

The *L. donovani* SGT (LdSGT) is also a part of complexes that include Hsp90, Hsp70 and Sti1. LdSGT, like Sti1, is essential for *L. donovani* viability, in spite of its lack of the eponymous glutamine-rich domain (Ommen *et al.*, 2010).

An X-ray diffraction analysis of *L. braziliensis* Hsp90 and Activator of Hsp90 ATPase 1 (Aha1) demonstrated the interaction of Aha1 with both the N-terminal domain and the middle domain of Hsp90, as well as an Aha1-dependent modulation of Hsp90 ATPase activity (Seraphim *et al.*, 2013)

Only indirect evidence exists for an interaction between Hsp90 and the non-TPR domain co-chaperone P23. Similar to the P23 homologue in yeast, Sba1 (Forafonov *et al.*, 2008), replacement of the putative *L. donovani P23* increases the GA sensitivity of Hsp90 (Ommen, 2009).

A similar effect was observed in *T. brucei* for the TPR domain-containing protein phosphatase 5 (PP5) that was shown to interact with Hsp90. PP5 overexpression protected Hsp90 against the inhibitor GA whilst its depletion by RNAi rendered the trypanosomes more sensitive to the drug (Jones *et al.*, 2008). *Leishmania* homologues of PP5 were already identified (Downing *et al.*, 2011; Johnson and Brown, 2009), and their function is probably similar.

From this one may deduce that the Hsp90 foldosome of *Leishmania* spp. comprises a variety of interacting co-chaperones and largely resembles its counterparts in the well-studied model eukaryotes such as yeast (Wandinger *et al.*, 2008).

The Hsp90 foldosome – effectors of signal transduction

For an organism devoid of regulated transcription, the leishmaniae express a surprising variety of protein kinases and protein phosphatases. Roughly 200 protein kinases have been identified (Parsons *et al.*, 2005) representing the well-known signal transduction cascades. However, whilst kinase cascades in higher eukaryotes are known to have proteins such as transcription factors as final substrates, the non-kinase substrate proteins in *Leishmania* are largely unknown. A phosphoproteome study resulted in the then surprising finding that a large portion of the phosphorylated proteins are chaperones and co-chaperones (Morales *et al.*, 2008). Moreover, several *Leishmania* chaperones were shown to be phosphorylated in a stage conversion-dependent manner (Morales *et al.*, 2010). This is especially true for the known components of the Hsp90 foldosome – Hsp90, Hsp70, Cyp40 and Sti1. A preliminary analysis of Sti1 phosphorylation showed at least one phosphorylation site to be essential for *Leishmania* viability (Morales *et al.*, 2010). Hsp90 phosphorylation also has a high impact on viability, proliferation and infectivity. Targeted mutagenesis at seven known or suspected phosphorylation sites in *L. donovani* Hsp90 revealed an amastigote-specific impact at two sites, whilst all mutations more or less affected *in vitro* promastigote growth (Hombach, 2013). These data implicate the *Leishmania* heat shock proteins as the so far elusive effectors and definitive substrates of protein phosphorylation pathways.

Open questions

In spite of the knowledge gained in recent years, many important issues are yet to be addressed, concerning the role of chaperones in signal transduction, the identification of client proteins, and a number of chaperones whose function in *Leishmania* has yet to be investigated.

Upstream kinases

Stage-specifically phosphorylated heat shock proteins may reflect modulated chaperone activity due to post-translational modifications. This begs the question of which kinase pathway is responsible for the modulation of the chaperones.

The studies performed with other biological systems may at least give some guidance for future research. Several protein kinases have been shown to target Hsp90, some of which are encoded by the *Leishmania* genomes.

Casein kinase 2 can phosphorylate the T_{22} moiety of *S. cerevisiae* Hsp90 which modulates the ATPase turnover rate and impedes the interaction with Aha1. We are currently investigating the role of the corresponding threonine in the *Leishmania* Hsp90 (Mollapour and Neckers, 2012; Mollapour *et al.*, 2011a,b).

There are several protein kinase A homologues in *L. major* (Parsons *et al.*, 2005). In humans, Hsp90α and Hsp90β are substrates of PKA phosphorylation at T_{90}, a modification that modulates ATPase turnover rates and the interactions with co-chaperones (Wang *et al.*, 2012).

Hsp90 is also a substrate for the mitogen-activated kinase 14 or P38 (Muller *et al.*, 2013). Although blast searches of the *Leishmania* spp. sequence databases show up to 44% sequence identity between human MPK14 and *Leishmania* spp. MPK3 and MPK4, no annotated MPK14 is listed. Both MPK4 and MPK3 were the subject of mutagenetic analyses in *L. mexicana*. Whilst MPK4 is essential for promastigote viability (Wang *et al.*, 2005), MPK3 impacts on the flagellar length of promastigotes (Erdmann *et al.*, 2006), a role similar to that of Hsp90 (Wiesgigl and Clos, 2001).

Clients of Hsp90

So far, only one client protein of *Leishmania* Hsp90 has been identified. A co-immune precipitation analysis showed that the silent information regulator 2 (SIR2)RP1 protein interacts with Hsp90 *in vivo* (Adriano *et al.*, 2007). In eukaryotic model organisms such as yeast, a vast variety of regulatory proteins are known to depend on Hsp90 chaperoning, including protein kinases, steroid hormone receptors, and transcription regulator including several oncogenes (Buchner, 1999; Picard, 2002; Rutherford, 1994). Unfortunately, this does not allow deductions for *Leishmania* spp. With the exception of protein kinases, none of the known classes of regulatory client proteins is expressed in the leishmaniae. Therefore, it seems worthwhile to analyse the functionality of known signal transduction kinases under Hsp90 inhibition (Wiesgigl and Clos, 2001) or in cells depending on mutated Hsp90 (Hombach *et al.*, 2012).

Another possible group of client proteins may be the cytoskeletal proteins. Both tubulin and actin have been shown to be clients of Hsp90 in mammalian cells (Koyasu *et al.*, 1986; Sanchez *et al.*, 1988; Wegele *et al.*, 2004; Weis *et al.*, 2010). Also, cytoskeletal proteins require massive reorganization and turnover during *Leishmania* stage conversion. It should be interesting to investigate interactions with Hsp90 either by immune electron microscopy or by high-resolution live imaging techniques.

Small heat shock proteins and their role in *Leishmania* survival

From our ongoing work with two *Leishmania* α-crystallin protein family members, a putative P23 co-chaperone and a putative Hsp23, it is already clear that both have strong impact on the intracellular amastigote stage, the latter through a pivotal effect on thermotolerance (A. Hombach and G. Ommen, unpublished). Both proteins show an unusual domain structure, with a short N-terminal region and an elongated C-terminal stretch. Another putative α-crystallin protein of 17 kDa has not yet been characterized, but it shows a more typical small HSP structure with a long N-terminal domain and a short C-terminal stretch. The role and function of the putative Hsp20, as well as the implications of the unusual domain structures of P23 and Hsp23, are worthy of further investigation.

All the Hsp70s and Hsp40s

Both Hsp70 and Hsp40 belong to the neglected chaperones in *Leishmania*. However, a recent study using antisense oligonucleotide lipofection-mediated RNA knockdown showed that Hsp70 encoded by gene *LinJ.28.2960* may play an important role as an antagonist of programmed cell death (Raina and Kaur, 2012). Unfortunately, the study did not show whether the RNA knockdown leads to a depletion of the Hsp70 chaperone. Other studies focussed on the role of Hsp70 as antigen in *Leishmania* infections (Quijada *et al.*, 1996a,b, 1998). Hsp70 is also known as amastigote stage-specific phosphoprotein (Morales *et al.*, 2010) and as part of the immune-modulatory exosome payload proteins (Silverman *et al.*, 2010a,b). Taken together, the information on Hsp70 is sketchy and requires dedicated efforts, given the essential role of Hsp70 chaperones in other eukaryotes.

Compared with Hsp70, the knowledge of Hsp40 is even sparser. It is known, however, that there are a large variety of Hsp40 gene family members (Folgueira and Requena, 2007), possibly reflecting diverse and distinct roles for this co-chaperone family. This diversity, however, makes it difficult to decide on targets for reverse genetic analyses. In other words – where to start? Until possible leads come out of system biology approaches ('omics'), the Hsp40s will be a blank spot on the chaperone map.

Organelle heat shock proteins – the glucose-regulated proteins (Grps) in lipophospoglycan (LPG) synthesis and antigenicity

A little more is known about compartmental HSPs. The best-known example is the *L. donovani* 94 kDa-glucose-regulated protein (Grp94) homologue that was identified under the moniker LPG3, since its gene emerged from a genetic complementation screen aimed at rescuing lipophosphoglycan synthesis in a deficient *L. donovani* strain (Descoteaux *et al.*, 2002). Grp94 was also implicated as an antigen during *L. infantum* infections (Larreta *et al.*, 2000, 2002). Grp94 family members belong to the Hsp90 family, but are localized to the endoplasmic reticulum and do not possess co-chaperone binding capacity. The ER-specific Hsp70 family member, Grp78, was also shown to evoke immune responses in experimental hosts and is indeed localized to the perinuclear region in *L. donovani* (Jensen *et al.*, 2001; Jensen *et al.*, 2002). Its abundance is induced under unfolded protein ER stress in good agreement with its expected function (Dolai *et al.*, 2011).

Other than their existence, nothing is known about the mitochondrial heat shock proteins mtHsp70 and TRAP1. Only the Hsp60 and Hsp10 family members CPN60.2 and CPN10 were shown to co-localize and interact inside the mitochondrion (Schlueter *et al.*, 2000; Zamora-Veyl *et al.*, 2005). Their abundance increases during promastigote-to-amastigote conversion (Rosenzweig *et al.*, 2008), and two Hsp60 chaperones, CPN60.2 and CPN60.3, are part of the exosome protein payload (Silverman *et al.*, 2010b).

Future trends

The *Leishmania* heat shock proteins have emerged as crucial players in the parasite's life cycle control and parasitic survival, impacting on cellular regulation pathways, immune modulation and general viability. However, our knowledge of their functions is still rudimentary and will require dedicated research efforts in the future. Most important in that context are the functions of heat shock proteins in *Leishmania* signal transduction. In this context, it will be crucial to identify both the upstream protein kinases and the client proteins affected by post-translational modifications of HSPs. Another important field of research will be the impact of HSPs on *Leishmania* survival under anti-parasitic drugs. For this, more must be learned about the mechanisms by which *Leishmania* spp. control gene expression, i.e. protein synthesis, in response to environmental adversity.

Acknowledgements

We wish to express our thanks to past and present members of our laboratory for the manifold contributions and discussions. Special thanks to Laura Jade Lee for the proofreading of the manuscript and to Gerald Späth for the constant exchange of ideas and concepts. The electron micrographs are the work of Manfred Krömer († 2009). We also much appreciate the numerous colleagues, inside and outside of the 'Leishmaniacs' community, who with their work and ideas added to our knowledge and understanding.

References

Adriano, M.A., Vergnes, B., Poncet, J., Mathieu-Daude, F., da Silva, A.C., Ouaissi, A., and Sereno, D. (2007). Proof of interaction between *Leishmania* SIR2RP1 deacetylase and chaperone HSP83. Parasitol. Res. *100*, 811–818.

Alcolea, P.J., Alonso, A., Gomez, M.J., Moreno, I., Dominguez, M., Parro, V., and Larraga, V. (2010a). Transcriptomics throughout the life cycle of *Leishmania infantum*: high down-regulation rate in the amastigote stage. Int. J. Parasitol. *40*, 1497–1516.

Alcolea, P.J., Alonso, A., Gomez, M.J., Sanchez-Gorostiaga, A., Moreno-Paz, M., Gonzalez-Pastor, E., Torano, A., Parro, V., and Larraga, V. (2010b). Temperature increase prevails over acidification in gene expression modulation of amastigote differentiation in *Leishmania infantum*. BMC Genomics *11*, 31.

Alexandratos, A., Clos, J., Samiotaki, M., Efstathiou, A., Panayotou, G., Soteriadou, K., and Smirlis, D. (2013). Restoration of Heat Shock Protein 83 expression in Histone H1 over-expressing *Leishmania donovani* attenuated parasites identifies HSP83 as a virulence factor. Mol. Microbiol. *88*, 1015–1031.

Aly, R., Argaman, M., Halman, S., and Shapira, M. (1994). A regulatory role for the 5′ and 3′ untranslated regions in differential expression of hsp83 in *Leishmania*. Nucleic Acids Res. *22*, 2922–2929.

Angeletti, P.C., Walker, D., and Panganiban, A.T. (2002). Small glutamine-rich protein/viral protein U-binding protein is a novel cochaperone that affects heat shock protein 70 activity. Cell Stress Chaperones *7*, 258–268.

Argaman, M., Aly, R., and Shapira, M. (1994). Expression of heat shock protein 83 in *Leishmania* is regulated post-transcriptionally. Mol Biochem. Parasitol. *64*, 95–110.

Banumathy, G., Singh, V., Pavithra, S.R., and Tatu, U. (2003). Heat shock protein 90 function is essential for Plasmodium falciparum growth in human erythrocytes. J. Biol. Chem. *278*, 18336–18345.

Barak, E., Amin-Spector, S., Gerliak, E., Goyard, S., Holland, N., and Zilberstein, D. (2005). Differentiation of *Leishmania donovani* in host-free system: analysis of signal perception and response. Mol. Biochem. Parasitol. *141*, 99–108.

Basha, E., O'Neill, H., and Vierling, E. (2012). Small heat shock proteins and alpha-crystallins: dynamic proteins with flexible functions. Trends Biochem. Sci. *37*, 106–117.

Bates, P.A. (1993). Axenic amastigote culture of *Leishmania* amastigotes. Parasitol. Today *9*, 143–146.

Bates, P.A. (1994). Complete developmental cycle of *Leishmania mexicana* in axenic culture. Parasitology *108*, 1–9.

Bates, P.A., Cobertson, C.D., Tetley, L., and Coombs, G.H. (1992). Axenic cultivation and characterization of *Leishmania mexicana* amastigote-like forms. Parasitology *105*, 193–202.

Bergerat, A., de Massy, B., Gadelle, D., Varoutas, P.C., Nicolas, A., and Forterre, P. (1997). An atypical topoisomerase II from Archaea with implications for meiotic recombination. Nature *386*, 414–417.

Beverley, S.M., and Turco, S.J. (1995). Identification of genes mediating lipophosphoglycan biosynthesis by functional complementation of *Leishmania donovani* mutants. Ann. Trop. Med. Parasitol. *89* (Suppl. 1), 11–17.

Bochtler, M., Ditzel, L., Groll, M., and Huber, R. (1997). Crystal structure of heat shock locus V (HslV) from *Escherichia coli*. Proc. Natl. Acad. Sci. U.S.A. *94*, 6070–6074.

Brandau, S., Dresel, A., and Clos, J. (1995). High constitutive levels of heat-shock proteins in human-pathogenic parasites of the genus *Leishmania*. Biochem. J. *310*, 225–232.

Brochu, C., Haimeur, A., and Ouellette, M. (2004). The heat shock protein HSP70 and heat shock cognate protein HSC70 contribute to antimony tolerance in the protozoan parasite leishmania. Cell Stress Chaperones *9*, 294–303.

Buchner, J. (1999). Hsp90 & Co. – a holding for folding. Trends Biochem. Sci. *24*, 136–141.

Byrd, C.A., Bornmann, W., Erdjument-Bromage, H., Tempst, P., Pavletich, N., Rosen, N., Nathan, C.F., and Ding, A. (1999). Heat shock protein 90 mediates macrophage activation by Taxol and bacterial lipopolysaccharide. Proc. Natl. Acad. Sci. U.S.A. *96*, 5645–5650.

Caplan, A.J., Ma'ayan, A., and Willis, I.M. (2007). Multiple kinases and system robustness: a link between Cdc37 and genome integrity. Cell Cycle *6*, 3145–3147.

Catelli, M.G., Binart, N., Jung-Testas, I., Renoir, J.M., Baulieu, E.E., Feramisco, J.R., and Welch, W.J. (1985). The common 90-kd protein component of non-transformed '8S' steroid receptors is a heat-shock protein. EMBO J. *4*, 3131–3135.

Chernoff, Y.O., Lindquist, S.L., Ono, B.-I., Inge-Vechtomov, S.G., and Liebman, S.W. (1995). Role of the chaperone protein Hsp104 in propagation of the yeast prion-like factor [psi^{+}]. Science *268*, 880–0884.

Cho, H.S., Shimazu, T., Toyokawa, G., Daigo, Y., Maehara, Y., Hayami, S., Ito, A., Masuda, K., Ikawa, N., Field, H.I., *et al.* (2012). Enhanced HSP70 lysine methylation promotes proliferation of cancer cells through activation of Aurora kinase B. Nat. Commun. 3, 1072.

Chrobak, M., Forster, S., Meisel, S., Pfefferkorn, R., Forster, F., and Clos, J. (2012). *Leishmania donovani* HslV does not interact stably with HslU proteins. Int. J. Parasitol. *42*, 329–339.

Chuang, S.E., Burland, V., Plunkett, G., 3rd, Daniels, D.L., and Blattner, F.R. (1993). Sequence analysis of four new heat-shock genes constituting the hslTS/ibpAB and hslVU operons in *Escherichia coli*. Gene *134*, 1–6.

Clayton, C.E. (2002). Life without transcriptional control? From fly to man and back again. EMBO J. *21*, 1881–1888.

Clos, J., Brandau, S., and and Hoyer, C. (1998). Chemical stress does not induce heat shock protein synthesis in *Leishmania donovani*. Protist *149*, 167–172.

Cohen-Freue, G., Holzer, T.R., Forney, J.D., and McMaster, W.R. (2007). Global gene expression in *Leishmania*. Int. J. Parasitol. 37, 1077–1086.

Couvreur, B., Wattiez, R., Bollen, A., Falmagne, P., Le Ray, D., and Dujardin, J.C. (2002). Eubacterial HslV and HslU subunits homologs in primordial eukaryotes. Mol. Biol. Evol. *19*, 2110–2117.

Cowen, L.E., Singh, S.D., Kohler, J.R., Collins, C., Zaas, A.K., Schell, W.A., Aziz, H., Mylonakis, E., Perfect, J.R., Whitesell, L., *et al.* (2009). Harnessing Hsp90 function as a powerful, broadly effective therapeutic strategy for fungal infectious disease. Proc. Natl. Acad. Sci. U.S.A. *106*, 2818–2823.

Czarnecka, E., Nagao, R.T., Key, J.L., and Gurley, W.B. (1988). Characterization of Gmhsp26-A, a stress gene encoding a divergent heat shock protein of soybean: heavy-metal-induced inhibition of intron processing. Mol. Cell Biol. *8*, 1113–1122.

Descoteaux, A., Avila, H.A., Zhang, K., Turco, S.J., and Beverley, S.M. (2002). *Leishmania* LPG3 encodes a GRP94 homolog required for phosphoglycan synthesis implicated in parasite virulence but not viability. EMBO J. *21*, 4458–4469.

Dillmann, W.H. (1999). Small heat shock proteins and protection against injury. Ann. N. Y. Acad. Sci. *874*, 66–68.

Dolai, S., Pal, S., Yadav, R.K., and Adak, S. (2011). Endoplasmic reticulum stress-induced apoptosis in *Leishmania* through Ca2+-dependent and caspase-independent mechanism. J. Biol. Chem. *286*, 13638–13646.

Donlin, L.T., Andresen, C., Just, S., Rudensky, E., Pappas, C.T., Kruger, M., Jacobs, E.Y., Unger, A., Zieseniss, A., Dobenecker, M.W., *et al.* (2012). Smyd2 controls cytoplasmic lysine methylation of Hsp90 and myofilament organization. Genes Dev. *26*, 114–119.

Downing, T., Imamura, H., Decuypere, S., Clark, T.G., Coombs, G.H., Cotton, J.A., Hilley, J.D., de Doncker, S., Maes, I., Mottram, J.C., *et al.* (2011). Whole genome sequencing of multiple *Leishmania donovani* clinical isolates provides insights into population structure and mechanisms of drug resistance. Genome Res. *21*, 2143–2156.

Dresel, A., and Clos, J. (1997). Transcription of the *Leishmania major* Hsp70-I gene locus does not proceed through the noncoding region. Exp. Parasitol. *86*, 206–212.

Erdmann, M., Scholz, A., Melzer, I.M., Schmetz, C., and Wiese, M. (2006). Interacting Protein Kinases Involved in the Regulation of Flagellar Length. Mol. Biol. Cell. *17*, 2035–2045.

Folgueira, C., and Requena, J.M. (2007). A postgenomic view of the heat shock proteins in kinetoplastids. FEMS Microbiol. Rev. *31*, 359–377.

Forafonov, F., Toogun, O.A., Grad, I., Suslova, E., Freeman, B.C., and Picard, D. (2008). p23/Sba1p protects against Hsp90 inhibitors independently of its intrinsic chaperone activity. Mol. Cell. Biol. *28*, 3446–3456.

Fraga, J., Montalvo, A.M., Van der Auwera, G., Maes, I., Dujardin, J.C., and Requena, J.M. (2013). Evolution and species discrimination according to the *Leishmania* heat-shock protein 20 gene. Infect. Genet. Evol. *18*, 229–237.

Gille, C., Goede, A., Schloetelburg, C., Preissner, R., Kloetzel, P.M., Gobel, U.B., and Frommel, C. (2003). A comprehensive view on proteasomal sequences: implications for the evolution of the proteasome. J. Mol. Biol. *326*, 1437–1448.

Glover, J.R., and Lindquist, S. (1998). Hsp104, Hsp70, and Hsp40: a novel chaperone system that rescues previously aggregated proteins. Cell *94*, 73–82.

Gottesman, S. (1996). Proteases and their targets in *Escherichia coli*. Annu. Rev. Genet. *30*, 465–506.

Gottesman, S. (2003). Proteolysis in bacterial regulatory circuits. Annu. Rev. Cell. Dev. Biol. *19*, 565–587.

Gottesmann, S., Squires, C., Pichersky, E., Carrington, M., Hobbs, M., Mattick, J.S., Dalrymple, B., Kuramitsu, H., Shiroza, T., Foster, T., Clark, W.P., Ross, B., Squires, C.L., and Maurizi, M.R. (1990). Conservation of the regulatory subunit for the Clp ATP-dependent protease in prokaryotes and eukaryotes. Proc. Natl. Acad. Sci. U.S.A. *87*, 3513–3517.

Graefe, S.E., Wiesgigl, M., Gaworski, I., Macdonald, A., and Clos, J. (2002). Inhibition of HSP90 in Trypanosoma cruzi Induces a Stress Response but No Stage Differentiation. Eukaryot. Cell *1*, 936–943.

Hilton, G.R., Lioe, H., Stengel, F., Baldwin, A.J., and Benesch, J.L. (2013). Small heat-shock proteins: paramedics of the cell. Top. Curr. Chem. *328*, 69–98.

Hombach, A. (2013). Mutational analysis of heat shock protein 90 from *Leishmania donovani*. In Faculty of Biology (Hamburg, University of Hamburg), pp. 77–83.

Hombach, A., Ommen, G., Chrobak, M., and Clos, J. (2012). The Hsp90-Sti1 interaction is critical for *Leishmania donovani* proliferation in both life cycle stages. Cell Microbiol. *15*, 585–600.

Hoyer, C., Mellenthin, K., Schilhabel, M., Platzer, M., and Clos, J. (2001). *Leishmania* and the leishmaniases: use of genetic complementation to identify gene(s) which specify species-specific organ tropism of *Leishmania*. Med. Microbiol. Immunol. *190*, 53–56.

Hubel, A., and Clos, J. (1996). The genomic organization of the HSP83 gene locus is conserved in three *Leishmania* species. Exp. Parasitol. *82*, 225–228.

Hubel, A., Brandau, S., Dresel, A., and Clos, J. (1995). A member of the ClpB family of stress proteins is expressed during heat shock in *Leishmania* spp. Mol. Biochem. Parasitol. *70*, 107–118.

Hubel, A., Krobitsch, S., Horauf, A., and Clos, J. (1997). *Leishmania major* Hsp100 is required chiefly in the mammalian stage of the parasite. Mol. Cell. Biol. *17*, 5987–5995.

Hunter, K.W., Cook, C.L., and Hayunga, E.G. (1984). Leishmanial differentiation *in vitro:* induction of heat shock proteins. Biochem. Biophys. Res. Commun. *125*, 755–760.

Ivens, A.C., Peacock, C.S., Worthey, E.A., Murphy, L., Aggarwal, G., Berriman, M., Sisk, E., Rajandream, M.A., Adlem, E., Aert, R., *et al.* (2005). The genome of the kinetoplastid parasite, *Leishmania major*. Science *309*, 436–442.

Jakobsson, M.E., Moen, A., Bousset, L., Egge-Jacobsen, W., Kernstock, S., Melki, R., and Falnes, P.O. (2013). Identification and characterization of a novel human methyltransferase modulating Hsp70 protein function through lysine methylation. J. Biol. Chem. *288*, 27752–27763.

Jensen, A.T., Curtis, J., Montgomery, J., Handman, E., and Theander, T.G. (2001). Molecular and immunological characterisation of the glucose regulated protein 78 of *Leishmania donovani*(1). Biochim. Biophys. Acta *1549*, 73–87.

Jensen, A.T., Ismail, A., Gaafar, A., El Hassan, A.M., and Theander, T.G. (2002). Humoral and cellular immune responses to glucose regulated protein 78 – a novel *Leishmania donovani* antigen. Trop. Med. Int. Health *7*, 471–476.

Johnson, J.L., and Brown, C. (2009). Plasticity of the Hsp90 chaperone machine in divergent eukaryotic organisms. Cell Stress Chaperones *14*, 83–94.

Jones, C., Anderson, S., Singha, U.K., and Chaudhuri, M. (2008). Protein phosphatase 5 is required for Hsp90 function during proteotoxic stresses in Trypanosoma brucei. Parasitol. Res. *102*, 835–844.

Kanemori, M., Nishihara, K., Yanagi, H., and Yura, T. (1997). Synergistic roles of HslVU and other ATP-dependent proteases in controlling *in vivo* turnover of sigma32 and abnormal proteins in *Escherichia coli*. J. Bacteriol. *179*, 7219–7225.

Kay, R.J., Russnak, R.H., Jones, D., Mathias, C., and Candido, E.P. (1987). Expression of intron-containing C. elegans heat shock genes in mouse cells demonstrates divergence of 3′ splice site recognition sequences between nematodes and vertebrates, and an inhibitory effect of heat shock on the mammalian splicing apparatus. Nucleic Acids Res. *15*, 3723–3741.

Kessel, M., Wu, W., Gottesman, S., Kocsis, E., Steven, A.C., and Maurizi, M.R. (1996). Six-fold rotational symmetry of ClpQ, the E. coli homolog of the 20S proteasome, and its ATP-dependent activator, ClpY. FEBS Lett. *398*, 274–278.

Kim, Y.E., Hipp, M.S., Bracher, A., Hayer-Hartl, M., and Hartl, F.U. (2013). Molecular chaperone functions in protein folding and proteostasis. Annu. Rev. Biochem. *82*, 323–355.

Koyasu, S., Nishida, E., Kadowaki, T., Matsuzaki, F., Iida, K., Harada, F., Kasuga, M., Sakai, H., and Yahara, I. (1986). Two mammalian heat shock proteins, HSP90 and HSP100, are actin-binding proteins. Proc. Natl. Acad. Sci. U.S.A. *83*, 8054–8058.

Krobitsch, S., and Clos, J. (1999). A novel role for 100 kD heat shock proteins in the parasite *Leishmania donovani*. Cell Stress Chaperones *4*, 191–198.

Krobitsch, S., and Lindquist, S. (2000). Aggregation of huntingtin in yeast varies with the length of polyglutamine expansion and the expression of chaperone proteins. Proc. Natl. Acad. Sci. U.S.A. *97*, 1583–1588.

Krobitsch, S., Brandau, S., Hoyer, C., Schmetz, C., Hübel, A., and Clos, J. (1998). *Leishmania donovani* heat shock protein 100: characterization and function in amastigote stage differentiation. J. Biol. Chem. *273*, 6488–6494.

Lahav, T., Sivam, D., Volpin, H., Ronen, M., Tsigankov, P., Green, A., Holland, N., Kuzyk, M., Borchers, C., Zilberstein, D., *et al.* (2011). Multiple levels of gene regulation mediate differentiation of the intracellular pathogen *Leishmania*. FASEB J. *25*, 515–525.

Lambertz, U., Silverman, J.M., Nandan, D., McMaster, W.R., Clos, J., Foster, L.J., and Reiner, N.E. (2012). Secreted virulence factors and immune evasion in visceral leishmaniasis. J. Leukoc. Biol. *91*, 887–899.

Landry, J., Chretien, P., Lambert, H., Hickey, E., and Weber, L.A. (1989). Heat shock resistance conferred by expression of the human HSP27 gene in rodent cells. J. Cell Biol. *109*, 7–15.

Larreta, R., Soto, M., Alonso, C., and Requena, J.M. (2000). *Leishmania infantum*: gene cloning of the GRP94 homologue, its expression as recombinant protein, and analysis of antigenicity. Exp. Parasitol. *96*, 108–115.

Larreta, R., Guzman, F., Patarroyo, M.E., Alonso, C., and Requena, J.M. (2002). Antigenic properties of the *Leishmania infantum* GRP94 and mapping of linear B-cell epitopes. Immunol. Lett. *80*, 199–205.

Lavoie, J.N., Gingras-Breton, G., Tanguay, R.M., and Landry, J. (1993). Induction of Chinese hamster HSP27 gene expression in mouse cells confers resistance to heat shock. HSP27 stabilization of the microfilament organization. J. Biol. Chem. *268*, 3420–3429.

Lawrence, F., and Robert-Gero, M. (1985). Induction of heat shock and stress proteins in promastigotes of three *Leishmania* species. Proc. Natl. Acad. Sci. U.S.A. *82*, 4414–4417.

Lee, M.G., Atkinson, B.L., Giannini, S.H., and Van der Ploeg, L.H. (1988). Structure and expression of the hsp 70 gene family of *Leishmania major* [published erratum appears in Nucleic Acids Res 1988 Dec 9;16(23):11400- 1]. Nucleic Acids Res. *16*, 9567–9585.

Lee, Y.R., Nagao, R.T., and Key, J.L. (1994). A soybean 101-kD heat shock protein complements a yeast HSP104 deletion mutant in acquiring thermotolerance. Plant Cell *6*, 1889–1897.

Li, J., Richter, K., and Buchner, J. (2011). Mixed Hsp90–co-chaperone complexes are important for the progression of the reaction cycle. Nat. Struct. Mol. Biol. *18*, 61–66.

Li, J., Soroka, J., and Buchner, J. (2012). The Hsp90 chaperone machinery: Conformational dynamics and regulation by co-chaperones. Biochim. Biophys. Acta *1823*, 624–635.

Lindquist, S. (1981). Regulation of protein synthesis during heat shock. Nature *293*, 311–314.

Lindquist, S. (1986). The heat-shock response. Annu. Rev. Biochem. *55*, 1151–1191.

Lindquist, S. (1992). Heat-shock proteins and stress tolerance in microorganisms. Curr. Opin. Genet. Dev. *2*, 748–755.

Liu, F.H., Wu, S.J., Hu, S.M., Hsiao, C.D., and Wang, C. (1999). Specific interaction of the 70-kDa heat shock cognate protein with the tetratricopeptide repeats. J. Biol. Chem. *274*, 34425–34432.

Lockwood, B.L., and Somero, G.N. (2011). Transcriptomic responses to salinity stress in invasive and native blue mussels (genus Mytilus). Mol. Ecol. *20*, 517–529.

Lockwood, B.L., Sanders, J.G., and Somero, G.N. (2010). Transcriptomic responses to heat stress in invasive and native blue mussels (genus Mytilus): molecular correlates of invasive success. J. Exp. Biol. *213*, 3548–3558.

Ma, J., and Lindquist, S. (1999). De novo generation of a PrPSc-like conformation in living cells. Nat. Cell Biol. *1*, 358–361.

Miller, J. (1988). Effects of temperature elevation on mRNA and protein synthesis in *Leishmania mexicana amazonensis*. Mol. Biochem. Parasitol. *30*, 175–184.

Miyata, Y. (2009). Protein kinase CK2 in health and disease: CK2: the kinase controlling the Hsp90 chaperone machinery. Cell. Mol. Life. Sci. *66*, 1840–1849.

Mollapour, M., and Neckers, L. (2012). Post-translational modifications of Hsp90 and their contributions to chaperone regulation. Biochim. Biophys. Acta *1823*, 648–655.

Mollapour, M., Tsutsumi, S., Kim, Y.S., Trepel, J., and Neckers, L. (2011a). Casein kinase 2 phosphorylation of Hsp90 threonine 22 modulates chaperone function and drug sensitivity. Oncotarget 2, 407–417.

Mollapour, M., Tsutsumi, S., Truman, A.W., Xu, W., Vaughan, C.K., Beebe, K., Konstantinova, A., Vourganti, S., Panaretou, B., Piper, P.W., *et al.* (2011b). Threonine 22 phosphorylation attenuates Hsp90 interaction with cochaperones and affects its chaperone activity. Mol. Cell *41*, 672–681.

Monnerat, S., Martinez-Calvillo, S., Worthey, E., Myler, P.J., Stuart, K.D., and Fasel, N. (2004). Genomic organization and gene expression in a chromosomal region of *Leishmania major*. Mol. Biochem. Parasitol. *134*, 233–243.

Morales, M.A., Watanabe, R., Laurent, C., Lenormand, P., Rousselle, J.C., Namane, A., and Spath, G.F. (2008). Phosphoproteomic analysis of *Leishmania donovani* pro- and amastigote stages. Proteomics *8*, 350–363.

Morales, M., Watanabe, R., Dacher, M., Chafey, P., Osorio y Fortéa, J., Beverley, S., Ommen, G., Clos, J., Hem, S., Lenormand, P., *et al.* (2010). Phosphoproteome dynamics reveals heat shock protein complexes specific to the *Leishmania* infectious stage. Proc. Natl. Acad. Sci. U.S.A. *107*, 8381–8386.

Morimoto, R.I. (1998). Regulation of the heat shock transcriptional response: cross-talk between a family of heat shock factors, molecular chaperones, and negative regulators. Genes Dev. *12*, 3788–3796.

Muhich, M.L., and Boothroyd, J.C. (1989). Synthesis of trypanosome hsp70 mRNA is resistant to disruption of trans- splicing by heat shock. J. Biol. Chem. *264*, 7107–7110.

Muller, P., Ruckova, E., Halada, P., Coates, P.J., Hrstka, R., Lane, D.P., and Vojtesek, B. (2013). C-terminal phosphorylation of Hsp70 and Hsp90 regulates alternate binding to co-chaperones CHIP and HOP to determine cellular protein folding/degradation balances. Oncogene 32, 3101–3110.

Munster, P.N., Srethapakdi, M., Moasser, M.M., and Rosen, N. (2001). Inhibition of heat shock protein 90 function by ansamycins causes the morphological and functional differentiation of breast cancer cells. Cancer Res. *61*, 2945–2952.

Myler, P.J., Beverley, S.M., Cruz, A.K., Dobson, D.E., Ivens, A.C., McDonagh, P.D., Madhubala, R., Martinez-Calvillo, S., Ruiz, J.C., Saxena, A., *et al.* (2001). The *Leishmania* genome project: new insights into gene organization and function. Med. Microbiol. Immunol. (Berl) *190*, 9–12.

Newnam, G.P., Wegrzyn, R.D., Lindquist, S.L., and Chernoff, Y.O. (1999). Antagonistic interactions between yeast chaperones hsp104 and hsp70 in prion curing. Mol. Cell. Biol. *19*, 1325–1333.

Ommen, G. (2009). Charakterisierung putativer Co-Chaperonen des Parasiten *Leishmania donovani* (Ross, 1903). In Fakultät für Mathematik, Informatik und Naturwissenschaften (Hamburg, University of Hamburg).

Ommen, G., Lorenz, S., and Clos, J. (2009). One-step generation of double-allele gene replacement mutants in *Leishmania donovani*. Int. J. Parasitol. 39, 541–546.

Ommen, G., Chrobak, M., and Clos, J. (2010). The co-chaperone SGT of *Leishmania donovani* is essential for the parasite's viability. Cell Stress Chaperones *39*, 541–546.

Parsell, D.A., and Lindquist, S. (1993). The function of heat-shock proteins in stress tolerance: degradation and reactivation of damaged proteins. Annu. Rev. Genet. 27, 437–496.

Parsell, D.A., and Lindquist, S. (1994). Heat shock proteins and stress tolerance. In The Biology of Heat Shock Proteins and Molecular Chaperones, Morimoto, R.I., Tissières, A., and Georgopoulos, C., eds. (Plainview, New York, Cold Spring Harbor Laboratory Press), pp. 457–494.

Parsell, D.A., Sanchez, Y., Stitzel, J.D., and Lindquist, S. (1991). Hsp104 is a highly conserved protein with two essential nucleotide-binding sites. Nature 353, 270–273.

Parsell, D.A., Kowal, A.S., and Lindquist, S. (1994a). *Saccharomyces cerevisiae* Hsp104 Protein. J. Biol. Chem. *269*, 4480–4487.

Parsell, D.A., Kowal, A.S., Singer, M.A., and Lindquist, S. (1994b). Protein disaggregation mediated by heat-shock protein Hsp104. Nature 372, 475–478.

Parsons, M., Worthey, E.A., Ward, P.N., and Mottram, J.C. (2005). Comparative analysis of the kinomes of three pathogenic trypanosomatids: *Leishmania major*, Trypanosoma brucei and Trypanosoma cruzi. BMC Genomics *6*, 127.

Patino, M.M., Liu, J.J., Glover, J.R., and Lindquist, S. (1996). Support for the prion hypothesis for inheritance of a phenotypic trait in yeast. Science *273*, 622–626.

Pearl, L.H., and Prodromou, C. (2006). Structure and mechanism of the Hsp90 molecular chaperone machinery. Annu. Rev. Biochem. *75*, 271–294.

Picard, D. (2002). Heat-shock protein 90, a chaperone for folding and regulation. Cell. Mol. Life Sci. *59*, 1640–1648.

van der Ploeg, L.H.T., Giannini, S.H., and Cantor, C.R. (1985). Heat shock genes: regulatory role for differentiation in parasitic protozoa. Science *228*, 1443–1446.

Przepiorka, D., and Srivastava, P.K. (1998). Heat shock protein–peptide complexes as immunotherapy for human cancer. Mol. Med. Today *4*, 478–484.

Queitsch, C., Sangster, T.A., and Lindquist, S. (2002). Hsp90 as a capacitor of phenotypic variation. Nature *417*, 618–624.

Quijada, L., Requena, J.M., Soto, M., and Alonso, C. (1996a). During canine viscero-cutaneous leishmaniasis the anti-Hsp70 antibodies are specifically elicited by the parasite protein. Parasitology *112 (Pt 3)*, 277–284.

Quijada, L., Requena, J.M., Soto, M., Gomez, L.C., Guzman, F., Patarroyo, M.E., and Alonso, C. (1996b). Mapping of the linear antigenic determinants of the *Leishmania infantum* hsp70 recognized by leishmaniasis sera. Immunol. Lett. *52*, 73–79.

Quijada, L., Soto, M., Alonso, C., and Requena, J.M. (1997). Analysis of post-transcriptional regulation operating on transcription products of the tandemly linked *Leishmania infantum* hsp70 genes. J. Biol. Chem. *272*, 4493–4499.

Quijada, L., Requena, J.M., Soto, M., and Alonso, C. (1998). Analysis of the antigenic properties of the L. infantum Hsp70: design of synthetic peptides for specific serodiagnosis of human leishmaniasis. Immunol. Lett. *63*, 169–174.

Raina, P., and Kaur, S. (2012). Knockdown of LdMC1 and Hsp70 by antisense oligonucleotides causes cell-cycle defects and programmed cell death in *Leishmania donovani*. Mol. Cell. Biochem. *359*, 135–149.

Reiling, L., Chrobak, M., Schmetz, C., and Clos, J. (2010). Overexpression of a single *Leishmania major* gene is sufficient to enhance parasite infectivity *in vivo* and *in vitro*. Mol. Microbiol. *76*, 1175–1190.

Renkawek, K., Bosman, G.J., and de Jong, W.W. (1994). Expression of small heat-shock protein hsp 27 in reactive gliosis in Alzheimer disease and other types of dementia. Acta Neuropathol. *87*, 511–519.

Retzlaff, M., Stahl, M., Eberl, H.C., Lagleder, S., Beck, J., Kessler, H., and Buchner, J. (2009). Hsp90 is regulated by a switch point in the C-terminal domain. EMBO Rep. *10*, 1147–1153.

Rohrwild, M., Coux, O., Huang, H.C., Moerschell, R.P., Yoo, S.J., Seol, J.H., Chung, C.H., and Goldberg, A.L. (1996). HslV-HslU: A novel ATP-dependent protease complex in *Escherichia coli* related to the eukaryotic proteasome. Proc. Natl. Acad. Sci. U.S.A. *93*, 5808–5813.

Rohrwild, M., Pfeifer, G., Santarius, U., Muller, S.A., Huang, H.C., Engel, A., Baumeister, W., and Goldberg, A.L. (1997). The ATP-dependent HslVU protease from *Escherichia coli* is a four-ring structure resembling the proteasome. Nat. Struct. Biol. *4*, 133–139.

Rosenzweig, D., Smith, D., Opperdoes, F., Stern, S., Olafson, R.W., and Zilberstein, D. (2008). Retooling *Leishmania* metabolism: from sand fly gut to human macrophage. FASEB J. *22*, 590–602.

Rutherford, S.L., and Zuker, C.S. (1994). Protein folding and the regulation of signaling pathways. Cell *79*, 1129–1132.

Rutherford, S.L., and Lindquist, S. (1998). Hsp90 as a capacitor for morphological evolution. Nature *396*, 336–342.

Sanchez, E.R., Redmond, T., Scherrer, L.C., Bresnick, E.H., Welsh, M.J., and Pratt, W.B. (1988). Evidence that the 90-kilodalton heat shock protein is associated with tubulin-containing complexes in L cell cytosol and in intact PtK cells. Mol. Endocrinol. *2*, 756–760.

Sanchez, Y., and Lindquist, S.L. (1990). Hsp104 required for induced thermotolerance. Science *248*, 1112–1115.

Sanchez, Y., Taulien, J., Borkovich, K.A., and Lindquist, S. (1992). Hsp104 is required for tolerance to many forms of stress. EMBO J. *11*, 2357–2364.

Sanchez, Y., Parsell, D.A., Taulien, J., Vogel, J.L., Craig, E.A., and Lindquist, S. (1993). Genetic evidence for a functional relationship between Hsp104 and Hsp70. J. Bacteriol. *175*, 6484–6491.

Sangster, T.A., Lindquist, S., and Queitsch, C. (2004). Under cover: causes, effects and implications of Hsp90-mediated genetic capacitance. Bioessays *26*, 348–362.

Sato, Y., and Yokoya, S. (2008). Enhanced tolerance to drought stress in transgenic rice plants overexpressing a small heat-shock protein, sHSP17.7. Plant Cell Rep. *27*, 329–334.

Schirmer, E.C., and Lindquist, S. (1997). Interactions of the chaperone Hsp104 with yeast Sup35 and mammalian PrP. Proc. Natl. Acad. Sci. U.S.A. *94*, 13932–13937.

Schirmer, E.C., Glover, J.R., Singer, M.A., and Lindquist, S. (1996). HSP100/Clp proteins: a common mechanism explains diverse functions. Trends Biochem. Sci. *21*, 289–296.

Schlueter, A., Wiesgigl, M., Hoyer, C., Fleischer, S., Klaholz, L., Schmetz, C., and Clos, J. (2000). Expression and Subcellular Localization of Cpn60 Protein Family Members in *Leishmania donovani*. Biochim. Biophys. Acta *1491*, 65–74.

Seraphim, T.V., Alves, M.M., Silva, I.M., Gomes, F.E., Silva, K.P., Murta, S.M., Barbosa, L.R., and Borges, J.C. (2013). Low resolution structural studies indicate that the activator of Hsp90 ATPase 1 (Aha1) of *Leishmania braziliensis* has an elongated shape which allows its interaction with both N- and M-domains of Hsp90. PLoS One *8*, e66822.

Shapira, M., and Pedraza, G. (1990). Sequence analysis and transcriptional activation of heat shock protein 83 of *Leishmania mexicana amazonensis*. Mol. Biochem. Parasitol. *42*, 247–256.

Shapira, M., Zilka, A., Garlapati, S., Dahan, E., Dahan, I., and Yavesky, V. (2001). Post transcriptional control of gene expression in *Leishmania*. Med. Microbiol. Immunol. (Berl) *190*, 23–26.

Sherman, M.Y., and Goldberg, A.L. (1993). Heat shock of *Escherichia coli* increases binding of dnaK (the hsp70 homolog) to polypeptides by promoting its phosphorylation. Proc. Natl. Acad. Sci. U.S.A. *90*, 8648–8652.

Sierra, J.M., and Zapata, J.M. (1994). Translational regulation of the heat shock response. Mol. Biol. Rep. *19*, 211–220.

Silva, K.P., Seraphim, T.V., and Borges, J.C. (2013). Structural and functional studies of *Leishmania braziliensis* Hsp90. Biochim. Biophys. Acta *1834*, 351–361.

Silverman, J.M., and Reiner, N.E. (2011a). Exosomes and other microvesicles in infection biology: organelles with unanticipated phenotypes. Cell. Microbiol. *13*, 1–9.

Silverman, J.M., and Reiner, N.E. (2011b). *Leishmania* exosomes deliver preemptive strikes to create an environment permissive for early infection. Front. Cell. Infect. Microbiol. *1*, 26.

Silverman, J.M., Chan, S.K., Robinson, D.P., Dwyer, D.M., Nandan, D., Foster, L.J., and Reiner, N.E. (2008). Proteomic analysis of the secretome of *Leishmania donovani*. Genome Biol. *9*, R35.

Silverman, J.M., Clos, J., de'Oliveira, C.C., Shirvani, O., Fang, Y., Wang, C., Foster, L.J., and Reiner, N.E. (2010a). An exosome-based secretion pathway is responsible for protein export from *Leishmania* and communication with macrophages. J. Cell Sci. *123*, 842–852.

Silverman, J.M., Clos, J., Horakova, E., Wang, A.Y., Wiesgigl, M., Kelly, I., Lynn, M.A., McMaster, W.R., Foster, L.J., Levings, M.K., *et al.* (2010b). *Leishmania* exosomes modulate innate and adaptive immune responses through effects on monocytes and dendritic cells. J. Immunol. *185*, 5011–5022.

Solomon, J.M., Rossi, J.M., Golic, K., McGarry, T., and Lindquist, S. (1991). Changes in Hsp70 alter thermotolerance and heat-shock regulation in *Drosophila*. New Biol. 3, 1106–1120.

Soroka, J., and Buchner, J. (2012). Mechanistic aspects of the Hsp90 phosphoregulation. Cell Cycle *11*, 1870–1871.

Soroka, J., Wandinger, S.K., Mausbacher, N., Schreiber, T., Richter, K., Daub, H., and Buchner, J. (2012). Conformational switching of the molecular chaperone Hsp90 via regulated phosphorylation. Mol. Cell *45*, 517–528.

Srethapakdi, M., Liu, F., Tavorath, R., and Rosen, N. (2000). Inhibition of Hsp90 function by ansamycins causes retinoblastoma gene product-dependent G1 arrest. Cancer Res. *60*, 3940–3946.

Srivastava, P. (2002). Interaction of heat shock proteins with peptides and antigen-presenting cells: chaperoning of the innate and adaptive immune responses. Annu. Rev. Immunol. *20*, 395–425.

Srivastava, P.K. (2006). Therapeutic cancer vaccines. Curr. Opin. Immunol. *18*, 201–205.

Truman, A.W., Kristjansdottir, K., Wolfgeher, D., Hasin, N., Polier, S., Zhang, H., Perrett, S., Prodromou, C., Jones, G.W., and Kron, S.J. (2012). CDK-dependent Hsp70 Phosphorylation controls G1 cyclin abundance and cell-cycle progression. Cell *151*, 1308–1318.

Vergnes, B., Gourbal, B., Girard, I., Sundar, S., Drummelsmith, J., and Ouellette, M. (2007). A proteomics screen implicates HSP83 and a small kinetoplastid calpain-related protein in drug resistance in *Leishmania donovani* clinical field isolates by modulating drug-induced programmed cell death. Mol. Cell. Proteomics *6*, 88–101.

Vogel, J.L., Parsell, D.A., and Lindquist, S. (1995). Heat-shock proteins Hsp104 and Hsp70 reactivate mRNA splicing after heat inactivation. Curr. Biol. *5*, 306–317.

Wallace, G.R., Ball, A.E., MacFarlane, J., el Safi, S.H., Miles, M.A., and Kelly, J.M. (1992). Mapping of a visceral leishmaniasis-specific immunodominant B-cell epitope of *Leishmania donovani* Hsp70. Infect. Immun. *60*, 2688–2693.

Wandinger, S.K., Richter, K., and Buchner, J. (2008). The Hsp90 chaperone machinery. J. Biol. Chem. *283*, 18473–18477.

Wang, Q., Melzer, I.M., Kruse, M., Sander-Juelch, C., and Wiese, M. (2005). LmxMPK4, a mitogen-activated protein (MAP) kinase homologue essential for promastigotes and amastigotes of *Leishmania mexicana*. Kinetoplastid Biol. Dis. *4*, 6.

Wang, X., Lu, X.A., Song, X., Zhuo, W., Jia, L., Jiang, Y., and Luo, Y. (2012). Thr90 phosphorylation of Hsp90alpha by protein kinase A regulates its chaperone machinery. Biochem. J. *441*, 387–397.

Webb, J.R., Campos-Neto, A., Skeiky, Y.A., and Reed, S.G. (1997). Molecular characterization of the heat-inducible LmSTI1 protein of *Leishmania major*. Mol. Biochem. Parasitol. *89*, 179–193.

Wegele, H., Muller, L., and Buchner, J. (2004). Hsp70 and Hsp90 – a relay team for protein folding. Rev. Physiol. Biochem. Pharmacol. *151*, 1–44.

Weis, F., Moullintraffort, L., Heichette, C., Chretien, D., and Garnier, C. (2010). The 90-kDa heat shock protein Hsp90 protects tubulin against thermal denaturation. J. Biol. Chem. *285*, 9525–9534.

Whitesell, L., and Cook, P. (1996). Stable and specific binding of heat shock protein 90 by geldanamycin disrupts glucocorticoid receptor function in intact cells. Mol. Endocrinol. *10*, 705–712.

Whitesell, L., and Lindquist, S.L. (2005). HSP90 and the chaperoning of cancer. Nat. Rev. *5*, 761–772.

Whitesell, L., Mimnaugh, E.G., De Costa, B., Myers, C.E., and Neckers, L.M. (1994). Inhibition of heat shock protein HSP90-pp60v-src heteroprotein complex formation by benzoquinone ansamycins: essential role for stress proteins in oncogenic transformation. Proc. Natl. Acad. Sci. U.S.A. *91*, 8324–8328.

Wiesgigl, M., and Clos, J. (1999). Uniform distribution of transcription complexes on the clpB gene locus of *Leishmania donovani.* Protist *150,* 369–373.

Wiesgigl, M., and Clos, J. (2001). Heat shock protein 90 homeostasis controls stage differentiation in *Leishmania donovani.* Mol. Biol. Cell. *12,* 3307–3316.

Wilson, M.E., Andersen, K.A., and Britigan, B.E. (1994). Response of *Leishmania chagasi* promastigotes to oxidant stress. Infect. Immun. *62,* 5133–5141.

Workman, P., Burrows, F., Neckers, L., and Rosen, N. (2007). Drugging the cancer chaperone HSP90: combinatorial therapeutic exploitation of oncogene addiction and tumor stress. Ann. N. Y. Acad. Sci. *1113,* 202–216.

Wu, C. (1995). Heat shock transcription factors: structure and regulation. Annu. Rev. Cell. Dev. Biol. *11,* 441–469.

Wu, C., Clos, J., Giorgi, G., Haroun, R.I., Kim, S.-J., Rabindran, S.K., Westwood, J.T., Wisniewski, J., and Yim, G. (1994). Structure and regulation of heat shock transcription factor. In The Biology of Heat Shock Proteins and Molecular Chaperones, Morimoto, R.I., Tissières, A., and Georgopoulos, C., eds. (Plainview, New York, Cold Spring Harbor Laboratory Press), pp. 395–416.

Xiao, H., Perisic, O., and Lis, J.T. (1991). Cooperative binding of Drosophila heat shock factor to arrays of a conserved 5 bp unit. Cell *64,* 585–593.

Xu, W., Mollapour, M., Prodromou, C., Wang, S., Scroggins, B.T., Palchick, Z., Beebe, K., Siderius, M., Lee, M.J., Couvillon, A., *et al.* (2012). Dynamic tyrosine phosphorylation modulates cycling of the HSP90-P50(CDC37)-AHA1 chaperone machine. Mol. Cell 47, 434–443.

Yau, W.L., Blisnick, T., Taly, J.F., Helmer-Citterich, M., Schiene-Fischer, C., Leclercq, O., Li, J., Schmidt-Arras, D., Morales, M.A., Notredame, C., *et al.* (2010). Cyclosporin A treatment of *Leishmania donovani* reveals stage-specific functions of cyclophilins in parasite proliferation and viability. PLoS Negl. Trop. Dis. *4,* e729.

Yin, H., Wang, H., Zong, H., Chen, X., Wang, Y., Yun, X., Wu, Y., Wang, J., and Gu, J. (2006). SGT, a Hsp90beta binding partner, is accumulated in the nucleus during cell apoptosis. Biochem. Biophys. Res. Commun. *343,* 1153–1158.

Yost, H.J., and Lindquist, S. (1986). RNA splicing is interrupted by heat shock and is rescued by heat shock protein synthesis. Cell *45,* 185–193.

Yost, H.J., and Lindquist, S. (1988). Translation of unspliced transcripts after heat shock. Science *242,* 1544–1548.

Yost, H.J., Petersen, R.B., and Lindquist, S. (1990). Posttranscriptional regulation of heat shock protein synthesis in *Drosophila.* In Stress proteins in biology and medicine, Morimoto, R.I., Tissières, A., and Georgopoulos, C., eds. (Plainview, New York, Cold Spring Harbor Laboratory Press), pp. 379–410.

Young, J.C. (2010). Mechanisms of the Hsp70 chaperone system. Biochem. Cell. Biol. *88,* 291–300.

Zamora-Veyl, F.B., Kroemer, M., Zander, D., and Clos, J. (2005). Stage-specific expression of the mitochondrial co-chaperonin of *Leishmania donovani,* CPN10. Kinetoplastid Biol. Dis. *4,* 3.

Zhang, G., Fang, X., Guo, X., Li, L., Luo, R., Xu, F., Yang, P., Zhang, L., Wang, X., Qi, H., *et al.* (2012). The oyster genome reveals stress adaptation and complexity of shell formation. Nature *490,* 49–54.

Zilberstein, D., and Shapira, M. (1994). The role of pH and temperature in the development of *Leishmania* parasites. Annu. Rev. Microbiol. *48,* 449–470.

Zilka, A., Garlapati, S., Dahan, E., Yaolsky, V., and Shapira, M. (2001). Developmental regulation of HSP83 in *Leishmania*: transcript levels are controlled by the efficiency of 3? RNA processing and preferential translation is directed by a determinant in the 3′ UTR. J. Biol. Chem. *11,* 11.

Role of Iron in *Leishmania*–Macrophage Interaction

3

Kavita Bharati, Saswat Kumar Bal, Shalini Saini, Vikash Bhardwaj and Chinmay K. Mukhopadhyay

Abstract

Iron is an essential micronutrient for most forms of life because of its vital role as a redox cofactor of proteins requiring for critical cellular processes. Being equally essential for survival of the host and pathogen, iron has taken its justified role in the immune status of the former and the virulence capacity of the later. Acquiring iron is a fundamental step in the development of a pathogen within the host, while host cells are known to implement several iron sequestration policies to encounter the different invasion strategies of pathogens. These illustrate the longstanding and ongoing battle for iron between host and pathogen in the evolution. This chapter describes the understanding of the role of iron in interaction between *Leishmania* and its mammalian host.

Introduction

Iron is the fourth most abundant element found on earth crust, forming 5% of the earth's inner and outer core. This is a metal of first transition series with atomic number 26 and molecular weight 55.85. It is a redox active element which is found in oxidation states between +2 and +6; among these +2 and +3 are two readily interconvertible redox states. Ferrous (Fe^{2+}) form is extremely water soluble at physiological pH, while ferric (Fe^{3+}) form is highly insoluble in water as the maximum solubility is 10^{-17} M at pH 7.0. Significant concentrations of water-soluble Fe^{3+} species can be attained only by strong complex formation, which compromises its bioavailability significantly. Iron can serve both as electron source and electron sink. It can bind to various ligands because of its coordination chemistry. Oxygen, nitrogen and sulfur atoms are the most preferred ligands to bind with biological molecules.

Because of its abundance and high redox potential, iron is involved in numerous biological reactions. For example, iron is required as a cofactor in many proteins involved in catalytic reactions. It serves as essential mineral element and trace element in biological systems. Iron is essential in all living organism ranging from archaea to mammals except non–pathogenic lactobacilli and pathogenic *Borrelia burgdorferi* (both of these use manganese in place of iron).

Aqueous chemistry and toxicity of iron

In acidic solution when pH is < 1.0, both ferrous and ferric iron exist as the hexaaquo complex, $[Fe(H_2O)_6]^{2+}$ or $[Fe(H_2O)_6]^{3+}$, but at pH above 2.0 aquated ferrous iron undergoes hydrolytic deprotonation and forms uncharged $[Fe(H_2O)_3(OH^-)_3]$ species, which is extremely insoluble. Insolubility of ferric hydroxide also enhances autoxidation of Fe^{2+} by mass action. Hydroxide iron complexes readily polymerize by dehydration to form polynuclear complexes with iron atoms linked by oxo- or hydroxo-bridges. Iron polymers in biological systems may range from two iron atoms to three-dimensional arrays containing more than 4000 iron atoms.

Excess of free reactive iron is toxic. In cytoplasm significant amount of iron is reduced and can participate in 'Fenton type' and Haber–Weiss redox chemistry. Iron transfers one electron to dioxygen and generates superoxide radicals ($O_2^{\cdot -}$)

that is precursor of hydrogen peroxide (H_2O_2). Hydrogen peroxide reacts with ferrous and produce hydroxyl radical ($OH^•$) (Fridovich, 1998; Liochev and Fridovich, 1994). These are known as reactive oxygen species (ROS) and mainly generated in mitochondria because of incomplete reduction of dioxygen but can also be generated in endoplasmic reticulum, peroxisomes and cytoplasm.

$$Fe(II) + H_2O_2 \rightarrow Fe(III) + OH^- + OH\cdot \text{ (Fenton)}$$

$$Fe(III) + O_2\cdot^- \rightarrow Fe(II) + O_2$$

Net reaction:

$$H_2O_2 + O_2\cdot^- \xrightarrow{Fe} OH^- + OH\cdot + O_2$$
(Haber-Weiss)

Reactive oxygen species (ROS) are also produced by membrane bound NADPH oxidase complex (Hampton *et al.*, 1998) that is primarily expressed in phagocytic macrophages and neutrophils and also in other cell types. NADPH oxidase contributes significantly in antimicrobial defence. Reactive oxygen species may react with nitric oxide (NO) and chloride to generate more potent oxidants like peroxynitrite ($ONOO^-$), hypochlorite (OCl^-) and reactive nitrogen species (RNS). Iron also involves in production of reactive species like peroxyl (ROO·), alkoxyl (RO·), thiyl (RS), or thiyl-peroxyl (RSOO·) radicals. These highly active radicals promote oxidation of proteins, modification of nucleic acid and peroxidation of membrane constituting lipids, induction of proteases, impaired synthesis of carbohydrates and altered cell proliferation (Halliwell, 1992; Halliwell and Gutteridge, 1990; McCord, 1998; Schaich, 1992).

Role of iron in physiology

Iron participates as cofactors in many diverse biological molecules because of its redox nature. There are four major classes of iron-containing proteins: haem-containing non-enzymic proteins (haemoglobin, myoglobin, cytochromes), iron–sulfur cluster-containing proteins (2Fe–2S, 4Fe–4S, flavoproteins), haem-containing enzymes (oxygenase, guanylate cyclase) and iron-containing enzymes devoid of iron–sulfur cluster or haem (oxo di-iron enzymes).

Iron is the key component of many cellular enzymes such as oxidase, catalase, peroxidase, cytochrome, ribonucleotide reductase, aconitase, cyclooxygenase, lipoxygenase, nitric oxide synthase (Boldt, 1999; Conrad *et al.*, 1999; Ponka, 1999), prolyl hydroxylase (PHD) (Vranka *et al.*, 2004) and tyrosine hydroxylase (Ramsey *et al.*, 1996). These enzymes are critical for many cellular processes and also contribute to several diseases like cancer, neurodegenerative disorders and ageing. Iron plays important role in many cellular processes such as electron transport, cellular respiration, ATP generation, synthesis of DNA, RNA and protein; cell proliferation and differentiation and regulation of gene expression (Andrews *et al.*, 1999; Boldt, 1999; Conrad *et al.*, 1999; Gerlach *et al.*, 1994; Wessling-Resnick, 1999). Cellular iron deficiency arrests cellular growth and may lead to cell death. In human about 80% of the total body iron is required for haemoglobin synthesis. Thus, iron deficiency may lead to anaemia. Iron has important role in neuronal functions as iron is required for myelin synthesis (Connor *et al.*, 1995), synthesis of monoamine transmitters (Beard *et al.*, 2003) and GABA (Li, 1998). It also affects the immune response in several ways as iron deficiency causes decreased bactericidal activity of macrophages (Hallquist *et al.*, 1992), reduced activity of reactive oxygen species producing enzyme myeloperoxidase in neutrophils (Britigan *et al.*, 1996) and decreased T-lymphocytes number (Kuvibidila *et al.*, 1999).

Role of iron in the host–parasite interaction

A major step in host–pathogen interaction is the multiplication of microbes within the cell to increase in number and disseminate. Microbes need to acquire nutrients from host to support their growth. Host tends to limit the availability of nutrient to the microbes. Thus, pathogens developed various sophisticated mechanisms to compete with host for their nutrients. Iron is one of the crucial micronutrients both for the host and microbes that mediate host–pathogen- interactions. Pathogen requires iron to establish infection, propagate pathogenesis and resist host defences. Host induces acute phase response to limit amount of circulating iron primarily through reduced dietary iron absorption and iron sequestration. The sequestration of iron involves binding with serum proteins such as transferrin, lactoferrin and ferritin and protects from iron induced cellular damage. Thus, during the course of microbial infections there is an arms race for iron between the host and pathogens. In the course of this arms race pathogens evolved various systems with very low iron-binding constants that allow binding and uptake iron from surrounding environment. Microorganisms usually require 0.4–4.0 μM iron for their metabolic functions (Lankford, 1973; Weinberg, 1978).

Microorganisms evolved specialized systems for iron uptake. Pathogens can acquire iron from the host mainly by two methods. One is using siderophores, the most widespread mode of iron acquisition. Second is using direct contact with the host iron sources. In the later case to acquire iron microorganisms release it from the host protein by enzymatic method (Philpott, 2006; Wandersman and Delepelaire, 2004). Siderophores are low molecular weight compounds with high affinity iron binding capacity. Approximately, 10^{-30} mol/l iron chelator can sequester Fe^{3+} from host protein–iron complexes. Siderophores are divided in three groups, either containing phenolic ring or hydroxamate structure or combination of both. Mycobacteria, nocardiae and rhodococci produce both membrane-associated mycobactin and extracellular siderophores, carboxymycobactin and exochelin (Ratledge, 2004). *Shigella* species produce aerobactin and catachol-type siderophore enterobactin (Payne, 1989).

Many pathogens directly use the host iron complexes for its requirement. *Listeria monocytogenes* expresses a surface ferrireductase which releases iron from iron complexes (Deneer *et al.*, 1995). Some pathogens, such as uropathogenic *E. coli* and *Vibrio cholerae,* release iron from host cell by production of exotoxins like haemolysin or cytolysin, which lyses red blood cells and releases haemoglobin. *Neisseria meningitidis, Haemophilus influenzae, Bordetella pertussis* and *Trypanosoma brucei* express surface receptor for transferrin, lactoferrin and other iron-binding molecules (Borst and Fairlamb, 1998; Cornelissen and Sparling, 1994; Schryvers and Morris, 1988). Some pathogens have receptors for haem and haemoglobin on their cell surfaces (Lee, 1995; Stojiljkovic *et al.*, 1995). *Serratia marcescens* expresses an extracellular haem-binding protein (Létoffé *et al.*, 1994).

In mammals, approximately 76% of body iron is contained in red blood cells as haemoglobin. Lysis of erythrocytes releases haemoglobin and its reactive ferric protoporphyrin-IX group (haemin), which is then rapidly bound by haptoglobin (Wagener *et al.*, 2003) and haemin-binding protein haemopexin (Delanghe and Langlois, 2001), respectively. Since, iron can generate reactive oxygen species, hydroxyl and other free radicals; thus, any free iron must be sequestered to prevent damage to DNA, protein and lipid. Moreover, other high-affinity iron-binding proteins such as lactoferrin (found in extracellular secretions), transferrin (found in serum) and ferritin (intracellular iron storage protein) bind to free iron that would otherwise can wreak oxidative havoc in the vasculature and in exposed tissues (Balla *et al.*, 2007). Thereby, these iron-binding proteins of host system limit the amount of free iron concentration of about 10^{-24} M (Doherty, 2007). Lactoferrin possess antimicrobial activity and is found at the mucosal surfaces that line the pathogen entry points. During microbial infection, usually transferrin iron saturation is reduced (Weinberg, 1999, 2000) and synthesis of ferritin goes high (Konijn *et al.*, 1981).

On the other hand, many organisms including *Escherichia coli* (Bullen *et al.*, 1968), *Klebsiella* (Ward *et al.*, 1986), *Listeria* (Martínez *et al.*, 1990), *Neisseria* (Bullen *et al.*, 1968), *Pasteurella* (Martínez *et al.*, 1990), *Shigella* (Payne, 1989), *Salmonella* (Griffiths, 1991), *Vibrio* (Karunasagar *et al.*, 1984) and *Yersinia* (Robins-Browne and Prpic, 1985) have displayed increased virulence in situations of increased iron availability. Prophylactic intramuscular iron-dextran injections in children caused an enhancement in *E. coli* bacteraemia and meningitis (Barry and Reeve, 1977). Non-lethal injections of *E. coli* in guinea pigs become lethal by the concomitant addition of either haem or iron (Bornside *et al.*, 1968; Bullen *et al.*, 1968) The virulence of *Yersinia enterocolitica* is enhanced by transfusion of blood stored for periods > 3 weeks, which is prone to haemolysis and subsequent availability of excess free iron (Leclercq *et al.*, 2005). Thus, iron overload or haemolysis situations may enhance bacteraemia and, subsequently, the patient may die from an infectious disease.

Pathogenic bacteria have evolved an array of intricate mechanisms to scavenge limited iron from host by using siderophore system and/or by using receptors that bind lactoferrin, transferrin, ferritin or haemoglobin (Lee, 1995). Both, gram positive and gram negative bacteria employ similar mechanisms to scavenge iron from the host. Bacteria of enteric genus and gram-positive *Streptomyces* species produce a siderophore, enterobactin (Fiedler *et al.*, 2001; O'Brien and Gibson, 1970; Pollack and Neilands, 1970) that has high affinity for iron. Haem and haemoproteins can also be used directly as a source of iron by *Shigella* (Wyckoff *et al.*, 2005), *Brucella* (Paulley *et al.*, 2007), *Yersinia* (Hornung *et al.*, 1996), *Neisseria* (Chen *et al.*, 1998; Zhu *et al.*, 2000) and *Haemophilus* species (Cope *et al.*, 1994). Intracellular pathogens like *Salmonella, Mycobacterium tuberculosis* and *Listeria* species reside within phagosomal compartments of macrophages and thereby evade the mammalian immune system. Host cells often withdraw iron from these intraphagosomal compartments in response to invasion. *Mycobacteria* acquire iron from transferrin, ferritin, and lactoferrin in the lung parenchyma by producing salicylic and citric acids and siderophores called mycobactins (Ratledge, 2004). Intraphagosomal pH and iron concentration are crucial factors for determining mycobacterial pathogenicity and both host and mycobacteria have developed competing mechanisms to adjust during an infection. Intracellular pathogens have to compete for iron with the host's pH-dependent metal transporter, natural resistance associated macrophage protein (NRAMP) (Canonne-Hergaux *et al.*, 1999; Gruenheid *et al.*, 1999). A NRAMP-like mycobacterial metal transporter, MNTH, has been implicated in iron acquisition by *M. tuberculosis* (Agranoff *et al.*, 1999). Iron restriction represents an important host mechanism of intracellular pathogen killing and NRAMP1 variants in humans have been associated with susceptibility or resistance to intracellular pathogens (Blackwell *et al.*, 2001).

Increased susceptibility to *Salmonella* and other intracellular pathogens is reported in iron-overload conditions, e.g. β-thalassemia (Wanachiwanawin, 2000). Dietary iron overload, common in certain African and Zimbabwean populations, is associated with increased risk of tuberculosis (Gangaidzo *et al.*, 2001). An autopsy series of African adults have demonstrated increased risk of death from tuberculosis in patients with increased splenic iron (Gordeuk *et al.*, 1996). Iron overload in the host cell from genetic predisposition, therapeutic intervention, or nutritional status can predispose them to a number of infections (Schaible and Kaufmann, 2004; Weinberg, 2000).

In HIV infection, iron serves as an important element during various steps of HIV life cycle such as regulation of HIV-1 transcription, translation of viral mRNA, and viral assembly (Drakesmith and Prentice, 2008). In hepatitis C virus (HCV) infection, excess iron in the liver acts a proinflammatory agent and is associated with increased morbidity and mortality of HCV disease (Fujita and Takei, 2007). HCV-induced reactive oxygen species promote hepatic iron level (Nishina *et al.*, 2008), which is associated with oxidative DNA damage in hepatocytes (Fujita *et al.*, 2007). There are reports where excess of iron is associated with an increased HIV-1 viral replication, whereas iron chelation with desferrioxamine and deferiprone correlates with lower viral replication (Traoré and Meyer, 2004). Depletion of cellular iron by over-expressing ferroportin, the sole iron exporter found in mammals has been reported to inhibit HIV-gene expression (Xu *et al.*, 2010). Feline panleukaemia virus (FPV) and canine parvovirus (CPV) use transferrin receptor-1 as a cellular receptor for viral entry (Parker *et al.*, 2001). Changes in iron metabolism accompanied by disease exacerbation have been reported in several viral infections.

The protozoan parasite *Trypanosoma brucei* possesses specific transferrin receptor (*Tb*TfR) for binding transferrin from host for utilization of host iron. *Tb*TfR is encoded by two of the expression-site associated genes (ESAG), *ESAG6* and *ESAG7*, of the variant surface glycoprotein

(VSG) that associate into a heterodimeric fashion to perform its function (Salmon *et al.*, 1994; Steverding *et al.*, 1995). *Tb*TfR captures transferrin and the ligand-bound receptor complex (*Tb*TfR-Tf) is delivered to the endosomes where acidic pH triggers the release of iron from TfR–Tf complex and then apo-Tf dissociates from the receptor. Transferrin is then delivered to the lysosomes where it is proteolytically degraded by the cathepsin-like protein, *Tb*catB (Steverding *et al.*, 1995). In contrast, *Trichomonas vaginalis* acquires iron either from holo-lactoferrin (holo-Lf) or haemoglobin (Lehker *et al.*, 1990; Peterson and Alderete, 1984). It possesses specific receptors for recognizing holo-Lf but not Tf. In iron-depleted condition, Lf-binding activity increases substantially. *Trichomonas* can also take up iron from haemoglobin in a receptor-mediated fashion. Both cytoadherence to the vaginal epithelium and the expression of adhesions, essential features of *Trichomonas vaginalis* pathogenesis, are enhanced by growth in medium supplemented with iron salts or with holo-Lf (Lehker *et al.*, 1991). Recent reports showed two iron regulatory element (IRE)-like hairpin-loop structures in mRNAs of differentially iron-regulated TVCP4 and TVCP12 cysteine proteinases and IRP-like trichomonad proteins in *T. vaginalis.* It suggests the existence of post-transcriptional iron regulation mechanism by an IRE/IRP-like (mainly reported as an exquisite iron sensing mechanism in mammalians) system in this parasite (Torres-Romero and Arroyo, 2009).

In *Plasmodium*-infected patients, despite the fact that intra-erythrocytic parasite is surrounded by haemoglobin; it is unable to utilize its iron molecules directly, and rather accumulate the haem as haemozoins (Goldberg *et al.*, 1990; Roth *et al.*, 1986). Iron supplementation of *plasmodium*-infected patients is associated with adverse outcomes and increases malaria morbidity (Oppenheimer, 1989; Prentice *et al.*, 2007). Infection with *Plasmodium* spp. is exacerbated by iron and its intracellular growth is inhibited by iron chelators, such as DFO (Mabeza *et al.*, 1999). Iron deficiency compromises intra-erythrocytic growth of *Plasmodium* and enhances death of infected erythrocytes hence influence the course of malaria (Koka *et al.*, 2007). There have been debates regarding presence of transferrin receptor like protein in plasmodium and parasitized erythrocytes and there are conflicting evidences regarding internalization of iron by the parasite (Haldar *et al.*, 1986; Pollack and Schnelle, 1988; Rodriguez and Jungery, 1986; Sanchez-Lopez and Haldar, 1992). Recently, a haem oxygenase-like protein has been identified in *P. falciparum* implicating the existence of haem degradation system in the parasite (Okada, 2009).

Special need of iron for *Leishmania*

Leishmania species are important pathogens of human and other vertebrate animals and may cause a complex disease leishmaniasis. They are digenetic parasite endemic in more than 98 countries of tropical and temperate regions (Desjeux, 2004). The 20 different species of *Leishmania* is carried by 30 species of female sandfly that belong to the genus *Phlebotomus* in the Old World and *Lutzomyia* in the New World. The extracellular form of *Leishmania* procyclic promastigote replicates in the digestive tract of phlebotominae sandfly and matures into infective metacyclic promastigote stage which is injected into dermis of host during blood meal. Inside host, metacyclic promastigotes are engulfed by macrophages. Within macrophage they transform into intracellular amastigotes and multiply within phagolysosomal parasitophorous vacuoles and finally burst the host to infect neighbouring macrophages (McConville *et al.*, 2007; Sacks and Kamhawi, 2001; Ueno and Wilson, 2012). Clinical manifestation depends on *Leishmania* species, endemic region and host immune status and ranges from cutaneous lesions (localized, mucosal, disseminated, and diffuse) to most fatal visceral disease (Murray *et al.*, 2005).

Like other organisms, *Leishmania* too needs to acquire iron from the harsh ambience of host macrophages for their intracellular growth (Huynh *et al.*, 2006; Marquis and Gros, 2007). *Leishmania* also requires the iron acquisition to combat oxidative stress because of the presence of iron-containing antioxidant enzyme superoxide dismutase (Fe-SOD) (Paramchuk *et al.*, 1997). Inactivation of Fe-SOD affects intracellular survival and virulence of the *Leishmania* (Ghosh *et al.*, 2003). Furthermore, intracellular iron status of *Leishmania* may affect its drug resistance ability (Wong and Chow, 2006). The role of intracellular iron in potentiating anti-leishmanial drug activity was also reported (Mehta and Shaha, 2006) as iron depletion by using iron chelator desferrioxamine (DFO) could inhibit antimonial or arsenical drug induced changes in mitochondrial membrane potential and ATPase activity. As a result DFO can reduce the level of cell death induced by treatment with Sb(III) or As(III), whereas addition of iron causes higher cell death. These evidences show that iron can potentiate parasite death induced by metalloids like arsenic and antimony (Mehta and Shaha, 2006).

Contrary to the effect of iron on *in vitro* survival and growth of the *Leishmania* parasite, intraperitoneal injections of iron chelator DFO in *L. major*-infected BALB/c mice exhibited only a modest effect on lesion development. While, when mice were given iron supplementation, in the form of intraperitoneal injections of iron-dextran, showed significant retardation in lesion development. Mice injected with 8 mg iron dextran showed no lesions up to 18 weeks after challenge (Bisti *et al.*, 2000). But the effect of iron supplementation appears to be due to its effect on the immune response of the host in terms of ROS generation, rather than to any direct effect on the parasite (Bisti *et al.*, 2000; Bisti and Soteriadou, 2006). These studies suggested that probably iron could mediate its effect via ROS signalling through NF-κB, leading to a sustained TH1 response against the parasites (Bisti *et al.*, 2006; Bisti and Soteriadou, 2006). In support of this hypothesis a very recent study also provides evidence that iron overload favours the elimination of *Leishmania infantum* from mouse tissues by interacting with reactive oxygen and nitrogen species (Vale-Costa *et al.*, 2013).

In contrast, effect of DFO generated iron deficiency was examined on the course of the infection of *Leishmania chagasi* in BALB/c mice. Result showed a significant reduction in haemoglobin concentration of DFO-treated mice and subsequent reduction in parasite load in spleen and liver. This study suggests that iron deficiency favours the host and reduced parasite multiplication to attenuate infection (Malafaia *et al.*, 2011). In another study, quercetin (Qr) was used in combination with serum albumin (SA) to interfere iron metabolism of amastigote form of *L. donovani* (Sen *et al.*, 2008). The effect of treatment with Qr and SA on the proliferation of amastigotes was determined by estimating splenic parasite load. Disintegration of parasites in response to combination treatment was assessed from ultrastructural analysis using a transmission electron microscope. Quenching of the tyrosyl radical of ribonucleotide reductase in Qr and SA treated amastigotes was detected by an electron paramagnetic resonance study. This report further provided evidence that iron metabolism of parasite as a potential target to exert leishmanicidal activity even in infected host (Sen *et al.*, 2008).

Mammalian iron homeostasis components important for sequestering iron

In order to avoid loss of iron to invading pathogens, mammals have evolved very well orchestrated and elaborate iron homeostasis mechanisms both at the cellular and systemic level. These regulations are controlled by different and well coordinated mechanism to avoid gaining iron by pathogens forming a distinct innate immune component of the host. A short description of the iron homeostasis components of mammalian hosts related to *Leishmania* infection has given below.

Transferrin, transferrin receptor and ferritin

Transferrin (Tf) is a single polypeptide of 80 kDa glycoprotein consisting of two globular domains each with high-affinity binding site for one Fe^{3+} atom (Yang *et al.*, 1984). Binding of iron to transferrin is pH dependent as pH below 6.5 releases iron from it. Transferrin mainly synthesized in liver (Morgan, 1983) but also significantly in testis, lactating mammary gland, brain, and in some foetal tissues (Dickson *et al.*, 1985; Takeda *et al.*, 1998). In normal conditions, the majority of host serum iron remains bound to host protein transferrin (6–15 g/l). Iron- transferrin complex enters cells via transferrin receptor

(TfR)-mediated endocytosis and Fe^{3+}–Tf-TfR1 complex enters in to endosomal compartment. In the acidic environment of endosome (pH 5.5), Fe^{3+} is released from Tf–TfR1 complex because of the conformational changes in Tf-TfR1 at low pH. The released Fe^{3+} is then reduced by a ferrireductase Steap3 (six-transmembrane epithelial antigen of the prostate) present on the endosomal membrane (Ohgami *et al.*, 2005). After conversion, Fe^{2+} exits from endosome via DMT1 (also known as NRAMP2) to cytoplasm and distributes to various organelles by still unknown mechanism. The apo-Tf-TfR1exocytoses from recycling endosomes to cell surface, where at neutral pH apo-Tf releases from TfR1. Other than transferrin bound form, iron is also detected as non-transferrin-bound iron (NTBI), circulating in the serum chelated by low molecular weight molecules such as citrate (Cook and Skikne, 1989).

Tf-TfR1 mediated iron uptake pathway is the major route of cellular iron entry in most of the cell types including monocytes and macrophages. TfR1 is ubiquitously expressed in all cell types. Hepatocytes, duodenal crypt cells and erythroid cells also possess a TfR1 homologue TfR2 that is thought to be more specialized for cellular need of iron; however, TfR2 is not regulated by cellular iron content. Polarized epithelial cells of kidney uses megalin-dependent, cubilin-mediated endocytosis of Tf-bound iron (Kozyraki *et al.*, 2001). There are reports of the presence of alternative cellular iron uptake pathways in certain cell types, i.e. enterobactin (low molecular weight siderophores) and neutrophil-derived protein neutral gelatinase-associated lipocalin (NGAL) (Goetz *et al.*, 2002; Yang *et al.*, 2002). Different forms of cellular iron uptake mechanism have been evolved for different forms of iron.

Internalized iron either used for metabolic functions by incorporation into iron containing proteins or transferred in to the mitochondrion (Richardson *et al.*, 1996). Initially, after entering into cytoplasm it becomes constituent of the labile iron pool (LIP). LIP is transit pool of iron in cytosol that binds to low molecular weight chelates, e.g. ATP, AMP or pyrophosphate (Kakhlon and Cabantchik, 2002; Petrat *et al.*, 2002). LIP is a source of redox active iron and may contribute to iron mediated toxicity. LIP levels contribute to the homeostatic response to iron availability.

Excess iron in cytosol is stored into ferritin, an iron storage protein composed of 24 subunits of H- and L-chains that assembles into a shell like structure forming a cavity that can accommodate up to 4500 Fe atoms (Harrison and Arosio, 1996). Ferritin also helps cells to detoxify iron-induced toxicity. H-chain of ferritin contains ferroxidase activity converting Fe^{2+} to Fe^{3+} and stores chemically less reactive ferrihydrite. Recently poly-C binding protein has been shown as an iron chaperon for storing iron into ferritin (Shi *et al.*, 2008).

Lactoferrin

Lactoferrin is an abundant, iron-binding protein and one of the most important components of innate immune system (Valenti *et al.*, 2004). It is present in the blood, in secreted fluids such as tears, semen, vaginal secretions and sweat but in greatest abundance in milk (Park *et al.*, 2011). It is an ~80 kDa member of the transferrin family of non-haem, iron-binding glycoproteins and was originally thought to function as an iron transport protein in blood. Detection of lactoferrin in neutrophils suggested later that this protein might have a greater role in innate immunity (Masson *et al.*, 1969). It can bind two ferric ions (Fe^{3+}) in two similar, but not identical, structural lobe regions. In comparison to transferrin, lactoferrin binds iron very tightly in each lobular cleft ($K_d = 10^{-20}$ M) suggesting that iron sequestration rather than transportation is the primary function of lactoferrin (Ammons and Copié, 2013). Within the cleft of each iron-binding lobe, a single ferric ion is surrounded by three side chains of Tyr, Asp and His residues, in addition to a carbonate ion providing two ligands to form an octahedral complex (Saponja and Vogel, 1996). The protein can exist in either an open, non-iron-binding form (apo-lactoferrin) or a closed, iron-bound form (holo-lactoferrin) (Baker and Baker, 2012).

Ferroportin–hepcidin axis of innate immunity

Ferroportin (Fpn-1), also known as Ireg1 (McKie *et al.*, 2000) or metal transport protein MTP1 (Abboud and Haile, 2000), is the only known mechanism for iron efflux from all mammalian cell types including enterocytes, hepatocytes, macrophages, placental syncytiotrophoblasts including central nervous system (Donovan *et al.*, 2005). Fpn-1 is negatively regulated by iron regulatory hormonal peptide hepcidin which binds to Fpn-1 for its internalization and degradation (Nemeth *et al.*, 2004). During extracellular infection hepcidin releases from hepatocytes to promote Fpn-1 degradation to restrict iron within macrophages or other phagocytic cells (Johnson *et al.*, 2010). There are evidences of release of hepcidin from macrophages involving TLR pathways (Layoun and Santos, 2012). However, role or fate of *Leishmania* infection on hepcidin-Fpn-1innate immunity axis is not reported so far.

Regulation of cellular iron metabolism of mammalian host

Iron is obligatory for metabolic processes. Scarcity of iron can lead to cell death while excess iron results in cellular toxicity. Thus, intracellular iron level must be sensitively and tightly regulated. In vertebrates cellular iron level is sensed and regulated post transcriptionally by iron regulatory proteins (IRP1 and IRP2). They are cytosolic proteins and bind to iron responsive elements (IREs) present in the untranslated regions of several iron homeostasis genes including TfR1 and ferritin. IRP1 and IRP2 both respond to cellular iron level via different mechanisms. In low iron conditions IRP1 binds to IRE but in normal iron conditions it forms Fe–S cluster containing cytosolic aconitase. In normal iron conditions IRP2 is down-regulated by iron mediated oxidative stress induced ubiquitination and subsequent proteosomal degradation (Guo *et al.*, 1995). IRPs may also respond to hypoxia (Hanson and Leibold, 1998; Schneider and Leibold, 2003), increased oxidative stress (Pantopoulos *et al.*, 1996), elevated levels of nitric oxide (Bouton and Drapier, 2003) and phosphorylation (Schalinske and Eisenstein, 1996).

Iron responsive elements regulate stability and translation of many proteins involved in iron homeostasis such as TfR1, H- and L-chains of ferritin, mitochondrial aconitase, ferroportin, and DMT1/NRAMP2 and 5-aminolaevulainate synthase (eALAS). IREs found either in the 5′ or 3′ UTRs of the iron-regulated mRNAs. IREs contain a heterogenic canonical IRE sequence producing a stem–loop or hairpin structure and also allow functional diversity (Gdaniec *et al.*, 1998; Ke *et al.*, 1998). Binding of IRPs to mRNA at the 3′ UTR protects mRNA from degradation, while binding of IRPs to 5′ IRE-containing mRNA prevents translation. TfR1 contains five IREs in its 3′ UTR. In low iron conditions IRPs bind to these IREs to stabilize TfR1 mRNAs. Up-regulation of TfR1 leads to increase in cellular iron uptake. Simultaneously, IRP binds to 5′UTR of ferritin mRNA to block its translation. In high-iron conditions IRP1 converts to Fe–S cluster containing cytosolic aconitase to decrease IRE binding resulting in decreased iron uptake and increased iron storage.

Hepcidin and its role in host–pathogen interaction

Hepcidin is a peptide hormone consisting of 25 amino acids. It is secreted by a wide variety of cell types, most predominantly by hepatocytes. It plays a central role in maintaining systemic iron homeostasis in humans and other mammals. It controls the plasma iron concentration and iron distribution in various tissues by inhibiting the intestinal absorption and recycling from the macrophages (Nemeth and Ganz, 2006). Hepcidin binds to the cell surface iron exporter ferroportin to induce the internalization and subsequent degradation of ferroportin (Nemeth *et al.*, 2004). Thus, in conditions when hepcidin level increases like in inflammatory states, serum iron level drops because of trapping iron inside macrophages and decreased iron transport through intestine leading to mild anaemia.

The expression of hepcidin in mammals is regulated transcriptionally in response to the level of iron and inflammatory stimulus. The supplementation of iron induces the hepcidin transcription whereas, the level of its transcript decreases in response to the need of iron (Lee and Beutler, 2009). During infection, transcription of hepcidin increases due to activation of TLRs as well as to inflammatory cytokines such as IL-6 mainly by STAT3 dependent mechanism (Drakesmith and Prentice, 2012).

Source of host iron and its acquisition systems in *Leishmania*

Iron is required for the growth of both intracellular and extracellular forms of *Leishmania* species (Chang and Hendricks, 1985) and they need to develop special mechanisms to acquire iron from immediate environment. In *in vitro* culture iron is mostly supplied in the form of haemin or other haem-containing compounds. However, the iron sources in mammalian host or insect vector have not been clearly defined so far. Similarly, only a little is understood in defining iron acquisition mechanisms in promastigote and amastigote forms of this parasite. In fact, recent studies identified a couple of members of iron acquisition machinery of *Leishmania* particularly in its amastigote form but far less is understood regarding iron acquisition at promastigote stage.

Iron circulating in blood is mostly in the form of haem attached with haemoglobin and reversibly bound to glycoprotein carriers like transferrin. It has been shown that *Leishmania* could utilize both haemoglobin and transferrin for its biological activities (Sengupta *et al.*, 1999; Voyiatzaki and Soteriadou, 1992). Haem or haemoglobin acquisition by the parasite might result from a special need to avail itself of haem rather than to utilize it as a source of iron because *Leishmania* species are defective in haem biosynthesis as they lack aminolaevulinate dehydratase and porphobilinogen deaminase (Chang and Chang, 1985). *Leishmania* can acquire haem by two mechanisms: (1) haemoglobin receptor-mediated endocytosis followed by haem salvage (Agarwal *et al.*, 2013; Campos-Salinas *et al.*, 2011; Krishnamurthy *et al.*, 2005) and (2) direct transmembrane transport (Campos-Salinas *et al.*, 2011; Huynh *et al.*, 2012).

Leishmania donovani promastigotes bind to haemoglobin via a 46 kDa specific receptors located near flagellar pocket that rapidly internalized through receptor-mediated endocytosis (Sengupta *et al.*, 1999). There are evidences for utilizing haemoglobin also as a source of iron for intracellular *L. infantum* amastigotes (Carvalho *et al.*, 2009). *L. infantum* axenic amastigotes unable to grow in iron-deficient medium but addition of haemoglobin can restore its growth, emphasizing the need for haem even in the amastigote stage of the parasite. Haemoglobin is internalized by the parasite utilizing a clathrin-mediated process (Agarwal *et al.*, 2013) and then it enters to the parasite's lysosome via endocytic pathway in a process mediated by the small GTPase Rab proteins Rab 5 (Singh *et al.*, 2003) and Rab 7 (Patel *et al.*, 2008). Once in the lysosome haemoglobin is degraded releasing haem, which has been proposed to translocate into the cytosol by the ATP-binding cassette protein, LABCG5 (Campos-Salinas *et al.*, 2011).

Leishmania mexicana amazonensis contains receptor that facilitates the binding of haem or other metalloporphyrin compounds, and this binding is regulated by the growth phase of the parasite (Galbraith and McElrath, 1988). *L. amazonensis* can acquire haem through a transmembrane protein LHR1 (Leishmania Haem Response-1), which is essential for parasite survival (Huynh *et al.*, 2012). Analysis of biological properties of the LHR1/Δlhr1 single knockout strain reveals that efficient haem uptake through LHR1 in *L. amazonensis* is important for growth, differentiation, and infectivity for mouse macrophages (Miguel *et al.*, 2013).

There are evidences of utilizing non-haem iron sources by *Leishmania*. It was reported earlier that three species of *Leishmania* might contain transferrin receptor (Schell *et al.*, 1991; Voyiatzaki and Soteriadou, 1990, 1992) to utilize transferrin-bound iron; however, later studies suggested the binding of transferrin to parasites was nonspecific in nature (Wilson *et al.*, 2002). *Leishmania* genome sequence also not revealed the presence of any gene like transferrin receptor (Ivens *et al.*, 2005). *Leishmania amazonensis* was found to subvert the transferrin trafficking of host macrophage to deliver iron- loaded transferrin to the phagolysosome, where amastigotes internalized transferrin through flagellar pocket via an endocytic pathway (Borges *et al.*, 1998). The same study also revealed that only holo-transferrin but not apo-transferrin was able to support parasite survival suggesting the ability of *Leishmania* to utilize iron loaded transferrin as source of iron. *Leishmania chagasi* promastigotes were also found capable of acquiring iron from ferrilactoferrin and other physiologic or non-physiologic iron chelates (Wilson *et al.*, 1994). There was suggestion of expression of surface receptors for ferrilactoferrin in *Leishmania donovani* plasma membrane (Britigan *et al.*, 1998). Microorganisms residing in the respiratory, intestinal and genitourinary tracts face iron deficiency on mucosal surface, because the available ferric iron is chelated by lactoferrin as part of the innate immune system of the host. However, parasitic protozoa have developed various mechanisms to acquire and utilize iron from host lactoferrin. *Toxoplasma gondii, Entamoeba histolytica, Trichomonas fetus* and *Trichomonas vaginalis* express lactoferrin-binding protein and use holo-lactoferrin as iron source for their growth *in vitro*. In some species, these lactoferrin-binding proteins are immunogenic, thus may serve as potential vaccine targets. There are reports of using surface reductase to identify and reduce lactoferrin bound ferric iron to accessible ferrous form by *Leishmania* spp. (Ortíz-Estrada *et al.*, 2012; Wilson *et al.*, 2002). Interestingly, a recent study also suggested that two antimicrobial cryptopeptides from the N1 domain of bovine lactoferrin, lactoferricin (LFcin17–30) and lactoferrampin (LFampin265–284), and a hybrid LF-peptide (LFchimera) are lethal to *Leishmania donovani* promastigotes (Silva *et al.*, 2012). Among these cryptopepetides, LF chimera showed higher leishmanicidal activity over its separated components and was reported as the only peptide showing activity against *Leishmania pifanoi* axenic amastigotes. No critical study has been reported so far defining the role of lactoferrin as iron source or anti-leishmanial *in vivo* condition.

Leishmania donovani infection also has the capacity to deplete intracellular labile iron pool for its intracellular growth (Das *et al.*, 2009). This mechanism allows the parasite to activate cellular iron sensing components iron regulatory proteins (IRP) to stabilize TfR1 mRNA and subsequent increase in host cellular iron pool by increasing uptake of transferrin. The advantage of this mechanism is at least twofold: to perpetuate iron uptake of host cells and to block ferritin translation to avoid sequestration of labile iron pool of the infected host (Das *et al.*, 2009). The host iron release component ferroportin (Fpn-1) also contains a single iron responsive element in its 5′ UTR. Activation of IRPs due to a decrease in host LIP should also block ferroportin translation to keep iron within the host unlike *Salmonella* or *Mycobacterium*-infected macrophages (Das *et al.*, 2009). Decreased ferroportin expression in *Leishmania donovani*-infected splenic macrophages in BALB/c mice was reported but the precise mechanism is not known so far (Das *et al.*, 2009). The ability of *Leishmania* to decrease host LIP also could activate a host transcription factor hypoxia inducible factor −1 (HIF-1) (Singh *et al.*, 2012). Interestingly, HIF-1 has been found as a survival factor for intracellular parasite (Arrais-Silva *et al.*, 2005; Singh *et al.*, 2012; Spear *et al.*, 2006) unlike most of the bacteria (Peyssonnaux *et al.*, 2005). In response to bacterial infections in phagocytes, usually HIF-1 promotes host immunity by inducing reactive nitrogen species and antimicrobial peptides (Nizet and Johnson, 2009; Peyssonnaux *et al.*, 2005) but *Leishmania* can avoid this innate immunity mechanism of host by an unknown mechanism.

Interestingly, compared with the amastigote form, the iron acquisition components in promastigotes are far less understood. Available data suggest that the promastigote acquires iron mainly by reduced form rather than in oxidized form. An NADPH-dependent iron reductase activity has been detected in *Leishmania*, suggesting a reductase-dependent mechanism of iron uptake from transferrin and lactoferrin. *Leishmania* contains a parasite-associated or -secreted reductase that reduces ferric to ferrous iron, decreasing its affinity for the extracellular chelate of the host and allowing it to be readily internalized by the parasite (Wilson *et al.*, 2002). The presence of plasma membrane associated ferric reductase LFR1 was confirmed later in *L. amazonenesis* (Flannery *et al.*, 2011). LFR1 is a transmembrane protein containing FAD-binding and NADPH-binding sites. It also contains a putative haem-binding site within its transmembrane regions. LFR1 uses cytosolic NADPH to reduce Fe^{3+} to Fe^{2+}, ferrous iron is then transported to parasite cytosol by ferrous iron transporter LIT1 (Jacques *et al.*, 2010). LIT1, a parasitic membrane protein member of ZIP family transporter in *Leishmania amazonensis* was reported earlier as an inducible ferrous iron transporter (Huynh *et al.*, 2006). LIT1 expression was only detected after intracellular invasion of the parasite in its amastigote stage. LIT1 knockout *L.*

amazonensis lost its virulence and replicating capacity within macrophages (Huynh *et al.*, 2006). Both LFR1 and LIT1 are up-regulated in low iron conditions further confirming their role in maintaining leishmanial iron homeostasis (Flannery *et al.*, 2011; Huynh *et al.*, 2006; Mittra *et al.*, 2013).

Recent evidence suggested that iron uptake capacity of the parasite might control the differentiation of *Leishmania amazonensis* as an independent factor other than temperature and pH. Iron depletion in culture medium up-regulated LIT1 expression and caused an increase in iron content of the parasites, growth arrest, and differentiation of wild-type (WT) promastigotes into infective *Leishmania amazonensis* amastigotes but not in the LIT1 null promastigotes (Mittra *et al.*, 2013). The same study showed that LIT1 up-regulation also increased the iron superoxide dismutase (FeSOD) activity in WT but not in LIT1-null parasites. FeSOD and reactive oxygen species contribute in differentiation of virulent *Leishmania* amastigotes in a process that is regulated by iron availability (Mittra *et al.*, 2013).

To survive, grow, replicate and residing inside the phagolysosomes of macrophages, *Leishmania* relies on its successful iron acquisition capability from host. Phagolysosomal environment is usually hostile for pathogens but also provide a ready access to nutrients that are being degraded by host cells. NRAMP1 or Slc11a1 is the main protein plays role in limiting iron availability to phagolysosomal *Leishmania*. NRAMP1 expresses on the phagolysosomal membrane of macrophage and granulocytes and pumps ferrous iron from phagosome lumen to cytosol. The NRAMP1 exhibits its antimicrobial properties through its iron transport property but the mechanism is not completely understood. NRAMP2, the other family member of the NRAMP1, contains an iron responsive element in its UTR and also may transport other divalent metal ions. However, any role or regulation of NRAMP2 in the host macrophages in *Leishmania*-infected macrophages has not been explored so far.

Summary and future aspects

Despite the crucial role of iron in host–pathogen interaction, far less has been understood regarding the iron homeostasis in *Leishmania* parasite and altered host iron homeostasis during its infection. There is recent advancement in understanding in iron and haem transport mechanism in *Leishmania* at the molecular level. However, intracellular iron transport, utilization and regulation of iron homeostasis genes of *Leishmania* remain almost unexplored, which needs to be addressed to understand the biology and mechanism of pathogenicity of this protozoan parasite. Given the limited availability of the effective and less toxic anti-leishmanial drugs as well as increased resistance of these drugs towards *Leishmania*, the parasite iron homeostasis could be a potential drug target that needs to be explored in future.

Acknowledgements

This work has been supported by Department of Biotechnology to CKM. KB and SS acknowledge CSIR for its support in terms of fellowship.

References

Abboud, S., and Haile, D.J. (2000). A novel mammalian iron-regulated protein involved in intracellular iron metabolism. J. Biol. Chem. 275, 19906–19912.

Agarwal, S., Rastogi, R., Gupta, D., Patel, N., Raje, M., and Mukhopadhyay, A. (2013). Clathrin-mediated hemoglobin endocytosis is essential for survival of *Leishmania*. Biochim. Biophys. Acta *1833*, 1065–1077.

Agranoff, D., Monahan, I.M., Mangan, J.A., Butcher, P.D., and Krishna, S. (1999). *Mycobacterium tuberculosis* expresses a novel pH-dependent divalent cation transporter belonging to the Nramp family. J. Exp. Med. *190*, 717–724.

Ammons, M.C., and Copié, V. (2013). Mini-review: Lactoferrin: a bioinspired, anti-biofilm therapeutic. Biofouling *29*, 443–455.

Andrews, N.C., Fleming, M.D., and Levy, J.E. (1999). Molecular insights into mechanisms of iron transport. Curr. Opin. Hematol. *6*, 61–64.

Andrews, S.C., Robinson, A.K., and Rodríguez-Quiñones, F. (2003). Bacterial iron homeostasis. FEMS Microbiol. Rev. *27*, 215–237.

Arrais-Silva, W.W., Paffaro, V.A., Jr., Yamada, A.T., and Giorgio, S. (2005). Expression of hypoxia-inducible factor-1alpha in the cutaneous lesions of BALB/c mice infected with *Leishmania amazonensis*. Exp. Mol. Pathol. *78*, 49–54.

Baker, H.M., and Baker, E.N. (2012). A structural perspective on lactoferrin function. Biochim. Biol. Cell. *90*, 320–328.

Balla, J., Vercellotti, G.M., Jeney, V., Yachie, A., Varga, Z., Jacob, H.S., Eaton, J.W., and Balla, G. (2007). Heme, heme oxygenase, and ferritin: how the vascular endothelium survives (and dies) in an iron-rich environment. Antioxid. Redox Signal. *9*, 2119–2137.

Barry, D.M., and Reeve, A.W. (1977). Increased incidence of gram-negative neonatal sepsis with intramuscula iron administration. Pediatrics *60*, 908–912.

Beard, J., Erikson, K.M., and Jones, B.C. (2003). Neonatal iron deficiency results in irreversible changes in dopamine function in rats. J. Nutr. *133*, 1174–1179.

Bisti, S., and Soteriadou, K. (2006). Is the reactive oxygen species-dependent-NF-kappaB activation observed in iron-loaded BALB/c mice a key process preventing growth of *Leishmania major* progeny and tissue-damage? Microbes Infect. *8*, 1473–1482.

Bisti, S., Konidou, G., Papageorgiou, F., Milon, G., Boelaert, J.R., and Soteriadou, K. (2000). The outcome of *Leishmania major* experimental infection in BALB/c mice can be modulated by exogenously delivered iron. Eur. J. Immunol. *30*, 3732–3740.

Bisti, S., Konidou, G., Boelaert, J., Lebastard, M., and Soteriadou, K. (2006). The prevention of the growth of *Leishmania major* progeny in BALB/c iron-loaded mice: a process coupled to increased oxidative burst, the amplitude and duration of which depend on initial parasite developmental stage and dose. Microbes Infect. *8*, 1464–1472.

Blackwell, J.M., Goswami, T., Evans, C.A., Sibthorpe, D., Papo, N., White, J.K., Searle, S., Miller, E.N., Peacock, C.S., Mohammed, H., and Ibrahim, M. (2001). SLC11A1 (formerly NRAMP1) and disease resistance. Cell. Microbiol. *3*, 773–784.

Boldt, D.H. (1999). New perspectives on iron: an introduction. Am. J. Med. Sci. *318*, 207–212.

Borges, V.M., Vannier-Santos, M.A., and de Souza, W. (1998). Subverted transferrin trafficking in *Leishmania*-infected macrophages. Parasitol. Res. *84*, 811–822.

Bornside, G.H., Bouis, P.J., Jr., and Cohn, I., Jr. (1968). Hemoglobin and *Escherichia coli*, a lethal intraperitoneal combination. J. Bacteriol. *95*, 1567–1571.

Borst, P., and Fairlamb, A.H. (1998). Surface receptors and transporters of *Trypanosoma brucei*. Annu. Rev. Microbiol. *52*, 745–778.

Bouton, C., and Drapier, J.C. (2003). Iron regulatory proteins as NO signal transducers. Science's STKE: signal transduction knowledge environment *2003*, pe17.

Britigan, B.E., Ratcliffe, H.R., Buettner, G.R., and Rosen, G.M. (1996). Binding of myeloperoxidase to bacteria: effect on hydroxyl radical formation and susceptibility to oxidant-mediated killing. Biochim. Biophys. Acta *1290*, 231–240.

Britigan, B.E., Lewis, T.S., McCormick, M.L., and Wilson, M.E. (1998). Evidence for the existence of a surface receptor for ferriclactoferrin and ferrictransferrin associated with the plasma membrane of the protozoan parasite *Leishmania donovani*. Adv. Exp. Med. Biol. *443*, 135–140.

Bullen, J.J., Leigh, L.C., and Rogers, H.J. (1968). The effect of iron compounds on the virulence of *Escherichia coli* for guinea-pigs. Immunology *15*, 581–588.

Campos-Salinas, J., Cabello-Donayre, M., García-Hernández R., Pérez-Victoria I., Castanys, S., Gamarro, F., and Pérez-Victoria, J.M. (2011). A new ATP-binding cassette protein is involved in intracellular haem trafficking in *Leishmania*. Mol. Microbiol. *79*, 1430–1444.

Canonne-Hergaux, F., Gruenheid, S., Govoni, G., and Gros, P. (1999). The Nramp1 protein and its role in resistance to infection and macrophage function. Proc. Assoc. Am. Phys. *111*, 283–289.

Carvalho, S., Cruz, T., Santarém, N., Castro, H., Costa, V., and Tomás, A.M. (2009). Heme as a source of iron to *Leishmania infantum* amastigotes. Acta Trop. *109*, 131–135.

Chang, C.S., and Chang, K.P. (1985). Heme requirement and acquisition by extracellular and intracellular stages of *Leishmania mexicana amazonensis*. Mol. Biochem. Parasitol. *16*, 267–276.

Chang, K.P., and Hendricks, L.D. (1985). Laboratory cultivation and maintenance of *Leishmania*. In Leishmaniasis, Chang, K-P., and Bray, R.S., eds. (Elsevier, New York), pp 214–244.

Chen, C.J., Elkins, C., and Sparling, P.F. (1998). Phase variation of hemoglobin utilization in *Neisseria gonorrhoeae*. Infect. Immun. *66*, 987–993.

Connor, J.R., Pavlick, G., Karli, D., Menzies, S.L., and Palmer, C. (1995). A histochemical study of iron-positive cells in the developing rat brain. J. Comp. Neurol. *355*, 111–123.

Conrad, M.E., Umbreit, J.N., and Moore, E.G. (1999). Iron absorption and transport. Am. J. Med. Sci. *318*, 213–229.

Cook, J.D., and Skikne, B.S. (1989). Iron deficiency: definition and diagnosis. J. Int. Med. *226*, 349–355.

Cope, L.D., Thomas, S.E., Latimer, J.L., Slaughter, C.A., Müller-Eberhard, U., and Hansen, E.J. (1994). The 100 kDa haem:haemopexin-binding protein of *Haemophilus influenzae*: structure and localization. Mol. Microbiol. *13*, 863–873.

Cornelissen, C.N., and Sparling, P.F. (1994). Iron piracy: acquisition of transferrin-bound iron by bacterial pathogens. Mol. Microbiol. *14*, 843–850.

Das, N.K., Biswas, S., Solanki, S., and Mukhopadhyay, C.K. (2009). *Leishmania donovani* depletes labile iron pool to exploit iron uptake capacity of macrophage for its intracellular growth. Cell. Microbiol. *11*, 83–94.

Delanghe, J.R., and Langlois, M.R. (2001). Hemopexin: a review of biological aspects and the role in laboratory medicine. Clin. Chim. Acta *312*, 13–23.

Deneer, H.G., Healey, V., and Boychuk, I. (1995). Reduction of exogenous ferric iron by a surface-associated ferric reductase of *Listeria* spp. Microbiology *141 (Pt 8)*, 1985–1992.

Desjeux, P. (2004). Leishmaniasis: current situation and new perspectives. Comp. Immunol. Microbiol. Infect. Dis. *27*, 305–318.

Dickson, P.W., Aldred, A.R., Marley, P.D., Tu, G.F., Howlett, G.J., and Schreiber, G. (1985). High prealbumin and transferrin mRNA levels in the choroid plexus of rat brain. Biochem. Biophys. Res. Comm. *127*, 890–895.

Doherty, C.P. (2007). Host–pathogen interactions: the role of iron. J. Nutr. *137*, 1341–1344.

Donovan, A., Lima, C.A., Pinkus, J.L., Pinkus, G.S., Zon, L.I., Robine, S., and Andrews, N.C. (2005). The iron exporter ferroportin/Slc40a1 is essential for iron homeostasis. Cell Metab. *1*, 191–200.

Drakesmith, H., and Prentice, A. (2008). Viral infection and iron metabolism. Nat. Rev. Microbiol. *6*, 541–552.

Drakesmith, H., and Prentice, A.M. (2012). Hepcidin and the iron–infection axis. Science *338*, 768–772.

Fiedler, H.P., Krastel, P., Müller, J., Gebhardt, K., and Zeeck, A. (2001). Enterobactin: the characteristic catecholate siderophore of Enterobacteriaceae is produced by *Streptomyces* species. FEMS Microbiol. Lett. *196*, 147–151.

Flannery, A.R., Huynh, C., Mittra, B., Mortara, R.A., and Andrews, N.W. (2011). LFR1 ferric iron reductase of *Leishmania amazonensis* is essential for the generation of infective parasite forms. J. Biol. Chem. *286*, 23266–23279.

Fridovich, I. (1998). Oxygen toxicity: a radical explanation. J. Exp. Biol. *201*, 1203–1209.

Fujita, N., and Takei, Y. (2007). Iron, hepatitis C virus, and hepatocellular carcinoma: iron reduction preaches the gospel for chronic hepatitis C. J. Gastroenterol. *42*, 923–926.

Fujita, N., Sugimoto, R., Urawa, N., Araki, J., Mifuji, R., Yamamoto, M., Horiike, S., Tanaka, H., Iwasa, M., Kobayashi, Y., *et al.* (2007). Hepatic iron accumulation is associated with disease progression and resistance to interferon/ribavirin combination therapy in chronic hepatitis C. J. Gastroenterol. Hepatol. *22*, 1886–1893.

Galbraith, R.A., and McElrath, M.J. (1988). Heme binding to *Leishmania mexicana amazonensis*. Mol. Biochem. Parasitol. *29*, 47–53.

Gangaidzo, I.T., Moyo, V.M., Mvundura, E., Aggrey, G., Murphree, N.L., Khumalo, H., Saungweme, T., Kasvosve, I., Gomo, Z.A., Rouault, T., *et al.* (2001). Association of pulmonary tuberculosis with increased dietary iron. J. Infect. Dis. *184*, 936–939.

Gdaniec, Z., Sierzputowska-Gracz, H., and Theil, E.C. (1998). Iron regulatory element and internal loop/bulge structure for ferritin mRNA studied by cobalt(III) hexammine binding, molecular modeling, and NMR spectroscopy. Biochemistry 37, 1505–1512.

Gerlach, M., Ben-Shachar, D., Riederer, P., and Youdim, M.B. (1994). Altered brain metabolism of iron as a cause of neurodegenerative diseases? J. Neurochem. *63*, 793–807.

Ghosh, S., Goswami, S., and Adhya, S. (2003). Role of superoxide dismutase in survival of *Leishmania* within the macrophage. Biochem. J. *369*, 447–452.

Goetz, D.H., Holmes, M.A., Borregaard, N., Bluhm, M.E., Raymond, K.N., and Strong, R.K. (2002). The neutrophil lipocalin NGAL is a bacteriostatic agent that interferes with siderophore-mediated iron acquisition. Mol. Cell *10*, 1033–1043.

Goldberg, D.E., Slater, A.F., Cerami, A., and Henderson, G.B. (1990). Hemoglobin degradation in the malaria parasite *Plasmodium falciparum*: an ordered process in a unique organelle. Proc. Natl. Acad. Sci. U.S.A. *87*, 2931–2935.

Gordeuk, V.R., McLaren, C.E., MacPhail, A.P., Deichsel, G., and Bothwell, T.H. (1996). Associations of iron overload in Africa with hepatocellular carcinoma and tuberculosis: Strachan's 1929 thesis revisited. Blood *87*, 3470–3476.

Griffiths, E. (1991). Iron and bacterial virulence-a brief overview. Biol. Metals *4*, 7–13.

Gruenheid, S., Canonne-Hergaux, F., Gauthier, S., Hackam, D.J., Grinstein, S., and Gros, P. (1999). The iron transport protein NRAMP2 is an integral membrane glycoprotein that colocalizes with transferrin in recycling endosomes. J. Exp. Med. *189*, 831–841.

Guo, B., Phillips, J.D., Yu, Y., and Leibold, E.A. (1995). Iron regulates the intracellular degradation of iron regulatory protein 2 by the proteasome. J. Biol. Chem. *270*, 21645–21651.

Haldar, K., Henderson, C.L., and Cross, G.A. (1986). Identification of the parasite transferrin receptor of *Plasmodium falciparum*-infected erythrocytes and its acylation via 1,2-diacyl-sn-glycerol. Proc. Natl. Acad. Sci. U.S.A. *83*, 8565–8569.

Halliwell, B. (1992). Oxygen radicals as key mediators in neurological disease: fact or fiction? Ann. Neurol. *32 Suppl*, S10–15.

Halliwell, B., and Gutteridge, J.M. (1990). Role of free radicals and catalytic metal ions in human disease: an overview. Methods Enzymol. *186*, 1–85.

Hallquist, N.A., McNeil, L.K., Lockwood, J.F., and Sherman, A.R. (1992). Maternal-iron-deficiency effects on peritoneal macrophage and peritoneal natural-killer-cell cytotoxicity in rat pups. Am. J. Clin. Nutr. *55*, 741–746.

Hampton, M.B., Zhivotovsky, B., Slater, A.F., Burgess, D.H., and Orrenius, S. (1998). Importance of the redox state of cytochrome c during caspase activation in cytosolic extracts. Biochem. J. *329 (Pt 1)*, 95–99.

Hanson, E.S., and Leibold, E.A. (1998). Regulation of iron regulatory protein 1 during hypoxia and hypoxia/reoxygenation. J. Biol. Chem. 273, 7588–7593.

Harrison, P.M., and Arosio, P. (1996). The ferritins: molecular properties, iron storage function and cellular regulation. Biochim. Biophys. Acta *1275*, 161–203.

Hornung, J.M., Jones, H.A., and Perry, R.D. (1996). The hmu locus of *Yersinia pestis* is essential for utilization of free haemin and haem–protein complexes as iron sources. Mol. Microbiol. *20*, 725–739.

Huynh, C., Sacks, D.L., and Andrews, N.W. (2006). A *Leishmania amazonensis* ZIP family iron transporter is essential for parasite replication within macrophage phagolysosomes. J. Exp. Med. *203*, 2363–2375.

Huynh, C., Yuan, X., Miguel, D.C., Renberg, R.L., Protchenko, O., Philpott, C.C., Hamza, I., and Andrews, N.W. (2012). Heme uptake by *Leishmania amazonensis* is mediated by the transmembrane protein LHR1. PLoS Pathog. *8*, e1002795.

Ivens, A.C., Peacock, C.S., Worthey, E.A., Murphy, L., Aggarwal, G., Berriman, M., Sisk, E., Rajandream, M.A., Adlem, E., Aert, R., *et al.* (2005). The genome of the kinetoplastid parasite, *Leishmania major*. Science *309*, 436–442.

Jacques, I., Andrews, N.W., and Huynh, C. (2010). Functional characterization of LIT1, the *Leishmania amazonensis* ferrous iron transporter. Mol. Biochem. Parasitol. *170*, 28–36.

Johnson, E.E., Sandgren, A., Cherayil, B.J., Murray, M., and Wessling-Resnick, M. (2010). Role of ferroportin in macrophage-mediated immunity. Infect. Immun. *78*, 5099–5106.

Kakhlon, O., and Cabantchik, Z.I. (2002). The labile iron pool: characterization, measurement, and participation in cellular processes(1). Free Radical Biol. Med. *33*, 1037–1046.

Karunasagar, I., Joseph, S.W., Twedt, R.M., Hada, H., and Colwell, R.R. (1984). Enhancement of *Vibrio parahaemolyticus* virulence by lysed erythrocyte factor and iron. Infect. Immun. *46*, 141–144.

Ke, Y., Wu, J., Leibold, E.A., Walden, W.E., and Theil, E.C. (1998). Loops and bulge/loops in iron-responsive element isoforms influence iron regulatory protein binding. Fine-tuning of mRNA regulation? J. Biol. Chem. *273*, 23637–23640.

Koka, S., Föller, M., Lamprecht, G., Boini, K.M., Lang, C., Huber, S.M., and Lang, F. (2007). Iron deficiency influences the course of malaria in *Plasmodium berghei* infected mice. Biochem. Biophys. Res. Comm. *357*, 608–614.

Konijn, A.M., Carmel, N., Levy, R., and Hershko, C. (1981). Ferritin synthesis in inflammation. II. Mechanism of increased ferritin synthesis. Br. J. Haematol. *49*, 361–370.

Kozyraki, R., Fyfe, J., Verroust, P.J., Jacobsen, C., Dautry-Varsat, A., Gburek, J., Willnow, T.E., Christensen, E.I., and Moestrup, S.K. (2001). Megalin-dependent cubilin-mediated endocytosis is a major pathway for the apical uptake of transferrin in polarized epithelia. Proc. Natl. Acad. Sci. U.S.A. *98*, 12491–12496.

Krishnamurthy, G., Vikram, R., Singh, S.B., Patel, N., Agarwal, S., Mukhopadhyay, G., Basu, S.K., and Mukhopadhyay, A. (2005). Hemoglobin receptor in *Leishmania* is a hexokinase located in the flagellar pocket. J. Biol. Chem. *280*, 5884–5891.

Kuvibidila, S.R., Kitchens, D., and Baliga, B.S. (1999). *In vivo* and *in vitro* iron deficiency reduces protein kinase C activity and translocation in murine splenic and purified T-cells. J. Cell. Biochem. *74*, 468–478.

Layoun, A., and Santos, M.M. (2012). Bacterial cell wall constituents induce hepcidin expression in macrophages through MyD88 signaling. Inflammation *35*, 1500–1506.

Leclercq, A., Martin, L., Vergnes, M.L., Ounnoughene, N., Laran, J.F., Giraud, P., and Carniel, E. (2005). Fatal *Yersinia enterocolitica* biotype 4 serovar O:3 sepsis after red blood cell transfusion. Transfusion *45*, 814–818.

Lee, B.C. (1995). Quelling the red menace: haem capture by bacteria. Mol. Microbiol. *18*, 383–390.

Lee, P.L., and Beutler, E. (2009). Regulation of hepcidin and iron-overload disease. Ann. Rev. Pathol. *4*, 489–515.

Lehker, M.W., Chang, T.H., Dailey, D.C., and Alderete, J.F. (1990). Specific erythrocyte binding is an additional nutrient acquisition system for *Trichomonas vaginalis*. J. Exp. Med. *171*, 2165–2170.

Lehker, M.W., Arroyo, R., and Alderete, J.F. (1991). The regulation by iron of the synthesis of adhesins and cytoadherence levels in the protozoan *Trichomonas vaginalis*. J. Exp. Med. *174*, 311–318.

Létoffé, S., Ghigo, J.M., and Wandersman, C. (1994). Iron acquisition from heme and hemoglobin by a *Serratia marcescens* extracellular protein. Proc. Natl. Acad. Sci. U.S.A. *91*, 9876–9880.

Li, D. (1998). Effects of iron deficiency on iron distribution and gamma-aminobutyric acid (GABA) metabolism in young rat brain tissues. Hokkaido J. Med. Sci. *73*, 215–225.

Liochev, S.I., and Fridovich, I. (1994). The role of $O_2^{\cdot-}$ in the production of HO·: *in vitro* and *in vivo*. Free Radical Biol. Med. *16*, 29–33.

Mabeza, G.F., Loyevsky, M., Gordeuk, V.R., and Weiss, G. (1999). Iron chelation therapy for malaria: a review. Pharmacol. Ther. *81*, 53–75.

McConville, M.J., de Souza, D., Saunders, E., Likic, V.A., and Naderer, T. (2007). Living in a phagolysosome; metabolism of *Leishmania* amastigotes. Trend. Parasitol. *23*, 368–375.

McCord, J.M. (1998). Oxidants and antioxidants: The concept of balance. Age *21*, 79–80.

McKie, A.T., Marciani, P., Rolfs, A., Brennan, K., Wehr, K., Barrow, D., Miret, S., Bomford, A., Peters, T.J., Farzaneh, F., Hediger, M.A., Hentze, M.W., and Simpson, R.J. (2000). A novel duodenal iron-regulated transporter, IREG1, implicated in the basolateral transfer of iron to the circulation. Mol. Cell *5*, 299–309.

Malafaia, G., Marcon Lde, N., Pereira Lde, F., Pedrosa, M.L., and Rezende, S.A. (2011). *Leishmania chagasi*: effect of the iron deficiency on the infection in BALB/c mice. Exp. Parasitol. *127*, 719–723.

Marquis, J.F., and Gros, P. (2007). Intracellular *Leishmania*: your iron or mine? Trend. Microbiol. *15*, 93–95.

Martínez, J.L., Delgado-Iribarren, A., and Baquero, F. (1990). Mechanisms of iron acquisition and bacterial virulence. FEMS Microbiol. Rev. *6*, 45–56.

Masson, P.L., Heremans, J.F., and Schonne, E. (1969). Lactoferrin, an iron-binding protein in neutrophilic leukocytes. J. Exp. Med. *130*, 643–658.

Mehta, A., and Shaha, C. (2006). Mechanism of metalloid-induced death in *Leishmania* spp.: role of iron, reactive oxygen species, Ca^{2+}, and glutathione. Free Radical Biol. Med. *40*, 1857–1868.

Miguel, D.C., Flannery, A.R., Mittra, B., and Andrews, N.W. (2013). Heme uptake mediated by LHR1 is essential for *Leishmania amazonensis* virulence. Infect. Immun. *81*, 3620–3626.

Mittra, B., Cortez, M., Haydock, A., Ramasamy, G., Myler, P.J., and Andrews, N.W. (2013). Iron uptake controls the generation of *Leishmania* infective forms through regulation of ROS levels. J. Exp. Med. *210*, 401–416.

Morgan, E.H. (1983). Effect of pH and iron content of transferrin on its binding to reticulocyte receptors. Biochim. Biophys. Acta *762*, 498–502.

Murray, H.W., Berman, J.D., Davies, C.R., and Saravia, N.G. (2005). Advances in leishmaniasis. Lancet *366*, 1561–1577.

Nemeth, E., and Ganz, T. (2006). Regulation of iron metabolism by hepcidin. Annu. Rev. Nutr. *26*, 323–342.

Nemeth, E., Tuttle, M.S., Powelson, J., Vaughn, M.B., Donovan, A., Ward, D.M., Ganz, T., and Kaplan, J. (2004). Hepcidin regulates cellular iron efflux by binding to ferroportin and inducing its internalization. Science *306*, 2090–2093.

Nishina, S., Hino, K., Korenaga, M., Vecchi, C., Pietrangelo, A., Mizukami, Y., Furutani, T., Sakai, A., Okuda, M., Hidaka, I., *et al.* (2008). Hepatitis C virus-induced reactive oxygen species raise hepatic iron level in mice by reducing hepcidin transcription. Gastroenterology *134*, 226–238.

Nizet, V., and Johnson, R.S. (2009). Interdependence of hypoxic and innate immune responses. Nat. Rev. Immunol. *9*, 609–617.

O'Brien, I.G., and Gibson, F. (1970). The structure of enterochelin and related 2,3-dihydroxy-N-benzoylserine conjugates from *Escherichia coli.* Biochim. Biophys. Acta *215*, 393–402.

Ohgami, R.S., Campagna, D.R., Greer, E.L., Antiochos, B., McDonald, A., Chen, J., Sharp, J.J., Fujiwara, Y., Barker, J.E., and Fleming, M.D. (2005). Identification of a ferrireductase required for efficient transferrin-dependent iron uptake in erythroid cells. Nat. Genet. *37*, 1264–1269.

Okada, K. (2009). The novel heme oxygenase-like protein from *Plasmodium falciparum* converts heme to bilirubin IX alpha in the apicoplast. FEBS Lett. *583*, 313–319.

Oppenheimer, S.J. (1989). Iron and malaria. Parasitol. Today *5*, 77–79.

Ortíz-Estrada, G., Luna-Castro, S., Piña-Vázquez, C., Samaniego-Barrón, L., León-Sicairos, N., Serrano-Luna, J., and de la Garza, M. (2012). Iron-saturated lactoferrin and pathogenic protozoa: could this protein be an iron source for their parasitic style of life? Future Microbiol. *7*, 149–164.

Pantopoulos, K., Weiss, G., and Hentze, M.W. (1996). Nitric oxide and oxidative stress (H_2O_2) control mammalian iron metabolism by different pathways. Mol. Cell. Biol. *16*, 3781–3788.

Paramchuk, W.J., Ismail, S.O., Bhatia, A., and Gedamu, L. (1997). Cloning, characterization and overexpression of two iron superoxide dismutase cDNAs from *Leishmania chagasi*: role in pathogenesis. Mol. Biochem. Parasitol. *90*, 203–221.

Park, J.H., Park, G.T., Cho, I.H., Sim, S.M., Yang, J.M., and Lee, D.Y. (2011). An antimicrobial protein, lactoferrin exists in the sweat: proteomic analysis of sweat. Exp. Dermatol. *20*, 369–371.

Parker, J.S., Murphy, W.J., Wang, D., O'Brien, S.J., and Parrish, C.R. (2001). Canine and feline parvoviruses can use human or feline transferrin receptors to bind, enter, and infect cells. J. Virol. *75*, 3896–3902.

Patel, N., Singh, S.B., Basu, S.K., and Mukhopadhyay, A. (2008). *Leishmania* requires Rab7-mediated degradation of endocytosed hemoglobin for their growth. Proc. Natl. Acad. Sci. U.S.A. *105*, 3980–3985.

Paulley, J.T., Anderson, E.S., and Roop, R.M., 2nd (2007). *Brucella abortus* requires the heme transporter BhuA for maintenance of chronic infection in BALB/c mice. Infect. Immun. *75*, 5248–5254.

Payne, S.M. (1989). Iron and virulence in *Shigella*. Mol. Microbiol. *3*, 1301–1306.

Peterson, K.M., and Alderete, J.F. (1984). Iron uptake and increased intracellular enzyme activity follow host lactoferrin binding by *Trichomonas vaginalis* receptors. J. Exp. Med. *160*, 398–410.

Petrat, F., de Groot, H., Sustmann, R., and Rauen, U. (2002). The chelatable iron pool in living cells: a methodically defined quantity. Biol. Chem. *383*, 489–502.

Peyssonnaux, C., Datta, V., Cramer, T., Doedens, A., Theodorakis, E.A., Gallo, R.L., Hurtado-Ziola, N., Nizet, V., and Johnson, R.S. (2005). HIF-1alpha expression regulates the bactericidal capacity of phagocytes. J. Clin. Invest. *115*, 1806–1815.

Philpott, C.C. (2006). Iron uptake in fungi: a system for every source. Biochim. Biophys. Acta *1763*, 636–645.

Pollack, J.R., and Neilands, J.B. (1970). Enterobactin, an iron transport compound from *Salmonella typhimurium*. Biochem. Biophys. Res. Comm. *38*, 989–992.

Pollack, S., and Schnelle, V. (1988). Inability to detect transferrin receptors on *P. falciparum* parasitized red cells. Br. J. Haematol. *68*, 125–129.

Ponka, P. (1999). Cellular iron metabolism. Kidney Int. Suppl. *69*, S2–11.

Prentice, A.M., Ghattas, H., Doherty, C., and Cox, S.E. (2007). Iron metabolism and malaria. Food Nutr. Bull. *28*, S524–539.

Ramsey, A.J., Hillas, P.J., and Fitzpatrick, P.F. (1996). Characterization of the active site iron in tyrosine hydroxylase. Redox states of the iron. J. Biol. Chem. *271*, 24395–24400.

Ratledge, C. (2004). Iron, mycobacteria and tuberculosis. Tuberculosis (Edinb) *84*, 110–130.

Ratledge, C., and Dover, L.G. (2000). Iron metabolism in pathogenic bacteria. Annu. Rev. Microbiol. *54*, 881–941.

Richardson, D.R., Ponka, P., and Vyoral, D. (1996). Distribution of iron in reticulocytes after inhibition of heme synthesis with succinylacetone: examination of the intermediates involved in iron metabolism. Blood *87*, 3477–3488.

Robins-Browne, R.M., and Prpic, J.K. (1985). Effects of iron and desferrioxamine on infections with *Yersinia enterocolitica*. Infect. Immun. *47*, 774–779.

Rodriguez, M.H., and Jungery, M. (1986). A protein on *Plasmodium falciparum*-infected erythrocytes functions as a transferrin receptor. Nature *324*, 388–391.

Roth, E.F., Jr., Brotman, D.S., Vanderberg, J.P., and Schulman, S. (1986). Malarial pigment-dependent error in the estimation of hemoglobin content in *Plasmodium falciparum*-infected red cells: implications for metabolic and biochemical studies of the erythrocytic phases of malaria. Am. J. Trop. Med. Hyg. *35*, 906–911.

Sacks, D., and Kamhawi, S. (2001). Molecular aspects of parasite–vector and vector–host interactions in leishmaniasis. Annu. Rev. Microbiol. *55*, 453–483.

Salmon, D., Geuskens, M., Hanocq, F., Hanocq-Quertier, J., Nolan, D., Ruben, L., and Pays, E. (1994). A novel heterodimeric transferrin receptor encoded by a pair of VSG expression site-associated genes in *T. brucei*. Cell *78*, 75–86.

Sanchez-Lopez, R., and Haldar, K. (1992). A transferrin-independent iron uptake activity in *Plasmodium falciparum*-infected and uninfected erythrocytes. Mol. Biochem. Parasitol. *55*, 9–20.

Saponja, J.A., and Vogel, H.J. (1996). Metal-ion binding properties of the transferrins: a vanadium-51 NMR study. J. Inorg. Biochem. *62*, 253–270.

Schaible, U.E., and Kaufmann, S.H. (2004). Iron and microbial infection. Nat. Rev. Microbiol. *2*, 946–953.

Schaich, K.M. (1992). Metals and lipid oxidation. Contemporary issues. Lipids *27*, 209–218.

Schalinske, K.L., and Eisenstein, R.S. (1996). Phosphorylation and activation of both iron regulatory proteins 1 and 2 in HL-60 cells. J. Biol. Chem. *271*, 7168–7176.

Schell, D., Evers, R., Preis, D., Ziegelbauer, K., Kiefer, H., Lottspeich, F., Cornelissen, A.W., and Overath, P. (1991). A transferrin-binding protein of *Trypanosoma brucei* is encoded by one of the genes in the variant surface glycoprotein gene expression site. EMBO J. *10*, 1061–1066.

Schneider, B.D., and Leibold, E.A. (2003). Effects of iron regulatory protein regulation on iron homeostasis during hypoxia. Blood *102*, 3404–3411.

Schryvers, A.B., and Morris, L.J. (1988). Identification and characterization of the transferrin receptor from *Neisseria meningitidis*. Mol. Microbiol. *2*, 281–288.

Sen, G., Mukhopadhyay, S., Ray, M., and Biswas, T. (2008). Quercetin interferes with iron metabolism in *Leishmania donovani* and targets ribonucleotide reductase to exert leishmanicidal activity. J. Antimicrob. Chemother. *61*, 1066–1075.

Sengupta, S., Tripathi, J., Tandon, R., Raje, M., Roy, R.P., Basu, S.K., and Mukhopadhyay, A. (1999). Hemoglobin endocytosis in *Leishmania* is mediated through a 46-kDa protein located in the flagellar pocket. J. Biol. Chem. *274*, 2758–2765.

Shi, H., Bencze, K.Z., Stemmler, T.L., and Philpott, C.C. (2008). A cytosolic iron chaperone that delivers iron to ferritin. Science *320*, 1207–1210.

Silva, T., Abengózar, M.Á., Fernández-Reyes, M., Andreu, D., Nazmi, K., Bolscher, J.G., Bastos, M., and Rivas, L. (2012). Enhanced leishmanicidal activity of cryptopeptide chimeras from the active N1 domain of bovine lactoferrin. Amino acids *43*, 2265–2277.

Singh, A.K., Mukhopadhyay, C., Biswas, S., Singh, V.K., and Mukhopadhyay, C.K. (2012). Intracellular pathogen *Leishmania donovani* activates hypoxia inducible factor-1 by dual mechanism for survival advantage within macrophage. PloS One *7*, e38489.

Singh, S.B., Tandon, R., Krishnamurthy, G., Vikram, R., Sharma, N., Basu, S.K., and Mukhopadhyay, A. (2003). Rab5-mediated endosome–endosome fusion regulates hemoglobin endocytosis in *Leishmania donovani*. EMBO J. *22*, 5712–5722.

Spear, W., Chan, D., Coppens, I., Johnson, R.S., Giaccia, A., and Blader, I.J. (2006). The host cell transcription factor hypoxia-inducible factor 1 is required for *Toxoplasma gondii* growth and survival at physiological oxygen levels. Cell. Microbiol. *8*, 339–352.

Steverding, D., Stierhof, Y.D., Fuchs, H., Tauber, R., and Overath, P. (1995). Transferrin-binding protein complex is the receptor for transferrin uptake in *Trypanosoma brucei*. J. Cell Biol. *131*, 1173–1182.

Stojiljkovic, I., Hwa, V., de Saint Martin, L., O'Gaora, P., Nassif, X., Heffron, F., and So, M. (1995). The *Neisseria meningitidis* haemoglobin receptor: its role in iron utilization and virulence. Mol. Microbiol. *15*, 531–541.

Takeda, A., Devenyi, A., and Connor, J.R. (1998). Evidence for non-transferrin-mediated uptake and release of iron and manganese in glial cell cultures from hypotransferrinemic mice. J. Neurosci. Res. *51*, 454–462.

Torres-Romero, J.C., and Arroyo, R. (2009). Responsiveness of *Trichomonas vaginalis* to iron concentrations: evidence for a post-transcriptional iron regulation by an IRE/IRP-like system. Infection, genetics and evolution. Infect. Genet. Evol. *9*, 1065–1074.

Traoré, H.N., and Meyer, D. (2004). The effect of iron overload on *in vitro* HIV-1 infection. J. Clin. Virol. *31 Suppl 1*, S92–98.

Ueno, N., and Wilson, M.E. (2012). Receptor-mediated phagocytosis of *Leishmania*: implications for intracellular survival. Trends Parasitol. *28*, 335–344.

Vale-Costa, S., Gomes-Pereira, S., Teixeira, C.M., Rosa, G., Rodrigues, P.N., Tomás, A., Appelberg, R., and Gomes, M.S. (2013). Iron overload favors the elimination of *Leishmania infantum from* mouse tissues through interaction with reactive oxygen and nitrogen species. PLoS Negl. Trop. Dis. *7*, e2061.

Valenti, P., Berlutti, F., Conte, M.P., Longhi, C., and Seganti, L. (2004). Lactoferrin functions: current status and perspectives. J. Clin. Gastroenterol. *38*, S127–129.

Voyiatzaki, C.S., and Soteriadou, K.P. (1990). Evidence of transferrin binding sites on the surface of *Leishmania* promastigotes. J. Biol. Chem. *265*, 22380–22385.

Voyiatzaki, C.S., and Soteriadou, K.P. (1992). Identification and isolation of the *Leishmania* transferrin receptor. J. Biol. Chem. *267*, 9112–9117.

Vranka, J.A., Sakai, L.Y., and Bächinger, H.P. (2004). Prolyl 3-hydroxylase 1, enzyme characterization and identification of a novel family of enzymes. J. Biol. Chem. *279*, 23615–23621.

Wagener, F.A., Volk, H.D., Willis, D., Abraham, N.G., Soares, M.P., Adema, G.J., and Figdor, C.G. (2003). Different faces of the heme-heme oxygenase system in inflammation. Pharmacol. Rev. *55*, 551–571.

Wanachiwanawin, W. (2000). Infections in E-beta thalassemia. J. Pediatr. Hematol. Oncol. *22*, 581–587.

Wandersman, C., and Delepelaire, P. (2004). Bacterial iron sources: from siderophores to hemophores. Annu. Rev. Microbiol. *58*, 611–647.

Ward, C.G., Hammond, J.S., and Bullen, J.J. (1986). Effect of iron compounds on antibacterial function of human polymorphs and plasma. Infect. Immun. *51*, 723–730.

Weinberg, E.D. (1978). Iron and infection. Microbiol. Rev. *42*, 45–66.

Weinberg, E.D. (1999). Iron loading and disease surveillance. Emerg. Infect. Dis. *5*, 346–352.

Weinberg, E.D. (2000). Modulation of intramacrophage iron metabolism during microbial cell invasion. Microbes Infect. *2*, 85–89.

Wessling-Resnick, M. (1999). Biochemistry of iron uptake. Crit. Rev. Biochem. Mol. Biol. *34*, 285–314.

Wilson, M.E., Vorhies, R.W., Andersen, K.A., and Britigan, B.E. (1994). Acquisition of iron from transferrin and lactoferrin by the protozoan *Leishmania chagasi*. Infect. Immun. *62*, 3262–3269.

Wilson, M.E., Lewis, T.S., Miller, M.A., McCormick, M.L., and Britigan, B.E. (2002). *Leishmania chagasi*: uptake of iron bound to lactoferrin or transferrin requires an iron reductase. Exp. Parasitol. *100*, 196–207.

Wong, I.L., and Chow, L.M. (2006). The role of *Leishmania enriettii* multidrug resistance protein 1 (LeMDR1) in mediating drug resistance is iron-dependent. Mol. Biochem. Parasitol. *150*, 278–287.

Wyckoff, E.E., Lopreato, G.F., Tipton, K.A., and Payne, S.M. (2005). *Shigella dysenteriae* ShuS promotes utilization of heme as an iron source and protects against heme toxicity. J. Bacteriol. *187*, 5658–5664.

Xu, M., Kashanchi, F., Foster, A., Rotimi, J., Turner, W., Gordeuk, V.R., and Nekhai, S. (2010). Hepcidin induces HIV-1 transcription inhibited by ferroportin. Retrovirology 7, 104.

Yang, F., Lum, J.B., McGill, J.R., Moore, C.M., Naylor, S.L., van Bragt, P.H., Baldwin, W.D., and Bowman, B.H. (1984). Human transferrin: cDNA characterization and chromosomal localization. Proc. Natl. Acad. Sci. U.S.A. *81*, 2752–2756.

Yang, J., Goetz, D., Li, J.Y., Wang, W., Mori, K., Setlik, D., Du, T., Erdjument-Bromage, H., Tempst, P., Strong, R., *et al.* (2002). An iron delivery pathway mediated by a lipocalin. Mol. Cell *10*, 1045–1056.

Zhu, W., Hunt, D.J., Richardson, A.R., and Stojiljkovic, I. (2000). Use of heme compounds as iron sources by pathogenic *Neisseriae* requires the product of the hemO gene. J. Bacteriol. *182*, 439–447.

Oxidative and Nitrosative Stress Response in *Leishmania*

4

Swati Pal and Subrata Adak

Abstract

The generation of reactive oxygen and reactive nitrogen species, in response to pathogen attacks, play a central role in host defence mechanism. *Leishmania* pathogens cleverly evade the assault by evolving several defence strategies to protect themselves against the toxic effects of these reactive species. In the arthropod host, reactive oxygen species (ROS) in *Leishmania* are generated as a result of cellular metabolism, uncoupled electron transfer in mitochondria, endoplasmic reticulum stress as well as unfavourable environment (limited nutrient and oxygen). In the human host, the parasite additionally encounters both the oxidative and nitrosative stress response; yet they remain successful in establishing and maintaining productive acute and persistent infections in the host. This chapter assembles the current knowledge regarding reactive oxygen and nitrogen species, and compares the adaptive mechanisms utilized by *Leishmania* species to resist the toxic effects resulting from exposure to various environmental stresses.

Introduction

> It is not the strongest of the species that survives, nor the most intelligent, but rather the one most adaptable to change.
>
> Leon C. Megginson

Leishmania, from the evolutionary perspective, provides one of the best examples of adroit adaptors in the living world. The life cycle of *Leishmania* allows it to thrive in the arthropod host as well as the mammalian host, presenting two completely different niches, thus necessitating a high degree of adaptation. This is well evident from the morphological as well as physiological changes occurring in the parasite with the change of hosts. In the alimentary canal of the sandfly vector, the parasite resides as elongated, flagellated promastigote, the non-infective procyclic form of which develops into the infective metacyclic form. These migrate from the midgut to the proboscis and are released on the mammalian host skin by the sandfly during its blood meal. On entering the second host, the parasites are taken up by the professional phagocytic cells, like macrophages and neutrophils, where they lose the flagella and readily revert to round, aflagellated amastigotes form. The phagocytes utilize receptor-mediated endocytosis for the uptake of the *Leishmania*, that involves opsonic or pattern recognition receptors like, CR3, CR1 or mannose fucose receptors depending on the species and stage of the parasite and produce a variety of reactive oxygen species (ROS) as well as reactive nitrogen species (RNS) to kill the microbe. This is accomplished mainly in two steps: first, during uptake of the promastigotes, the phagocytosis event initiates an oxidative burst response in the mammalian cell generating toxic ROS. Second, once infection with amastigotes is established, the quiescent macrophage can be activated by the cytokines to produce RNS in an effort to kill the intracellular *Leishmania*. The events are tightly regulated by efficient feedback mechanisms. In order to evade the host's defence mechanism, the parasite adopts a biologically much complicated physiology which in turn confers the ability to bypass the host's immune response as well as development of drug resistance; this has made parasite treatment most challenging. This chapter will provide a detailed discussion on the oxidative and nitrosative onslaught experienced by this organism and the intelligent redox control that follows.

Sources of superoxide and hydrogen peroxide in *Leishmania*

Research on reactive oxygen species (ROS) dates back to 1818, when H_2O_2 was first identified by Louis Jacques Thenard (Thénard, 1818). Biological ROS production was first reported by Commoner *et al.* (1954) and high concentrations of it were soon recognized to be responsible for damage of key cellular components like DNA, proteins and lipids, thus leading to apoptosis and carcinogenesis (Jeffree, 1958). In aerobic organisms, ROS are generated as by-products of normal cellular metabolism, primarily in the mitochondria. During oxidative phosphorylation, leakage of electrons occurs from the mitochondrial electron transport chain, mainly at complexes I and III, producing superoxide, which is readily converted to H_2O_2 in the cell by the enzyme, superoxide dismutase. The second major source of ROS in eukaryotic organisms is the pathogen-induced respiratory burst in the phagocytic cells. ROS in *Leishmania* are generated as a result of cellular metabolism, uncoupled electron transfer in mitochondria, endoplasmic reticular stress, drug metabolism as well as various exogenous agents (Adak and Pal, 2013).

Mitochondrial reactive oxygen species generation

Leishmaniae and other trypanosomatids are unique in that they contain a single large mitochondrion per cell that ramifies extensively, occupying about 12% of the cell volume and along with the kinetoplast DNA, forms the kinetoplast–mitochondrial complex (Paulin, 1975; Simpson, 1972). Kinetoplast resides in the mitochondrial matrix and remains perpendicular to the flagellar axis. It consists of a dense network of two types of circular DNA molecules, the maxicircles and minicircles, representing many copies of the mitochondrial genome. In the past decades literature has been enriched by research on ROS generation by mitochondrial complexes of *Leishmania*. The data from several laboratories (Bermudez *et al.*, 1997; Luque-Ortega and Rivas, 2007; Martin and Mukkada, 1979a,b; Mukherjee *et al.*, 2012) reported evidences for the presence of all the complexes in the respiratory chain of *Leishmania* spp. Employing respiratory chain inhibitors, like rotenone, thenoyltrifluoroacetone and antimycin A, Mehta and Shaha (2004) presented compelling evidence for the generation of ROS by complexes I, II and III respectively. They showed that the ROS generated by inhibition of complex II resulted in increased levels of intracellular Ca^{2+}, which subsequently led to cellular apoptosis. Inhibition of complex I only increased the Ca^{2+} content marginally.

Endoplasmic reticulum stress-mediated reactive oxygen species generation

Endoplasmic reticulum (ER) is a multifunctional organelle that hosts the important steps in folding and post-translational modification of proteins as well as their transport to other cellular compartments. In the ER, numerous thiol-disulphide oxidoreductases mediate disulfide bond formation in myriads of secretory and membrane proteins (Appenzeller-Herzog and Ellgaard, 2008; Hatahet and Ruddock, 2009; Sevier and Kaiser, 2008). The sophisticated redox networks permit the formation and isomerization of disulphide bonds until the native state of the protein is attained. The major players in oxidative reactions are protein disulphide isomerases (PDI) (Ben Khalaf *et al.*, 2012; Hong and Soong, 2008; Kushawaha *et al.*, 2012; Padilla *et al.*, 2003) and putative Ero1 (Gene Bank accession number CAJ 04154); both highly conserved in *Leishmania*. Biochemical studies established that yeast Ero1 (Ero1p) oxidizes the CXXC motif in the N-terminal thioredoxin fold domain (a-domain) of Pdi1p, a pathway preferred for oxidizing the ER thiol pool (Vitu *et al.*, 2010). The generation of each disulphide bond by Ero1p is accompanied by the generation of one molecule of H_2O_2, a potential source of reactive oxygen species (Gross *et al.*, 2006). Usually a fine balance is maintained between the production of peroxides during routine oxidative protein folding and their detoxification or utilization as a messenger in the cell-signalling cascades (Veal *et al.*, 2007) or as a direct protein disulphide introducer (Karala *et al.*, 2009); however, ROS production beyond the capacity of the cellular antioxidant defence systems could induce harmful ER oxidative stress. Dolai *et al.* (2011) had shown that upon induction of ER stress by treatment with tunicamycin, an elevation in the level of intracellular ROS and Ca^{2+} is observed, which subsequently contributes to the mitochondrial membrane potential depolarization and ATP loss. This, in turn, leads to the release of cytochrome c and endonuclease G (Endo G) from mitochondria to cytosol and subsequent translocation of Endo G to the nucleus, which then causes DNA fragmentation and phosphatidylserine exposure, ultimately resulting in programmed cell death.

Drug mediated reactive oxygen species generation

Killing of the parasites by producing reactive oxygen intermediates has largely been the basis of parasitic chemotherapy. The hosts, including humans, have effective antioxidative machinery, which include the enzymes like catalase, superoxide dismutase, glutathione peroxidase and the low-molecular-weight non-enzyme molecules like α-tocopherol, ascorbate, glutathione and urate. In contrast, lack of one or other of these enzymes or molecules renders the parasites more susceptible to oxidative damage than the host (Callahan *et al.*, 1988; Docampo and Moreno, 1986; Dolai *et al.*, 2009; Fonseca-Silva *et al.*, 2011; Mookerjee Basu *et al.*, 2006; Vennerstrom and Eaton, 1988). Some of the currently used drugs lead to the increase in intracellular ROS content either by targeting the mitochondrial electron transport chain (ETC) complexes, thus causing electron leakage, or by inhibiting the ROS scavenging enzymes and preventing them from eliminating the toxicity. The most widely used drug for treatment of all types of leishmaniasis, that perturbs the redox homeostasis, has been the antimonials, available as sodium stibogluconate (Pentostam) for intravenous or intramuscular administration and meglumine antimoniate (Glucantime) for intramuscular administration (Ameen, 2007; Herwaldt, 1999; Murray *et al.*, 2005). Wyllie *et al.* (2004) have demonstrated the role of the antimonial compounds in deranging the thiol-redox potential of both the promastigote as well as amastigote stages of the protozoon in two distinct ways. First, the Sb^{III} form induces rapid efflux of intracellular trypanothione and glutathione, thus reducing the thiol buffering capacity. Secondly, Sb^{III} also inhibits the enzyme trypanothione reductase and prevents regeneration of the reduced molecules (Wyllie *et al.*, 2008). Apart from these, the adverse effects of the drug have also been demonstrated on DNA replication by dose dependent inhibition of DNA topoisomerase I of *L. donovani* (Chakraborty and Majumder, 1988; Walker and Saravia, 2004). Treatment with other widely used drugs like pentamidine, amphotericin B, miltefosine and paromomycin has also been shown to induce ROS accumulation in the *Leishmania* mitochondria (Moreira *et al.*, 2011; Sundar and Chatterjee, 2006), though the exact modes of action of these drugs need further elucidation. Pentamidine has been shown to accumulate in the mitochondria (Basselin *et al.*, 1997; Croft and Brazil, 1982; Ouellette *et al.*, 2004) and has a role as potential enhancer of the mitochondrial ETC complex II inhibitors (Mehta and Shaha, 2004). A number of model drugs like paraquat, quercetin, menadione, staurosporine, 3,3′-di-indolylmethane, tafenoquine and curcumin among others, also rely on ROS generation as the principal mode of action against the parasite. Fumarate reductase (FRD) constitutes another unique enzyme in the respiratory chain of trypanosomatids that catalyses the reduction of fumarate to succinate. FRD donates electrons to the electron transport chain via complex II and plays the crucial role in anaerobic energy metabolism. Chen *et al.* (2001) have shown that chalcones, for example, licochalcone A, 2,4-dimethoxy-4′-allyloxychalcone (24m4ac) and 2,4-dimethoxy-4′-butoxychalcone (24mbc) specifically inhibit FRD, leading to alterations in the function and ultrastructure of the mitochondria, ultimately killing the parasite.

Oxidative burst mediated reactive oxygen species generation in the host

The parasite, probably, faces the greatest assault due to oxidative stress in the mammalian phagocytic cells. Inside the neutrophil or the macrophage, receptor-dependent activation of nicotinamide adenine dinucleotide (NADPH)-oxidase enzyme (NOX) results in transfer of one electron from NADPH to O_2, with the concomitant generation of superoxide ($O_2{\cdot}^-$), a mechanism called oxidative burst (Fig. 4.1) (Lapouge *et al.*, 2000; Leto *et al.*, 2009; Morel *et al.*, 1991). $O_2{\cdot}^-$ and its dismutation product H_2O_2 have widely been thought to be responsible for the killing of phagocytosed micro-organisms, either directly (Babior *et al.*, 1973) or by acting as substrates for Fenton reactions or myeloperoxidase (MPO)-mediated halogenations (Klebanoff, 1970).

This oxidase complex consists of membrane-associated flavocytochrome b_{558} (composed of two subunits, gp91phox and p22phox) and four cytosolic components, p47phox, p67phox, p40phox, and the GTPase Rac, all of which assemble during phagocytosis or upon activation (Bylund *et al.*, 2010; Minakami and Sumimotoa, 2006; Park *et al.*, 1994; Wientjes *et al.*, 1993). gp91phox in cyt b_{558}, representing the only catalytic component in NADPH oxidase, includes the redox centres, FAD, and two non-identical haems. On receiving appropriate stimuli, Cyt b_{558}, an active site of the enzyme, binds tightly to the flavine adenine dinucleotide (FAD), which in turn allows it to complex with the cytosolic components, thus facilitating electron transfer between FAD and haem, ultimately generating superoxide anion (Babior, 1999; Hashida *et al.*, 2004). NOX can be activated by various cytokines, like TNF-α and IL-1 (Gauss *et al.*, 2007), oxidized low density lipoproteins as well as exogenous stimulators, for example phorbol myristate acetate (Rueckschloss *et al.*, 2001). Activated NOX has been shown to assemble within discrete subcellular compartments, like caveolae lipid rafts or endosomes, thus facilitating toxic ROS production within a restricted subcellular location. Phagocytes are estimated to produce relatively high concentrations of $O_2{\cdot}^-$, which then undergoes dismutation to H_2O_2 by superoxide dismutase (Fridovich, 1998). H_2O_2 is thought to be acted upon by MPO, and released into the vacuole from the cytoplasmic granules, to produce hypochlorous acid (HOCl), a potent antimicrobial (Klebanoff, 1967, 1968) oxidant. In *Leishmania*, different life cycle stages exhibit different levels of ROS sensitivity, and modulate the oxidative responses of the macrophages in different ways. Although, both the promastigote and amastigote stages are taken up by phagocytosis, the phagocytic $O_2{\cdot}^-$ production is much higher in response to promastigotes than to the amastigotes.

Targets of reactive oxygen species in *Leishmania*

Carbohydrates, lipids, amino acids and nucleic acids form the basic components of a cell and are essentially non-reactive with molecular oxygen (O_2). In 1954, Gerschman *et al.*, for the first time, hypothesized that oxygen poisoning and radiation injuries have at least one common mode of action, that is, through the formation of oxidizing free radicals like OH·, $HO_2{\cdot}$ and $O_2{\cdot}^-$ (Gerschman *et al.*, 1954). Oxic conditions can become toxic if there is uncontrolled production of reactive oxygen species or ineffective mechanisms of ROS scavenging. Oxidative damage may lead to loss

$$2O_2 + NADPH \xrightarrow{\text{(NADPH)-oxidase}} 2O_2{\cdot}^- + NADP^+ + H^+$$

$$2O_2{\cdot}^- + 2H^+ \xrightarrow{\text{superoxide dismutase}} H_2O_2 + O_2$$

$$H_2O_2 + Cl^- \xrightarrow{\text{myeloperoxidase}} HOCl + H_2O$$

Figure 4.1 Oxidative burst mediated ROS generation in the host.

of cell integrity and/or viability. In the following sections the molecular targets of oxidative stress in *Leishmania* spp. will be discussed in details (Table 4.1).

Lipid peroxidation

The plasma membrane as well as the other cellular biomembranes surrounding the different organelles of the cell consist of phospholipid

Table 4.1 Targets as well as scavengers of reactive oxygen species (ROS) and reactive nitrogen species (RNS)

Source ROS/ RNS	ROS/RNS	Scavengers	Specific targets	Consequence of oxidation/nitration
Mitochondria, endoplasmic reticulum, drug, oxidative burst	Hydroxyl radical (OH·), hydrogen peroxide (H_2O_2), superoxide ($O_2{\cdot}^-$), hypochlorous acid (HOCl),	Superoxide dismutase, tryparedoxin (TXN), TXN-dependent peroxidases, ascorbate peroxidase, peroxiredoxin, non-selenium glutathione peroxidase, ascorbate, trypanothione, glutathione, and ovothiol A	Lipid	Polyunsaturated fatty acids are easily oxidized to produce toxic lipid peroxides and peroxyl radicals (Haliwell and Chirico, 1993). Extensive lipid peroxidation leads to membrane damage causing increased membrane fluidity, inactivating membrane bound proteins and receptors and promoting efflux of solute from different compartments (Avery, 2011; West and Marnett, 2006)
			DNA	Oxidation of DNA results in the formation of lesions including oxidized bases (purines and pyrimidines), abasic sites (also called apurinic/apyrimidinic (AP) sites), and DNA single- and/or double-strand breaks, some of which may lead to miscoding and hence, be detrimental to the cell (Amesh, 1989; Marnett, 2000)
			Protein	Loss of sulfhydryl (–SH) groups, targeting of Fe–S clusters, formation of carbonyls, disulphide crosslinks, methionine sulfoxide, dityrosine cross-links. Oxidation-sensitive proteins include phosphatases, kinases, transcription factors, metabolic enzymes, chaperones and members of the ubiquitin proteasome system (Deavall *et al.*, 2012; Avery, 2011)
Phagocytic cell	Nitric oxide (NO) peroxynitrite ($ONOO^-$)	Peroxiredoxin1, pseudo-peroxidase, thiols	Lipid	Lipid peroxidation in membranes, liposomes, and lipoproteins by abstracting a hydrogen atom from polyunsaturated fatty acids (PUFA) producing lipid hydroperoxy radicals, conjugated dienes, and aldehydes (Hogg and Kalyanaraman, 1999)
			DNA	Oxidative modifications in both nucleobases and sugar–phosphate backbone; 8-oxoguanine is the major oxidation product, which further reacts with peroxynitrite, yielding cyanuric acid and oxazolone (Burney *et al.*, 1999; Niles *et al.*, 2006)
			Protein	Iron–sulfur (Fe/S) centre alterations, S nitrosylation in cysteine residue, cysteine oxidation, tyrosine nitration, tryptophan oxidation, methionine oxidation and histidine oxidation (Manuel and Ransijn, 1997; Fang, 2004)

bilayer containing embedded proteins. These are responsible for maintaining the integrity of cell and the organelles in it and allow selective passage of ions and organic molecules. Lipid peroxidation appears to be the most obvious consequence of oxidative stress. While the monounsaturated and saturated fatty acids (SFA) exhibit high resistance to free radical attack, the polyunsaturated fatty acids (PUFA) are easily oxidized by the protonated form of the superoxide radical ($HO_2\cdot$) and the hydroxyl radical ($OH\cdot$) producing toxic lipid peroxides (Halliwell and Gutteridge, 1995; Howlett and Avery, 1997); the latter are further decomposed by $HO_2\cdot$ to form peroxyl radicals, which then attack more PUFA side chains (Aikens and Dix, 1991; Wang *et al.*, 2006). Extensive lipid peroxidation leads to membrane damage causing increased membrane fluidity, inactivating membrane bound proteins and receptors and promoting efflux of solute from different compartments or the cell as a whole (Avery, 2011; Stubbs and Smith, 1984). Among the trypanosomatids, *Trypanosoma brucei* provides the example of FA synthesis by altogether a different mechanism involving endoplasmic reticulum-based elongases (Lee *et al.*, 2006). Further studies demonstrated that promastigotes of *Leishmania* spp. can grow in chemically defined medium that lacks lipid moieties (Steiger and Steiger, 1977), clearly highlighting the parasite's capacity to synthesize FA *de novo*. This was corroborated by *L. major* genome database that revealed the presence of genes encoding *T. brucei* Elongases 1–4 (TbELO1–4) orthologues as well as additional ELO-like open reading frames (ORFs). While the additional ELOs are engaged in elongating saturated FAs, the PUFAs in *L. major* are formed by TbELO1–4 orthologues and desaturase enzymes (Livore *et al.*, 2007). In *T. brucei, T. cruzi* and *L. major* the stearate that is produced by the ELO pathway can be desaturated to make C18:1 and C18:2 molecules; *L. major* possess the capacity to convert C18:2 to arachidonate and/or to even larger and more unsaturated PUFAs, up to 22:6 (Paul *et al.*, 2001). PUFAs may form part of the membrane bound free glycosylphosphatidylinositol (GPI) molecules (Morita *et al.*, 2000) as well as GPI-anchored variant surface glycoproteins (VSGs) attached to the GPI, which are responsible for promastigote uptake by the phagocytic cells, antigenic variations allowing the parasite to escape the host's immune system or even modification of the host's immune response. Nevertheless, the presence of PUFA in the parasite has made it more vulnerable to ROS. Moreover, the ability of the mitochondria to utilize oxygen at high rate and the presence of large quantities of PUFA along with peroxidation catalysts, like haem and non-haem iron, in the mitochondrial membrane have put this organelle at high risks of lipid peroxidation-induced injuries (Bindoli, 1988; Dix and Aikens, 1993). Indeed, studies with exogenous H_2O_2 as well as a number of ROS generating drugs like TTFA, camptothecin, curcumine, etc. have shown that *Leishmania* spp. are prone to oxidative stress mediated lipid peroxidation resulting in increased Ca^{2+} release from the ER compartment, loss of mitochondrial membrane permeabilization, release of cytochrome c to the cytoplasm and ultimately apoptosis (Das *et al.*, 2001, 2008, 2009; Mukherjee *et al.*, 2002; Roy *et al.*, 2008; Sen *et al.*, 2004). *In vivo* models of mice also support this, where it has been shown that increase in phagocytic ROS production leads to increased parasite death (Murray and Nathan, 1999). Lipid peroxidation may lead to the generation of a variety of oxidized products including reactive electrophiles, such as epoxides and aldehydes, which target the DNA, protein, and other macromolecules. Some examples are malondialdehyde (MDA), 4-hydroxy-2-nonenal (HNE), 2-propenal (acrolein), and isoprostanes, which can be measured as an indirect index of oxidative stress (Requena *et al.*, 2003; Wang *et al.*, 2006). MDA reacts with nucleic acid bases to form dG, dA, and dC adducts (Chaudhary *et al.*, 1994; Marnett, 1999) and is mutagenic (Basu and Marnett, 1983).

DNA oxidation

Oxidation of DNA results in the formation of lesions including oxidized bases (purines and pyrimidines), abasic sites (also called apurinic/apyrimidinic (AP) sites), and DNA single- and/or double-strand breaks, some of which may lead to miscoding and hence, be detrimental to the cell (Ames, 1989; Marnett, 2000). $HO\cdot$ and singlet oxygen ($^1O_2$) are the principal ROS directly affecting DNA. Owing to low oxidation potential, guanine is the most vulnerable site for oxidative damage on DNA, specifically at the position C-8 (Johnson *et al.*, 1996). This site, not being normally involved in bonding between guanosine and cytosine, is easily prone to ROS-mediated modifications in both single- and double-stranded DNA. The process is initiated by addition of the hydroxyl radical to the C-8 position of the guanine ring, producing an 8-hydroxy-7,8-dihydroguanyl radical, which can then be either oxidized to 8-oxo-7,8-dihydroguanine (8-oxoGua) or reduced to give the open-ringed 2,6-diamino-4-hydroxy-5-formamidopyrimidine (FapyGua) (Shibutani *et al.*, 1991; Spassky and Angelov, 1997). 8-oxoGua has even lower oxidation potential and is considered a clinical biomarker for oxidative DNA damage (Faure *et al.*, 1998). The other purine, adenosine, can be oxidized on either C-2 or -8, producing 2-hydroxyadenine (2-OH-Ade) as well as 8-oxo-7,8-dihydroadenine (8-oxoAde) and 4,6-diamino-5-formamidopyrimidine (FapyAde). Further purine lesions due to oxidative stress include 8,5′-cyclo-2′-deoxyguanosine (cyclo-dG) and the adenine equivalent cyclo-dA and the deamination products xanthine and hypoxanthine formed from guanine and adenine, respectively (Evans *et al.*, 2004). dGTP in the nucleotide pool can also be attacked by ROS and undergo conversion to form 8-oxoGua. Hence, during DNA replication, 8-oxoGua can be incorporated into DNA opposite both dC and dA on the template strand, resulting in AT to CG transversions (Cheng *et al.*, 1992; Moriya, 1993). The pyrimidines typically are hydroxylated on C-5 and C-6. The most common DNA lesions that arise due to oxidative damage of the pyrimidines are due to the formation of 5,6-dihydroxy-5,6-dihydrothymine (thymine glycol) and 5,6-dihydroxy-5,6-dihydrocytosine (cytosine glycol). Products of cytosine may deaminate and dehydrate producing uracil glycol, 5-hydroxycytosine (5-OH-Cyt), and 5-hydroxyuracil (5-OHUra). Among all the DNA oxidation products, 8-oxo-Gua is the most potent promutagen. During DNA replication, the oxidized guanosine occasionally rotates on the sugar–phosphate backbone, presenting N-7 for bonding, a condition that mimics double hydrogen bonding by thymidine, resulting in the insertion of adenosine into the complementary strand. This induces a GC to TA transversion in the subsequent replication cycle (Cooke *et al.*, 2003; Einolf *et al.*, 1998). The pyrimidine-derived products, 5-OH-Cyt and 5-OH-Ura are capable of producing mutations by causing GC to AT transitions and GC to CG transversions (Evans *et al.*, 2004; Purmal *et al.*, 1994). Left unrepaired, DNA mutations in critical genes can be detrimental to the cell.

Protein oxidation

The deleterious effects of ROS are largely mediated through protein oxidation. Oxidation-susceptible proteins are mainly transcription factors, metabolic enzymes, chaperones, kinases, phosphatases, and members of the ubiquitin proteasome system (Deavall *et al.*, 2012). Thus, protein oxidation can have a large impact on cellular homeostasis by directly affecting cell signalling as well as cellular respiration and metabolism (Diao *et al.*, 2010). Some proteins are very sensitive to oxidation than others. This is dependent on the relative content of oxidation-susceptible amino acid residues, protein localization within the cell, presence of metal-binding sites, molecular conformation, and rate of degradation (Avery, 2011). Newly synthesized proteins were found to be easy victims of ROS-induced ubiquitin-dependent degradation indicating that proper folding and inclusion of proteins in complexes account for protection against damage (Medicherla and Goldberg, 2008). Several types of ROS-induced protein modifications have been demonstrated, some examples include loss of sulfhydryl groups, targeting of Fe–S clusters, formation of carbonyls, disulphide cross-links, methionine sulfoxide, dityrosine cross-links, nitrotyrosine as well as glyoxidation and lipid peroxidation adducts (Cecarini *et al.*, 2007). Among the metabolic enzymes, the oxidatively targeted proteins include glyceraldehyde-3-phosphate dehydrogenase (GAPDH) and several citric acid cycle enzymes, like α-oxoglutarate dehydrogenase, isocitrate dehydrogenase, succinate dehydrogenase and aconitase, the last two being FeS enzymes. Among these, GAPDH has been found to be inactivated even by mild H_2O_2 stress (Schmalhausen *et al.*, 2003). The high reactivity of ROS with the Fe–S clusters makes a number of proteins that use these as co-factors, extremely ROS sensitive. Not only the functions of these proteins are perturbed, ROS attack causes release of the labile iron from the clusters into the intracellular milieu. Such an increase in cellular iron availability can induce catalysis of Fenton reaction that can lead to additional oxidative damage and killing. ROS can attack peptides by directly modifying amino acid side chains. The most vulnerable amino acids are those with aromatic side chain groups, for example, phenylalanine and histidine, and those containing sulfhydryl groups, for example, methionine and cysteine. Cysteine is also susceptible to oxidant-induced cross-linking. Oxidative modifications of sulfur-containing amino acids can be reversible. For example, methionine sulfoxide reductase enzymes can reverse the oxidation of methionine, which is one of the most oxidation-prone amino acid residues. Perhaps, the most common measure of ROS-induced protein alteration is the irreversible formation of carbonyl products by several amino acids, like arginine, proline, histidine and lysine. Carbonylation enables the proteins to be inactivated by proteasomal degradation. The carbonyls can further react with the α-amino groups of lysine residues on other protein molecules, leading to the formation of intra- or intermolecular cross-links which can promote the formation of high-molecular-mass aggregates.

Defence against reactive oxygen species

Ironically, despite all the antimicrobial approaches taken by the mammalian system, the parasites have evolved the capability to survive and multiply within the hosts, eventually establishing the disease. *Leishmania* spp. try to modulate the host's immune system at every step of infection, for example, right after the entry into the mammalian system, by evading complement activation, the parasites try to infect macrophages successfully (Puentes *et al.*, 1990); once taken up by the phagocytes, they try to modulate the host cytokine production (Reiner *et al.*, 1994; Sartori *et al.*, 1997), delay the fusion of the phagosome with the endosome (Desjardins and Descoteaux, 1997) and apoptosis of the phagocytes (Moore and Matlashewski, 1994) as well as prevent the activation of the NADPH oxidase enzyme (Lodge and Descoteaux, 2006) and hence the onset of oxidative burst; lastly, if the oxidative burst is initiated by the mammalian phagocytes, *Leishmania* uses its redox armamentarium to combat the adverse effects of the ROS. Not all promastigotes which enter macrophages survive. The level and extent of the oxidative stress dictates the parasitic cells to establish long or short term response (Krauth-Siegel and Comini, 2008). Laboratory infections of macrophages with *L. tropica* have indicated that the percentage of infected macrophages drops from a maximum of 63% at 30 minutes to 25% at 8 hours (Meshnick and Eaton, 1981). This decline is closely associated with a fall in reactive oxygen species production by the macrophage. Therefore, most of the killing of parasites occurs in the first of eight hours of infection and parasites which can survive the stress exhibit resistance to ROS. In mammalian cells, oxidative stress is prevented by the following antioxidant enzymes:

1. Cu,Zn-superoxide dismutase, expressed in the cytoplasm and the mitochondrial intermembrane space and responsible for the conversion of superoxide anion into hydrogen peroxide (Fridovich, 1995; McCord and Fridovich, 1969; Okado-Matsumoto and Fridovich, 2001);
2. Mn-SOD, with the same function as Cu, Zn-SOD, but found only in the mitochondrial matrix (Fridovich, 1995);
3. glutathione peroxidase enzyme responsible for the reduction of hydrogen peroxide and hydroperoxides utilizing reduced glutathione as hydrogen donor (Ursini *et al.*, 1995). Mammalian system contains several isoenzymes of this enzyme, of which GPX 1–4 contain selenium (Se) atom as selenocysteine at the catalytic site;
4. catalase, one of the most efficient enzymes with high turnover number, which catalyses the dismutation of hydrogen peroxide to oxygen and water (Chance *et al.*, 1979); and
5. peroxiredoxins, which include glutathione-dependent 1-cys peroxiredoxins; and thioredoxin-dependent 2-cys peroxiredoxins (Poole, 2007).

Although the parasite lacks potent ROS scavenging enzymes, like catalase and Se-containing glutathione peroxidase, the redox system in *Leishmania* is efficiently regulated. This is well documented by their perfect adaptation to diverse metabolic and environmental conditions implemented by their digenetic lifestyle, including the phagocytic oxidative burst (Table 4.1). The redox equilibrium in the parasite is maintained by the enzymes superoxide dismutase, tryparedoxin (TXN), TXN-dependent peroxidases (TXNPxs) and ascorbate peroxidase (APX). In addition, the low-molecular-weight thiols such as trypanothione, glutathione, and ovothiol A also effectively maintain the redox balance. Besides acting as ROS-scavengers, these molecules indirectly take part in regulating the H_2O_2-mediated intracellular signalling processes, such as cell differentiation, proliferation, and apoptosis.

Superoxide dismutase

Trypanosomatids like *Leishmania* and *Trypanosoma* possess unique iron-containing SOD. The first evidence for the presence of Fe-SOD in *Leishmania* came from the works of Meshnick and Eaton (1981). They demonstrated that *Leishmania* contain 60% more SOD activity than the free-living trypanosomatid, *Crithidia fasciculata*, and almost four times as much activity as in the extracellular parasitic trypanosomatid, *T. brucei*, thus establishing the importance of SOD in contributing to the parasitic adaptation against phagocytic oxidative burst. Later, Lashitew Gedamu's lab, from *L. chagasi*, cloned and characterized two distinct isoforms of the enzyme, Fe-SODA and Fe-SODB, sharing only 37% identity in amino acid sequence (Paramchuk *et al.*, 1997). Fe-SODB has been found to be constitutively expressed from multiple-copy gene, localized to the cytoplasm and functioning as an initial defence mechanism. On the other hand, Fe-SODA has been shown to be encoded by a single-copy gene and bears a 31 amino acids extension at the amino-terminal end. This N-terminal extension is hydrophobic in nature that bears sequence homology to mitochondrial transit peptide, suggesting a mitochondrial location for the enzyme. Ghosh *et al.* (2003) showed that, in *L. tropica* and *L. donovani* promastigotes, reducing the level of SOD enzyme by antisense RNA expression renders the parasites more sensitive to $O_2^{\cdot-}$ and H_2O_2. *L. donovani* amastigotes with reduced SOD enzyme exhibit markedly reduced survival within mouse macrophages. Data from Datta's lab point to the presence of an endogenous Cu/Zn-SOD in a specialized organelle, the glycosome, of *Leishmania* spp. that is similar to the peroxisomal enzyme found in the mammals (Dey and Datta, 1994).

Trypanothione and trypanothione reductase (TR)

Trypanosomatids present a peculiar example of thiol redox homeostasis based on the dithiol trypanothione [N1,N8-bis(glutathionyl)spermidine] (Fig. 4.2), which largely replaces glutathione in these organisms. It is produced by the unique ability of this group of protozoa to conjugate two molecules of the sulfur-containing tripeptide, glutathione (glutamic acid–cysteine–glycine), and one molecule of the polyamine, spermidine. The NADPH-dependent flavoenzyme, trypanothione reductase (TR), helps to maintain the reduced state of trypanothione [$T(SH)_2$]. Although $T(SH)_2$ and reduced glutathione (GSH) display similar redox potential, the former is a much more efficient reducing agent. Formation of an intramolecular disulfide is kinetically favoured in the dithiol $T(SH)_2$ when compared with GSH, where intermolecular oxidation of two molecules producing GS-SG is observed. In addition, the positive charge in the spermidine bridge shifts the thiol pK value to the physiological pH range that accounts for the high efficiency of the molecule in thiol disulfide exchange reactions in the parasite. TR forms the only link between NADPH and thiol-based redox system in these organisms. TR belongs to the FAD disulfide oxidoreductase family, which also includes glutathione reductase, lipoamide dehydrogenase, thioredoxin reductase, and the flavoprotein AhpF, a component of the

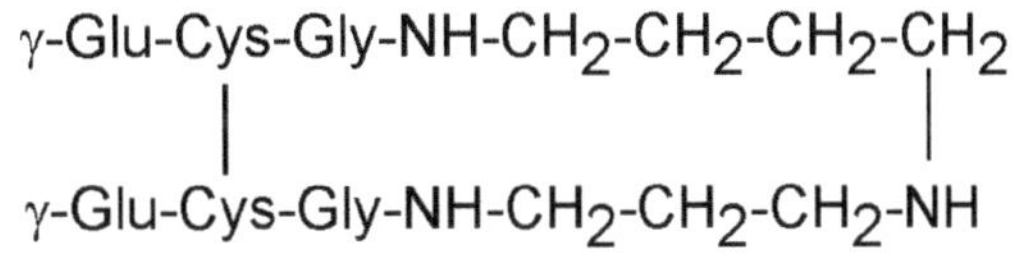

Figure 4.2 Structure of trypanothione.

alkylhydroperoxide reductase system found in *S. typhimurium*. Although TR shares many physical and chemical properties with glutathione reductase, the enzymes exhibit mutually exclusive specificity towards their disulfide substrates. Conditional knockdown of TR in trypanosomes rendered the parasites highly sensitive to oxidative stress accompanied by loss of virulence (Krieger *et al.*, 2000). In *L. donovani*, deletion of the first allele by homologous recombination was easily achieved. But attempts to delete the second allele proved to be futile, accompanied by a genomic rearrangement involving the translocation of the TR locus to a larger chromosome, generating an unaffected third copy of the gene (Dumas *et al.*, 1997). TR heterozygotes of *L. donovani* and *L. major* show attenuated infectivity and markedly reduced survivability within the host macrophages. This proved the essential role of the enzyme in the parasite. Trypanothione and TR, along with tryparedoxin (TXN) and tryparedoxin peroxidase (TXNPx) comprise the trypanothione peroxidase system (Fig. 4.3).

Tryparedoxin (TXN)

TXN, a distant member of the thioredoxin family (Ludemann *et al.*, 1998), is characterized by a typical WCPPCR active site motif that is thought to be responsible for the specificity of the enzyme towards oxidized trypanothione (TS_2). TXNs are specifically reduced by trypanothione; the reducing equivalents are transferred to different thiol-containing molecules that participate in various physiological activities of the parasite. Some examples include TXN peroxidase and non-selenium glutathione peroxidase-like enzymes (nsGPXs), which establish the role of TXN in parasitic peroxide detoxification. Other examples include the universal minicircle sequence binding protein, ribonucleotide reductase, that is required for nuclear and mitochondrial DNA replication and a monothiol glutaredoxin (1-Cys GRX) involved in iron–sulfur cluster biosynthesis (Comini *et al.*, 2008). Kinetoplastid genome presents several tryparedoxin-like sequences. Castro *et al.* (2004) characterized two TXNs from *L. infantum*: cytosolic LiTXN1 and mitochondrial LiTXN2. The former is up-regulated in stationary-phase promastigotes and in amastigotes (Castro *et al.*, 2004) and has been shown to be essential for parasite survival (Romao *et al.*, 2009). The absence of TXN in mammalian hosts as well as its importance in the parasite survival has marked this enzyme as a potential candidate for drug target.

Tryparedoxin peroxidase (TXNPx)

TXNPxs belong to the large family of peroxiredoxins found in different organisms and taking part in a wide variety of functions ranging from general cellular ROS detoxification to mediating signal transduction pathways during proliferation or differentiation processes. In trypanosomatids, TXNPxs catalyse the reduction of H_2O_2 or small-chain organic hydroperoxides using tryparedoxin as the electron donor (Hillebrand *et al.*, 2003; Schmidt and Krauth-Siegel, 2003). The unique characteristics of the TXNPxs not only render them a good drug target, but are also used to produce vaccines. *Leishmania* spp. possess more than one TXNPxs, eight in *L. major* and three in *L. infantum* and *L. brasiliensis*, that are distributed in the cytosol as well as the mitochondria. Both the cytosolic and the mitochondrial isoforms

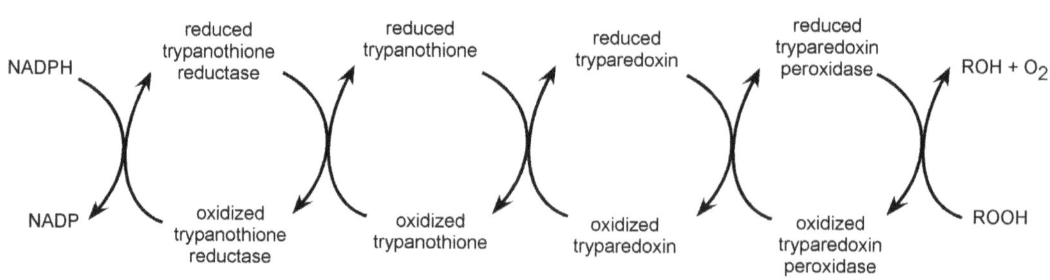

Figure 4.3 Trypanothione-dependent pathway for transfer of reducing equivalents.

present two different domains, I and II, which are common to the 2-Cys peroxiredoxins and rely on the Cys residues for the peroxidase activity. Although, the isoforms share identical amino acid sequence for the domain I, they differ for the domain II. TXNPxs include members of the peroxiredoxin (PRX) and of the non-selenium glutathione peroxidase (nsGPX) families.

Peroxiredoxin (PRX)

Peroxiredoxins or peroxidoxins were the first enzymes reported to display tryparedoxin peroxidase (TXNPx) activity (Nogoceke *et al.*, 1997), catalysing the reduction of hydroperoxides and peroxynitrites. PRXs can be classified based on their mitochondrial or cytosolic distribution. Mitochondrial PRXs differ from the typical 2-Cys PRXs in having the second redox active Cys embedded in an Ile–Pro–Cys motif, instead of the usual Val–Cys–Pro sequence. The enzyme forms an obligate head-to-tail homodimer, whose active sites are formed by the N-proximal peroxidatic cysteine from one subunit and a C-proximal resolving cysteine from the other. Mitochondrial PRXs are encoded by single copy genes, whereas closely related genes clustered within the same chromosomal locus, codes for the cytosolic enzymes. In *Leishmania* spp., one of these genes specifies a glycosomal targeting signal, suggesting the role of this enzyme in this peroxisome-like organelle. Barr and Gedamu (2001, 2003) have identified two PRXs from *L. chagasi*, which are differentially regulated. LcPRX1was found to be expressed more in the amastigote stage, while LcPRX2 transcript level was more abundant in the promastigote stage. They have also demonstrated that although both the isoforms participate in ROS detoxification, LcPRX1 is capable of scavenging reactive nitrogen species (RNS) as well.

Non-selenium glutathione peroxidase (nsGPX)

GPXs are among the front-line enzymes that reduce H_2O_2 or other organic hydroperoxides with high efficiency (Schlecker *et al.*, 2005). The catalytic core usually consists of selenocysteine (SeCys), tryptophan, and a glutamine residue. In trypanosomatids, the SeCys is replaced by a Cys residue, a substitution that accounts for reduced peroxidase activity exhibited by the nsGPXs. The nsGPXs form the second class of peroxidases in trypanosomatids that obtain their reducing equivalents from NADPH via the $T(SH)_2$/TXN couple. These nsGPXs also lack the glutathione-binding residues, consequently displaying extremely low affinities for the thiol, with K_M values within the millimolar range. Tryparedoxin or thioredoxin may act as alternative reductants for these enzymes (Castro *et al.*, 2002; Diechtierow and Krauth-Siegel, 2011).

Ovothiol A

Ovothiols (OSH) are histidine-derived thiols, first observed in marine invertebrates. Members of the ovothiol family differ from each other depending on the number of methyl substituents on the α-amino group ($n = 0$, 1 or 2 for ovothiol A, B and c, respectively). The thiol can react with H_2O_2 in a non-enzymatic way forming water and ovothiol disulfide (Ariyanayagam and Fairlamb, 2001). The latter is brought back to the reduced state via thiol-disulfide exchange with glutathione (GSH), which is oxidized in the process to produce glutathione disulfide (GSSG). Glutathione reductase (GR) acts as the reducing agent for GSSG (Shapiro and Hopkins, 1991; Shapiro and Turner, 1988). In addition to this, OSHs also exhibit free radical scavenging properties (Holler and Hopkins, 1990; Marjanovic *et al.*, 1995). The thiol participates in one-electron transfer reactions and although, is a better free radical scavenger than aliphatic thiols such as glutathione, it has been proved to be less efficient than trypanothione [$T(SH)_2$]. Ovothiol A (N-methyl-4-mercaptohistidine) has long been detected in *Crithidia fasciculata* and *L. donovani* (Spies and Steenkamp, 1994; Steenkamp and Spies, 1994). Later, results from A.H. Fairlamb's laboratory demonstrated that the ovothiol A content varies depending on the species and the life cycle stage of the parasites (Ariyanayagam and Fairlamb, 2001). It was detected in the insect stage of *Leishmania* spp., and, particularly in the late log phase or stationary phase of growth, the level can even exceed the $T(SH)_2$ content. However, only *L. major* has been found to maintain substantial levels of OSH in the amastigote stage. No ovothiol reductase activity has been detected in trypanosomatids compelling the redox reaction involving OSH to be mainly dependent upon the trypanothione/trypanothione reductase activity.

Ascorbate

L-Ascorbic acid or vitamin C acts as a pivotal antioxidant molecule donating electrons in various enzymatic and non-enzymatic reactions. The chief redox functions include donating electrons to APX family of enzymes as well as scavenging free radicals, such as RO·, RO_2·, HO·, NO_2·, several non-radical reactive species, e.g. for hypochlorous acid, ozone, and nitrating agents derived from peroxynitrite, and maintenance of α-tocopherol (vitamin E) in its reduced form. The presence of ascorbate has been detected in trypanosomatids as early as in 1994, but only recently the biosynthetic pathway has been described in the *T. brucei*, *T. cruzi* and *L. donovani* (Biyani and Madhubala, 2011; Clark *et al.*, 1994; Wilkinson *et al.*, 2005). Humans are not able to synthesize ascorbate because they lack the enzyme gulonolactone oxidase, which is required to produce ascorbate from glucose. This imparts the ascorbate and ascorbate peroxidase system in the parasite the probability to become potent drug targets.

Ascorbate peroxidase (APX)

Detection of ascorbate-dependent peroxidase activity in *T. cruzi* epimastigote extracts dates back to 1980. Much later, in 2002, an endoplasmic reticulum-located unusual plant-like APX enzyme (TcAPX) has been discovered in the organism (Wilkinson *et al.*, 2002) and in 2005, the enzyme (LmAPX) was detected in the inner mitochondrial membrane of *L. major* with its catalytic domain facing the inter-membrane space (Adak and Datta, 2005; Dolai *et al.*, 2008). LmAPX is the only haem-containing peroxidase discovered in *Leishmania* till date and is more related to TcAPX. However, the primary sequence of LmAPX bears high level of sequence identity with both yeast cytochrome C Peroxidase (CCP) (~35%) and the pea APX (~36%) and is a functional hybrid of both the enzymes. Interestingly, although all the key residues on both distal and proximal sites of the haem are conserved among the three enzymes, LmAPX lacks both the cytochrome c-binding residues (Asp34/Glu35, Tyr39, and Glu290 in yeast CCP) and the critical ascorbate binding residue (Arg 172 in pea APX). LmAPX contains similar proximal cation (K^+) binding loop as found in pea APX, whereas the c-terminal insertion and the proximal Trp radical stabilizing Met residues of LmAPX are identical to CCP (Jasion *et al.*, 2012; Jasion *et al.*, 2011; Yadav *et al.*, 2008). This renders LmAPX an unique mechanism for catalysing the redox reaction. In presence of electron donors, the resting form of LmAPX can form two types of intermediates: haem ferryl protoporphyrin π-cation radical ($^{+}$·P $Fe^{IV}=O$ Trp) intermediate, containing oxyferryl haem $Fe^{IV}=O$ and a porphyrin π-cation radical as well as haem ferryl tryptophan radical ($Fe^{IV}=O$,P,Trp·$^{+}$) intermediate, where a Trp^+ radical cation located on the indole ring of Trp-208 is formed instead of the π-cation. Low-molecular-mass (micromolecule) electron donors like ascorbate act as the electron donor for the former, while the latter intermediate accepts reducing equivalents from macromolecules like cytochrome C (Dolai *et al.*, 2008; Jasion and Poulos, 2012). Like well-known peroxidases, LmAPX also shows pseudocatalase activity and catalyses the degradation of H_2O_2 with concomitant formation of O_2 under low ascorbate concentration at physiological pH (Dolai *et al.*, 2007). Dolai et. al (Dolai *et al.*, 2008, 2009) demonstrated that overexpression of the enzyme protects the promastigote cells against the oxidative-stress induced apoptosis. Knocking the gene out of the cell by homologous recombination results in elevated levels of intracellular H_2O_2, thus throwing the parasite under constant oxidative stress (Pal *et al.*, 2010). This signals for the early onset of metacyclogenesis producing infective metacyclic promastigotes from non-infective procyclic ones. The metacyclics can infect host cells and eventually transform into amastigotes. But they proceed towards apoptosis if the favourable condition (high temperature and low pH, characteristic of the condition prevalent in the mammalian host) is not reached; in this respect, metacyclogenesis is comparable to terminal differentiation. The infectious parasite population in the midgut of infected sandfly or *in vitro* stationary phase cultures contain both metacyclic and apoptotic parasites and a combination of these two types of cells are needed for infectivity of the parasite within host macrophages (van Zandbergen *et al.*, 2006). Macrophage activation is suppressed by the apoptotic cells with a concomitant reduction in the production of inflammatory cytokines such as TNF-α and an increase in the expression of anti-inflammatory cytokines such as transforming growth factor-β, while the virulent metacyclics proceed to invade macrophages and cause disease. Deprived of LmAPX, the stationary phase culture was found to contain a higher number of metacyclic and apoptotic parasites that exhibited higher infectivity compared to both wild-type and LmAPX-overexpressing cells (Pal *et al.*, 2010). Thus, LmAPX represents a distinct class of genes whose deletion resulted in enhanced infectivity of the parasite.

Nitrosative stress in *Leishmania*

In addition to generation of oxidative stress, for effective killing of the parasites, the first evidence that the phagocytic cells also depend on non-oxidative mechanisms came from the observations of Scott *et al.* (1985), who demonstrated *L. major* elimination by a murine macrophage cell line, IC-21, which is deficient in the production of oxygen metabolites. Later it was shown that the leishmanicidal activity of the macrophage is largely mediated by nitric oxide (NO) generated from L-arginine (Liew *et al.*, 1990). Indeed, the parasiticidal activity can be abolished on treatment of the macrophage *in vitro* with L-N^G-monomethyl arginine (L-NMMA), a specific inhibitor of L-arginine: nitric oxide pathway (Nathan and Xie, 1994). Disease severity was found to be attributable to resistance of *Leishmania* spp. to reactive nitrogen species (RNS).

Generation of reactive nitrogen species

Development of leishmaniasis not only depends on the species initiating the disease, but also on the immunological competence of the host involving a cell-mediated response to combat the disease (Reiner and Locksley, 1995). Although, immediately after a microbial infection, phagocytes are typically stimulated to produce ROS, the production of RNS requires *de novo* protein synthesis. Studies using mice, susceptible (BALB/c) and resistant (C57BL/6, C3H/HeN or B10.D2) to leishmaniasis, have greatly contributed to the understanding of the underlying mechanisms of the disease development. The intrinsic susceptibility or resistance has been shown to be attributable to distinct subtypes of $CD4^+$ T-helper (Th) cells. Th1 cells respond to antigens and IL-12 and secret IL-2, IFN-γ and TNF-β, whereas, Th2 cells are triggered by IL-4 and their effector cytokines are IL-4, IL-5, and IL-13. In both the susceptible and resistant mice, the initial cytokine response to leishmanial infection is of Th0 type, with both IFN- γ and IL-4 detected in the draining lymph nodes within the first week of infection (Reiner *et al.*, 1994; Scott *et al.*, 1996). In the BALB/c mice, the high level of IL-4 is maintained, a condition that renders the mice unable to produce or respond to IL-12 and consequently accounts for their failure to mount a Th1 response (Guler *et al.*, 1996). In infected, resistant C57BL/6 mice, the initial high level of IL-4 is quickly controlled and the macrophages, acting as antigen-presenting cells, produce IL-12 that leads to differentiation and proliferation of Th1 cells. IL-12 induces both natural killer (NK) cells and Th1 cells to produce IFN- γ, which results in macrophage activation. IFN-γ induces phosphorylation of the tyrosine kinases Jak 1 and Jak 2, which in turn phosphorylate p91 STAT protein. The latter then promotes the expression of interferon regulatory factor-1 (IRF-1) gene that is thought to be involved in the inducible nitric oxide synthase (iNOS) gene transcription. Several cytokines, like TNF-α, IL-2, IL-4 as well as IL-7, act synergistically with IFN-γ to enhance NO production by iNOS (Belosevic *et al.*, 1990; Bogdan *et al.*, 1991; de Kossodo *et al.*, 1994; Gessner *et al.*, 1993).

Nitric oxide synthases (NOS) are widely distributed in mammalian tissues that catalyses the generation of nitric oxide (NO) from the guanidine nitrogen of L-arginine. NOS isoforms can be categorized according to their mode of expression: (i) constitutive NOS (NOS1 and NOS3) that are dependent on Ca^{2+}/calmodulin and (ii) inducible NOS (NOS2 or iNOS), the expression of which is independent of Ca^{2+}, but is increased by cytokines and other inflammatory stimuli. The latter is expressed in macrophages and neutrophils, which show species-dependent variations in the amount of NO synthesis. For example, rodent cells produce more NO than the human cells. NO is highly unstable and quickly reacts with superoxide yielding peroxynitrite ($ONOO^-$). The latter, being highly reactive, can react with various biological targets as well as other molecules forming additional reactive nitrogen species (RNS). Even modest increases in the levels of $O_2{\cdot}^-$ and NO will greatly enhance the $ONOO^-$ content. This is depicted in Fig. 4.4. Interestingly, generation of ROS and RNS are efficiently regulated by fine-tuning of negative feedback mechanisms. The excessive amount of NO produced, represses NADPH-oxidase activity protecting the cell against superoxide and peroxynitrite overload (Selemidis *et al.*, 2007).

Targets of reactive nitrogen species (RNS) in *Leishmania*

RNS toxicity is likely to be complex, involving several targets. Nitric oxide is highly diffusible and hydrophobic, and thus partitions readily into biological membranes affecting the cellular organelles. Work by Mauel and Ransijn with *L. enrietti* and *L. major* has thrown light on the deleterious effects of the toxic molecules in the organisms (Mauel and Ransijn, 1997). They have demonstrated that both the promastigote and the amastigote stages are sensitive to NO, although *L. major* exhibited less sensitivity than the other species. The cellular transformation from promastigote to amastigote and vice versa is delayed by RNS. Various other laboratories have shown apoptotic death in *Leishmania* spp. in response to NO (Lemesre *et al.*, 1997; Liew *et al.*, 1990). Later, Murray and Nathan (1999) provided evidence that RNS alone is sufficient and necessary to control *L. donovani* infection in mice. The following sections summarize the RNS toxicity in the parasite (Table 4.1). Although the effect of RNS on parasite life cycles has been well investigated, detailed studies to identify macromolecular targets and the mode of inactivation in the parasite still remain to be done. Notably, the concentration of NO required to kill the promastigote and axenic amastigotes of *Leishmania* spp. *in vitro* was comparable to that required for tumour cell killing (Hibbs *et al.*, 1988) and hence, works in the mammalian system to identify RNS targets can throw some light on the probable targets in the organism.

Lipid peroxidation

Lipid peroxidation comprises one of the major outcomes of RNS-mediated cytotoxicity. $ONOO^-$ can trigger lipid peroxidation in membranes, liposomes, and lipoproteins by abstracting a hydrogen atom from polyunsaturated fatty acids (PUFA) producing lipid hydroperoxy radicals, conjugated dienes, and aldehydes (Hogg and Kalyanaraman, 1999). The reaction proceeds as a chain-reaction with the concomitant attack on neighbouring PUFAs generating additional radicals capable of propagating the reaction. Degeneration of membrane lipids results in

$$\text{L-Arg} + O_2 + \text{NADPH} \xrightarrow{\text{NO synthase}} \text{NO} + \text{Citrulline} + \text{NADP}^+$$

$$\text{NO} + O_2 \longrightarrow N_2O_3$$

$$N_2O_3 + \text{RSH} \longrightarrow \text{RSNO}$$

$$\text{NO} + O_2{\cdot}^- \longrightarrow \text{ONOO}^-$$

$$\text{ONOO}^- + H^+ \longrightarrow \text{HNOOH} \longrightarrow {\cdot}\text{HO} + NO_2$$

$${\cdot}\text{HO} + \text{NO} \longrightarrow HNO_2$$

$$Fe^{3+} + \text{NO} \longrightarrow Fe^{2+}(\text{NO}{\cdot})$$

Figure 4.4 Generation of reactive nitrogen species. NO, nitric oxide; N_2O_3, nitrous anhydride; RSH, thiol; RSNO, nitroso thiol; $ONOO^-$, peroxinitrite; ONOOH, peroxynitrous acid; HNO_2, nitrous acid; $Fe^{2+}(NO{\cdot})$, complex formed with the iron-containing proteins.

altered membrane permeability and fluidity with significant biological consequences (Richter, 1987) including loss of membrane potential and even apoptotic cell death. Mitochondria, being extensively membranous in structure and bearing high content of PUFA, are readily prone to ROS- and RNS-mediated damage. This poses serious problems especially for kinetoplastids, where the fully functional single large mitochondrion occupies most of the intracellular space.

Nucleic acids damage

RNS has been reported to induce DNA damage and apoptosis in a variety of cell types. $ONOO^-$ is capable of introducing oxidative modifications in both nucleobases and sugar–phosphate backbone (Burney *et al.*, 1999; Niles *et al.*, 2006). Among the four DNA bases, guanine has the lowest reduction potential, and hence exhibits the highest reactivity with peroxynitrite. The reaction starts with the formation of G radical cation ($G^{\cdot+}$), that is subsequently converted to several oxidation and nitration products in presence of $ONOO^-$. 8-Oxoguanine is the major oxidation product, which further reacts with peroxynitrite, yielding cyanuric acid and oxazolone (Niles *et al.*, 2006). The obvious outcomes of these reactions comprise base transition as well as transversion. Guanine oxidation may ultimately lead to guanine fragmentation. Guanine can also be nitrated by $ONOO^-$ with the generation of 8-nitroguanine. The latter depurinates easily, forming an important source of apurinic sites that are prone to cleavage by endonucleases *in vivo* to give DNA single-strand breaks. Peroxynitrite may also attack the sugar–phosphate backbone by abstracting a hydrogen atom from the deoxyribose moiety; as a result the sugar ring opens up, leading to DNA strand breaks. Holzmuller and co-workers demonstrated that NO-mediated cell death in *Leishmania amazonensis* was associated with extensive nuclear DNA fragmentation (Holzmuller *et al.*, 2002). Their results revealed that this DNA fragmentation was independent of caspase activities, but could be prevented by administering proteasome inhibitors such as lactacysteine and calpain inhibitor I, thus suggesting the involvement of proteasome activity in NO-mediated apoptosis in the parasite.

Protein oxidation

RNS targets only a few chemical groups in proteins and is known to cause iron–sulfur (Fe/S) centre alterations or S nitrosylation, and hence, can interfere with the activity of certain enzymes, particularly those that depend on the presence of free thiols or Fe–S clusters for activity. This is corroborated by *in vitro* experiments with *Leishmania* promastigotes and amastigotes that involved addition of FeSO4 and L-cysteine immediately after NO flushing. An excess of these molecules greatly reversed the leishmanicidal effect of NO previously observed after a post-flushing incubation of 12 h (Lemesre *et al.*, 1997). It is to be noted here that this combination maximally reconstituted cis-aconitase activity in the tumour cells (Drapier and Hibbs, 1986). Upon exposure of the parasites or their extracts to NO has resulted in drastic reductions in the activities of enzymes like glyceraldehyde-3-phosphate dehydrogenase (GAPDH) and cis-aconitase (Mauel and Ransijn, 1997). The former is functional in the glycolytic pathway, and bears a sulfhydryl group in its catalytic site), while cis-aconitase is a citric acid cycle enzyme which contains an iron–sulfur centre. Addition of citric acid cycle substrates such as α-ketoglutarate and cis-aconitate after NO flushing reduced the inhibitory effects of NO in promastigotes and axenic amastigotes of *Leishmania* (Lemesre *et al.*, 1997). Indeed, NO-resistant amastigotes have been found to up-regulate the expression of these two enzymes. Another enzyme that has also been found in increased concentration in the resistant strain is 6-phosphogluconate dehydrogenase, a pentose phosphate pathway candidate, although NO inactivation of this enzyme in *Leishmania* has not been reported yet. All these indicate an RNS-mediated induction of altered metabolism in the parasite. Successful survival and proliferation of the parasite in the macrophage requires effective metabolite uptake from the harbouring phagocyte. *Leishmania* promastigotes, when exposed to activated murine macrophages, exhibit reduced capability for nutrient uptake. NO donors have been shown to confer S-nitrosylation-mediated inactivation of cruzipain, the major cysteine proteinase from *T. cruzi* epimastigotes (Salvati *et al.*, 2001), and falcipain, the cruzipain-homologous cysteine proteinase from *Plasmodium falciparum* trophozoites (Venturini *et al.*, 2000). Cysteine proteinases appear to be involved in several life cycle stages of *Leishmania*, influencing the host–parasite relationship that involves parasitic nutritional acquisition at the expense of the host, penetration of the parasite into host cells, and escaping mechanisms of the parasite from the host's immune system. Cysteine proteinase inhibitors were found to block *Leishmania* replication and differentiation, allowing the protein to be identified as potential a drug target. The work in *T. cruzi* and *P. falciparum* was extended by Salvati *et al.* (2001) in *L. infantum* promastigotes using NO donors that demonstrated a dose-dependent inactivation of the protein in the parasite accompanied by an increase in the loss of cell viability. The protein inactivation was shown to be due to nitrosylation of the Sγ atom of the Cys-reactive residue.

$ONOO^-$ and its protonated form ONOOH can exert direct oxidative modifications by one- or two-electron oxidation mechanism. $ONOO^-$ may also alter protein structure by attacking different amino acids in the peptide chains. This is mainly achieved in three ways: (1) cysteine oxidation, (2) tyrosine nitration and (3) tryptophan, methionine and histidine oxidation. In mammals, these three mechanisms are responsible for RNS-mediated inactivation of several enzymes, including GAPDH, creatine kinase, complex I (NADH dehydrogenase), complex II (succinate dehydrogenase), complex III (cytochrome c reductase) as well as complex V (ATP synthase) from the mitochondrial respiratory chain. Cysteine oxidation is the most prevalent modification, where $ONOO^-$ reacts with thiols (particularly with the anionic form, RS^-) generating the intermediate sulfenic acid (RSOH), which when reacts with another thiol, forms the disulfide RSSR. The latter may act with other reactive oxygen species (RNS) forming thiyl radicals ($RS^·$). Cysteine oxidation may also inactivate several protein tyrosine phosphatases and modulate phosphotyrosine-dependent signalling. Apart from these, peroxynitrite can directly oxidize low molecular weight thiols, particularly reduced glutathione (GSH) or trypanothione [T(SH)2]. Accordingly, the susceptibility of *Leishmania* cells to $ONOO^{--}$-mediated toxicity largely depends on the amount of intracellular GSH or [$T(SH)_2$] (Romao *et al.*, 2006).

In mammals, extensive studies on RNS-exposed mitochondria has revealed the respiratory complexes like complex I, II, III and IV to be the potential targets for RNS, mainly because they all contain catalytically active Fe-S prosthetic group. Inactivation of these complexes is mediated by formation of nitrosyl–Fe complexes as well as cysteine oxidation and tyrosine nitration of amino acids, as mentioned in the previous section. To determine the effects of RNS on cellular respiration, the kinetics of parasite oxygen consumption was measured as a function of concentration and time of exposure to NO donors (Mauel and Ransijn, 1997). The results revealed a dose-dependent drop in the respiration rate. $ONOO^-$ can potentiate oxidative injury by accelerating flavine reduction and promoting Fenton chemistry. ROS and RNS orchestrate the destabilization of iron–sulphur clusters, releasing free iron, which further exacerbates this process.

Cellular defences against nitrosative stress in *Leishmania*

As stated earlier, despite the ability of the phagocytic cells to control *Leishmania* infection by producing ROS and RNS, strains causing cutaneous and visceral leishmaniasis persist long enough to propagate the disease. Although the parasitic defences against oxidative stress has been studied extensively, the details of defence mechanisms against reactive nitrogen species remain to be elucidated. The available literature indicates the role of thiols as well as enzymes such as peroxiredoxin 1 and a haem-containing pseudoperoxidase in mediating RNS detoxification (Table 4.1).

Thiols

Reduced glutathione (GSH) has been reported to confer protection of mammalian macrophages against the cytotoxic effects of NO. Romao *et al.* d1999, 2006) demonstrated a protective role of GSH in *Leishmania* spp. against NO generators like S-nitroso-*N*-acetyl-D,L-penicillamine (SNAP). The sensitivity of distinct species of *Leishmania* to SNAP was found to be inversely correlated with their glutathione concentration. The major antioxidant thiol pool in these organisms is trypanothione $T(SH)_2$, which is maintained in the reduced state by trypanothione reductase (TR). Owing to the presence of two sulfhydryl groups with low pK_a values, $T(SH)_2$ is a faster reagent in redox reactions than GSH. Interestingly, in comparison to control parasites, those with reduced expression of TR, were found to be more sensitive to the cytotoxic effect of SNAP (Romao *et al.*, 2006). In mammalian cells, the interaction among NO, Fe^{II} and thiols are known to spontaneously produce dinitrosyl–iron complexes (DNIC) that remain bound to proteins, for example glutathione transferase (GST) isoenzymes. *In vitro*, free DNICs are known to show various toxic effects like competitive inhibition of glutathione reductase; this is prevented once the protein complex is formed with GST (Becker *et al.*, 1998; Pedersen *et al.*, 2007). Interestingly, Bocedi *et al.* (2010) have shown that $T(SH)_2$ intercepts NO in a dinitrosyl–trypanothionyl–iron complex (DNTIC) with about 600-fold higher propensity than GSH. Moreover, this complex was found to be harmless to the parasite, even in the free form, and, hence, needs no sequestration by specific proteins.

Peroxiredoxin 1

Peroxidoxins (or peroxiredoxins) are highly conserved enzymes implicated to confer resistance against RNS in bacteria, yeast and human cells. The first reports on enzymatic defence in *Leishmania* against RNS came from Lashitew Gedamu's lab (Barr and Gedamu, 2003). They reported the presence of two peroxidoxins, peroxidoxin 1 (LcPxn1) and peroxidoxin 2 (LcPxn2), in the cytoplasm of *L. chagasi*. The genes are differentially expressed, with LcPxn1 being expressed more in the amastigote stage while the LcPxn2 transcripts were more abundant in the promastigote stage. Although, initially, both the isoforms were shown to be active against ROS, LcPxn1 was found to confer additional resistance against RNS. This is significant enough if we consider the fact that the other isoform or the peroxiredoxins characterized from *L. major* and *L. infantum* so far, have been shown to lack this ability. Overexpression of LcPxn1 in *L. chagasi* promastigotes resulted in enhanced survival upon exposure to H_2O_2, tBOOH, OH, $ONOO^-$, and NO as well as within the macrophages. Structure-function analysis by site-directed mutagenesis revealed that the conserved Cys-52 residue is essential for detoxifying hydrogen peroxide, t-butyl hydroperoxide, and hydroxyl radicals, whereas the conserved Cys-173 residue is essential for detoxifying t-butyl hydroperoxide and peroxynitrite.

Pseudoperoxidase

In the biological world, the first evidence of haem-containing enzyme-mediated defence against toxic RNS came from *L. major* with the discovery of a pseudoperoxidase (LmPP) by Bose *et al.* (2012). The enzyme sequence has been found to bear ~14% identical and ~40% conserved residues when compared to the class I peroxidases like horseradish peroxidase, but lacks a distal site His residue, that in most of the peroxidases plays a crucial role in the formation of active intermediates (compounds I and II). The enzyme has been shown to be expressed in amastigote stage as a cell surface protein and protects cells from $ONOO^-$ but not from H_2O_2 under both *in vitro* and *in vivo* conditions. Lack of the distal His is thought to enable LmPP to catalyse the breakdown of $ONOO^-$ in the absence of any electron donor. The protein exists in dimeric form with the spectral characteristics of six coordinate low-spin haem environment that is rapidly converted by $ONOO^-$ to a ferric high-spin form at neutral pH. Lack of the distal His is thought to enable LmPP to catalyse the breakdown of $ONOO^-$ in the absence of any electron donor. Overexpression of the enzyme in *Leishmania* conferred increased protection in presence of RNS and consequently helped survivability of the parasite upon infection in macrophage cell cultures as well as in BALB/c mice. In contrast, LmPP knocked-out parasites were very much susceptible to NO-mediated killing and displayed decreased virulence.

Conclusion

Successful parasitism requires fulfilment of at least two possible strategies: first, suppression of the microbicidal activities of the host and, second, acquisition of the host's nutritive pool for the growth and proliferation of the parasite. For the past few decades, studies on *Leishmania* have enriched our understanding of the oxidative stress problems with which all organisms contend, the peculiar mechanisms adopted by the parasites to break the host defence and how the unique molecules and pathways in the parasite can be intelligently exploited in order to create new drug targets. While various other parasites can also infect macrophages, *Leishmania* is distinguished in that it not only manipulates the host's antimicrobial strategies, but also relies on its own defence mechanisms to survive and reproduce within the macrophage phagolysosome. The proteome analysis in *L. donovani* promastigotes treated with sub-lethal dose of ROS, RNS or a combination of both revealed differential expressions of redox homeostasis proteins (Sardar *et al.*, 2013). The enzymes trypanothione reductase, trypanothione synthetase, tryparedoxin peroxidase, pteridine reductase, glutathione peroxidase-like protein, peroxidoxin, protein disulfide isomerase, endoplasmic reticulum oxidoreductin and dihydrolipoamide dehydrogenase exhibited a ~1.2- to 2-fold increase in expression levels. Ascorbate dependent peroxidase, another ROS-scavenging enzyme, and γ-glutamyl cysteine synthetase, an enzyme involved in GSH synthesis, were up-regulated by ~2- to 3.5-fold, whereas thioredoxin-like protein, thiol-dependent reductase 1, glutathione S-transferase/glutaredoxin and glutaredoxin-like protein were down-regulated by ~0.5- to 0.8-fold. The meta-containing domain protein was found to be down-regulated by ~2-fold.

Living cells require ROS for their normal growth and proliferation, but at the same time must possess appropriate scavenging system to get rid of the excess ROS, which, otherwise, might be detrimental. In order to maintain the redox homeostasis, cells constantly undergo a wide variety of adaptations that are honed through evolution, resulting in the redundancy of antioxidant systems as well as intelligent use of a single protein in diverse physiological processes. This has been appropriately addressed by H.G. Wells in *The Time Machine*: 'We are kept keen on the grindstone of pain and necessity'.

Future perspectives

Given the rapid spread of drug resistance against *Leishmania* parasites in > 98 countries (www.who.int/gb/ebwha/pdf_files/EB118/B118_4-en.pdf; accessed on 19 December 2008) it is very important promote research in this particular area. The emergence of *Leishmania*-HIV co-infection cases also made the situation more challenging and warrants newer drug discovery. Hence, the future perspective towards this is to investigate the oxidative as well as nitrosative stress response in the trypanosomatid parasites those are important for its growth/survival with the focus on discovering new therapeutic targets.

Acknowledgments

This work was supported by Council of Scientific and Industrial Research (CSIR) Project BSC 0114.

References

Adak, S., and Datta, A.K. (2005). *Leishmania major* encodes an unusual peroxidase that is a close homologue of plant ascorbate peroxidase: a novel role of the transmembrane domain. Biochem. J. *390*, 465–474.

Adak, S., and Pal, S. (2013). Ascorbate peroxidase acts as a novel determiner of redox homeostasis in *Leishmania*. Antioxid. Redox Signal. *19*, 746–754.

Aikens, J., and Dix, T.A. (1991). Perhydroxyl radical (HOO.) initiated lipid peroxidation. The role of fatty acid hydroperoxides. J. Biol. Chem. *266*, 15091–15098.

Ameen, M. (2007). Cutaneous leishmaniasis: therapeutic strategies and future directions. Expert Opin. Pharmacother. *8*, 2689–2699.

Ames, B.N. (1989). Endogenous DNA damage as related to cancer and aging. Mutat. Res. *214*, 41–46.

Appenzeller-Herzog, C., and Ellgaard, L. (2008). In vivo reduction-oxidation state of protein disulfide isomerase: the two active sites independently occur in the reduced and oxidized forms. Antioxid. Redox Signal *10*, 55–64.

Ariyanayagam, M.R., and Fairlamb, A.H. (2001). Ovothiol and trypanothione as antioxidants in trypanosomatids. Mol. Biochem. Parasitol. *115*, 189–198.

Avery, S.V. (2011). Molecular targets of oxidative stress. Biochem. J. *434*, 201–210.

Babior, B.M. (1999). NADPH oxidase: an update. Blood *93*, 1464–1476.

Babior, B.M., Kipnes, R.S., and Curnutte, J.T. (1973). Biological defense mechanisms. The production by leukocytes of superoxide, a potential bactericidal agent. J. Clin. Invest. *52*, 741–744.

Barr, S.D., and Gedamu, L. (2001). Cloning and characterization of three differentially expressed peroxidoxin genes from *Leishmania chagasi*. Evidence for an enzymatic detoxification of hydroxyl radicals. J. Biol. Chem. *276*, 34279–34287.

Barr, S.D., and Gedamu, L. (2003). Role of peroxidoxins in *Leishmania chagasi* survival. Evidence of an enzymatic defense against nitrosative stress. J. Biol. Chem. *278*, 10816–10823.

Basselin, M., Lawrence, F., and Robert-Gero, M. (1997). Altered transport properties of pentamidine-resistant *Leishmania donovani* and *L. amazonensis* promastigotes. Parasitol. Res. *83*, 413–418.

Basu, A.K., and Marnett, L.J. (1983). Unequivocal demonstration that malondialdehyde is a mutagen. Carcinogenesis *4*, 331–333.

Becker, K., Savvides, S.N., Keese, M., Schirmer, R.H., and Karplus, P.A. (1998). Enzyme inactivation through sulfhydryl oxidation by physiologic NO-carriers. Nat. Struct. Biol. *5*, 267–271.

Belosevic, M., Finbloom, D.S., Meltzer, M.S., and Nacy, C.A. (1990). IL-2. A cofactor for induction of activated macrophage resistance to infection. J. Immunol. *145*, 831–839.

Ben Khalaf, N., De Muylder, G., Louzir, H., McKerrow, J., and Chenik, M. (2012). *Leishmania major* protein disulfide isomerase as a drug target: enzymatic and functional characterization. Parasitol. Res. *110*, 1911–1917.

Bermudez, R., Dagger, F., D'Aquino, J.A., Benaim, G., and Dawidowicz, K. (1997). Characterization of mitochondrial electron-transfer in *Leishmania mexicana*. Mol. Biochem. Parasitol. *90*, 43–54.

Bindoli, A. (1988). Lipid peroxidation in mitochondria. Free Radic. Biol. Med. *5*, 247–261.

Biyani, N., and Madhubala, R. (2011). *Leishmania donovani* encodes a functional enzyme involved in vitamin C biosynthesis: arabino-1,4-lactone oxidase. Mol. Biochem. Parasitol. *180*, 76–85.

Bocedi, A., Dawood, K.F., Fabrini, R., Federici, G., Gradoni, L., Pedersen, J.Z., and Ricci, G. (2010). Trypanothione efficiently intercepts nitric oxide as a harmless iron complex in trypanosomatid parasites. FASEB J. *24*, 1035–1042.

Bogdan, C., Stenger, S., Rollinghoff, M., and Solbach, W. (1991). Cytokine interactions in experimental cutaneous leishmaniasis. Interleukin 4 synergizes with interferon-gamma to activate murine macrophages for killing of *Leishmania major* amastigotes. Eur. J. Immunol. *21*, 327–333.

Bose, M., Saha, R., Sen Santara, S., Mukherjee, S., Roy, J., and Adak, S. (2012). Protection against peroxynitrite by pseudoperoxidase from *Leishmania major*. Free Radic. Biol. Med. *53*, 1819–1828.

Burney, S., Caulfield, J.L., Niles, J.C., Wishnok, J.S., and Tannenbaum, S.R. (1999). The chemistry of DNA damage from nitric oxide and peroxynitrite. Mutat. Res. *424*, 37–49.

Bylund, J., Brown, K.L., Movitz, C., Dahlgren, C., and Karlsson, A. (2010). Intracellular generation of superoxide by the phagocyte NADPH oxidase: how, where, and what for? Free Radic. Biol. Med. *49*, 1834–1845.

Callahan, H.L., Crouch, R.K., and James, E.R. (1988). Helminth anti-oxidant enzymes: a protective mechanism against host oxidants? Parasitol. Today *4*, 218–225.

Castro, H., Budde, H., Flohe, L., Hofmann, B., Lunsdorf, H., Wissing, J., and Tomas, A.M. (2002). Specificity and kinetics of a mitochondrial peroxiredoxin of *Leishmania infantum*. Free Radic. Biol. Med. *33*, 1563–1574.

Castro, H., Sousa, C., Novais, M., Santos, M., Budde, H., Cordeiro-da-Silva, A., Flohe, L., and Tomas, A.M. (2004). Two linked genes of *Leishmania infantum* encode tryparedoxins localised to cytosol and mitochondrion. Mol. Biochem. Parasitol. *136*, 137–147.

Cecarini, V., Gee, J., Fioretti, E., Amici, M., Angeletti, M., Eleuteri, A.M., and Keller, J.N. (2007). Protein oxidation and cellular homeostasis: Emphasis on metabolism. Biochim. Biophys. Acta *1773*, 93–104.

Chakraborty, A.K., and Majumder, H.K. (1988). Mode of action of pentavalent antimonials: specific inhibition of type I DNA topoisomerase of *Leishmania donovani*. Biochem. Biophys. Res. Commun. *152*, 605–611.

Chance, B., Sies, H., and Boveris, A. (1979). Hydroperoxide metabolism in mammalian organs. Physiol. Rev. *59*, 527–605.

Chaudhary, A.K., Nokubo, M., Reddy, G.R., Yeola, S.N., Morrow, J.D., Blair, I.A., and Marnett, L.J. (1994). Detection of endogenous malondialdehyde–deoxyguanosine adducts in human liver. Science *265*, 1580–1582.

Chen, M., Zhai, L., Christensen, S.B., Theander, T.G., and Kharazmi, A. (2001). Inhibition of fumarate reductase in *Leishmania major* and *L. donovani* by chalcones. Antimicrob. Agents Chemother. *45*, 2023–2029.

Cheng, K.C., Cahill, D.S., Kasai, H., Nishimura, S., and Loeb, L.A. (1992). 8-Hydroxyguanine, an abundant form of oxidative DNA damage, causes G – – T and A – – C substitutions. J. Biol. Chem. *267*, 166–172.

Clark, D., Albrecht, M., and Arevalo, J. (1994). Ascorbate variations and dehydroascorbate reductase activity in Trypanosoma cruzi epimastigotes and trypomastigotes. Mol. Biochem. Parasitol. *66*, 143–145.

Comini, M.A., Rettig, J., Dirdjaja, N., Hanschmann, E.M., Berndt, C., and Krauth-Siegel, R.L. (2008). Monothiol glutaredoxin-1 is an essential iron–sulfur protein in the mitochondrion of African trypanosomes. J. Biol. Chem. *283*, 27785–27798.

Commoner, B., Townsend, J., and Pake, G.E. (1954). Free radicals in biological materials. Nature *174*, 689–691.

Cooke, M.S., Evans, M.D., Dizdaroglu, M., and Lunec, J. (2003). Oxidative DNA damage: mechanisms, mutation, and disease. FASEB J. *17*, 1195–1214.

Croft, S.L., and Brazil, R.P. (1982). Effect of pentamidine isethionate on the ultrastructure and morphology of *Leishmania mexicana amazonensis in vitro*. Ann. Trop. Med. Parasitol. *76*, 37–43.

Das, M., Mukherjee, S.B., and Shaha, C. (2001). Hydrogen peroxide induces apoptosis-like death in *Leishmania donovani* promastigotes. J. Cell Sci. *114*, 2461–2469.

Das, R., Roy, A., Dutta, N., and Majumder, H.K. (2008). Reactive oxygen species and imbalance of calcium homeostasis contributes to curcumin induced programmed cell death in *Leishmania donovani*. Apoptosis *13*, 867–882.

Deavall, D.G., Martin, E.A., Horner, J.M., and Roberts, R. (2012). Drug-induced oxidative stress and toxicity. J. Toxicol. *2012*, 645460.

Desjardins, M., and Descoteaux, A. (1997). Inhibition of phagolysosomal biogenesis by the *Leishmania* lipophosphoglycan. J. Exp. Med. *185*, 2061–2068.

Dey, R., and Datta, S.C. (1994). Leishmanial glycosomes contain superoxide dismutase. Biochem. J. *301 (Pt 2)*, 317–319.

Diao, Y., Liu, W., Wong, C.C., Wang, X., Lee, K., Cheung, P.Y., Pan, L., Xu, T., Han, J., Yates, J.R., 3rd, *et al.* (2010). Oxidation-induced intramolecular disulfide bond inactivates mitogen-activated protein kinase kinase 6 by inhibiting ATP binding. Proc. Natl. Acad. Sci. U.S.A. *107*, 20974–20979.

Diechtierow, M., and Krauth-Siegel, R.L. (2011). A tryparedoxin-dependent peroxidase protects African trypanosomes from membrane damage. Free Radic. Biol. Med. *51*, 856–868.

Dix, T.A., and Aikens, J. (1993). Mechanisms and biological relevance of lipid peroxidation initiation. Chem. Res. Toxicol. *6*, 2–18.

Docampo, R., and Moreno, S.N. (1986). Free radical metabolism of antiparasitic agents. Fed. Proc. *45*, 2471–2476.

Dolai, S., Yadav, R.K., Datta, A.K., and Adak, S. (2007). Effect of thiocyanate on the peroxidase and pseudocatalase activities of *Leishmania major* ascorbate peroxidase. Biochim. Biophys. Acta *1770*, 247–256.

Dolai, S., Yadav, R.K., Pal, S., and Adak, S. (2008). *Leishmania major* ascorbate peroxidase overexpression protects cells against reactive oxygen species-mediated cardiolipin oxidation. Free Radic. Biol. Med. *45*, 1520–1529.

Dolai, S., Yadav, R.K., Pal, S., and Adak, S. (2009). Overexpression of mitochondrial *Leishmania major* ascorbate peroxidase enhances tolerance to oxidative stress-induced programmed cell death and protein damage. Eukaryot. Cell *8*, 1721–1731.

Dolai, S., Pal, S., Yadav, R.K., and Adak, S. (2011). Endoplasmic Reticulum Stress-induced Apoptosis in *Leishmania* through Ca2+-dependent and Caspase-independent Mechanism. J. Biol. Chem. *286*, 13638–13646.

Drapier, J.C., and Hibbs, J.B., Jr. (1986). Murine cytotoxic activated macrophages inhibit aconitase in tumor cells. Inhibition involves the iron–sulfur prosthetic group and is reversible. J. Clin. Invest. *78*, 790–797.

Dumas, C., Ouellette, M., Tovar, J., Cunningham, M.L., Fairlamb, A.H., Tamar, S., Olivier, M., and Papadopoulou, B. (1997). Disruption of the trypanothione reductase gene of *Leishmania* decreases its ability to survive oxidative stress in macrophages. EMBO J. *16*, 2590–2598.

Einolf, H.J., Schnetz-Boutaud, N., and Guengerich, F.P. (1998). Steady-state and pre-steady-state kinetic analysis of 8-oxo-7,8-dihydroguanosine triphosphate incorporation and extension by replicative and repair DNA polymerases. Biochemistry *37*, 13300–13312.

Evans, M.D., Dizdaroglu, M., and Cooke, M.S. (2004). Oxidative DNA damage and disease: induction, repair and significance. Mutat. Res. *567*, 1–61.

Fang, F.C. (2004). Antimicrobial reactive oxygen and nitrogen species: concepts and controversies. Nat. Rev. Microbiol. *2*, 820–832.

Faure, H., Mousseau, M., Cadet, J., Guimier, C., Tripier, M., Hida, H., and Favier, A. (1998). Urine 8-oxo-7,8-dihydro-2-deoxyguanosine vs. 5-(hydroxymethyl) uracil as DNA oxidation marker in adriamycin-treated patients. Free Radic. Res. *28*, 377–382.

Fonseca-Silva, F., Inacio, J.D., Canto-Cavalheiro, M.M., and Almeida-Amaral, E.E. (2011). Reactive oxygen species production and mitochondrial dysfunction contribute to quercetin induced death in *Leishmania amazonensis*. PLoS One *6*, e14666.

Fridovich, I. (1995). Superoxide radical and superoxide dismutases. Annu. Rev. Biochem. *64*, 97–112.

Fridovich, I. (1998). Oxygen toxicity: a radical explanation. J. Exp. Biol. *201*, 1203–1209.

Gauss, K.A., Nelson-Overton, L.K., Siemsen, D.W., Gao, Y., DeLeo, F.R., and Quinn, M.T. (2007). Role of NF-kappaB in transcriptional regulation of the phagocyte NADPH oxidase by tumor necrosis factor-alpha. J. Leukoc. Biol. *82*, 729–741.

Gerschman, R., Gilbert, D.L., Nye, S.W., Dwyer, P., and Fenn, W.O. (1954). Oxygen poisoning and x-irradiation: a mechanism in common. Science *119*, 623–626.

Gessner, A., Vieth, M., Will, A., Schroppel, K., and Rollinghoff, M. (1993). Interleukin-7 enhances antimicrobial activity against *Leishmania major* in murine macrophages. Infect. Immun. *61*, 4008–4012.

Ghosh, S., Goswami, S., and Adhya, S. (2003). Role of superoxide dismutase in survival of *Leishmania* within the macrophage. Biochem. J. *369*, 447–452.

Gross, E., Sevier, C.S., Heldman, N., Vitu, E., Bentzur, M., Kaiser, C.A., Thorpe, C., and Fass, D. (2006). Generating disulfides enzymatically: reaction products and electron acceptors of the endoplasmic reticulum thiol oxidase Ero1p. Proc. Natl. Acad. Sci. U.S.A. *103*, 299–304.

Guler, M.L., Gorham, J.D., Hsieh, C.S., Mackey, A.J., Steen, R.G., Dietrich, W.F., and Murphy, K.M. (1996). Genetic susceptibility to *Leishmania*: IL-12 responsiveness in TH1 cell development. Science *271*, 984–987.

Halliwell, B., and Chirico, S. (1993). Lipid peroxidation: its mechanism, measurement, and significance. Am. J. Clin. Nutr. *57*, 715S–724S; discussion 724S–725S.

Halliwell, B., and Gutteridge, J.M. (1995). The definition and measurement of antioxidants in biological systems. Free Radic. Biol. Med. *18*, 125–126.

Hashida, S., Yuzawa, S., Suzuki, N.N., Fujioka, Y., Takikawa, T., Sumimoto, H., Inagaki, F., and Fujii, H. (2004). Binding of FAD to cytochrome b558 is facilitated during activation of the phagocyte NADPH oxidase, leading to superoxide production. J. Biol. Chem. *279*, 26378–26386.

Hatahet, F., and Ruddock, L.W. (2009). Protein disulfide isomerase: a critical evaluation of its function in disulfide bond formation. Antioxid. Redox Signal. *11*, 2807–2850.

Herwaldt, B.L. (1999). Leishmaniasis. Lancet *354*, 1191–1199.

Hibbs, J.B., Jr., Taintor, R.R., Vavrin, Z., and Rachlin, E.M. (1988). Nitric oxide: a cytotoxic activated macrophage effector molecule. Biochem. Biophys. Res. Commun. *157*, 87–94.

Hillebrand, H., Schmidt, A., and Krauth-Siegel, R.L. (2003). A second class of peroxidases linked to the trypanothione metabolism. J. Biol. Chem. *278*, 6809–6815.

Hogg, N., and Kalyanaraman, B. (1999). Nitric oxide and lipid peroxidation. Biochim. Biophys. Acta *1411*, 378–384.

Holler, T.P., and Hopkins, P.B. (1990). Ovothiols as free-radical scavengers and the mechanism of ovothiol-promoted NAD(P)H-O2 oxidoreductase activity. Biochemistry *29*, 1953–1961.

Holzmuller, P., Sereno, D., Cavaleyra, M., Mangot, I., Daulouede, S., Vincendeau, P., and Lemesre, J.L. (2002). Nitric oxide-mediated proteasome-dependent oligonucleosomal DNA fragmentation in *Leishmania amazonensis* amastigotes. Infect. Immun. *70*, 3727–3735.

Hong, B.X., and Soong, L. (2008). Identification and enzymatic activities of four protein disulfide isomerase (PDI) isoforms of *Leishmania amazonensis*. Parasitol. Res. *102*, 437–446.

Howlett, N.G., and Avery, S.V. (1997). Induction of lipid peroxidation during heavy metal stress in Saccharomyces cerevisiae and influence of plasma membrane fatty acid unsaturation. Appl. Environ. Microbiol. *63*, 2971–2976.

Jasion, V.S., and Poulos, T.L. (2012). *Leishmania major* peroxidase is a cytochrome c peroxidase. Biochemistry *51*, 2453–2460.

Jasion, V.S., Polanco, J.A., Meharenna, Y.T., Li, H., and Poulos, T.L. (2011). Crystal structure of *Leishmania major* peroxidase and characterization of the compound I tryptophan radical. J. Biol. Chem. *286*, 24608–24615.

Jasion, V.S., Doukov, T., Pineda, S.H., Li, H., and Poulos, T.L. (2012). Crystal structure of the *Leishmania major* peroxidase-cytochrome c complex. Proc. Natl. Acad. Sci. U.S.A. *109*, 18390–18394.

Jeffree, G.M. (1958). Hydrogen peroxide and cancer. Nature *182*, 892.

Johnson, T.M., Yu, Z.X., Ferrans, V.J., Lowenstein, R.A., and Finkel, T. (1996). Reactive oxygen species are downstream mediators of p53-dependent apoptosis. Proc. Natl. Acad. Sci. U.S.A. *93*, 11848–11852.

Karala, A.R., Lappi, A.K., Saaranen, M.J., and Ruddock, L.W. (2009). Efficient peroxide-mediated oxidative refolding of a protein at physiological pH and implications for oxidative folding in the endoplasmic reticulum. Antioxid. Redox Signal. *11*, 963–970.

Klebanoff, S.J. (1967). Iodination of bacteria: a bactericidal mechanism. J. Exp. Med. *126*, 1063–1078.

Klebanoff, S.J. (1968). Myeloperoxidase-halide-hydrogen peroxide antibacterial system. J. Bacteriol. *95*, 2131–2138.

Klebanoff, S.J. (1970). Myeloperoxidase: contribution to the microbicidal activity of intact leukocytes. Science *169*, 1095–1097.

de Kossodo, S., Grau, G.E., Louis, J.A., and Muller, I. (1994). Tumor necrosis factor alpha (TNF-alpha) and TNF-beta and their receptors in experimental cutaneous leishmaniasis. Infect. Immun. *62*, 1414–1420.

Krauth-Siegel, R.L., and Comini, M.A. (2008). Redox control in trypanosomatids, parasitic protozoa with trypanothione-based thiol metabolism. Biochim. Biophys. Acta *1780*, 1236–1248.

Krieger, S., Schwarz, W., Ariyanayagam, M.R., Fairlamb, A.H., Krauth-Siegel, R.L., and Clayton, C. (2000). Trypanosomes lacking trypanothione reductase are avirulent and show increased sensitivity to oxidative stress. Mol. Microbiol. 35, 542–552.

Kushawaha, P.K., Gupta, R., Tripathi, C.D., Sundar, S., and Dube, A. (2012). Evaluation of *Leishmania donovani* protein disulfide isomerase as a potential immunogenic protein/vaccine candidate against visceral Leishmaniasis. PLoS One 7, e35670.

Lapouge, K., Smith, S.J., Walker, P.A., Gamblin, S.J., Smerdon, S.J., and Rittinger, K. (2000). Structure of the TPR domain of p67phox in complex with Rac. GTP. Mol. Cell 6, 899–907.

Lee, S.H., Stephens, J.L., Paul, K.S., and Englund, P.T. (2006). Fatty acid synthesis by elongases in trypanosomes. Cell *126*, 691–699.

Lemesre, J.L., Sereno, D., Daulouede, S., Veyret, B., Brajon, N., and Vincendeau, P. (1997). *Leishmania* spp.: nitric oxide-mediated metabolic inhibition of promastigote and axenically grown amastigote forms. Exp. Parasitol. *86*, 58–68.

Leto, T.L., Morand, S., Hurt, D., and Ueyama, T. (2009). Targeting and regulation of reactive oxygen species generation by Nox family NADPH oxidases. Antioxid. Redox Signal. *11*, 2607–2619.

Liew, F.Y., Millott, S., Parkinson, C., Palmer, R.M., and Moncada, S. (1990). Macrophage killing of *Leishmania* parasite *in vivo* is mediated by nitric oxide from L-arginine. J. Immunol. *144*, 4794–4797.

Livore, V.I., Tripodi, K.E., and Uttaro, A.D. (2007). Elongation of polyunsaturated fatty acids in trypanosomatids. FEBS J. *274*, 264–274.

Lodge, R., and Descoteaux, A. (2006). Phagocytosis of *Leishmania donovani* amastigotes is Rac1 dependent and occurs in the absence of NADPH oxidase activation. Eur. J. Immunol. *36*, 2735–2744.

Ludemann, H., Dormeyer, M., Sticherling, C., Stallmann, D., Follmann, H., and Krauth-Siegel, R.L. (1998). *Trypanosoma brucei* tryparedoxin, a thioredoxin-like protein in African trypanosomes. FEBS Lett. *431*, 381–385.

Luque-Ortega, J.R., and Rivas, L. (2007). Miltefosine (hexadecylphosphocholine) inhibits cytochrome c oxidase in *Leishmania donovani* promastigotes. Antimicrob. Agents Chemother. *51*, 1327–1332.

McCord, J.M., and Fridovich, I. (1969). Superoxide dismutase. An enzymic function for erythrocuprein (hemocuprein). J. Biol. Chem. *244*, 6049–6055.

Marjanovic, B., Simic, M.G., and Jovanovic, S.V. (1995). Heterocyclic thiols as antioxidants: why ovothiol C is a better antioxidant than ergothioneine. Free Radic. Biol. Med. *18*, 679–685.

Marnett, L.J. (1999). Lipid peroxidation-DNA damage by malondialdehyde. Mutat. Res. *424*, 83–95.

Marnett, L.J. (2000). Oxyradicals and DNA damage. Carcinogenesis *21*, 361–370.

Martin, E., and Mukkada, A.J. (1979a). Identification of the terminal respiratory chain in kinetoplast. mitochondrial complexes of *Leishmania tropica* promastigotes. J. Biol. Chem. *254*, 12192–12198.

Martin, E., and Mukkada, A.J. (1979b). Respiratory chain components of *Leishmania tropica* promastigotes. J. Protozool. *26*, 138–142.

Mauel, J., and Ransijn, A. (1997). *Leishmania* spp.: mechanisms of toxicity of nitrogen oxidation products. Exp. Parasitol. *87*, 98–111.

Medicherla, B., and Goldberg, A.L. (2008). Heat shock and oxygen radicals stimulate ubiquitin-dependent degradation mainly of newly synthesized proteins. J. Cell Biol. *182*, 663–673.

Mehta, A., and Shaha, C. (2004). Apoptotic death in *Leishmania donovani* promastigotes in response to respiratory chain inhibition: complex II inhibition results in increased pentamidine cytotoxicity. J. Biol. Chem. *279*, 11798–11813.

Meshnick, S.R., and Eaton, J.W. (1981). Leishmanial superoxide dismutase: a possible target for chemotherapy. Biochem. Biophys. Res. Commun. *102*, 970–976.

Minakami, R., and Sumimotoa, H. (2006). Phagocytosis-coupled activation of the superoxide-producing phagocyte oxidase, a member of the NADPH oxidase (nox) family. Int. J. Hematol. *84*, 193–198.

Mookerjee Basu, J., Mookerjee, A., Sen, P., Bhaumik, S., Banerjee, S., Naskar, K., Choudhuri, S.K., Saha, B., Raha, S., and Roy, S. (2006). Sodium antimony gluconate induces generation of reactive oxygen species and nitric oxide via phosphoinositide 3-kinase and mitogen-activated protein kinase activation in *Leishmania donovani*-infected macrophages. Antimicrob. Agents Chemother. *50*, 1788–1797.

Moore, K.J., and Matlashewski, G. (1994). Intracellular infection by *Leishmania donovani* inhibits macrophage apoptosis. J. Immunol. *152*, 2930–2937.

Moreira, W., Leprohon, P., and Ouellette, M. (2011). Tolerance to drug-induced cell death favours the acquisition of multidrug resistance in *Leishmania*. Cell Death Dis. *2*, e201.

Morel, F., Doussiere, J., and Vignais, P.V. (1991). The superoxide-generating oxidase of phagocytic cells. Physiological, molecular and pathological aspects. Eur. J. Biochem. *201*, 523–546.

Morita, Y.S., Paul, K.S., and Englund, P.T. (2000). Specialized fatty acid synthesis in African trypanosomes: myristate for GPI anchors. Science *288*, 140–143.

Moriya, M. (1993). Single-stranded shuttle phagemid for mutagenesis studies in mammalian cells: 8-oxoguanine in DNA induces targeted G.C – >T.A transversions in simian kidney cells. Proc. Natl. Acad. Sci. U.S.A. *90*, 1122–1126.

Mukherjee, S.B., Das, M., Sudhandiran, G., and Shaha, C. (2002). Increase in cytosolic Ca^{2+} levels through the activation of non-selective cation channels induced by oxidative stress causes mitochondrial depolarization leading to apoptosis-like death in *Leishmania donovani* promastigotes. J. Biol. Chem. *277*, 24717–24727.

Mukherjee, S., Sen Santara, S., Das, S., Bose, M., Roy, J., and Adak, S. (2012). NAD(P)H cytochrome b5 oxidoreductase deficiency in *Leishmania major* results in impaired linoleate synthesis followed by increased oxidative stress and cell death. J. Biol. Chem. *287*, 34992–35003.

Murray, H.W., and Nathan, C.F. (1999). Macrophage microbicidal mechanisms *in vivo:* reactive nitrogen versus oxygen intermediates in the killing of intracellular visceral *Leishmania donovani*. J. Exp. Med. *189*, 741–746.

Murray, H.W., Berman, J.D., Davies, C.R., and Saravia, N.G. (2005). Advances in leishmaniasis. Lancet *366*, 1561–1577.

Nathan, C., and Xie, Q.W. (1994). Regulation of biosynthesis of nitric oxide. J. Biol. Chem. *269*, 13725–13728.

Niles, J.C., Wishnok, J.S., and Tannenbaum, S.R. (2006). Peroxynitrite-induced oxidation and nitration products of guanine and 8-oxoguanine: structures and mechanisms of product formation. Nitric Oxide *14*, 109–121.

Nogoceke, E., Gommel, D.U., Kiess, M., Kalisz, H.M., and Flohe, L. (1997). A unique cascade of oxidoreductases catalyses trypanothione-mediated peroxide metabolism in *Crithidia fasciculata*. Biol. Chem. *378*, 827–836.

Okado-Matsumoto, A., and Fridovich, I. (2001). Subcellular distribution of superoxide dismutases (SOD) in rat liver: Cu,Zn-SOD in mitochondria. J. Biol. Chem. *276*, 38388–38393.

Ouellette, M., Drummelsmith, J., and Papadopoulou, B. (2004). Leishmaniasis: drugs in the clinic, resistance and new developments. Drug Resist. Updat. *7*, 257–266.

Padilla, A., Noiva, R., Lee, N., Mohan, K.V., Nakhasi, H.L., and Debrabant, A. (2003). An atypical protein disulfide isomerase from the protozoan parasite *Leishmania* containing a single thioredoxin-like domain. J. Biol. Chem. *278*, 1872–1878.

Pal, S., Dolai, S., Yadav, R.K., and Adak, S. (2010). Ascorbate peroxidase from *Leishmania major* controls the virulence of infective stage of promastigotes by regulating oxidative stress. PLoS One *5*, e11271.

Paramchuk, W.J., Ismail, S.O., Bhatia, A., and Gedamu, L. (1997). Cloning, characterization and overexpression of two iron superoxide dismutase cDNAs from *Leishmania chagasi*: role in pathogenesis. Mol. Biochem. Parasitol. *90*, 203–221.

Park, J.W., Benna, J.E., Scott, K.E., Christensen, B.L., Chanock, S.J., and Babior, B.M. (1994). Isolation of a complex of respiratory burst oxidase components from resting neutrophil cytosol. Biochemistry *33*, 2907–2911.

Paul, K.S., Jiang, D., Morita, Y.S., and Englund, P.T. (2001). Fatty acid synthesis in African trypanosomes: a solution to the myristate mystery. Trends Parasitol. *17*, 381–387.

Paulin, J.J. (1975). The chondriome of selected trypanosomatids. A three-dimensional study based on serial thick sections and high voltage electron microscopy. J. Cell. Biol. *66*, 404–413.

Pedersen, J.Z., De Maria, F., Turella, P., Federici, G., Mattei, M., Fabrini, R., Dawood, K.F., Massimi, M., Caccuri, A.M., and Ricci, G. (2007). Glutathione transferases sequester toxic dinitrosyl–iron complexes in cells. A protection mechanism against excess nitric oxide. J. Biol. Chem. *282*, 6364–6371.

Poole, L.B. (2007). The catalytic mechanism of peroxiredoxins. Subcell. Biochem. *44*, 61–81.

Puentes, S.M., Da Silva, R.P., Sacks, D.L., Hammer, C.H., and Joiner, K.A. (1990). Serum resistance of metacyclic stage *Leishmania major* promastigotes is due to release of C5b-9. J. Immunol. *145*, 4311–4316.

Purmal, A.A., Kow, Y.W., and Wallace, S.S. (1994). 5-Hydroxypyrimidine deoxynucleoside triphosphates are more efficiently incorporated into DNA by exonuclease-free Klenow fragment than 8-oxopurine deoxynucleoside triphosphates. Nucleic Acids Res. *22*, 3930–3935.

Reiner, S.L., and Locksley, R.M. (1995). The regulation of immunity to *Leishmania major*. Annu. Rev. Immunol. *13*, 151–177.

Reiner, S.L., Zheng, S., Wang, Z.E., Stowring, L., and Locksley, R.M. (1994). *Leishmania* promastigotes evade interleukin 12 (IL-12) induction by macrophages and stimulate a broad range of cytokines from $CD4^+$ T-cells during initiation of infection. J. Exp. Med. *179*, 447–456.

Requena, J.R., Levine, R.L., and Stadtman, E.R. (2003). Recent advances in the analysis of oxidized proteins. Amino Acids *25*, 221–226.

Richter, C. (1987). Biophysical consequences of lipid peroxidation in membranes. Chem. Phys. Lipids *44*, 175–189.

Romao, P.R., Fonseca, S.G., Hothersall, J.S., Noronha-Dutra, A.A., Ferreira, S.H., and Cunha, F.Q. (1999). Glutathione protects macrophages and *Leishmania major* against nitric oxide-mediated cytotoxicity. Parasitology *118* (Pt 6), 559–566.

Romao, P.R., Tovar, J., Fonseca, S.G., Moraes, R.H., Cruz, A.K., Hothersall, J.S., Noronha-Dutra, A.A., Ferreira, S.H., and Cunha, F.Q. (2006). Glutathione and the redox control system trypanothione/trypanothione reductase are involved in the protection of *Leishmania* spp. against nitrosothiol-induced cytotoxicity. Braz. J. Med. Biol. Res. *39*, 355–363.

Romao, S., Castro, H., Sousa, C., Carvalho, S., and Tomas, A.M. (2009). The cytosolic tryparedoxin of *Leishmania infantum* is essential for parasite survival. Int. J. Parasitol. *39*, 703–711.

Roy, A., Ganguly, A., BoseDasgupta, S., Das, B.B., Pal, C., Jaisankar, P., and Majumder, H.K. (2008). Mitochondria-dependent reactive oxygen species-mediated programmed cell death induced by 3,3′-diindolylmethane through inhibition of F0F1-ATP synthase in unicellular protozoan parasite *Leishmania donovani*. Mol. Pharmacol. *74*, 1292–1307.

Rueckschloss, U., Galle, J., Holtz, J., Zerkowski, H.R., and Morawietz, H. (2001). Induction of NAD(P)H oxidase by oxidized low-density lipoprotein in human endothelial cells: antioxidative potential of hydroxymethylglutaryl coenzyme A reductase inhibitor therapy. Circulation *104*, 1767–1772.

Salvati, L., Mattu, M., Colasanti, M., Scalone, A., Venturini, G., Gradoni, L., and Ascenzi, P. (2001). NO donors inhibit *Leishmania infantum* cysteine proteinase activity. Biochim. Biophys. Acta *1545*, 357–366.

Sardar, A.H., Kumar, S., Kumar, A., Purkait, B., Das, S., Sen, A., Kumar, M., Sinha, K.K., Singh, D., Equbal, A., *et al.* (2013). Proteome changes associated with *Leishmania donovani* promastigote adaptation to oxidative and nitrosative stresses. J. Proteomics. *81*, 185–199.

Sartori, A., Oliveira, M.A., Scott, P., and Trinchieri, G. (1997). Metacyclogenesis modulates the ability of *Leishmania* promastigotes to induce IL-12 production in human mononuclear cells. J. Immunol. *159*, 2849–2857.

Schlecker, T., Schmidt, A., Dirdjaja, N., Voncken, F., Clayton, C., and Krauth-Siegel, R.L. (2005). Substrate specificity, localization, and essential role of the glutathione peroxidase-type tryparedoxin peroxidases in Trypanosoma brucei. J. Biol. Chem. *280*, 14385–14394.

Schmalhausen, E.V., Pleten, A.P., and Muronetz, V.I. (2003). Ascorbate-induced oxidation of glyceraldehyde-3-phosphate dehydrogenase. Biochem. Biophys. Res. Commun. *308*, 492–496.

Schmidt, H., and Krauth-Siegel, R.L. (2003). Functional and physicochemical characterization of the thioredoxin system in Trypanosoma brucei. J. Biol. Chem. *278*, 46329–46336.

Scott, P., James, S., and Sher, A. (1985). The respiratory burst is not required for killing of intracellular and extracellular parasites by a lymphokine-activated macrophage cell line. Eur. J. Immunol. *15*, 553–558.

Scott, P., Eaton, A., Gause, W.C., di Zhou, X., and Hondowicz, B. (1996). Early IL-4 production does not predict susceptibility to *Leishmania major*. Exp. Parasitol. *84*, 178–187.

Selemidis, S., Dusting, G.J., Peshavariya, H., Kemp-Harper, B.K., and Drummond, G.R. (2007). Nitric oxide suppresses NADPH oxidase-dependent superoxide production by S-nitrosylation in human endothelial cells. Cardiovasc. Res. *75*, 349–358.

Sen, N., Das, B.B., Ganguly, A., Mukherjee, T., Tripathi, G., Bandyopadhyay, S., Rakshit, S., Sen, T., and Majumder, H.K. (2004). Camptothecin induced mitochondrial dysfunction leading to programmed cell death in unicellular hemoflagellate *Leishmania donovani*. Cell Death Differ. *11*, 924–936.

Sevier, C.S., and Kaiser, C.A. (2008). Ero1 and redox homeostasis in the endoplasmic reticulum. Biochim. Biophys. Acta *1783*, 549–556.

Shapiro, B.M., and Hopkins, P.B. (1991). Ovothiols – biological and chemical perspectives. Adv. Enzymol. Relat. Areas Mol. Biol. *64*, 291–316.

Shapiro, B.M., and Turner, E. (1988). Oxidative stress and the role of novel thiol compounds at fertilization. Biofactors *1*, 85–88.

Shibutani, S., Takeshita, M., and Grollman, A.P. (1991). Insertion of specific bases during DNA synthesis past the oxidation-damaged base 8-oxodG. Nature *349*, 431–434.

Simpson, L. (1972). The kinetoplast of the hemoflagellates. Int. Rev. Cytol. *32*, 139–207.

Spassky, A., and Angelov, D. (1997). Influence of the local helical conformation on the guanine modifications generated from one-electron DNA oxidation. Biochemistry *36*, 6571–6576.

Spies, H.S., and Steenkamp, D.J. (1994). Thiols of intracellular pathogens. Identification of ovothiol A in *Leishmania donovani* and structural analysis of a novel thiol from Mycobacterium bovis. Eur. J. Biochem. *224*, 203–213.

Steenkamp, D.J., and Spies, H.S. (1994). Identification of a major low-molecular-mass thiol of the trypanosomatid *Crithidia fasciculata* as ovothiol A. Facile isolation and structural analysis of the bimane derivative. Eur. J. Biochem. *223*, 43–50.

Steiger, R.F., and Steiger, E. (1977). Cultivation of *Leishmania donovani* and *Leishmania braziliensis* in defined media: nutritional requirements. J. Protozool. *24*, 437–441.

Stubbs, C.D., and Smith, A.D. (1984). The modification of mammalian membrane polyunsaturated fatty acid composition in relation to membrane fluidity and function. Biochim. Biophys. Acta *779*, 89–137.

Sundar, S., and Chatterjee, M. (2006). Visceral leishmaniasis – current therapeutic modalities. Indian J. Med. Res. *123*, 345–352.

Thénard, L.J. (1818). Observations sur des nouvelles combinaisons entre l'oxigène et divers acides. Ann. Chim. Phys. *8*, 306–312.

Ursini, F., Maiorino, M., Brigelius-Flohe, R., Aumann, K.D., Roveri, A., Schomburg, D., and Flohe, L. (1995). Diversity of glutathione peroxidases. Methods Enzymol. *252*, 38–53.

Veal, E.A., Day, A.M., and Morgan, B.A. (2007). Hydrogen peroxide sensing and signaling. Mol. Cell *26*, 1–14.

Vennerstrom, J.L., and Eaton, J.W. (1988). Oxidants, oxidant drugs, and malaria. J. Med. Chem. *31*, 1269–1277.

Venturini, G., Salvati, L., Muolo, M., Colasanti, M., Gradoni, L., and Ascenzi, P. (2000). Nitric oxide inhibits cruzipain, the major papain-like cysteine proteinase from Trypanosoma cruzi. Biochem. Biophys. Res. Commun. *270*, 437–441.

Vitu, E., Kim, S., Sevier, C.S., Lutzky, O., Heldman, N., Bentzur, M., Unger, T., Yona, M., Kaiser, C.A., and Fass, D. (2010). Oxidative activity of yeast Ero1p on protein disulfide isomerase and related oxidoreductases of the endoplasmic reticulum. J. Biol. Chem. *285*, 18155–18165.

Walker, J., and Saravia, N.G. (2004). Inhibition of *Leishmania donovani* promastigote DNA topoisomerase I and human monocyte DNA topoisomerases I and II by antimonial drugs and classical antitopoisomerase agents. J. Parasitol. *90*, 1155–1162.

Wang, G., Hong, Y., Johnson, M.K., and Maier, R.J. (2006). Lipid peroxidation as a source of oxidative damage in Helicobacter pylori: protective roles of peroxiredoxins. Biochim. Biophys. Acta *1760*, 1596–1603.

West, J.D., and Marnett, L.J. (2006). Endogenous reactive intermediates as modulators of cell signaling and cell death. Chem. Res. Toxicol. *19*, 173–194.

Wientjes, F.B., Hsuan, J.J., Totty, N.F., and Segal, A.W. (1993). p40phox, a third cytosolic component of the activation complex of the NADPH oxidase to contain src homology 3 domains. Biochem. J. *296 (Pt 3)*, 557–561.

Wilkinson, S.R., Obado, S.O., Mauricio, I.L., and Kelly, J.M. (2002). *Trypanosoma cruzi* expresses a plant-like ascorbate-dependent hemoperoxidase localized to the endoplasmic reticulum. Proc. Natl. Acad. Sci. U.S.A. *99*, 13453–13458.

Wilkinson, S.R., Prathalingam, S.R., Taylor, M.C., Horn, D., and Kelly, J.M. (2005). Vitamin C biosynthesis in trypanosomes: a role for the glycosome. Proc. Natl. Acad. Sci. U.S.A. *102*, 11645–11650.

Wyllie, S., Cunningham, M.L., and Fairlamb, A.H. (2004). Dual action of antimonial drugs on thiol redox metabolism in the human pathogen *Leishmania donovani*. J. Biol. Chem. *279*, 39925–39932.

Wyllie, S., Vickers, T.J., and Fairlamb, A.H. (2008). Roles of trypanothione S-transferase and tryparedoxin peroxidase in resistance to antimonials. Antimicrob. Agents Chemother. *52*, 1359–1365.

Yadav, R.K., Dolai, S., Pal, S., and Adak, S. (2008). Role of tryptophan-208 residue in cytochrome c oxidation by ascorbate peroxidase from *Leishmania major*-kinetic studies on Trp208Phe mutant and wild type enzyme. Biochim. Biophys. Acta *1784*, 863–871.

van Zandbergen, G., Bollinger, A., Wenzel, A., Kamhawi, S., Voll, R., Klinger, M., Muller, A., Holscher, C., Herrmann, M., Sacks, D., *et al.* (2006). *Leishmania* disease development depends on the presence of apoptotic promastigotes in the virulent inoculum. Proc. Natl. Acad. Sci. U.S.A. *103*, 13837–13842.

Cell Death in a Kinetoplastid Parasite, the *Leishmania* spp.

5

Radhika Mathur and Chandrima Shaha

Abstract

Apoptosis is a distinct form of cell death that plays an important role in various physiological processes in mammals. Earlier it was thought that apoptosis evolved with multicellular organisms; however, studies over the past years have shown that single-celled organisms such as *Leishmania* can also undergo programmed cell death. It is now believed that apoptosis plays an important role in the life cycle of *Leishmania* parasites. The process enables selection of fittest cells amongst the parasites within the sandfly vector and within the mammalian cells for necessary control of parasite numbers and for the evasion of immune responses. Like mammalian cells, *Leishmania* parasites also show typical features of apoptotic death, like cell shrinkage, nuclear condensation and DNA damage in response to heat stress, serum deprivation and a range of anti-leishmanial drugs. Understanding the molecular processes involved in apoptotic death in trypanosomatids might help in defining the cell death machinery and thus provide new targets for chemotherapeutic drug development.

Introduction

The parasites of the genus *Leishmania* belongs to a large complex of vector borne organisms, the kinetoplastid haemoflagellates. They belong to the trypanosomatidae family and are responsible for causing a group of diseases called as leishmaniasis which ranges from asymptomatic infections, cutaneous and mucocutaneous lesions to fatal visceral involvement. There are 30 species of *Leishmania* that infect mammals, of which 21 are involved in human infection (Hommel, 1999).

Visceral leishmaniasis is characterized by prolonged fever, splenomegaly, anaemia and excessive weight loss. The symptoms of cutaneous leishmaniasis are considered to be dependent on the species of the parasite and immunological response of the patients. The species of the Old World (the eastern hemisphere) cause self-limiting cutaneous disease while New World (America) species cause a wide range of symptoms that also involve mucosal involvement (Sundar and Rai, 2002). The disease is considered endemic in 88 countries, 72 of which are developing countries.

Leishmaniasis is linked to immunosuppressive conditions and this has become a problem because of prevalence of HIV in similar areas (Hommel, 1999). Although HIV–*Leishmania* co-infection has been reported in 35 countries, the association of VL with HIV is bringing this severe form of leishmaniasis to entirely new areas and changing its epidemiology (Desjeux, 1999).

Life cycle of *Leishmania* spp.

Leishmania parasites exist as extracellular motile promastigote forms in the invertebrate host and as intracellular non-motile amastigotes in the vertebrate host. Dipteran sandfly of the genus *Phlebotomus* (Old World leishmaniasis) and *Lutzomyia* (New World) act as vectors for the these parasites (Lainson *et al.*, 1987). The amastigote form lives inside phagolysosomes present within phagocytic cells in the mammalian hosts. The sandfly vector acquires amastigote forms when it feeds on the blood of infected individuals. The parasite continues its development inside the sandfly, where it undergoes transformation to the procyclic promastigotes that attach to the gut wall and differentiate to non-dividing metacyclic promastigotes incapable of attaching to the midgut and as a result migrate to the mouth parts. Metacyclic promastigotes enter skin of the vertebrate host through bite of the infected sandfly where eventually they infect macrophages (van Zandbergen *et al.*, 2004). The replicated amastigotes are released into the bloodstream or interstitial spaces, where they are phagocytosed by macrophages invading other tissues including liver, spleen and bone marrow. Parasites are again transferred to sandfly from the blood of an infected individual during a blood meal to continue the cycle of infection (Chang *et al.*, 1990).

Treatment of leishmaniasis

In the absence of suitable vaccines, the treatment of leishmaniasis depends mainly on chemotherapy. Pentavalent antimonial compounds such as sodium stibogluconate (SSG) and meglumine antimoniate have long been the primary drugs for anti-leishmanial therapy. It is believed that pentavalent antimony has to be converted to trivalent form to be active (Shaked-Mishan *et al.*, 2001) and acts by compromising the thiol redox potential of the cell (Wyllie *et al.*, 2004). Amongst other drugs, amphotericin B, a polyene antibiotic, has been widely used to treat visceral leishmaniasis. The drug binds with ergosterol, the major sterol present on *Leishmania* membrane, leading to formation of membrane pores, alters membrane permeability and finally causes cell lysis (Ouellette *et al.*, 2004). Miltefosine, an alkyl phospholipid, alters glycosylphosphatidylinositol (GPI) anchor synthesis, ether-lipid metabolism, signal transduction and alkyl-specific acyl-coenzyme A acyl-transferase (Paris *et al.*, 2004). The anti-leishmanial drug pentamidine is used to treat antimonial-resistant cutaneous leishmaniasis. It acts by inhibiting polyamine biosynthesis and disruption of mitochondrial membrane potential (Croft *et al.*, 2006).

Studies have shown that paromomycin, an aminoglycoside antibiotic produced by *Streptomyces riomosus*, is a suitable anti-leishmanial drug acting by altering membrane permeability and RNA synthesis (Maarouf *et al.*, 1997). Ergosterol biosynthesis pathway has been explored over the past few years for development of drugs against trypanosomatids because unlike mammals, trypanosomatids have ergosterol and not cholesterol as the major sterol (de and Rodrigues, 2009). Various inhibitors of the ergosterol biosynthetic pathway such as azoles, azasterols, statins, etc. are being studied to explore the clinical value of these drugs in different forms of leishmaniasis (Ghannoum and Rice, 1999; Zhou *et al.*, 2006; Lange *et al.*, 2000; Istvan and Deisenhofer, 2001). Sitamaquine, targeting mitochondria, and azithromycin, targeting protein synthesis, are two other drugs that are being explored. These drugs have shown promising results as anti-leishmanial agents (Carvalho *et al.*, 2011; Dinos *et al.*, 2001).

Programmed cell death (PCD)

The term 'apoptosis', which describes the phenotype of programmed cell death pathway (PCD), was originally coined by Kerr and coworkers in 1972. The process plays an important role in various physiological processes during development including embryogenesis and maintenance of the immune system (Lockshin *et al.*, 2001). Different types of programmed cell death have been described, the most investigated being apoptosis and autophagy (Kroemer *et al.*, 2009). It was initially assumed that during evolution PCD emerged along with multicellular organisms as death of single celled organism would be against the evolutionary principle. However, increasing experimental evidence suggests that this process also occurs in a variety of single celled organisms.

PCD can be initiated by various internal and external stimuli and is coordinated by network of regulators and effectors. Irrespective of the type of stimulus experienced by the cell, the end result of apoptosis is change in membrane permeability, nuclear condensation, rounding of cells, cell shrinkage and blebbing followed by dismantling of the cell in small bodies called apoptotic bodies that are engulfed by macrophages. Various biochemical features associated with apoptosis include DNA fragmentation, phosphatidylserine externalization, mitochondrial dysfunction, cleavage of poly (ADP) ribosyl polymerase, caspase activation and proteolytic cleavage of various intracellular substrates (Kerr *et al.*, 1972).

However, some of these features are associated with normal cellular function as well and therefore need to be explored in detail. In mammalian cells there are two major apoptotic pathways. The intrinsic pathway involves mitochondrial membrane permeabilization regulated by the members of BCl2/Bax protein family that results in the release of cytochrome c in the cytosol that activates caspase-9 and eventually caspase-3, leading to cell death. The extrinsic pathway involves activation of death receptors via the TNF superfamily that activates death domain proteins, in turn resulting in caspase-8 cleavage, activating caspase-3 (Movassagh and Foo, 2008).

Autophagy, on the other hand, is a type of cell death that lacks chromatin condensation but is characterized by formation of large number of autophagic vacuoles in the cytoplasm (Levine and Klionsky, 2004). Autophagy is a catabolic process and involves degradation of damaged cellular components through lysosomal machinery. Regulated autophagy ensures synthesis, degradation and recycling of cellular components; however, very high autophagy induced by adverse stress situations results in cell death (Patel *et al.*, 2012). Autophagy is also considered to play important role in cellular differentiation, tissue remodelling, growth control, size regulation, mitochondrial homeostasis, cellular immunity, adaptation to stresses, and unconventional protein secretion (Williams *et al.*, 2013).

Programmed cell death in protozoan parasites

Protozoan parasites that diverged around 1–2 billion years ago express cell death phenotypes and at least 10 species have been shown to do so (Ameisen, 1996). Kinetoplastid parasites like the *Leishmania* and the *Trypanosoma* have been shown to demonstrate apoptotic phenotype (Fig. 5.1) that in many ways resembles mammalian apoptotic features (Das *et al.*, 2001). Since *Leishmania* is one of the earliest eukaryotes with

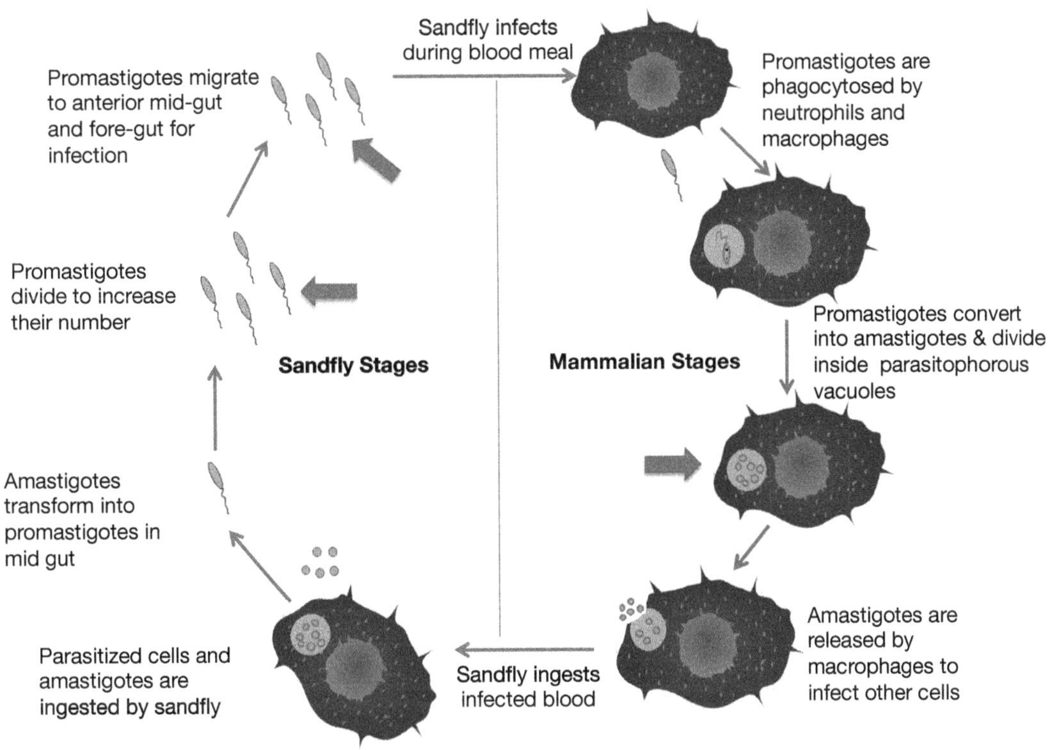

Figure 5.1 Life cycle of *L. donovani*. Arrows indicate stages where apoptosis plays an important role to control parasite numbers.

mitochondria, the mitochondrial death pathway may have evolved after its emergence. The biological relevance of PCD in *Leishmania* may rest in maximizing fitness of the parasites that is required both within the sandfly gut and inside the host cell. The origin of PCD in the early eukaryotes could have arisen due to conflict between two different genomes after acquisition of the mitochondria. In addition, the possibility of the evolution of PCD in these organisms may have their origin when host–parasite relationship was established with the appearance of the vector and the host. The *Leishmania* parasites show autophagic phenotype as well as the apoptotic phenotype during their normal life cycle and when they are exposed to stress.

Relevance of studying cell death in *Leishmania*

The process of apoptosis is very essential for the survival of an organism. It leads to clearing of damaged cells and allows the survival of fittest cells in a colony (Welburn *et al.*, 1997). Many studies have shown that apoptosis-like cell death occurs in *Leishmania*. These studies primarily demonstrate cell death under various conditions of drug treatment or during infection. Cell death by apoptosis is considered essential as it allows the removal of unhealthy cells cleanly so that only fit cells that are capable of propagation and infection survive in the sandfly and migrate to pharynx for infection (Welburn and Maudlin, 1999). For establishment of a successful infection the parasites also have to survive in the harsh conditions present inside the macrophages and maintain optimum numbers (Zangger *et al.*, 2002).

The study of apoptosis in organisms such as *Leishmania* can provide new insights into the process of infection and might offer new possibilities in controlling the disease. In addition, the study of this process in *Leishmania* may reveal novel effector molecules that are present in both mammals and the parasite. Also, increased knowledge of such an important pathway in lower eukaryotes such as *Leishmania* can be used to study how processes in higher eukaryotes evolved.

Apoptosis in *Leishmania*

Unlike metazoan apoptotic pathways, apoptotic or other cell death pathways in trypanosomatids such as *Leishmania* is not fully understood, therefore features such as cell shrinkage etc. are not considered to be reliable markers for apoptosis in these organisms. The final outcome of apoptosis in trypanosomatids is similar to those observed in mammals. However, absence of key regulatory molecules of apoptosis such as members of BCl2, caspase and TNF family suggest that pathway of apoptosis is different in these divergent eukaryotes (Lee *et al.*, 2002). Various studies over the past years have led to the identification of markers of apoptosis in *Leishmania*. These include exposure of phosphatidylserine on cell surface (de Freitas Balanco *et al.*, 2001), release of cytochrome c (Gannavaram *et al.*, 2008), caspase like and metacaspase activities (Lee *et al.*, 2002; Paris *et al.*, 2004), nuclease activity (Gannavaram *et al.*, 2008; BoseDasgupta *et al.*, 2008), mitochondrial membrane potential loss (Mukherjee *et al.*, 2002) and changes in ion homeostasis (Sudhandiran and Shaha, 2003; Lindoso *et al.*, 2004). Wide range of stimuli such as heat shock, drugs, reactive oxygen species, antimicrobial peptides and prostaglandins lead to apoptotic features in trypanosomatids (Lee *et al.*, 2002; Duszenko *et al.*, 2006; Jimenez-Ruiz *et al.*, 2010). A summary of different biochemical changes that occur in *Leishmania* during its life cycle or due to drug exposure is given below.

Phosphatidylserine exposure

The two leaflets of plasma membrane bilayer are asymmetrical, that is, the composition of the two layers is not identical. The outer leaflet contains phosphatidylcholine and sphingomyelin while lipids such as phosphatidylserine and phosphatidylethanolamine are more abundant in the inner leaflet (Bevers and Williamson, 2010). However, during apoptosis the asymmetry is lost and phosphatidylserine is exposed at the outer surface. This exposure of phosphatidylserine can be identified by its reactivity towards Annexin V and is therefore considered as an early event in cells undergoing apoptosis (Martin *et al.*, 1995). Various studies have used Annexin V binding to show exposure of PS in *Leishmania* in response to heat shock, serum deprivation and a range of anti-leishmanial drugs (Jimenez-Ruiz *et al.*, 2010; de Freitas Balanco *et al.*, 2001). It has also been proposed that PS exposure plays an important role during establishment of infection as the parasites use this feature for allowing recognition and engulfment by the macrophages and thus evade immune response (Martin *et al.*, 1995). Recent studies have shown that ABC transporter LABCG2 is involved in PS exposure and hence regulates infectivity and pathogenicity of the parasites (Campos-Salinas *et al.*, 2013). However, there are other reports that suggest that PS is not the only lipid that can bind to Annexin V, other anionic lipids such as phosphatidic acid, phosphatidylinositol also interact with Annexin V during drug treatment or membrane permeabilization (Weingartner *et al.*, 2012). It is therefore apparent that PS externalization should be considered as an apoptotic marker when combined with other apoptotic-like features.

Cytochrome C release

The mitochondrial death pathway in higher eukaryotes is initiated by the permeabilization of the mitochondrial outer membrane by the proapoptotic BCl2 members. Subsequently, cytochrome c is released from the mitochondria and it binds to Apaf1 present in the cytosol leading to activation of caspase-9 followed by caspase-3 (Brenner and Mak, 2009; Pradelli *et al.*, 2010; Huttemann *et al.*, 2011). Release of cytochrome c is a well-studied phenomenon in *Leishmania*, in response to different apoptotic signals (Gannavaram *et al.*, 2008). Homology studies have revealed a possible homologue of Apaf1 in *Leishmania* (Gannavaram and Debrabant, 2012b). However, the downstream process, i.e. cytochrome c binding to Apaf1 and further activation of caspases are not well characterized in *Leishmania* partly because of absence of caspase-9 component.

Caspase and metacaspase activity

Caspases are important regulators of apoptotic processes in metazoans. There are extensive reports that provide evidence for the presence of 'caspase like activities' in *Leishmania*. Caspase genes have not been reported in any of the *Leishmania* genomes; however, studies suggest that proteases of the parasites undergoing apoptosis can degrade classical substrates of mammalian caspases (Lee *et al.*, 2002; Paris *et al.*, 2004; Bose-Dasgupta *et al.*, 2008). Metacaspases are cysteine proteases with structural similarity to caspases. All *Leishmania* species except *L. infantum* and *L. donovani* express metacaspase (Lee *et al.*, 2002). In *L. donovani* metacaspases are expressed in both promastigote and amastigote forms. They express arginine lysine specific activity and do not require any proteolytic activation. They are localized in acidocalcisomes as inactive enzymes and are released during the process of apoptosis. Unlike *L. donovani* metacaspases, the *L. major* metacaspases are activated after autoprocessing. Overexpression of metacaspases increased the sensitivity of parasites towards oxidative stress and caused a rapid loss of mitochondrial membrane potential, suggesting a role of metacaspases during apoptosis (Luder *et al.*, 2010; Zalila *et al.*, 2011).

Nuclease activity

In metazoans extensive DNA degradation during apoptosis is caused by a number of nucleases. These nucleases can be divided into two categories depending on whether they require caspases for activation or not. Like caspases, *Leishmania* genome does not show any homologue of caspase-derived DNase (CAD). Two endonucleases, apoptosis inducing factor (AIF) and endonuclease G (Endo G) have been known to induce DNA fragmentation in a caspase independent manner in the mammalian system (Li *et al.*, 2001). A weak homologue of AIF has been recently reported in *Leishmania* genomes. The role of this homologue in apoptosis is not characterized. Endonuclease G is a mitochondrial enzyme and is released in response to stimuli inducing apoptosis. Many studies have reported that *Leishmania* Endo G is activated during oxidative stress and mediates DNA degradation (BoseDasgupta *et al.*, 2008; Gannavaram *et al.*, 2008). The overexpression of Endo G led to spontaneous DNA fragmentation in amastigotes but not in promastigotes suggesting requirement of additional factors expressed in a stage specific manner. Oxidative stress-induced high levels of apoptosis in Endo G overexpressing cells suggest a role of this gene in *Leishmania* apoptosis (Gannavaram *et al.*, 2008). Other endonucleases, such as TatD related nuclease and flap endonuclease, have also been reported have a role in DNA degradation during apoptosis in *Leishmania* (BoseDasgupta *et al.*, 2008). Treatment of parasites with inducers of apoptosis led to the migration and subsequent interaction of TatD with Endo G in the nucleus, suggesting a role of these endonucleases during apoptosis (Gannavaram and Debrabant, 2012a).

Mitochondrial membrane potential (Ψ)

Unlike mammalian cells that have numerous mitochondria *Leishmania* has a single branched mitochondrion, therefore the functioning of this organelle is vital for the survival of the parasite. The functioning of mitochondria is essential as they are sites for respiratory chain and generates ATP for the cells (Gottlieb, 2000, 2003). Oxidative stress induced apoptosis leads to changes in the mitochondria and causes disruption of mitochondrial membrane potential (Ψ) in *L. donovani* (Mukherjee *et al.*, 2002). Total loss of Ψ suddenly would lead to the process of necrosis in *Leishmania* as the parasite does not have any other mitochondria. Therefore, single mitochondrion of the parasite adapts itself to the process of apoptosis and produces variable energy generating state to complete the process of apoptosis (Mukherjee *et al.*, 2002). Various studies in metazoans have identified a megachannel in mitochondria to be involved in apoptosis (Kroemer, 2003). However, in *Leishmania* inhibition of this megachannel did not reverse oxidative stress-induced dissipation of mitochondrial membrane potential, suggesting that type of apoptotic stimulus might decide the involvement of this megachannel (Mukherjee *et al.*, 2002). Various anti-leishmanial drugs that induce apoptosis also act by dissipation of Ψ. Inhibition of the respiratory chain complex I, II and III caused apoptotic death of the *L. donovani* promastigotes. While treatment with complex I inhibitor leads to an increase in mitochondrial potential, inhibition of complex II and III lead to a fall in the potential, suggesting that single mitochondrion of *Leishmania* plays an different roles during apoptosis depending on the type of apoptotic inducer used (Mehta and Shaha, 2004).

Changes in ion homeostasis

Changes in the ion homeostasis are related to the apoptotic process in *Leishmania*. Ca^{2+} levels play an important role in *Leishmania* apoptosis pathway similar to metazoans. The changes in mitochondrial membrane potential during apoptosis have been linked to increase in the levels of cytosolic Ca^{2+}. Inhibition of this increase by chelating extracellular and intracellular calcium levels rescues oxidative stress induced dissipation of Ψ and subsequent apoptosis-like cell death. Studies from this laboratory have indicated that elevated cytosolic Ca^{2+} levels is a result of influx of extracellular Ca^{2+} through non-selective ion channels as well as release of Ca^{2+} from intracellular acidocalciosomes (Mehta and Shaha, 2004). Studies have also shown that Ca^{2+} induces apoptotic death in *L. amazonensis* promastigotes subjected to heat shock. Apart from Ca^{2+}, K^+ also plays an important role in apoptosis (Sen *et al.*, 2004). Apoptotic death induced by agents such as camptothecin led to the impairment of Na^+,K^+-ATPase pump and therefore decreases intracellular K^+ levels and pH. This is followed by activation of caspase like protease and cytochrome c release leading to apoptosis. Results also suggest that low levels of K^+ inside the cell promote activation of nuclease leading to DNA fragmentation. These studies indicate that ions play a crucial role during programmed cell death (Gottlieb, 2000).

Natural cell death during cell cycle

Various reports show that apoptosis-like cell death pathway serve in the life cycle of unicellular parasites like *Leishmania*. Apoptotic cell death is considered important during growth and survival as it is useful in controlling cell number in sandfly and mammalian host, to avoid hyperparasitism (Al-Olayan *et al.*, 2002; Arnoult *et al.*, 2002). The process of apoptosis is vital in various stages of life cycle to eliminate unfit cell and control cell numbers for survival of the parasite. As the promastigotes grow within the sandfly gut, their number increases gradually and this can lead to nutrient deprivation (Lee *et al.*, 2002).

Leishmania can be grown in *in vitro* cultures. Stationary phase cultures of various *Leishmania* species show features corresponding to apoptosis. These include DNA fragmentation, nuclear condensation and presence of endonuclease activity. These features increase with an increase in the number of days the cells are in culture. Also the intensity of apoptotic features is more prominent in promastigote stage of the parasite as compared to the amastigote stage suggesting that the promastigote stage is more sensitive to nutrient deprivation (Lee *et al.*, 2002; Zangger *et al.*, 2002). The induction of caspase like activity in *Leishmania* has been observed in both promastigotes and axenic amastigotes by checking the cleavage of PPL, a caspase-specific cell permeable substrate. It fluoresces upon cleavage and therefore helps in monitoring total protease activity inside the cell. In *Leishmania* the induction of PPL cleavage activity was partially inhibited by mammalian caspase inhibitors suggesting that some protease activity in *Leishmania* during nutrient deprivation is caspase like. The basal level of PPL cleavage was higher in axenic amastigotes of *L. donovani* than in promastigotes. However, the pattern was reversed during stationary phase, when promastigotes showed a considerably higher activity than the axenic amastigotes. The induction of PPL cleavage correlated with DNA fragmentation (Lee *et al.*, 2002). Unlike *L. donovani,* in which caspase-like activity has been reported in both promastigotes and amastigotes, *L. major* and *L. mexicana* show this kind of activity only in amastigotes. No activity has been observed in the promastigote stage (Zangger *et al.*, 2002). These discrepancies suggest that among *Leishmania* species there could be both caspase independent and caspase dependent apoptosis. The increase in number of parasites and their gradual migration into the stationary phase also leads to dissipation of the mitochondrial potential in *L. donovani* parasites. This is followed by changes in membrane integrity (Nicholson and Thornberry, 1997). It is still unclear whether similar set of events occur in amastigotes, but increased DNA fragmentation and nuclear condensation suggest that stationary phase of amastigotes also show apoptosis-like features (Lee *et al.*, 2002).

Induction of cell death by anti-leishmanial drugs

Pentavalent antimonial compounds such as pentostam still play a leading role in the treatment of leishmaniasis. These compounds are thought to act by inhibition of various metabolic pathways, though the mode of action is still not very clear. Studies indicate that antimony induces efflux of intracellular thiols and inhibits trypanothione reductase to compromise the overall thiol redox potential of the cell (Wyllie *et al.*, 2004). The use of antimonial drugs in treating leishmaniasis has reduced considerably over the years due to emergence of drug resistance and various side effects associated with high doses of antimony. Various drugs cause apoptosis-like cell death in *Leishmania*. Treatment of *L. donovani* amastigotes with increasing doses of pentavalent antimony leads to induction of caspase like activity (Sereno *et al.*, 2001). However, no such activity was observed in promastigotes even with higher dose of pentavalent antimony. Both amastigotes and promastigotes however showed DNA laddering after antimony exposure (Lee *et al.*, 2002). Similar to *L. donovani* promastigotes, axenic amastigotes of *L. infantum* also show caspase independent apoptotic features on treatment with trivalent antimonial compounds. Caspase independent cell death mechanism is known in multicellular organisms where various factors released from mitochondria cause apoptotic cell death. Studies from this laboratory have shown that trivalent antimonial compound (PAT) induces apoptotic cell death in *L. donovani* amastigotes. These parasites show nuclear DNA fragmentation and phosphatidylserine externalization, hallmarks of apoptotic cell death (Sudhandiran and Shaha, 2003).

Amphotericin B and its formulations are increasingly being used for the treatment of visceral leishmaniasis. This drug has been proven to be quite effective in treating antimonial-resistant leishmaniasis. Amphotericin B is a polyene antibiotic and it is believed to be leishmanicidal owing to its ability to bind to ergosterol (Bolard, 1986). This leads to formation of transmembrane channels that alters membrane permeability to cations, glucose, and water and also affects the activity of membrane bound enzymes (Hartsel and Bolard,

1996; Baginski *et al.*, 1997). Recent reports have shown that Amphotericin B also binds to cholesterol with significant affinity inhibiting *Leishmania* entry into the macrophages by sequestering cholesterol and inducing changes in the receptor pattern on the macrophage surface such that infection by parasites can be avoided. Treatment of *L. donovani* amastigotes and promastigotes with amphotericin B shows DNA ladder formation, a hallmark of apoptosis. These parasites also show PPL cleavage activity, suggesting that caspase like enzymes are activated on treatment with amphotericin B. However, promastigotes and amastigotes show differential sensitivity towards the drug suggesting that apoptotic machinery in both the parasite forms may be differentially sensitive towards the inducers of apoptosis (Lee *et al.*, 2002). The drug also induces a dissipation of mitochondrial membrane potential in promastigotes, which is another hallmark of apoptosis in these parasites. Rapid changes in membrane permeability were observed following PPL cleavage activity, suggesting that parasites undergo apoptosis on treatment with this drug. Treatment of *L. donovani* promastigotes with caspase inhibitors prior to treatment with amphotericin B inhibited PPL cleavage significantly and also reduced the extent of DNA damage (Lee *et al.*, 2002).

Miltefosine, an alkylphosphocholine, was originally developed for the treatment of cancer. However, it was later approved as first effective and safe oral treatment for human visceral leishmaniasis (Paris *et al.*, 2004). The mode of action of the drug is not very well understood, though studies indicate that miltefosine may act by damaging the flagellar movements (Santa-Rita *et al.*, 2000), perturbation of alkyl lipid metabolism (Lira *et al.*, 2001), interference with ether lipid remodelling by inhibiting alkyl lysophosphatidylcholine specific acyl coenzyme A acyltransferase (Lux *et al.*, 2000), inhibition of *de novo* synthesis of phosphatidylcholine and disruption of glycosylphosphatidylinositol anchor biosynthesis pathway (Lux *et al.*, 1996). The anticancer activity of miltefosine has been linked to its capacity to induce apoptosis in numerous tumour cell lines (Engelmann *et al.*, 1996; Henke *et al.*, 1999). It is believed that the drug accumulates in the plasma membrane due to its amphiphilic properties, but the molecular mechanisms implicated in the initiation of apoptotic signalling are not well understood. Miltefosine has been reported to cause morphological changes in *L. donovani* promastigotes and amastigotes. It causes cell shrinkage and phosphatidylserine externalization at higher doses. Treatment with miltefosine also induced DNA fragmentation in a concentration and time dependent manner suggesting that parasites undergo programmed cell death upon treatment with the drug. Studies have also revealed that miltefosine induces cell cycle arrest at sub-G_0/G_1 phase and activates caspase-like and other proteases upon treatment. However, only caspase inhibitors could rescue the parasite from apoptosis, inhibitors of other proteases did not have much effect on cell shrinkage and DNA fragmentation caused by miltefosine (Paris *et al.*, 2004). The efficiency of the drug in causing apoptosis has been found to be dependent on the strain of the parasite. Results similar to *L. donovani* have been reported in *L. amazonensis* and *L. tropica*. However, *L. mexicana* and *L. major* have been reported to have low sensitivity towards miltefosine (Escobar *et al.*, 2002). Various other groups have shown that the intracellular form of the parasite shows apoptotic features on exposure to other anti-leishmanial agents such as nitric oxide, heat shock, serum deprivation etc., suggesting that the parasite can undergo programmed cell death in response to different stimuli.

Oxidative stress induced cell death

Oxidative stress has been implicated in apoptosis in mammalian and yeast cells. Studies indicate that apoptosis induced by H_2O_2 changes the levels of intracellular Ca^{2+}, and induces lipid peroxidation in mammalian cell lines. It also decreases the levels of BCl2 and activates Ca^{2+}, Mg^{2+}-dependent endonucleases in higher organisms (Hildeman *et al.*, 2003; Singh *et al.*, 2007). Many antioxidants act as inhibitors of apoptosis suggesting that oxidative stress can indeed act as an apoptotic stimulus.

During its life cycle, *Leishmania* parasites infect mammalian macrophages that produce large amounts of reactive oxygen and nitrogen intermediates to control infection. The mode of action of various anti-leishmanial drugs also involves production of ROS to facilitate killing of the parasite. High doses of H_2O_2 are lethal to both amastigotes and promastigote form of *L. donovani*. Studies have showed that *L. donovani* promastigotes undergo apoptosis on treatment with H_2O_2 (Das *et al.*, 2001). Oxidative stress induced by H_2O_2 induced changes in cellular metabolism of the promastigotes, permeability of the cellular membrane and reduced promastigote motility. H_2O_2 also promoted cell shrinkage and gradual cytoplasmic condensation followed by nuclear breakdown. DNA laddering, another prominent hallmark of cells undergoing apoptosis was also observed in *L. donovani* promastigotes treated with H_2O_2. Oxidative stress induced apoptosis also activated caspase like proteases in these parasites thus promoting programmed cell death pathway (Mukherjee *et al.*, 2002).

Treatment of *L. amazonensis* parasites with LQB118 induces ROS production in a concentration dependent manner. Promastigotes treated with this drug showed several features typical of apoptosis. These included reduction of mitochondrial membrane potential, DNA fragmentation and sub-G_0/G_1 cell cycle phenotype. Similar features were also observed in the amastigote stage with variation in intensities suggesting that LQB118 induced ROS triggers mitochondria dependent apoptosis in the parasites (Ribeiro *et al.*, 2013). Treatment of parasites with antioxidants inhibits ROS production and promotes cell survival. Antioxidants also inhibited features corresponding to apoptosis in both H_2O_2- and LQB118-treated parasites confirming that oxidative stress acts as an apoptotic stimulus in these parasites (Ribeiro *et al.*, 2013; Mukherjee *et al.*, 2002). Similar studies have also revealed that other *Leishmania* species such as *L. chagasi*, *L. major* also undergo apoptosis on treatment with agents that produce high amounts of oxidative stress (Dolai *et al.*, 2011).

Role of apoptotic mimicry and apoptosis in *Leishmania*

The idea that apoptosis is only a feature of multicellular organisms has been challenged with the discovery of features corresponding to PCD in unicellular organisms such as *Leishmania*. The identification of such a cell death programme in lower eukaryotes questions our understanding of the evolution of the cell death pathway and its players.

Apoptotic mimicry by the parasites helps *Leishmania* to evade immune responses of the host and establish themselves as intracellular parasites. Exposure of phosphatidylserine allows internalization of the amastigotes and inhibits inflammatory responses, thus promoting parasite survival (de Freitas Balanco *et al.*, 2001). Observations suggest that the process of apoptosis may be quite relevant in trypanosomatids. Investigations show that inoculum of *Leishmania* parasites containing dead parasites are more effective infective agents (van Zandbergen *et al.*, 2006). Apoptosis can act as a mechanism to select fittest cells from a colony to facilitate survival in harsh conditions present in the host. It can therefore be used by the parasite as a means to maximize their biological fitness. The process of apoptosis may also contribute in controlling population of parasites inside host to promote host survival and subsequent parasite transmission. The process may therefore allow a tight regulation of cell cycle in response to different environmental challenges such as nutrient deprivation, heat shock, oxidative stress etc. for adaption and survival. Programmed cell death pathway may therefore play an important role in establishment of stable host parasite relationship.

Autophagy in *Leishmania*

Autophagy is a regular cellular housekeeping process that is activated in response to nutrient deprivation or chemotherapeutic stress, leading to recycling of cellular macromolecules to provide metabolic precursors. Failure to activate autophagy in response to stress can result in cell death. On the contrary, autophagy in excess may also result in cell death. Trypanosomatids possess a functional autophagic system (Herman *et al.*, 2006). It is known that autophagy in these parasites is essential for differentiation, parasite maintenance and survival (Alvarez *et al.*, 2008a,b; Besteiro *et al.*, 2006). Recent studies indicate that autophagy in trypanosomatids such as *Leishmania* is activated during starvation induced stress (Besteiro *et al.*, 2006; Alvarez *et al.*, 2008a,b), differentiation, treatment with different drugs (Menna-Barreto *et al.*, 2009b; Menna-Barreto *et al.*, 2007, 2009a; Merkel *et al.*, 2007) and antimicrobial peptides (Bera *et al.*, 2003; McGwire and Kulkarni, 2010). Features corresponding to autophagy have been observed in *Leishmania* promastigotes. Treatment with antimicrobial peptides or starvation leads to degradation of cytoplasmic components and appearance of autophagic vacuoles. Cell death induced by these stresses lack typical apoptotic features like loss of nuclear and mitochondrial integrity. *Leishmania* promastigotes treated with antimicrobial peptides show preservation of microtubules suggesting that the parasites have machinery for the process of autophagy. Cytoskeletal proteins are considered important during autophagy as they play an important role in sequestration and formation of autophagosomes and their fusion with lysosomes (Bera *et al.*, 2003). Autophagy can serve as a survival mechanism through protective action on cell function or it can induce death.

Future trends

Pathways of apoptosis peculiar to the *Leishmania* parasite where specific molecules can be identified and can potentially be targeted would be of great value for development of drugs. Fig. 5.2 shows the points in the life cycle where apoptosis is important for the *Leishmania* parasite. This can be in the sandfly gut or within the macrophages. While inducing *Leishmania* death in the vector will be difficult, it could be possible to target the parasite within the macrophages of the host through small molecule inhibitors. Our understanding of protozoal parasites has increased with genome sequencing of multiple kinetoplastid parasites and this information reveals that typical apoptotic proteins identified in the higher eukaryotes are not present. It is possible that the existing molecules within the parasites can perform functions

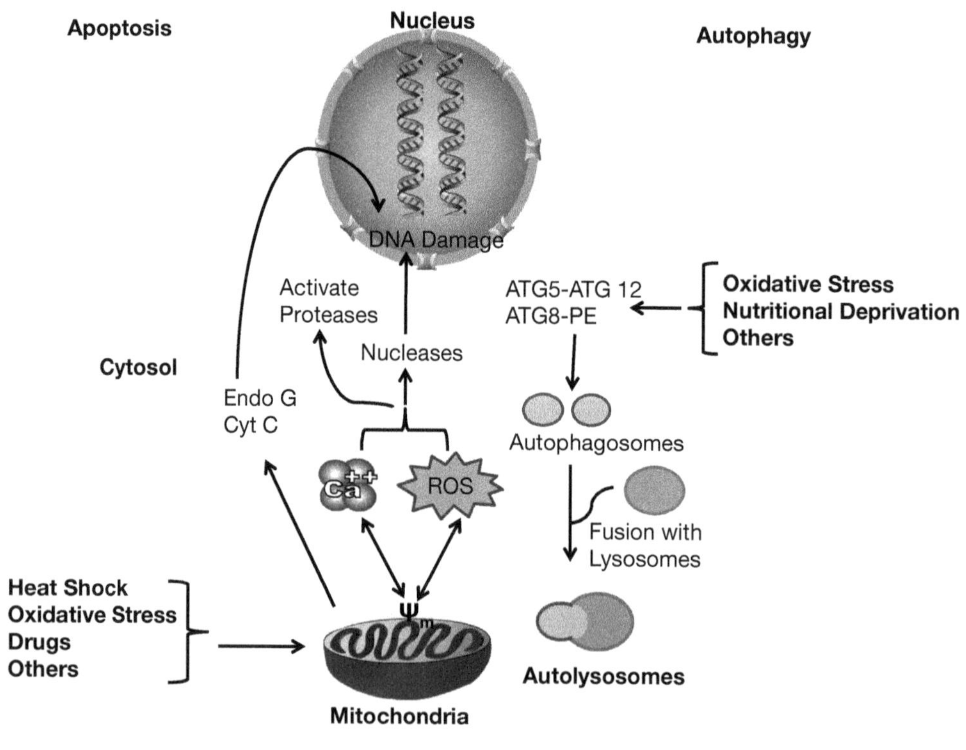

Figure 5.2 Programmed cell death pathways in *Leishmania* spp. The cartoon describes different events that occur during apoptosis and autophagy.

related to apoptosis, for example the cysteine proteases. Therefore, the search for unique death inducers should continue for identifying exclusive proteins.

Conclusion

PCD is a molecular event that is essential for development and for cell death under stress that helps to maintain homeostasis. The recognition that unicellular organisms such as trypanosomatid parasites possess endogenous machinery that helps the cells to die in a regulated manner has raised possibilities to identify molecules related to the cell death pathway peculiar to these parasites that would help in drug development. Our knowledge on the biology of these parasites were limited; however, with the solving of the genome sequences of multiple trypanosomatids, it is now possible to look realistically into the possibilities of molecules that can be investigated further for their role in cell death. In the post-genomic era since 2005, many studies have come out on a medley of genes correlated to a given change. However, that has provided a description rather than an understanding. Although data mining of the genomes has provided valuable data by way of comparative analysis and functional correlation of the *Leishmania* proteome, we do have to get back to basic biological studies to understand much of what the genome sequence has revealed. Ideally, molecules related to successes of host–pathogen interaction should also be explored for their role in the parasite survival at the time of contacting the host. Substantial insights will only be achieved through further explorations linking the causality.

Acknowledgements

Financial support came from funds of the Department of Biotechnology, Govt. of India (http://dbtindia.nic.in/index.asp) towards creating a Centre for Molecular Medicine at the National Institute of Immunology.

References

Al-Olayan, E.M., Williams, G.T., and Hurd, H. (2002). Apoptosis in the malaria protozoan, *Plasmodium berghei*: a possible mechanism for limiting intensity of infection in the mosquito. Int. J. Parasitol. 32, 1133–1143.

Alvarez, V.E., Kosec, G., Sant'Anna, C., Turk, V., Cazzulo, J.J., and Turk, B. (2008a). Autophagy is involved in nutritional stress response and differentiation in *Trypanosoma cruzi*. J. Biol. Chem. *283*, 3454–3464.

Alvarez, V.E., Kosec, G., Sant, A.C., Turk, V., Cazzulo, J.J., and Turk, B. (2008b). Blocking autophagy to prevent parasite differentiation: a possible new strategy for fighting parasitic infections? Autophagy. *4*, 361–363.

Ameisen, J.C. (1996). The origin of programmed cell death. Science 272, 1278–1279.

Arnoult, D., Akarid, K., Grodet, A., Petit, P.X., Estaquier, J., and Ameisen, J.C. (2002). On the evolution of programmed cell death: apoptosis of the unicellular eukaryote *Leishmania major* involves cysteine proteinase activation and mitochondrion permeabilization. Cell Death. Differ. *9*, 65–81.

Baginski, M., Resat, H., and McCammon, J.A. (1997). Molecular properties of amphotericin B membrane channel: a molecular dynamics simulation. Mol. Pharmacol. *52*, 560–570.

Bera, A., Singh, S., Nagaraj, R., and Vaidya, T. (2003). Induction of autophagic cell death in *Leishmania donovani* by antimicrobial peptides. Mol. Biochem. Parasitol. *127*, 23–35.

Besteiro, S., Williams, R.A., Morrison, L.S., Coombs, G.H., and Mottram, J.C. (2006). Endosome sorting and autophagy are essential for differentiation and virulence of *Leishmania major*. J. Biol. Chem. *281*, 11384–11396.

Bevers, E.M., and Williamson, P.L. (2010). Phospholipid scramblase: an update. FEBS Lett. *584*, 2724–2730.

Bolard, J. (1986). How do the polyene macrolide antibiotics affect the cellular membrane properties? Biochim. Biophys. Acta *864*, 257–304.

BoseDasgupta, S., Das, B.B., Sengupta, S., Ganguly, A., Roy, A., Dey, S., Tripathi, G., Dinda, B., and Majumder, H.K. (2008). The caspase-independent algorithm of programmed cell death in *Leishmania* induced by baicalein: the role of LdEndoG, LdFEN-1 and LdTatD as a DNA 'degradesome'. Cell Death. Differ. *15*, 1629–1640.

Brenner, D., and Mak, T.W. (2009). Mitochondrial cell death effectors. Curr. Opin. Cell Biol. *21*, 871–877.

Campos-Salinas, J., Leon-Guerrero, D., Gonzalez-Rey, E., Delgado, M., Castanys, S., Perez-Victoria, J.M., and Gamarro, F. (2013). LABCG2, a new ABC transporter implicated in phosphatidylserine exposure, is involved in the infectivity and pathogenicity of *Leishmania*. PLoS. Negl. Trop. Dis. 7, e2179.

Carvalho, L., Luque-Ortega, J.R., Lopez-Martin, C., Castanys, S., Rivas, L., and Gamarro, F. (2011). The 8-aminoquinoline analogue sitamaquine causes oxidative stress in *Leishmania donovani* promastigotes by targeting succinate dehydrogenase. Antimicrob. Agents Chemother. *55*, 4204–4210.

Chang, K.P., Chaudhuri, G., and Fong, D. (1990). Molecular determinants of *Leishmania* virulence. Annu. Rev. Microbiol. *44*, 499–529.

Croft, S.L., Sundar, S., and Fairlamb, A.H. (2006). Drug resistance in leishmaniasis. Clin. Microbiol. Rev. *19*, 111–126.

Das, M., Mukherjee, S.B., and Shaha, C. (2001). Hydrogen peroxide induces apoptosis-like death in *Leishmania donovani* promastigotes. J. Cell Sci. *114*, 2461–2469.

Desjeux, P. (1999). Global control and *Leishmania* HIV co-infection. Clin. Dermatol. *17*, 317–325.

Dinos, G.P., Michelinaki, M., and Kalpaxis, D.L. (2001). Insights into the mechanism of azithromycin interaction with an *Escherichia coli* functional ribosomal complex. Mol. Pharmacol. *59*, 1441–1445.

Dolai, S., Pal, S., Yadav, R.K., and Adak, S. (2011). Endoplasmic reticulum stress-induced apoptosis in *Leishmania* through Ca2+-dependent and caspase-independent mechanism. J. Biol. Chem. *286*, 13638–13646.

Duszenko, M., Figarella, K., Macleod, E.T., and Welburn, S.C. (2006). Death of a trypanosome: a selfish altruism. Trends Parasitol. *22*, 536–542.

Engelmann, J., Henke, J., Willker, W., Kutscher, B., Nossner, G., Engel, J., and Leibfritz, D. (1996). Early stage monitoring of miltefosine induced apoptosis in KB-cells by multinuclear NMR spectroscopy. Anticancer Res. *16*, 1429–1439.

Escobar, P., Matu, S., Marques, C., and Croft, S.L. (2002). Sensitivities of *Leishmania* species to hexadecylphosphocholine (miltefosine), ET-18-OCH(3) (edelfosine) and amphotericin B. Acta Trop. *81*, 151–157.

de Freitas Balanco, J.M., Moreira, M.E., Bonomo, A., Bozza, P.T., Amarante-Mendes, G., Pirmez, C., and Barcinski, M.A. (2001). Apoptotic mimicry by an obligate intracellular parasite down-regulates macrophage microbicidal activity. Curr. Biol. *11*, 1870–1873.

Gannavaram, S., and Debrabant, A. (2012a). Involvement of TatD nuclease during programmed cell death in the protozoan parasite *Trypanosoma brucei*. Mol. Microbiol. *83*, 926–935.

Gannavaram, S., and Debrabant, A. (2012b). Programmed cell death in *Leishmania*: biochemical evidence and role in parasite infectivity. Front Cell Infect. Microbiol. *2*, 95.

Gannavaram, S., Vedvyas, C., and Debrabant, A. (2008). Conservation of the pro-apoptotic nuclease activity of endonuclease G in unicellular trypanosomatid parasites. J. Cell Sci. *121*, 99–109.

Ghannoum, M.A., and Rice, L.B. (1999). Antifungal agents: mode of action, mechanisms of resistance, and correlation of these mechanisms with bacterial resistance. Clin. Microbiol. Rev. *12*, 501–517.

Gottlieb, R.A. (2000). Role of mitochondria in apoptosis. Crit Rev. Eukaryot. Gene Expr. *10*, 231–239.

Gottlieb, R.A. (2003). Mitochondrial signaling in apoptosis: mitochondrial daggers to the breaking heart. Basic Res. Cardiol. *98*, 242–249.

Hartsel, S., and Bolard, J. (1996). Amphotericin B: new life for an old drug. Trends Pharmacol. Sci. *17*, 445–449.

Henke, J., Engelmann, J., Kutscher, B., Nssner, G., Engel, J., Voegeli, R., and Leibfritz, D. (1999). Changes of intracellular calcium, fatty acids and phospholipids during miltefosine-induced apoptosis monitored by fluorescence- and 13C NMR-spectroscopy. Anticancer Res. *19*, 4027–4032.

Herman, M., Gillies, S., Michels, P.A., and Rigden, D.J. (2006). Autophagy and related processes in trypanosomatids: insights from genomic and bioinformatic analyses. Autophagy. *2*, 107–118.

Hildeman, D.A., Mitchell, T., Aronow, B., Wojciechowski, S., Kappler, J., and Marrack, P. (2003). Control of Bcl-2 expression by reactive oxygen species. Proc. Natl. Acad. Sci. U.S.A. *100*, 15035–15040.

Hommel, M. (1999). Visceral leishmaniasis: biology of the parasite. J. Infect. *39*, 101–111.

Huttemann, M., Pecina, P., Rainbolt, M., Sanderson, T.H., Kagan, V.E., Samavati, L., Doan, J.W., and Lee, I. (2011). The multiple functions of cytochrome c and their regulation in life and death decisions of the mammalian cell: From respiration to apoptosis. Mitochondrion. *11*, 369–381.

Istvan, E.S., and Deisenhofer, J. (2001). Structural mechanism for statin inhibition of HMG-CoA reductase. Science *292*, 1160–1164.

Jimenez-Ruiz, A., Alzate, J.F., Macleod, E.T., Luder, C.G., Fasel, N., and Hurd, H. (2010). Apoptotic markers in protozoan parasites. Parasit. Vectors. *3*, 104.

Kerr, J.F., Wyllie, A.H., and Currie, A.R. (1972). Apoptosis: a basic biological phenomenon with wide-ranging implications in tissue kinetics. Br. J. Cancer *26*, 239–257.

Kroemer, G. (2003). Mitochondrial control of apoptosis: an introduction. Biochem. Biophys. Res. Commun. *304*, 433–435.

Kroemer, G., Galluzzi, L., Vandenabeele, P., Abrams, J., Alnemri, E.S., Baehrecke, E.H., Blagosklonny, M.V., El-Deiry, W.S., Golstein, P., Green, D.R., *et al.* (2009). Classification of cell death: recommendations of the Nomenclature Committee on Cell Death 2009. Cell Death. Differ. *16*, 3–11.

Lainson, R., Ryan, L., and Shaw, J.J. (1987). Infective stages of *Leishmania* in the sandfly vector and some observations on the mechanism of transmission. Mem. Inst. Oswaldo Cruz *82*, 421–424.

Lange, B.M., Rujan, T., Martin, W., and Croteau, R. (2000). Isoprenoid biosynthesis: the evolution of two ancient and distinct pathways across genomes. Proc. Natl. Acad. Sci. U.S.A. *97*, 13172–13177.

Lee, N., Bertholet, S., Debrabant, A., Muller, J., Duncan, R., and Nakhasi, H.L. (2002). Programmed cell death in the unicellular protozoan parasite *Leishmania*. Cell Death. Differ. *9*, 53–64.

Levine, B., and Klionsky, D.J. (2004). Development by self-digestion: molecular mechanisms and biological functions of autophagy. Dev. Cell *6*, 463–477.

Li, L.Y., Luo, X., and Wang, X. (2001). Endonuclease G is an apoptotic DNase when released from mitochondria. Nature *412*, 95–99.

Lindoso, J.A., Cotrim, P.C., and Goto, H. (2004). Apoptosis of *Leishmania (Leishmania) chagasi* amastigotes in hamsters infected with visceral leishmaniasis. Int. J. Parasitol. *34*, 1–4.

Lira, R., Contreras, L.M., Rita, R.M., and Urbina, J.A. (2001). Mechanism of action of anti-proliferative lysophospholipid analogues against the protozoan parasite *Trypanosoma cruzi*: potentiation of *in vitro* activity by the sterol biosynthesis inhibitor ketoconazole. J. Antimicrob. Chemother. *47*, 537–546.

Lockshin, R.A., Facey, C.O., and Zakeri, Z. (2001). Cell death in the heart. Cardiol. Clin. *19*, 1–11.

Luder, C.G., Campos-Salinas, J., Gonzalez-Rey, E., and van Zandbergen, G. (2010). Impact of protozoan cell death on parasite–host interactions and pathogenesis. Parasit. Vectors. *3*, 116.

Lux, H., Hart, D.T., Parker, P.J., and Klenner, T. (1996). Ether lipid metabolism, GPI anchor biosynthesis, and signal transduction are putative targets for anti-leishmanial alkyl phospholipid analogues. Adv. Exp. Med. Biol. *416*, 201–211.

Lux, H., Heise, N., Klenner, T., Hart, D., and Opperdoes, F.R. (2000). Ether – lipid (alkyl-phospholipid) metabolism and the mechanism of action of ether – lipid analogues in *Leishmania*. Mol. Biochem. Parasitol. *111*, 1–14.

Maarouf, M., Lawrence, F., Brown, S., and Robert-Gero, M. (1997). Biochemical alterations in paromomycin-treated *Leishmania donovani* promastigotes. Parasitol. Res. *83*, 198–202.

McGwire, B.S., and Kulkarni, M.M. (2010). Interactions of antimicrobial peptides with *Leishmania* and trypanosomes and their functional role in host parasitism. Exp. Parasitol. *126*, 397–405.

Martin, S.J., Reutelingsperger, C.P., McGahon, A.J., Rader, J.A., van Schie, R.C., LaFace, D.M., and Green, D.R. (1995). Early redistribution of plasma membrane phosphatidylserine is a general feature of apoptosis regardless of the initiating stimulus: inhibition by overexpression of Bcl-2 and Abl. J. Exp. Med. *182*, 1545–1556.

Mehta, A., and Shaha, C. (2004). Apoptotic death in *Leishmania donovani* promastigotes in response to respiratory chain inhibition: complex II inhibition results in increased pentamidine cytotoxicity. J. Biol. Chem. *279*, 11798–11813.

Menna-Barreto, R.F., Correa, J.R., Pinto, A.V., Soares, M.J., and de Castro, S.L. (2007). Mitochondrial disruption and DNA fragmentation in *Trypanosoma cruzi* induced by naphthoimidazoles synthesized from beta-lapachone. Parasitol. Res. *101*, 895–905.

Menna-Barreto, R.F., Correa, J.R., Cascabulho, C.M., Fernandes, M.C., Pinto, A.V., Soares, M.J., and de Castro, S.L. (2009a). Naphthoimidazoles promote different death phenotypes in *Trypanosoma cruzi*. Parasitology *136*, 499–510.

Menna-Barreto, R.F., Salomao, K., Dantas, A.P., Santa-Rita, R.M., Soares, M.J., Barbosa, H.S., and de Castro, S.L. (2009b). Different cell death pathways induced by drugs in T*rypanosoma cruzi:* an ultrastructural study. Micron. *40*, 157–168.

Merkel, P., Beck, A., Muhammad, K., Ali, S.A., Schonfeld, C., Voelter, W., and Duszenko, M. (2007). Spermine isolated and identified as the major trypanocidal compound from the snake venom of Eristocophis macmahoni causes autophagy in *Trypanosoma brucei*. Toxicon *50*, 457–469.

Movassagh, M., and Foo, R.S. (2008). Simplified apoptotic cascades. Heart Fail. Rev. *13*, 111–119.

Mukherjee, S.B., Das, M., Sudhandiran, G., and Shaha, C. (2002). Increase in cytosolic Ca^{2+} levels through the activation of non-selective cation channels induced by oxidative stress causes mitochondrial depolarization leading to apoptosis-like death in *Leishmania donovani* promastigotes. J. Biol. Chem. *277*, 24717–24727.

Nicholson, D.W., and Thornberry, N.A. (1997). Caspases: killer proteases. Trends Biochem. Sci. *22*, 299–306.

Ouellette, M., Drummelsmith, J., and Papadopoulou, B. (2004). Leishmaniasis: drugs in the clinic, resistance and new developments. Drug Resist. Updat. *7*, 257–266.

Paris, C., Loiseau, P.M., Bories, C., and Breard, J. (2004). Miltefosine induces apoptosis-like death in *Leishmania donovani* promastigotes. Antimicrob. Agents Chemother. *48*, 852–859.

Patel, A.S., Lin, L., Geyer, A., Haspel, J.A., An, C.H., Cao, J., Rosas, I.O., and Morse, D. (2012). Autophagy in idiopathic pulmonary fibrosis. PLoS. One. *7*, e41394.

Pradelli, L.A., Beneteau, M., and Ricci, J.E. (2010). Mitochondrial control of caspase-dependent and -independent cell death. Cell Mol. Life Sci. *67*, 1589–1597.

Ribeiro, G.A., Cunha-Junior, E.F., Pinheiro, R.O., da-Silva, S.A., Canto-Cavalheiro, M.M., da Silva, A.J., Costa, P.R., Netto, C.D., Melo, R.C., Almeida-Amaral, E.E., and Torres-Santos, E.C. (2013). LQB-118, an orally active pterocarpanquinone, induces selective oxidative stress and apoptosis in *Leishmania amazonensis*. J. Antimicrob. Chemother. *68*, 789–799.

Santa-Rita, R.M., Santos, B.H., Meirelles, M.N., and de Castro, S.L. (2000). Effect of the alkyl-lysophospholipids on the proliferation and differentiation of *Trypanosoma cruzi*. Acta Trop. *75*, 219–228.

Sen, N., Das, B.B., Ganguly, A., Mukherjee, T., Bandyopadhyay, S., and Majumder, H.K. (2004). Camptothecin-induced imbalance in intracellular cation homeostasis regulates programmed cell death in unicellular hemoflagellate *Leishmania donovani*. J. Biol. Chem. *279*, 52366–52375.

Sereno, D., Holzmuller, P., Mangot, I., Cuny, G., Ouaissi, A., and Lemesre, J.L. (2001). Antimonial-mediated DNA fragmentation in *Leishmania infantum* amastigotes. Antimicrob. Agents Chemother. *45*, 2064–2069.

Shaked-Mishan, P., Ulrich, N., Ephros, M., and Zilberstein, D. (2001). Novel Intracellular SbV reducing activity correlates with antimony susceptibility in *Leishmania donovani*. J. Biol. Chem. *276*, 3971–3976.

Singh, M., Sharma, H., and Singh, N. (2007). Hydrogen peroxide induces apoptosis in HeLa cells through mitochondrial pathway. Mitochondrion. *7*, 367–373.

de Souza, W., and Rodrigues, J.C. (2009). Sterol biosynthesis pathway as target for anti-trypanosomatid drugs. Interdiscip. Perspect. Infect. Dis. *2009*, 642502.

Sudhandiran, G., and Shaha, C. (2003). Antimonial-induced increase in intracellular Ca2+ through non-selective cation channels in the host and the parasite is responsible for apoptosis of intracellular *Leishmania donovani* amastigotes. J. Biol. Chem. *278*, 25120–25132.

Sundar, S., and Rai, M. (2002). Laboratory diagnosis of visceral leishmaniasis. Clin. Diagn. Lab. Immunol. *9*, 951–958.

Weingartner, A., Kemmer, G., Muller, F.D., Zampieri, R.A., Gonzaga dos, S.M., Schiller, J., and Pomorski, T.G. (2012). *Leishmania* promastigotes lack phosphatidylserine but bind annexin V upon permeabilization or miltefosine treatment. PLoS. One. *7*, e42070.

Welburn, S.C., and Maudlin, I. (1999). Tsetse–trypanosome interactions: rites of passage. Parasitol. Today *15*, 399–403.

Welburn, S.C., Barcinski, M.A., and Williams, G.T. (1997). Programmed cell death in trypanosomatids. Parasitol. Today *13*, 22–26.

Williams, R.A., Mottram, J.C., and Coombs, G.H. (2013). Distinct roles in autophagy and importance in infectivity of the two ATG4 cysteine peptidases of *Leishmania major*. J. Biol. Chem. *288*, 3678–3690.

Wyllie, S., Cunningham, M.L., and Fairlamb, A.H. (2004). Dual action of antimonial drugs on thiol redox metabolism in the human pathogen *Leishmania donovani*. J. Biol. Chem. *279*, 39925–39932.

Zalila, H., Gonzalez, I.J., El-Fadili, A.K., Delgado, M.B., Desponds, C., Schaff, C., and Fasel, N. (2011). Processing of metacaspase into a cytoplasmic catalytic domain mediating cell death in *Leishmania major*. Mol. Microbiol. *79*, 222–239.

van Zandbergen, G., Klinger, M., Mueller, A., Dannenberg, S., Gebert, A., Solbach, W., and Laskay, T. (2004). Cutting edge: neutrophil granulocyte serves as a vector for *Leishmania* entry into macrophages. J. Immunol. *173*, 6521–6525.

van Zandbergen, G., Bollinger, A., Wenzel, A., Kamhawi, S., Voll, R., Klinger, M., Muller, A., Holscher, C., Herrmann, M., Sacks, D., *et al.* (2006). *Leishmania* disease development depends on the presence of apoptotic promastigotes in the virulent inoculum. Proc. Natl. Acad. Sci. U.S.A. *103*, 13837–13842.

Zangger, H., Mottram, J.C., and Fasel, N. (2002). Cell death in *Leishmania* induced by stress and differentiation: programmed cell death or necrosis? Cell Death. Differ. *9*, 1126–1139.

Zhou, W., Lepesheva, G.I., Waterman, M.R., and Nes, W.D. (2006). Mechanistic analysis of a multiple product sterol methyltransferase implicated in ergosterol biosynthesis in *Trypanosoma brucei*. J. Biol. Chem. *281*, 6290–6296.

Elucidating the Strategies of Immune Evasion by *Leishmania*

6

Supriya Srivastav, Anindita Ukil and Pijush K. Das

Abstract

The protozoa *Leishmania* species are obligate intracellular parasites that harbour in the macrophages of their host. In order to inhabit an intracellular niche, they have evolved various means to attenuate and/or subvert how their host cell integrates signals from the external 'immune' environment. Since macrophages being the first line of defence in the immune system are specialized for the identification and destruction of invading pathogens by triggering an innate immune response, *Leishmania* have evolved a range of mechanisms for suppressing some critical macrophage activities. This is achieved either by employing strategies to inhibit proteins that play a positive role in immune cell activation or by activation of molecules that govern the negative regulation of immune cell signalling and function. This chapter highlights the strategies employed by *Leishmania* to suppress macrophage defence mechanism to create a favourable niche which aids the parasite in intracellular growth and virulence.

Introduction

Leishmania is a genus of protozoan parasites that are transmitted by the bite of phlebotomine sandflies and give rise to a range of diseases (collectively known as leishmaniases) that affect over 150 million people worldwide (Kaye and Scott, 2011). In the mammalian host, *Leishmania* survives and multiplies in macrophages, which are the first line of defence against any invading pathogen. To establish a successful infection, *Leishmania* parasites must counter the immune responses evoked by macrophages. *Leishmania* invariably develop mechanisms to neutralize the macrophage microbicidal machinery and the outcome of infection depends on the balance between the host's ability to activate macrophage killing and the parasite's ability to escape or evade this host immune response (Gantt *et al.*, 2001). Intracellular survival of *Leishmania* parasites has additional pleiotropic effects on macrophages, adversely interfering with the function of these cells in innate and acquired immune response. This includes inhibition of microbicidal activity, inhibition of respiratory burst, prevention of apoptosis, impairment of antigen presentation and protective cytokine responses etc. Although identification of macrophages as host cells for the mammalian stage of *Leishmania* species traces back to many decades ago, several questions concerning the strategies by which these parasites establish themselves in macrophages, which are endowed with potent innate microbicidal mechanisms, are still to be answered. This chapter focuses on how *Leishmania* selectively manipulate diverse macrophage functions, including pathogen recognition, production of cytokines and other microbicidal molecules and other immune signalling pathways of the host to ensure their survival.

Defying recognition by the host: modulation of Toll-like receptor-mediated signalling

Recognition of a pathogen through various pathogen-associated molecular patterns (PAMPs) by macrophages triggers a diverse array of signalling cascades which initiate on the cell surface and ultimately culminate in the production of various microbicidal agents thereby providing immunity to infections (Thoma-Uszynski *et al.*, 2001; Diaz *et al.*, 2003). Pattern recognition receptors (PRRs) present on host cells are evolutionarily conserved proteins that are able to distinguish molecular patterns present on microorganisms but absent in the host. (Kolli *et al.*, 2013). In this context, there exists a role of Toll-like receptors (TLRs) present on the surface of macrophages, which upon countering a pathogen, elicit an array of signalling events resulting in the activation of NF-κB and production of proinflammatory cytokines which ultimately lead to the effective clearance of the parasite (Underhill and Ozinsky, 2002). *Leishmania* species present unusual challenges to the immune system with their capacity to down-regulate TLR-mediated inflammatory response and therefore enter and reside within macrophages by stealth (Cameron *et al.*, 2004; Olivier *et al.*, 2005; Tuon *et al.*, 2008). All the species of *Leishmania* possess a cell surface glycophospholipid LPG (lipophosphoglycan), which is a ligand for TLR2 (McConville *et al.*, 1995). *Leishmania* LPG has been reported to activate natural killer (NK) cells and monocytes through TLR2. Binding of *L. major* LPG to TLR2 initiates IFN-γ and TNF-α production and nuclear translocation of NF-κB in human NK cells (Becker *et al.*, 2003). *L. donovani* LPG has been shown to elicit IL-12 production in macrophages (Balaraman *et al.*, 2005; de Veer *et al.*, 2003). However, it is interesting to note that infection by *L. donovani* promastigotes suppresses proinflammatory response in macrophages despite the presence of LPG. Ligand recognition by TLR leads to the assembly of signalling complex consisting of MyD88, IRAK and TRAF (tumour necrosis factor receptor-associated factor) 6. TRAF6 is an E3 ubiquitin ligase, which in association with an ubiquitin editing enzyme complex causes the addition of Lys 63-linked polyubiquitin chains (Megas *et al.*, 2011). These chains are recognized by TAB2, which functions as an adaptor linking TGF-β-associated kinase (TAK) 1 to TRAF6, thus facilitating TAK1 activation and subsequent NF-κB-triggered proinflammatory response (Kanayama *et al.*, 2004). However, *L. donovani* causes discrepancy in TLR signalling by interfering with the proper assembly of these signalling complexes which may be attributed to the de-ubiquitination of TRAF6 in *L. donovani*-infected macrophages brought about by host de-ubiquitinating enzyme A20 (Srivastav *et al.*, 2012). A recent study suggests the role of TRAF3 in impairing TLR4-mediated host response by *L. donovani*. It has been demonstrated that *Leishmania* alters the ubiquitination pattern of TRAF3 thereby leading to the defective cytosolic translocation of the TLR4-anchored multiprotein complex (Gupta *et al.*, 2014).

Hence, In the light of the ongoing host–*Leishmania* interaction, the implication of *Leishmania*-triggered TLR response depends upon the ability of cells of the immune system to eliminate the parasite versus the ability of parasites to counteract TLR detection by interfering with TLR signalling cascade thereby rendering them refractory to subsequent TLR stimulation. Subverting the TLR signalling cascade allows the parasite to successfully enter the host macrophages and avoid the otherwise pathogen-recognition-triggered immune response.

Deactivation of host signalling: mitogen-activated protein (MAP) kinases and phosphatases

Mitogen-activated protein kinases (MAPKs) are an important group of serine/threonine kinases that through phosphorylation, and hence activation status of transcription factors, link transmembrane signalling with gene activation in the nucleus (Su and Karin, 1996). MAPK family includes extracellular signal-related kinase 1 and 2 (ERK1/2), p38 and c-jun N-terminal kinase (JNK). The phosphorylation of these MAP kinases is mediated by seven MAPK kinases (MAPKKs) that have specificity for individual MAPK isoforms. Thus, ERK1 and ERK2 are activated by MEK1 and MEK2, ERK5 is activated by MEK5, JNK is activated by MKK4 and MKK7, and p38 is activated by MKK3 and MKK6 (Kyriakis and Avruch, 2001), (Awasthi *et al.*, 2003). *Leishmania*-induced macrophage deactivation during intracellular infection may be linked to defects in the signalling pathway, which in turn facilitate the parasite's infection and propagation within the cell (Reiner *et al.*, 1994). The antimicrobial state of macrophages is dependent on the appropriate phosphorylation of MAP kinases, and there are several lines of evidence that *Leishmania* interferes with signalling transduction in macrophages (Nandan and Reiner, 1995). In naive bone marrow-derived macrophages, *Leishmania* parasites fail to induce the phosphorylation of p38, JNK and ERK1/2 and there is a down-modulation of basal p38 phosphorylation levels occurring within the first 30 min of infection (Prive and Descoteaux, 2000). They further implicated the important role of *Leishmania* LPG in MAPK inactivation. *Leishmania*-induced ceramide generation was linked with inactivation of ERK1/2 which subsequently led to inhibition of AP-1 and NF-κB-dependent NO production during infection (Ghosh *et al.*, 2002). Cystatin, a small molecular protein, could also very well suppress *Leishmania* infection through p105-tumour progression locus 2 (Tpl2)-ERK mediated NF-κB activation (Kar *et al.*, 2009). The activation of p38 MAPK pathway to resolve infection has also been documented by several researchers (Awasthi *et al.*, 2003; Ukil *et al.*, 2011). An independent study showed that specific inhibition of p38 MAPK increased *L. donovani* survival in human peripheral blood mononuclear macrophages (Ono and Han, 2000). This notion was corroborated by the observation that treatment of macrophages with anisomycin, an activator of p38 reduces the survival of *L. donovani* (Junghae and Raynes, 2002) and 18β-glycyrrhetinic (GRA), an herbal immunomodulator, could act as an efficient anti-leishmanial compound by activating p38-mediated NF-κB activation (Ukil *et al.*, 2005, 2011). However, some contradictory findings are also there. Chandra and Naik (2008) had demonstrated an increase in ERK1/2 phosphorylation in *L. donovani*-infected THP-1 cells and human and Feng *et al.* (1999) have shown similar findings in J774 macrophage cells. These could be due to discrepancies between cell lines, as discussed in the review (Shadab and Ali, 2011) that differences in the cell lines contribute to the contradictory findings in *L. donovani* infection. Apart from this, strain difference could be another factor as *L. major*, causative agent of cutaneous leishmaniasis, is known to increase ERK phosphorylation whereas it impairs p38 phosphorylation (Mathur *et al.*, 2004). Infection with *L. amazonensis* has also been shown to activate macrophage ERK (Yang *et al.*, 2007). As the general term leishmaniasis includes a wide spectrum of diseases varying from self healing cutaneous to fatal visceral infection, therefore it is desirable to consider the MAPK activation status for a particular type of infection rather than generalize it. MAP kinases are primarily inactivated by a group of dual-specificity protein phosphatases by dephosphorylation of the critical tyrosine and threonine residues of activated MAP kinases (Kondoh and Nishida, 2007). In this capacity, this group of protein phosphatases may serve as pivotal feedback control regulators in the innate immune response during microbial infection, and several studies have shown that host tyrosine phosphatases, specially SHP-1 (Src homology 2 domain-containing tyrosine phosphatase-1) are up-regulated during *L. donovani* infection (Forget *et al.*, 2006; Abu-Dayyeh *et al.*, 2008). *L. donovani* infection markedly up-regulated the expression of three phosphatases, MAP kinase phosphatase (MKP)1, MKP3 and protein phosphatase 2A (PP2A). Inhibition of these phosphatases prior to infection leads to preferential induction of Th2 response through deactivation of p38 by MKP1. On the contrary, MKP3 and PP2A might play important roles in the inhibition of inducible nitric oxide synthase (iNOS) expression through deactivation of ERK1/2. From all these observations it may be hypothesized that the fate of infection depends on cellular kinase/phosphatase balance, rather than either kinases or phosphatases individually.

Disarming the enemy: inhibition of host microbicidal molecules

One of the dangers encountered by *Leishmania* upon entering macrophages is the ability of these cells to produce microbicidal agents which include deadly free radicals such as reactive nitrogen intermediates (RNIs) and reactive oxygen species (ROS). Microbial phagocytosis triggers production of ROS/RNI which function in microbial killing as well as antigen processing and presentation (Horta *et al.*, 2012). Although *Leishmania* species are susceptible to exogenous superoxide radical, hydrogen peroxide, nitric oxide, and peroxynitrite *in vitro*, they manage to survive the endogenous oxidative burst during phagocytosis and the subsequent elevated NO production in the macrophage.

NO derived from L-arginine by the catalytic action of inducible NO synthase (iNOS) plays an important role in parasite killing. Macrophages express high levels of iNOS when activated by a number of immunological stimuli which include interferon-gamma (IFN-γ), tumour necrosis factor alpha (TNF-α) and lipopolysaccharide (LPS) (Segal, 2008), (Ellson *et al.*, 2006). Mice with a disrupted iNOS gene are highly susceptible to *L. major* infection and develop a strong Th1-cell response compared to control mice, suggesting that NO is likely to be the effector molecule in controlling intracellular parasitic infections. NO is produced by nitric oxide synthase (NOS) which converts one of the terminal nitrogens of the guanidino group of L-arginine to NO, producing L-citrulline (Stuehr, 1999). To ensure their survival, the *Leishmania* parasites have evolved effective means to inhibit NO synthesis (Proudfoot *et al.*, 1995). A mechanism involved in the down-regulation of NO production is by conversion of arginine to ornithine and urea via the arginase pathway. In support of this mechanism, recently, it has been shown that arginase, as well as expression of polyamine genes are up-regulated by *L. amazonensis* amastigotes. Major surface glycolipids, glycoinositol-phospholipids and lipophosphoglycan (LPG), of *Leishmania* are potent inhibitors of NO synthesis. LPG can inhibit IL-12 synthesis, thereby indirectly blocking the induction of iNOS. *Leishmania* has the ability to rapidly activate protein tyrosine phosphatases (PTPs) in infected macrophages and thus can interfere with several molecules involved in NO production including JAK2, ERK1/2, and the transcription factors NF-κB and AP-1 (Kar *et al.*, 2009, 2010).

The generation of ROS by macrophages, also known as respiratory burst is a crucial event in parasite killing, which depends on the assembling of NADPH oxidase complex to produce superoxide anion. This complex contains the membrane-bound cytochrome b558, which is composed of at least two polypeptides ($gp91^{phox}$ and $p22^{phox}$) and two non-identical haem groups that are associated with $gp91^{phox}$ (Nauseef, 2004). Four additional components of the NADPH oxidase enzyme, p40, p47, p67 and Rac2 are mostly found in the cytosol and associate with the membrane-bound components upon activation (Nauseef, 2004). Assembling of NADPH oxidase complex is achieved by translocation of cytosolic components of NADPH oxidase complex to the plasma membrane which depends on intracellular calcium, IP3-mediated signalling, and protein kinase C (PKC) activation (Iles and Forman, 2002; Hoyal *et al.*, 1998). PKC activation results in the phosphorylation of the p47 subunit of the NADPH oxidase, thereby resulting in oxidase activation. This utilizes cytosolic NADPH to reduce extracellular O_2 to $O_2{\cdot}^-$ and the superoxide anion thus produced is enzymatically converted to hydrogen peroxide by superoxide dismutase (SOD). Hydrogen peroxide generated can serve as a precursor for hydroxyl radical (·OH) generation. Superoxide anion can react with several cellular molecules, including DNA, thus causing DNA damage.

Interestingly, LPG of *L. donovani* promastigotes is able to block NADPH oxidase assembly at the phagosome membrane. *L. donovani* amastigotes can block superoxide release by inhibiting the phosphorylation of the NADPH oxidase component p47(phox), leading to defective recruitment of p47(phox) and p67(phox) to the phagosome (Lodge *et al.*, 2006). Several studies have suggested the involvement of surface molecules LPG and gp63 of *Leishmania* and abnormal protein kinase C (PKC) activity in suppressing the ROS generation by the host macrophages (Olivier *et al.*, 1992; Descoteaux and Turco, 1999). Uncoupling proteins (UCPs) belong to the family of transporters present in the inner membrane of mitochondria and are the major regulator of ROS generation in the macrophage (Negre-Salvayre *et al.*, 1997). Several studies have emphasized the role of mitochondrial inner membrane UCPs in regulating the mitochondrial ROS generation in diverse cellular contexts and disease conditions. Mitochondrial uncoupling proteins are associated with mitochondrial ROS generation, which is the major contributor of total cellular ROS generation. *L. donovani* infection is associated with strong up-regulation of uncoupling protein

2 (UCP2), a negative regulator of mitochondrial ROS generation located at the inner membrane of mitochondria, possibly preventing ROS-mediated PTP inactivation to suppress macrophage defence mechanisms (Basu Ball *et al.*, 2011).

Delaying phagosome maturation: buying time for promastigote to amastigote conversion

Once internalized, microorganisms are sequestered in phagosomes, which play a key role in macrophages' innate ability to restrict the spread of infectious diseases (Desjardins and Griffiths, 2003). *Leishmania* engage a variety of receptors (CR3, CR1, and Fc-γR) on the surface of host cells and are internalized by the conventional phagocytic pathway (Antoine *et al.*, 2004). The newly formed phagosomes subsequently mature as a result of highly regulated fusion and fission events with early and late endosomes and lysosomes resulting in the formation of an acidified parasitophorous vacuoles containing the markers of a mature phagolysosome (Muraille *et al.*, 2010). Phagosomes are initially inapt at killing and degrading pathogens and acquire their microbicidal properties through a complex maturation process, involving sequential fusion with endocytic organelles, leading to the formation of phagolysosomes (McConville and Naderer, 2011). Through a series of evolutionary processes, pathogens have developed strategies to adapt and survive in their hosts by avoiding the harsh environment of phagolysosomes. The intracellular pathogen *Leishmania* relies on the use of its cell surface LPG to alter the functional properties of the host phagosomes where they reside to survive in mammalian macrophages (Dermine *et al.*, 2005). *Leishmania* LPG has been reported to inhibit the ability of phagosomes to fuse with late endocytic organelles and lysosomes. *Leishmania* promastigotes use LPG to directly alter the formation of phagosome lipid microdomains and/or disorganize these structures after phagocytosis. This process is likely to be beneficial for the parasite because phagosome microdomains contain proteins involved in important aspects of phagolysosome biogenesis. Unlike $LPG^{-}/^{-}$ mutants, wild-type *L. donovani* reside within phagosomes not displaying flotillin-1-enriched microdomains, suggesting that the parasite might target these structures for its survival in macrophages. Since microdomains have been shown to play key roles in various cellular functions, including signal transduction, targeting of these structures in phagosomes by pathogens might confer a survival advantage (Dermine *et al.*, 2001; Simons and Vaz, 2004). When promastigotes attach to the macrophage surface, LPG is transferred from the parasite to the outer leaflet of the macrophage plasma membrane and, upon internalization, to the inner leaflet of the phagosomal membrane. Insertion of LPG in one leaflet of a membrane bilayer alters the biophysical properties of the entire membrane, making it less fusogenic (Kaneshiro *et al.*, 1982), (Denny *et al.*, 2001). Inhibition of phagolysosome formation is characterized by a defective recruitment of rab7 to the phagosome (Vieira *et al.*, 2003). Delaying phagolysosome maturation gives the engulfed promastigotes time to evolve into amastigotes which are resistant to lysosomal enzymes and the acidic pH inside the phagolysosome (Holm *et al.*, 2001).

Creating a favourable environment: polarization of Th (T-helper cell) response

When *Leishmania* parasites, in the form of the flagellated promastigotes, infect an individual, they are phagocytosed into the macrophages. Depending on the genetic constitution of the host, there can be preferential induction of IL-12, which drives the development of Th1 cells (Alexander and Brombacher, 2012). Early studies particularly on *L. major* largely defined the Th1/Th2 paradigm of resistance/susceptibility to infection and the role of IL-12 and IL-4 respectively in driving Th1 and Th2 cell development (Proudfoot *et al.*, 1995). The Th1 cells produce IFN-γ which activates macrophages to produce NO, and this kills the parasites thereby leading to disease resistance (Huang *et al.*, 1998). On the contrary, in susceptible individuals, there is preferential induction of IL-4 which triggers the differentiation of Th2 cells which produce more IL-4 that can inhibit macrophage activation and the synthesis of NO. Hence, polarization of Th responses to either Th1 or Th2 may lead to life or death consequences in *Leishmania* infections as balance of pro and anti-inflammatory cytokines released from macrophages and other immune cells is an important mediator in the establishment of infection. The control of *Leishmania* infection requires generation of a strong Th1 response, which is driven by the presence of IL-12 at the initiation of the immune response. The cytokine IL-12 plays a critical role in the regulation of cellular immune responses. A growing body of evidences support the host protective role of IL-12-induced differentiation of Th1 cells that produce IFN-γ, thereby triggering NO production in macrophages (Ukil *et al.*, 2005; Diefenbach *et al.*, 1999). However, infection of macrophages with either the amastigote or promastigote forms of *Leishmania* does not activate macrophages to produce IL-12 (Cameron *et al.*, 2004). IL-12 inhibition has been also reported to occur in *L. major*-infected mice (Belkaid *et al.*, 1998). The intracellular mechanism is still unclear. Macrophage complement receptors and Fcγ receptor, which are known to interact with *Leishmania* during phagocytosis, have been shown to repress IL-12. Recently, the IL-12 related cytokine IL-27 has been shown to counter-regulate the early disease execratory effects of IL-4 in this mouse strain during *L. major* infection by promoting a Th1 response. In the absence of WSX-1 signalling early lesion growth is greatly exacerbated although healing under the influence of IL-12 occurs late in infection (Alexander and Brombacher, 2012).

Moreover, *Leishmania* prevents the activation of an effective immune response by inhibiting production of a number of cytokines, particularly those involved in the inflammatory response (IL-1) and tumour necrosis factor alpha (TNF-α) or in T-lymphocyte activation (IL-12) (Reiner *et al.*, 1987, 1990). For instance, LPS-induced IL-1β secretion is inhibited in *L. donovani*-infected and LPG-exposed macrophages (Olivier *et al.*, 2005). The inhibition of TNF-α production in infected macrophages may be attributed to IL-10 and PKC inhibition (Bhattacharyya *et al.*, 2001). The capacity of *Leishmania* to infect macrophages without inducing proinflammatory cytokines, and then to inhibit their induction in response to various agonists, probably represents a survival mechanism whereby the parasites can inhibit a harmful inflammatory reaction.

Th2 cytokines promote survival of the parasite by down-regulating host-mediated oxidative and inflammatory pathways. Pathogenesis and failure to check the proliferation of the intracellular parasites in leishmaniasis has been ascribed to polarized Th2 response, but the precise mechanism resulting in the inability to control disease progression is not very well documented. Several studies indicate that amastigote interactions with mammalian cells promote the production of anti-inflammatory cytokines such as IL-10 and TGF-β while suppressing the production of IL-12, superoxide and nitric oxide (Kima, 2007). In case of murine visceral leishmaniasis Th1 response is suppressed at least in part by TGF-β and IL-10 without Th2 cytokine production (Wilson *et al.*, 2005). This contrasts with the cutaneous species *L. major*, in which a Th2 response suppresses Th1 cytokines and leads to murine disease progression. IL-10 is an anti-inflammatory cytokine that inhibits both antigen presentation and subsequent release of proinflammatory cytokines. It promotes the Th2 immune response by down-regulating IL-12 production and plays a major role in disease susceptibility during *Leishmania* infection (Khan *et al.*, 2002). An inactivation of IL-10 in mice results in an increased production of IL-12 and IFN-γ (Trinchieri, 2007). Recent report suggests that induction of arginase 1 by an IL-10-mediated STAT3-dependent pathway activating IL-4Ra expression could be a potential mechanism by which the pathogen *L. donovani* escapes the host immune response (Biswas *et al.*, 2011). When these parasites are internalized, the macrophages are induced to secrete IL-10 and TGF-β, which in turn block the induction of iNOS, therefore inhibiting the production of NO. It is quite likely that this strategy for parasite uptake can contribute significantly to the persistence of the infection. A recent study reported that infection with *L. amazonensis* amastigotes results in the activation of the MAP kinase ERK1/2. These authors showed that activation of ERK1/2 was required for the parasite-induced secretion of IL-10 by the infected macrophage (Yang *et al.*, 2007).

Suppression of antigen presentation: escaping cytotoxic T-cell response

One of the many roles of macrophages in the immune system includes acting as antigen-presenting cells (APCs). However, despite the role of macrophages in antigen presentation, many intracellular pathogens, such as *Leishmania* seek residence in macrophages. Although dendritic cells participate in the initial priming of the anti-leishmanial immune response and may sustain short-term survival of *Leishmania* parasites, macrophages are the major site for parasite replication (Meier *et al.*, 2003). Upon activation, infected macrophages mobilize potent microbicidal mechanisms that eliminate the intracellular pathogen. This transition from a resting to an activated state is mediated by the interaction with specific T-cells that recognize pathogen-derived peptides complexed to major histocompatibility complex (MHC) molecules at the surface of host cells (Overath and Aebischer, 1999). Following the presentation of antigens via MHC class I or class II molecules expressed at the macrophage plasma membrane, microbicidal activities of macrophages can be enhanced or induced by effector T-lymphocytes (Antoine *et al.*, 2004). However, *Leishmania* have evolved mechanisms to evade or to interfere with antigen presentation processes, allowing parasites to partially resist these T-cell-mediated immune responses. Consequently, the ability of macrophages to process and present *Leishmania* antigens is necessary for their efficient interaction with effector T-cells and the focused delivery of cytokines which enhance leishmanicidal activity (Meier *et al.*, 2003). A number of studies have, therefore, investigated the antigen-presenting potential of *Leishmania*-infected macrophages.

Several reports have linked defective antigen presentation to increased membrane fluidity and it has been demonstrated that the defective antigen presentation could be corrected by a single injection of cholesterol liposome in infected hamster (Banerjee *et al.*, 2009). Serum cholesterol content is found to be reduced in patients of visceral leishmaniasis at the active stage of the disease and more interestingly bears an inverse correlation to splenic parasite load (Ghosh *et al.*, 2014). Membrane cholesterol content is also reported to be reduced in infected condition resulting in the disruption of membrane raft. A recent study demonstrates that *L. donovani* infection down-regulates microRNA (miR)-122 and genes involved in cholesterol biosynthesis in infected mouse livers to reduce serum cholesterol (Ghosh *et al.*, 2013). Physiological stimulation of T-cells by antigenic peptides presented in the context of MHC molecules requires a stable contact between T-cells and APCs to enable the generation of efficient activation signals. Recent studies on the membrane features that affect presentation of antigens by APCs to T-cells indicates that lipid rafts in the outer leaflet of the membrane lipid bilayer induce clustering of the peptide–MHC complex on the surface of APCs. Membrane lipid raft are micro-domains consisting of cholesterol and sphingolipid which acts as important pathogen portal. Disruption of lipid rafts leads to faulty antigen presentation.

The processing of exogenous antigens for cross-presentation within phagosomes is a highly complex process that requires limited proteolytic activities regulated in part by the intraphagosomal pH and the levels of hydrolytic enzymes. Phagolysosomes contribute to host immunity by linking microbial killing within these organelles with antigen processing for presentation on MHC class I or II molecules to T-cells. An intracellular pathogen can directly target and cleave components of the membrane fusion machinery of its host cells to inhibit antigen presentation. The *Leishmania* cell surface metalloprotease gp63 cleaves a subset of soluble N-ethylmaleimide-sensitive factor attachment protein receptors (SNAREs), including VAMP8 (Matheoud *et al.*, 2013). gp63-mediated VAMP8 inactivation or disruption prevents the NADPH oxidase complex from assembling on phagosomes, thus altering their pH and degradative properties. Consequently, the presentation of exogenous *Leishmania* antigens on MHC class I molecules, also known as cross-presentation, is inhibited, resulting in reduced T-cell activation. Recent findings indicated that SNAP-23 regulates the recruitment of gp91phox to phagosomes (Sakurai *et al.*, 2012). Certain reports indicate that VAMP8 is required for the early recruitment of gp91phox to phagosomes, which is crucial for the establishment of the conditions required for the processing of exogenous antigens for cross-presentation. The observations that gp63-mediated cleavage of VAMP8 (and other SNAREs) and the consequential inhibition of cross-presentation are specific to the promastigote stage indicate that these events are limited to the early phase of the establishment of *Leishmania* infection. These findings indicate that *Leishmania* subverts immune recognition by altering phagosome function and highlight the importance of VAMP8 in phagosome biogenesis and antigen cross-presentation.

Protection of niche: inhibition of macrophage apoptosis by *Leishmania* parasites

Programmed cell death or apoptosis, is a signal-dependent physiological suicide mechanism that preserves homeostasis by maintaining the delicate balance between cell proliferation and cell death (Carratelli *et al.*, 2002). However, apart from serving all these diverse spectrum of functions, it serves as a defence mechanism against viruses and probably other infectious agents, such as intracellular bacteria and parasites (Vaux *et al.*, 1994). Conversely, as pathogens are under great selective pressure to defeat the host defence systems, they have evolved a variety of ways to specifically antagonize apoptotic death of the invaded host cell, allowing them more time to replicate (Heussler *et al.*, 2001). For the successful survival of intracellular pathogens, protection of their niche, i.e. the host cell is a necessary pre-requisite. As apoptosis of infected cells is one of the classical defence mechanisms resulting in the elimination of the host cell along with the pathogen, many pathogens including *Leishmania* escape immune surveillance by developing mechanisms to suppress host cell apoptosis (Gao and Kwaik, 2000; Lisi *et al.*, 2005).

Apoptotic processes occur through different pathways, but all finally activate effector caspase-3 and -7, triggering disassembly of the nucleus and other organelles and encapsulation of the products in membrane-bound apoptotic bodies (Mallat and Tedgui, 2000). The extrinsic pathway of apoptosis is triggered by binding of cell surface receptors, like CD95 [Fas/Apo1, tumour necrosis factor receptor superfamily member 6 (TNFRSF6)] or CD120 [(tumour necrosis factor receptor superfamily (TNFRS)], by FasL (Fas ligand, TNF superfamily member 6) or TNF-α (tumour necrosis factor-α), followed by caspase-8 activation. The intrinsic pathway, on the other hand, initiated by DNA damage, involves the release of cytochrome c and other mitochondrial components into the cytosol, due to loss of integrity of the mitochondrial outer membrane and caspase-9 activation. Activation of the caspase cascade, observed in apoptotic process, leads to cleavage of a variety of target proteins including poly (ADP-ribose) polymerase (PARP), nuclear lamins, and protein kinase C (PKC), triggering disassembly of the cell (Goebel *et al.*, 2001; Kaufmann *et al.*, 1993). PKC isozymes play an important role in regulating cell survival and cell death during intracellular microbial infectious (Olivier *et al.*, 1992).

L. donovani was the first parasite reported to enhance host cell viability by inhibiting growth factor deprivation-induced apoptosis (Moore and Matlashewski, 1994). One potential mechanism behind this inhibition has been through the activation of NF-κB and PI-3 Kinase/Akt pathways (Ruhland *et al.*, 2007). In this context, it is interesting to note that *L. donovani* infection leads to an inhibition of PKC-mediated c-fos gene expression in macrophages (Descoteaux *et al.*, 1992; Moore *et al.*, 1993). Since the transcription factor c-fos triggers the expression of pro-apoptotic molecules, impairment of its expression might prevent apoptosis. It is of interest to note that though infected cells are capable of ROS production during early hours, there is complete abrogation of the downstream caspase cascade leading to oxidative stress-mediated apoptosis. In this context it has been reported that *L. donovani* employs differential induction of host SOCS proteins to subvert macrophage apoptotic machinery triggered by parasite internalization-mediated oxidative burst, thus establishing its replicative niche inside the host (Srivastav *et al.*, 2013). SOCS proteins play an important role in stabilizing the survival machinery of infected cells in the course of phagocytosis and their down-regulation leads to increased cell death and diminished persistence of infection.

Conclusion

In order to inhabit the phagocytes, intracellular pathogens have developed strategies permitting their rapid physiological adaptation in the immunologically hostile environment, of their host cells. Amastigote survival within infected cells is undoubtedly dependent on the outcome of the parasite's interaction with the host cell at multiple points such as escape from first-line defence systems, and the capacity to manipulate several macrophage functions (Fig. 6.1).

In the early stages of infection, these parasites must either avoid inappropriate activation of their host cell or elaborate mechanisms that inactivate anti-parasitic effectors at the time of infection. Further, parasites have to reduce proinflammatory signals such as cytokines or chemokines emanating from the infected cell, which can promote anti-leishmanial responses in their vicinity. Moreover, it would be prudent for the parasite to prevent its own molecules from initiating antigen presentation pathways and therefore limit T-cell recognition of the infected cell. Finally, parasites must promote survival of the infected cell in a milieu that is rich with immunological activity. In learning to evade innate host defences,

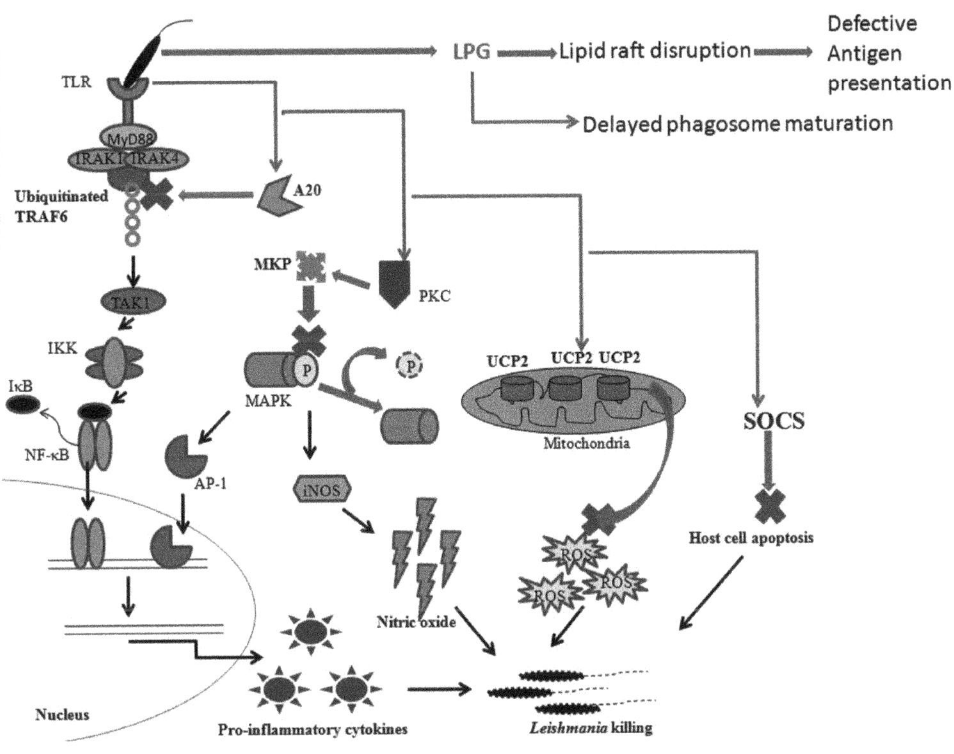

Figure 6.1 Modulation of macrophage immune-defensive signalling cascades by *Leishmania*: attachment of *Leishmania* parasites on macrophage surface results in interfering with TLR-mediated proinflammatory response through de-ubiquitinating enzyme A20 which disrupts the upstream TRAF6-TAK-TAB assembly thereby inhibiting NF-κB activation. This allows the parasite to silently enter macrophages without being recognized by pathogen recognition receptors. The *Leishmania* surface molecule LPG depletes membrane cholesterol hence disrupting lipid rafts and suppressing antigen presentation. This molecule also leads to delayed phagosome maturation thus allowing the promastigotes more time to convert to amastigotes which are better adapted to survive in the harsh acidic environment of the phagolysosome. Upon entering macrophages *Leishmania* activate a number of host molecules like PKC, MAPK and UCP2 which help the pathogen to inhibit harmful microbicidal molecules like nitric oxide and ROS. Further, activation of host SOCS proteins help the parasite to suppress macrophage apoptosis, thereby protecting its niche for survival. Normal macrophage signalling cascades are shown with black arrows. Red arrows indicate the modulations caused by *Leishmania* parasites or their surface molecule LPG. A colour version of this figure is available in the plate section at the back of the book.

protozoan parasites appear to have mastered the intricacies of both cell biology and cellular immunology. However, our current understanding of the mechanisms that *Leishmania* parasites deploy to modulate these processes is incomplete. The evidence that is currently available suggests that *Leishmania* parasites selectively engage and exploit host cell processes from the point of parasite entry and promote the release of anti-inflammatory cytokines.

Reviewing the available literature that falls within the scope of this study, we have tried to gain insight into some of the mechanisms by which *Leishmania* modulates signalling pathways of the host macrophages. Such strategies of host immune response modulation may be true for a number of other intracellular pathogens, which have also evolved means to inhibit or attenuate immune cell functions, contributing to their survival and the development of disease progression. Further, the mechanistic detail of this study is likely to be of great importance in explaining the widely different clinical manifestations of leishmaniasis. Since signalling cascades are vulnerable to pharmacological manipulation, a sound knowledge of the modulation of host cell functions by *Leishmania* may permit the development of new therapies to control a diverse spectrum of infectious agents.

References

Abu-Dayyeh, I., Shio, M.T., Sato, S., Akira, S., Cousineau, B., and Olivier, M. (2008). *Leishmania*-induced IRAK-1 inactivation is mediated by SHP-1 interacting with an evolutionarily conserved KTIM motif. PLoS Negl. Trop. Dis. *2*, e305.

Alexander, J., and Brombacher, F. (2012). T-helper1/T-helper2 cells and resistance/susceptibility to *Leishmania* infection: is this paradigm still relevant? Front. Immunol. *3*, 80.

Antoine, J.C., Prina, E., Courret, N., and Lang, T. (2004). *Leishmania* spp.: on the interactions they establish with antigen-presenting cells of their mammalian hosts. Adv. Parasitol. *58*, 1–68.

Awasthi, A., Mathur, R., Khan, A., Joshi, B.N., Jain, N., Sawant, S., Boppana, R., Mitra, D., and Saha, B. (2003). CD40 signaling is impaired in *L. major*-infected macrophages and is rescued by a p38MAPK activator establishing a host-protective memory T-cell response. J. Exp. Med. *197*, 1037–1043.

Balaraman, S., Singh, V.K., Tewary, P., and Madhubala, R. (2005). *Leishmania* lipophosphoglycan activates the transcription factor activating protein 1 in J774A.1 macrophages through the extracellular signal-related kinase (ERK) and p38 mitogen-activated protein kinase. Mol. Biochem. Parasitol. *139*, 117–127.

Banerjee, S., Ghosh, J., Sen, S., Guha, R., Dhar, R., Ghosh, M., Datta, S., Raychaudhury, B., Naskar, K., Haldar, A.K., *et al.* (2009). Designing therapies against experimental visceral leishmaniasis by modulating the membrane fluidity of antigen-presenting cells. Infect. Immun. *77*, 2330–2342.

Basu Ball, W., Kar, S., Mukherjee, M., Chande, A.G., Mukhopadhyaya, R., and Das, P.K. (2011). Uncoupling protein 2 negatively regulates mitochondrial reactive oxygen species generation and induces phosphatase-mediated anti-inflammatory response in experimental visceral leishmaniasis. J. Immunol. *187*, 1322–1332.

Becker, I., Salaiza, N., Aguirre, M., Delgado, J., Carrillo-Carrasco, N., Kobeh, L.G., Ruiz, A., Cervantes, R., Torres, A.P., Cabrera, N., *et al.* (2003). *Leishmania* lipophosphoglycan (LPG) activates NK cells through Toll-like receptor-2. Mol. Biochem. Parasitol. *130*, 65–74.

Belkaid, Y., Butcher, B., and Sacks, D.L. (1998). Analysis of cytokine production by inflammatory mouse macrophages at the single-cell level: selective impairment of IL-12 induction in *Leishmania*-infected cells. Eur. J. Immunol. *28*, 1389–1400.

Bhattacharyya, S., Ghosh, S., Jhonson, P.L., Bhattacharya, S.K., and Majumdar, S. (2001). Immunomodulatory role of interleukin-10 in visceral leishmaniasis: defective activation of protein kinase C-mediated signal transduction events. Infect. Immun. *69*, 1499–1507.

Biswas, A., Bhattacharya, A., Kar, S., and Das, P.K. (2011). Expression of IL-10-triggered STAT3-dependent IL-4Ralpha is required for induction of arginase 1 in visceral leishmaniasis. Eur. J. Immunol. *41*, 992–1003.

Cameron, P., McGachy, A., Anderson, M., Paul, A., Coombs, G.H., Mottram, J.C., Alexander, J., and Plevin, R. (2004). Inhibition of lipopolysaccharide-induced macrophage IL-12 production by *Leishmania mexicana* amastigotes: the role of cysteine peptidases and the NF-kappaB signaling pathway. J. Immunol. *173*, 3297–3304.

Carratelli, C.R., Rizzo, A., Catania, M.R., Galle, F., Losi, E., Hasty, D.L., and Rossano, F. (2002). Chlamydia pneumoniae infections prevent the programmed cell death on THP-1 cell line. FEMS Microbiol. Lett. *215*, 69–74.

Chandra, D., and Naik, S. (2008). *Leishmania donovani* infection down-regulates TLR2-stimulated IL-12p40 and activates IL-10 in cells of macrophage/monocytic lineage by modulating MAPK pathways through a contact-dependent mechanism. Clin. Exp. Immunol. *154*, 224–234.

Denny, P.W., Field, M.C., and Smith, D.F. (2001). GPI-anchored proteins and glycoconjugates segregate into lipid rafts in Kinetoplastida. FEBS Lett. *491*, 148–153.

Dermine, J.F., Duclos, S., Garin, J., St-Louis, F., Rea, S., Parton, R.G., and Desjardins, M. (2001). Flotillin-1-enriched lipid raft domains accumulate on maturing phagosomes. J. Biol. Chem. *276*, 18507–18512.

Dermine, J.F., Goyette, G., Houde, M., Turco, S.J., and Desjardins, M. (2005). *Leishmania donovani* lipophosphoglycan disrupts phagosome microdomains in J774 macrophages. Cell Microbiol. *7*, 1263–1270.

Descoteaux, A., and Turco, S.J. (1999). Glycoconjugates in *Leishmania* infectivity. Biochim. Biophys. Acta *1455*, 341–352.

Descoteaux, A., Matlashewski, G., and Turco, S.J. (1992). Inhibition of macrophage protein kinase C-mediated protein phosphorylation by *Leishmania donovani* lipophosphoglycan. J. Immunol. *149*, 3008–3015.

Desjardins, M., and Griffiths, G. (2003). Phagocytosis: latex leads the way. Curr. Opin. Cell Biol. *15*, 498–503.

Diaz, N.L., Fernandez, M., Figueira, E., Ramirez, R., Monsalve, I.B., and Tapia, F.J. (2003). Nitric oxide and cellular immunity in experimental cutaneous leishmaniasis. Clin. Exp. Dermatol. *28*, 288–293.

Diefenbach, A., Schindler, H., Rollinghoff, M., Yokoyama, W.M., and Bogdan, C. (1999). Requirement for type 2 NO synthase for IL-12 signaling in innate immunity. Science *284*, 951–955.

Ellson, C.D., Davidson, K., Ferguson, G.J., O'Connor, R., Stephens, L.R., and Hawkins, P.T. (2006). Neutrophils from p40phox-/- mice exhibit severe defects in NADPH oxidase regulation and oxidant-dependent bacterial killing. J. Exp. Med. *203*, 1927–1937.

Feng, G.J., Goodridge, H.S., Harnett, M.M., Wei, X.Q., Nikolaev, A.V., Higson, A.P., and Liew, F.Y. (1999). Extracellular signal-related kinase (ERK) and p38 mitogen-activated protein (MAP) kinases differentially regulate the lipopolysaccharide-mediated induction of inducible nitric oxide synthase and IL-12 in macrophages: *Leishmania* phosphoglycans subvert macrophage IL-12 production by targeting ERK MAP kinase. J. Immunol. *163*, 6403–6412.

Forget, G., Gregory, D.J., Whitcombe, L.A., and Olivier, M. (2006). Role of host protein tyrosine phosphatase SHP-1 in *Leishmania donovani*-induced inhibition of nitric oxide production. Infect. Immun. *74*, 6272–6279.

Gantt, K.R., Goldman, T.L., McCormick, M.L., Miller, M.A., Jeronimo, S.M., Nascimento, E.T., Britigan, B.E., and Wilson, M.E. (2001). Oxidative responses of human and murine macrophages during phagocytosis of *Leishmania chagasi*. J. Immunol. *167*, 893–901.

Gao, L.Y., and Kwaik, Y.A. (2000). The modulation of host cell apoptosis by intracellular bacterial pathogens. Trends Microbiol. *8*, 306–313.

Ghosh, J., Bose, M., Roy, S., and Bhattacharyya, S.N. (2013). *Leishmania donovani* targets Dicer1 to down-regulate miR-122, lower serum cholesterol, and facilitate murine liver infection. Cell Host Microbe *13*, 277–288.

Ghosh, J., Guha, R., Das, S., and Roy, S. (2014). Liposomal cholesterol delivery activates the macrophage innate immune arm to facilitate intracellular *Leishmania donovani* killing. Infect. Immun. *82*, 607–617.

Ghosh, S., Bhattacharyya, S., Sirkar, M., Sa, G.S., Das, T., Majumdar, D., Roy, S., and Majumdar, S. (2002). *Leishmania donovani* suppresses activated protein 1 and NF-kappaB activation in host macrophages via ceramide generation: involvement of extracellular signal-regulated kinase. Infect. Immun. *70*, 6828–6838.

Goebel, S., Gross, U., and Luder, C.G. (2001). Inhibition of host cell apoptosis by *Toxoplasma gondii* is accompanied by reduced activation of the caspase cascade and alterations of poly(ADP-ribose) polymerase expression. J. Cell Sci. *114*, 3495–3505.

Gupta, P., Giri, J., Srivastav, S., Chande, A.G., Mukhopadhyaya, R., Das, P.K., and Ukil, A. (2014). *Leishmania donovani* targets tumor necrosis factor receptor-associated factor (TRAF) 3 for impairing TLR4-mediated host response. FASEB J. *28*, 1756–1768.

Heussler, V.T., Kuenzi, P., and Rottenberg, S. (2001). Inhibition of apoptosis by intracellular protozoan parasites. Int. J. Parasitol. *31*, 1166–1176.

Holm, A., Tejle, K., Magnusson, K.E., Descoteaux, A., and Rasmusson, B. (2001). *Leishmania donovani* lipophosphoglycan causes periphagosomal actin accumulation: correlation with impaired translocation of PKCalpha and defective phagosome maturation. Cell Microbiol. 3, 439–447.

Horta, M.F., Mendes, B.P., Roma, E.H., Noronha, F.S., Macedo, J.P., Oliveira, L.S., Duarte, M.M., and Vieira, L.Q. (2012). Reactive oxygen species and nitric oxide in cutaneous leishmaniasis. J. Parasitol. Res. *2012*, 203818.

Hoyal, C.R., Giron-Calle, J., and Forman, H.J. (1998). The alveolar macrophage as a model of calcium signaling in oxidative stress. J. Toxicol. Environ. Health B Crit. Rev. *1*, 117–134.

Huang, F.P., Xu, D., Esfandiari, E.O., Sands, W., Wei, X.Q., and Liew, F.Y. (1998). Mice defective in Fas are highly susceptible to *Leishmania major* infection despite elevated IL-12 synthesis, strong Th1 responses, and enhanced nitric oxide production. J. Immunol. *160*, 4143–4147.

Iles, K.E., and Forman, H.J. (2002). Macrophage signaling and respiratory burst. Immunol. Res. *26*, 95–105.

Junghae, M., and Raynes, J.G. (2002). Activation of p38 mitogen-activated protein kinase attenuates *Leishmania donovani* infection in macrophages. Infect. Immun. *70*, 5026–5035.

Kanayama, A., Seth, R.B., Sun, L., Ea, C.K., Hong, M., Shaito, A., Chiu, Y.H., Deng, L., and Chen, Z.J. (2004). TAB2 and TAB3 activate the NF-kappaB pathway through binding to polyubiquitin chains. Mol. Cell *15*, 535–548.

Kaneshiro, E.S., Gottlieb, M., and Dwyer, D.M. (1982). Cell surface origin of antigens shed by *Leishmania donovani* during growth in axenic culture. Infect. Immun. *37*, 558–567.

Kar, S., Ukil, A., and Das, P.K. (2009). Signaling events leading to the curative effect of cystatin on experimental visceral leishmaniasis: involvement of ERK1/2, NF-kappaB and JAK/STAT pathways. Eur. J. Immunol. *39*, 741–751.

Kar, S., Ukil, A., Sharma, G., and Das, P.K. (2010). MAPK-directed phosphatases preferentially regulate pro- and anti-inflammatory cytokines in experimental visceral leishmaniasis: involvement of distinct protein kinase C isoforms. J. Leukoc. Biol. *88*, 9–20.

Kaufmann, S.H., Desnoyers, S., Ottaviano, Y., Davidson, N.E., and Poirier, G.G. (1993). Specific proteolytic cleavage of poly(ADP-ribose) polymerase: an early marker of chemotherapy-induced apoptosis. Cancer Res. *53*, 3976–3985.

Kaye, P., and Scott, P. (2011). Leishmaniasis: complexity at the host–pathogen interface. Nat. Rev. Microbiol. *9*, 604–615.

Khan, A.Q., Shen, Y., Wu, Z.Q., Wynn, T.A., and Snapper, C.M. (2002). Endogenous pro- and anti-inflammatory cytokines differentially regulate an *in vivo* humoral response to Streptococcus pneumoniae. Infect. Immun. *70*, 749–761.

Kima, P.E. (2007). The amastigote forms of *Leishmania* are experts at exploiting host cell processes to establish infection and persist. Int. J. Parasitol. *37*, 1087–1096.

Kolli, D., Velayutham, T.S., and Casola, A. (2013). Host–viral interactions: role of pattern recognition receptors (PRRs) in human pneumovirus infections. Pathogens *2*, 232–263.

Kondoh, K., and Nishida, E. (2007). Regulation of MAP kinases by MAP kinase phosphatases. Biochim. Biophys. Acta *1773*, 1227–1237.

Kyriakis, J.M., and Avruch, J. (2001). Mammalian mitogen-activated protein kinase signal transduction pathways activated by stress and inflammation. Physiol. Rev. *81*, 807–869.

Lisi, S., Sisto, M., Acquafredda, A., Spinelli, R., Schiavone, M., Mitolo, V., Brandonisio, O., and Panaro, M. (2005). Infection with *Leishmania infantum* Inhibits actinomycin D-induced apoptosis of human monocytic cell line U-937. J. Eukaryot. Microbiol. *52*, 211–217.

Lodge, R., Diallo, T.O., and Descoteaux, A. (2006). *Leishmania donovani* lipophosphoglycan blocks NADPH oxidase assembly at the phagosome membrane. Cell Microbiol. *8*, 1922–1931.

McConville, M.J., and Naderer, T. (2011). Metabolic pathways required for the intracellular survival of *Leishmania*. Annu. Rev. Microbiol. *65*, 543–561.

McConville, M.J., Schnur, L.F., Jaffe, C., and Schneider, P. (1995). Structure of *Leishmania* lipophosphoglycan: inter- and intra-specific polymorphism in Old World species. Biochem. J. *310 (Pt 3)*, 807–818.

Mallat, Z., and Tedgui, A. (2000). Apoptosis in the vasculature: mechanisms and functional importance. Br. J. Pharmacol. *130*, 947–962.

Matheoud, D., Moradin, N., Bellemare-Pelletier, A., Shio, M.T., Hong, W.J., Olivier, M., Gagnon, E., Desjardins, M., and Descoteaux, A. (2013). *Leishmania* evades host immunity by inhibiting antigen cross-presentation through direct cleavage of the SNARE VAMP8. Cell Host Microbe *14*, 15–25.

Mathur, R.K., Awasthi, A., Wadhone, P., Ramanamurthy, B., and Saha, B. (2004). Reciprocal CD40 signals through p38MAPK and ERK-1/2 induce counteracting immune responses. Nat. Med. *10*, 540–544.

Megas, C., Hatzivassiliou, E.G., Yin, Q., Marinopoulou, E., Hadweh, P., Vignali, D.A., and Mosialos, G. (2011). Mutational analysis of TRAF6 reveals a conserved functional role of the RING dimerization interface and a potentially necessary but insufficient role of RING-dependent TRAF6 polyubiquitination towards NF-kappaB activation. Cell Signal. *23*, 772–777.

Meier, C.L., Svensson, M., and Kaye, P.M. (2003). *Leishmania*-induced inhibition of macrophage antigen presentation analyzed at the single-cell level. J. Immunol. *171*, 6706–6713.

Moore, K.J., and Matlashewski, G. (1994). Intracellular infection by *Leishmania donovani* inhibits macrophage apoptosis. J. Immunol. *152*, 2930–2937.

Moore, K.J., Labrecque, S., and Matlashewski, G. (1993). Alteration of *Leishmania donovani* infection levels by selective impairment of macrophage signal transduction. J. Immunol. *150*, 4457–4465.

Muraille, E., Gounon, P., Cazareth, J., Hoebeke, J., Lippuner, C., Davalos-Misslitz, A., Aebischer, T., Muller, S., Glaichenhaus, N., and Mougneau, E. (2010). Direct visualization of peptide/MHC complexes at the surface and in the intracellular compartments of cells infected *in vivo* by *Leishmania major*. PLoS Pathog. *6*, e1001154.

Nandan, D., and Reiner, N.E. (1995). Attenuation of gamma interferon-induced tyrosine phosphorylation in mononuclear phagocytes infected with *Leishmania donovani*: selective inhibition of signaling through Janus kinases and Stat1. Infect. Immun. *63*, 4495–4500.

Nauseef, W.M. (2004). Assembly of the phagocyte NADPH oxidase. Histochem. Cell Biol. *122*, 277–291.

Negre-Salvayre, A., Hirtz, C., Carrera, G., Cazenave, R., Troly, M., Salvayre, R., Penicaud, L., and Casteilla, L. (1997). A role for uncoupling protein-2 as a regulator of mitochondrial hydrogen peroxide generation. FASEB J. *11*, 809–815.

Olivier, M., Brownsey, R.W., and Reiner, N.E. (1992). Defective stimulus–response coupling in human monocytes infected with *Leishmania donovani* is associated with altered activation and translocation of protein kinase C. Proc. Natl. Acad. Sci. U.S.A. *89*, 7481–7485.

Olivier, M., Gregory, D.J., and Forget, G. (2005). Subversion mechanisms by which *Leishmania* parasites can escape the host immune response: a signaling point of view. Clin. Microbiol. Rev. *18*, 293–305.

Ono, K., and Han, J. (2000). The p38 signal transduction pathway: activation and function. Cell Signal. *12*, 1–13.

Overath, P., and Aebischer, T. (1999). Antigen presentation by macrophages harboring intravesicular pathogens. Parasitol. Today *15*, 325–332.

Prive, C., and Descoteaux, A. (2000). *Leishmania donovani* promastigotes evade the activation of mitogen-activated protein kinases p38, c-Jun N-terminal kinase, and extracellular signal-regulated kinase-1/2 during infection of naive macrophages. Eur. J. Immunol. *30*, 2235–2244.

Proudfoot, L., O'Donnell, C.A., and Liew, F.Y. (1995). Glycoinositolphospholipids of *Leishmania major* inhibit nitric oxide synthesis and reduce leishmanicidal activity in murine macrophages. Eur. J. Immunol. *25*, 745–750.

Reiner, N.E., Ng, W., and McMaster, W.R. (1987). Parasite-accessory cell interactions in murine leishmaniasis. II. *Leishmania donovani* suppresses macrophage expression of class I and class II major histocompatibility complex gene products. J. Immunol. *138*, 1926–1932.

Reiner, N.E., Ng, W., Wilson, C.B., McMaster, W.R., and Burchett, S.K. (1990). Modulation of *in vitro* monocyte cytokine responses to *Leishmania donovani*. Interferon-gamma prevents parasite-induced inhibition of interleukin 1 production and primes monocytes to respond to *Leishmania* by producing both tumor necrosis factor-alpha and interleukin 1. J. Clin. Invest. *85*, 1914–1924.

Reiner, S.L., Zheng, S., Wang, Z.E., Stowring, L., and Locksley, R.M. (1994). *Leishmania* promastigotes evade interleukin 12 (IL-12) induction by macrophages and stimulate a broad range of cytokines from CD4$^+$ T-cells during initiation of infection. J. Exp. Med. *179*, 447–456.

Ruhland, A., Leal, N., and Kima, P.E. (2007). *Leishmania promastigotes* activate PI3K/Akt signalling to confer host cell resistance to apoptosis. Cell Microbiol. *9*, 84–96.

Sakurai, C., Hashimoto, H., Nakanishi, H., Arai, S., Wada, Y., Sun-Wada, G.H., Wada, I., and Hatsuzawa, K. (2012). SNAP-23 regulates phagosome formation and maturation in macrophages. Mol. Biol. Cell *23*, 4849–4863.

Segal, A.W. (2008). The function of the NADPH oxidase of phagocytes and its relationship to other NOXs in plants, invertebrates, and mammals. Int. J. Biochem. Cell Biol. *40*, 604–618.

Shadab, M., and Ali, N. (2011). Evasion of Host Defence by *Leishmania donovani*: Subversion of Signaling Pathways. Mol. Biol. Int. *2011*, 343961.

Simons, K., and Vaz, W.L. (2004). Model systems, lipid rafts, and cell membranes. Annu. Rev. Biophys. Biomol. Struct. *33*, 269–295.

Srivastav, S., Kar, S., Chande, A.G., Mukhopadhyaya, R., and Das, P.K. (2012). *Leishmania donovani* exploits host deubiquitinating enzyme A20, a negative regulator of TLR signaling, to subvert host immune response. J. Immunol. *189*, 924–934.

Srivastav, S., Basu Ball, W., Gupta, P., Giri, J., Ukil, A., and Das, P.K. (2014). *Leishmania donovani* prevents oxidative burst mediated apoptosis of host macrophages through selective induction of suppressors of cytokine signaling (SOCS) proteins. J. Biol. Chem. *289*, 1092–1105.

Stuehr, D.J. (1999). Mammalian nitric oxide synthases. Biochim. Biophys. Acta *1411*, 217–230.

Su, B., and Karin, M. (1996). Mitogen-activated protein kinase cascades and regulation of gene expression. Curr. Opin. Immunol. *8*, 402–411.

Thoma-Uszynski, S., Stenger, S., Takeuchi, O., Ochoa, M.T., Engele, M., Sieling, P.A., Barnes, P.F., Rollinghoff, M., Bolcskei, P.L., Wagner, M., *et al.* (2001). Induction of direct antimicrobial activity through mammalian Toll-like receptors. Science *291*, 1544–1547.

Trinchieri, G. (2007). Interleukin-10 production by effector T-cells: Th1 cells show self control. J. Exp. Med. *204*, 239–243.

Tuon, F.F., Amato, V.S., Bacha, H.A., Almusawi, T., Duarte, M.I., and Amato Neto, V. (2008). Toll-like receptors and leishmaniasis. Infect. Immun. *76*, 866–872.

Ukil, A., Biswas, A., Das, T., and Das, P.K. (2005). 18 Beta-glycyrrhetinic acid triggers curative Th1 response and nitric oxide up-regulation in experimental visceral leishmaniasis associated with the activation of NF-kappa B. J. Immunol. *175*, 1161–1169.

Ukil, A., Kar, S., Srivastav, S., Ghosh, K., and Das, P.K. (2011). Curative effect of 18beta-glycyrrhetinic acid in experimental visceral leishmaniasis depends on phosphatase-dependent modulation of cellular MAP kinases. PLoS One *6*, e29062.

Underhill, D.M., and Ozinsky, A. (2002). Toll-like receptors: key mediators of microbe detection. Curr. Opin. Immunol. *14*, 103–110.

Vaux, D.L., Haecker, G., and Strasser, A. (1994). An evolutionary perspective on apoptosis. Cell *76*, 777–779.

de Veer, M.J., Curtis, J.M., Baldwin, T.M., DiDonato, J.A., Sexton, A., McConville, M.J., Handman, E., and Schofield, L. (2003). MyD88 is essential for clearance of *Leishmania major*: possible role for lipophosphoglycan and Toll-like receptor 2 signaling. Eur. J. Immunol. 33, 2822–2831.

Vieira, O.V., Bucci, C., Harrison, R.E., Trimble, W.S., Lanzetti, L., Gruenberg, J., Schreiber, A.D., Stahl, P.D., and Grinstein, S. (2003). Modulation of Rab5 and Rab7 recruitment to phagosomes by phosphatidylinositol 3-kinase. Mol. Cell Biol. *23*, 2501–2514.

Wilson, M.E., Jeronimo, S.M., and Pearson, R.D. (2005). Immunopathogenesis of infection with the visceralizing *Leishmania* species. Microb. Pathog. *38*, 147–160.

Yang, Z., Mosser, D.M., and Zhang, X. (2007). Activation of the MAPK, ERK, following *Leishmania amazonensis* infection of macrophages. J. Immunol. *178*, 1077–1085.

Role of STAT Signalling in Immunity to Leishmaniasis

7

Steve Oghumu*, James Stock*, Cesar Terrazzas*, Gayathri Natarajan, Sanjay Varikuti and Abhay R. Satoskar

Abstract

The JAK/STAT signalling pathway is a principal mediator of the effects of cytokines on immune cells and therefore plays a major role in the orchestration of immunological responses to an infectious challenge. The STAT family is comprised of seven genes that code for STAT1, STAT2, STAT3, STAT4, STAT5A, STAT5B and STAT6, and most of these STATs as well as their associated signalling mediators have been shown to be critical in immunity and/or pathogenesis of various forms of *Leishmania* infection. *Leishmania* are also known to exploit the JAK/STAT signalling pathway to favour parasite establishment, growth and replication. Understanding how *Leishmania* modulates this host intracellular pathway could lead to the development of more effective therapeutic approaches against leishmaniasis. In this chapter we will discuss the importance of STATs and their signalling ligands in various forms of *Leishmania* infection and how the parasite has exploited this intracellular network to evade anti-parasitic immune responses and facilitate survival within the host. We will also briefly address how our knowledge of these mechanisms can aid in the development of effective therapeutic strategies in the management of leishmaniasis.

Introduction

Leishmaniasis is a major public health problem present mainly in tropical countries. Over 12 million people currently suffer from the disease, and approximately 2 million people are infected annually. This pathogen is an obligate kinetoplastid protozoan parasite of the genus *Leishmania*, and is transmitted by a sand fly vector *Phlebotomus* and *Lutzomyia* in the Old and New World, respectively (Hughes and Piontkivska, 2003; Walters *et al.*, 1989a,b). Several species are spread around the world and the infection presents with different clinical manifestations such as cutaneous (CL), mucocutaneous (MCL) and visceral leishmaniasis (VL) (Herwaldt, 1999; Mougneau *et al.*, 2011; Reithinger *et al.*, 2007; Schonian *et al.*, 2011).

The pathology of leishmaniasis is dependent on the parasite species. There are approximately 30 species of *Leishmania* of which 20 are confirmed to be pathogenic to humans. Based on their geographical distribution, *Leishmania* species are designated as 'Old World' or 'New World' species. Old World leishmaniasis occurs in parts of Asia, the Middle East, North Africa and southern Europe. Among the Old World species, *Leishmania major*, *Leishmania tropica* and *Leishmania aethiopica* cause CL while *Leishmania donovani* and *Leishmania infantum* cause VL. New World leishmaniasis on the other hand is found in regions of Mexico, Central and South America (Reithinger *et al.*, 2007). New World CL is caused by *Leishmania mexicana*, *Leishmania amazonensis* and *Leishmania braziliensis* and VL is caused by *Leishmania chagasi*(Herwaldt, 1999; McMahon-Pratt and Alexander, 2004; Reithinger *et al.*, 2007).

Leishmania parasites exist in different stages. In the sandfly, parasites are found as extracellular flagellate procyclic promastigotes (Bates and Rogers, 2004; Dostalova and Volf, 2012), and upon a sandfly bite, metacyclic promastigotes

* Equal contributors.

are introduced into the mammalian host, which are subsequently phagocytosed by neutrophils, macrophages and dendritic cells(DCs) present in the skin (Sacks and Perkins, 1985). Within the intracellular environment, *Leishmania* parasites transform into the amastigote stage and localize in parasitophorous vacuoles, which display features of phagolysosomes.

The immune system plays an important role in determining the outcome of *Leishmania* infection. Although macrophages have broad spectra of microbicidal functions, *Leishmania* parasites have developed strategies to evade these immune mechanisms and survive in these cells (Gupta *et al.*, 2013; Terrazas *et al.*, 2010). Most parasites are therefore able to establish non-healing lesions in CL or persist in the spleen, liver and bone marrow during VL (Gupta *et al.*, 2013).

The immune response generated during leishmaniasis has been the focus of research for several years. Importantly, it has been shown that the dichotomy of the immune response plays a role in infection or immunity. Understanding the factors that enhance host immune protection can lead to the generation of improved prophylactic and therapeutic strategies. It must be noted that different *Leishmania* species elicit different immune responses in the host, resulting in a more complex scenario for generalized therapeutic approaches (McGwire and Satoskar, 2013).

In mouse models of leishmaniasis, genetic differences contribute to the type of immune response generated which ultimately determine the outcome of infection. For example, in CL, resistant mice develop a strong Th1 response which is accompanied by the production of interferon gamma (IFN-γ) by T-cells and nitric oxide (NO) by macrophages. This immune response is associated with low parasitic burdens and increased clearance of infection. In contrast, genetic backgrounds prone to induce a Th2 response show increased susceptibility to CL (Alexander and Bryson, 2005; Kropf *et al.*, 2003; Rosas *et al.*, 2005; Satoskar *et al.*, 1995). On the other hand, this dichotomy is less clear in VL. *L. donovani* infection requires a Th1 response for protection, but a Th2 response also contributes to control of infection in the liver and helps in granuloma formation. Both Th1 and Th2 responses are mainly regulated by hallmark interleukins (IL) which are soluble factors derived mainly from T-cells (Alexander *et al.*, 2000; Stager *et al.*, 2003a).

The Th1 response is characterized by enhanced IFN-γ production by T-cells which in turn activate macrophages to produce NO and reactive oxygen species (ROS), enhancing its microbicidal capability (Kima and Soong, 2013; Mosser and Edwards, 2008; Rosas *et al.*, 2006). This response requires T-cell polarization by antigen-presenting DCs and macrophages via IL-12 production after parasite antigen recognition. Additionally IL-12 production can be enhanced by CD40–CD40L interactions and is a requirement for IFN-γ production by T-cells (Steinman, 2007). Th1 responses further help in the development of $CD8^+$ cytotoxic T-cells which kill parasite-infected cells (Kima and Soong, 2013). Conversely, Th2 responses which are associated with the cytokines IL-4, IL-5, and IL-13 produced by T lymphocytes, contribute to the maturation of granulomas as well as antibody class switching to IgE. Further, the Th2 response is required for effective drug therapy during *L. donovani* infection (Alexander *et al.*, 2000; Kropf *et al.*, 2003; Stager *et al.*, 2003a).

Cytokines produced by activated cells during the initiation of an immune response following a *Leishmania* infectious challenge require recognition by target cells in order to induce their biological activity. Such recognition is mediated by specific cell surface receptors which trigger complex intracellular signalling pathways resulting in cellular activation, differentiation and proliferation, and ultimately leading to inflammation, generation of immune responses and disease resolution.

The intracellular signalling pathways initiated by cytokine recognition are mediated by intracytoplasmic proteins known as Janus kinases (JAKs) and signal transducers and activators of transcription (STATs) and are responsible for the induction of cytokine responsive genes. The JAK/STAT pathway is the principal signalling mechanism for cytokines and is involved in cell proliferation, differentiation, migration and apoptosis. These cellular events are critical to haematopoiesis, immune development and other processes. In immunity, the JAK/STAT pathway is important for inflammation as well as the generation and specialization of effector responses. The STAT family is comprised of seven genes that code for STAT1, STAT2, STAT3, STAT4, STAT5A, STAT5B and STAT6. Homodimeric or heterodimeric combinations of STATs are activated by different cytokines, and redundancy in STAT activation by multiple cytokines is known to occur (Adamson *et al.*, 2009; Delgoffe *et al.*, 2011; Murray, 2007).

The JAK/STAT pathway is initiated by ligand recognition which induces receptor multimerization. Such receptors can form homodimers or heterodimers. Their cytoplasmic domains are associated with JAKs which are activated when receptor multimerization occurs, initiating transphosphorylation between JAKs (Behrmann *et al.*, 2004; Murray, 2007). STATs are present in the cytoplasm and are activated by JAKs which phosphorylate tyrosine residues on STAT molecules. This event leads to STAT homo- or hetero-dimerization and subsequent nuclear translocation. STATs are imported to the nucleus through the nuclear pore complex or by carrier molecules. Once the STAT has translocated into the nucleus, it binds to the promoter region via specific DNA binding domains and activates or represses the transcription of target genes (Bromberg and Darnell, 2000; Imada and Leonard, 2000). STAT signalling is therefore crucial for translating extracellular signals and controlling gene expression, thereby ultimately regulating cell function.

Leishmania parasites have the ability to interfere with several host immune response pathways including STAT signalling, thereby decreasing the effectiveness of macrophage activation and subsequent parasite killing (Bhardwaj *et al.*, 2005; Contreras *et al.*, 2010; Cummings *et al.*, 2010; Matte and Descoteaux, 2010; McDowell and Sacks, 1999; Nandan and Reiner, 1995; Olivier *et al.*, 2005; Osorio *et al.*, 2012; Xin *et al.*, 2008). Understanding how *Leishmania* modulates host intracellular pathways could lead to the development of more effective therapeutic approaches against leishmaniasis. In this chapter we will discuss the importance of STATs and their signalling ligands in various forms of *Leishmania* infection and how the parasite has exploited this intracellular network to evade anti-parasitic immune responses and facilitate survival within the host. We will also briefly address how our knowledge of these mechanisms can aid in the development of effective therapeutic strategies in the management of leishmaniasis.

STAT1 and leishmaniasis

STAT1 is a 91 kDa protein which serves as an intermediate to signalling from IFN-γ, a type II interferon, and type I interferons (IFN-α and IFN-β). Expression of STAT1 occurs commonly in T-cells, macrophages, and DCs and is present in the cytosol as a non-active monomer. Active STAT1 homodimers form the gamma interferon activated (GAF) complex which translocate to the nucleus and induce the transcription of cytokines such as IFN-γ, IL-12, chemokines such as CXCL9, and cell surface receptors such as FcγR1 (Najjar and Fagard, 2010). Studies have hinted at other roles for STAT1 besides its role as a gamma induced signal transducer; it has been shown to play a role in phagosome acidification and natural antimicrobial resistance during the early phase of *L. major* infection (Spath *et al.*, 2009).

Prior to translocation of phosphorylated STAT1 dimers to the nucleus, activation of STAT1 occurs through the JAK/STAT signalling pathway. Binding of IFN-γ to the interferon gamma receptor-1 (IFNGR1) subunit causes recruitment of the IFNGR2 subunit which is required for subsequent downstream signalling. Each IFNGR subunit has an associated JAK protein, with JAK1 localized to inactive IFNGR1 and JAK2 localized to inactive IFNGR2. Following the assembly of the IFNGR complex, JAKs 1 and 2 become activated, resulting in autophosphorylation of JAK2 and subsequent phosphorylation of JAK1. These activated JAK kinases together phosphorylate IFNGR subunits at unique tyrosine residues in the cytoplasmic tail, leading to binding of STAT1. STAT1 is subsequently phosphorylated and released upon interaction with another activated STAT1–IFNGR complex leading to dimerization of phosphorylated STAT1 monomers. Active STAT1 homodimers are translocated to the nucleus, and bind to unique gamma interferon activated sequence (GAS) promoter elements to induce transcription of its target genes(Najjar and Fagard, 2010). Recently, it has been shown that STAT1 forms dimers with STAT4, and bind to promoters of the genes encoding the IL-12 subunits p35 and Ebi3 in response to IL-35. Interestingly, naive T-cells targeted by IL-35 transform into regulatory T-cells (Collison *et al.*, 2012). However, the role of IL-35 during leishmaniasis remains elusive.

STAT1 has a complex role in protective immunity because its expression occurs in a variety of cell types, including (but not limited to) DCs, monocytes, macrophages, and T-cells. STAT1 has been shown to be important in mediating IFN-γ signalling, a cytokine which activates macrophages, thereby enhancing the killing of phagocytosed *Leishmania* parasites. STAT1 also contributes to the development of a Th1-type response, characterized by IFN-γ production and activity, leading to a resistant phenotype in most *Leishmania* infections (Roberts, 2005).

Extensive studies using STAT1-deficient mice have demonstrated variability in the outcome of infection depending on the infective species of *Leishmania*. Research in our laboratory showed that $stat1^{-/-}$ C57BL/6 mice lose their natural protective immunity during *L. major* infection. Lack of STAT1 resulted not only in higher parasite loads but also in a complete change in the immune response reflected by the inability to produce IFN-γ (Rosas *et al.*, 2003). The impaired activity of STAT1-deficient DCs (Johnson and Scott, 2007) and T-cells (Barbi *et al.*, 2009) have been shown to contribute to this susceptible phenotype. In this regard, DCs deficient in STAT1 have a significantly reduced ability to present antigen and demonstrate an impaired up-regulation of costimulatory molecules, making them poor inducers of a Th1 response (Johnson and Scott, 2007). Moreover, STAT1-deficient $CD4^+$ T-cells lose their ability to home to the infection site where they exert their effector functions (Barbi *et al.*, 2009). Although STAT1 can be activated in response to different cytokines, the IFN-γ/STAT1 pathway seems to play a main role during *L. major* infection as reflected by the similar inability of IFNGR-deficient mice to control infection (Rosas *et al.*, 2003).

Surprisingly, studies of VL caused by *L. donovani* have shown that STAT1-deficient BALB/c mice are resistant to *L. donovani* (Rosas *et al.*, 2006). Further, STAT1-deficient T-cells induce more pathology when transferred to T-cell-deficient mice infected with *L. donovani*, suggesting that STAT1 signalling in T-cells likely does not contribute to susceptibility to VL. Other mechanisms could therefore be responsible for the resistance observed in STAT1-deficient mice. For example, STAT1 could mediate migration of monocytes to the site of infection, by interfering with the up-regulation of chemokine receptors (Cummings *et al.*, 2010). The possibility of STAT1-mediated recruitment of disease promoting monocytes favouring parasite establishment will require additional research.

Given the importance of STAT1 in immune responses to CL and VL, it is not surprising that the *Leishmania* parasite has targeted this pathway to evade intracellular killing caused by IFN-γ activation of phagocytes. It has been shown that intermediates in the JAK/STAT pathway are targets for *L. mexicana* and *L. major* virulence factors. Studies using cell lines derived from mouse macrophages and monocytes showed that STAT1 activation is inhibited by parasites from both species. *L. major* and *L. mexicana* also interfere with IFNGR receptor expression as well as phosphorylation of JAKs 1 and 2 to facilitate escape from the host's cellular killing machinery (Bhardwaj *et al.*, 2005). Similarly, *L. donovani* has been shown to attenuate IFNGR expression and tyrosine phosphorylation of JAK1, JAK2 and STAT1 in monocytes and macrophages (Blanchette *et al.*, 1999; Nandan and Reiner, 1995). The effects of modulation of the STAT1 signalling pathway in host cells by *Leishmania* results in the alteration of macrophage functions such as the suppression of IFN-γ inducible molecules including the production of the primary mediator of intracellular killing, NO, and the expression of the antigen-presenting molecule, MHC class II.

Multiple mechanisms are employed by *Leishmania* parasites in interfering with the STAT1 pathway. *L. donovani*, *L. mexicana* and *L. major* have been shown to decrease IFNGR expression on macrophages (Bhardwaj *et al.*, 2005; Ray *et al.*, 2000). However, this mechanism may not be employed by all *Leishmania* parasites, as this effect was not found in *L. amazonensis*-infected cells (Kima *et al.*, 1997). *L. mexicana* also interferes with STAT1 mediated IFN-γ signalling by enhancing the phosphorylation of STAT1β, a dominant negative mediator of STAT1 signal transduction in macrophages (Bhardwaj *et al.*, 2005). Another mechanism utilized by *L. donovani* in attenuating STAT1 signalling in infected macrophages is the induction of the Src homology-2 domain-containing phosphatase-1 (SHP-1), a negative regulator of STAT signalling. The *L. donovani* parasite-derived molecule, elongation factor-1α (EF-1α), has been shown to activate SHP-1, and attenuate IFN-γ induced iNOS production in infected macrophages (Nandan *et al.*, 2002). SHP-1 seems to play a key role in *Leishmania* survival. Infected macrophages exposed to tyrosine phosphatase inhibitors have a normal response to IFN-γ stimulation, suggesting a potential therapeutic target of the STAT1 signalling pathway (Forget *et al.*, 2006; Nandan *et al.*, 1999). Indeed, mice deficient in SHP-1 have been shown to be resistant to *Leishmania* infection (Forget *et al.*, 2001).

STAT2 and leishmaniasis

STAT2 mediates the signalling of type I interferons, IFN-α and IFN-β, and is important in innate immunity as well as in immunity to viral pathogens (Park *et al.*, 2000; Perry *et al.*, 2011). STAT2 associates with STAT1 and, in combination with interferon regulatory factor 9 (IRF9), form a heterotrimeric transcription factor complex known as interferon stimulated response gene factor 3 (ISGF3), the master regulator of type I interferon induced transcriptional activation (Au-Yeung *et al.*, 2013). Numerous studies using STAT1 and STAT2-deficient mice demonstrate that these transcription factors are essential mediators of the proinflammatory activity of type I interferons (Brierley and Fish, 2005; Chowdhury and Farrar, 2013; Rauch *et al.*, 2013).

Although studies utilizing STAT2-deficient mice in *Leishmania* infection are presently lacking, a number of studies have demonstrated a role for this transcription factor in innate immunity to CL caused by *L. major* and *L. amazonensis* (Diefenbach *et al.*, 1998; Xin *et al.*, 2008). Infection of DCs with *L. amazonenesis* was shown to cause degradation of STAT2 as well as suppression in STAT2 phosphorylation, resulting in impaired DC function and Th1 activation (Xin *et al.*, 2008). IFN-α and IFN-β, which signal via STAT1 and STAT2, have also been shown to be involved in innate immunity to *L. major* through the induction of NO, an essential leishmanicidal molecule produced by macrophages and NK cells (Diefenbach *et al.*, 1998). These studies emphasize the vital role of type I IFN and STAT2 in innate immune responses to CL and the subsequent priming of Th1 adaptive immune responses.

STAT3 and leishmaniasis

STAT3 is a 92 kDa protein which is present in the cytoplasm as an inactive monomer. The methods of classical activation for STAT3 closely resemble other STATS discussed in this chapter, although STAT3 can be activated by different signals, including the cytokines IL-6 (Heinrich *et al.*, 1998), IL-10 (Biswas *et al.*, 2011) and IL-12 (Jacobson *et al.*, 1995), as well as other cytokines and growth factors. While other STATs require a limited number of signals to be activated, one study lists as many as 39 potential activators for STAT3. Some of these factors activate STAT3 via different pathways outside of the classical JAK/STAT activation pathway (Aggarwal *et al.*, 2009). Activation can occur through phosphorylation or acetylation of three different residues: a serine residue at position 727, tyrosine 705, and lysine 685. Acetylation of the latter position stabilizes the dimer and increases the DNA-binding capability of STAT3.

During the classical JAK/STAT activation pathway for STAT3, IL-6 binds to its transmembrane receptor causing the formation of a protein complex which phosphorylates STAT3, releasing it into the cytoplasm. Dimerization of STAT3 and translocation into the nucleus results in the transcription of its many target genes, including those which promote the development of Th17 type helper T-cells (Chen and O'Shea, 2008; Harris *et al.*, 2007; Rebe *et al.*, 2013).

Given the large number of ligands which signal through STAT3, it is not surprising that STAT3 has been implicated in a number of autoimmune and neoplastic diseases. (Bournazou and Bromberg, 2013; Camporeale and Poli, 2012; Rebe *et al.*, 2013). In these diseases and their model systems, activation by IL-6 appears to be the main contributor to pathology, with subsequent IL-17 production, which causes the proliferation of Th17 cells.

Mice genetically deficient in STAT3 are not viable, so studies examining the role of STAT3 in *Leishmania* infections are difficult to perform. However, Cre-lox based conditional cell-specific knock-out models have demonstrated a role for STAT3 signalling in IL-6 dependent T-cell proliferation (Takeda *et al.*, 1998). Based on the already well-established requirement for STAT3 in the development of Th17 cells, we can only speculate on the role it plays in immunity to Leishmaniasis. Th17 responses are less polarized and often occur in concert with the development of other subsets of helper T-cells (Nylen and Gautam, 2010). Th17 cells are proinflammatory and have been shown to attract neutrophils to the site of infection (Boaventura *et al.*, 2010). In the context of cutaneous infections by *L. major*, Th17 cells are associated with increased susceptibility to the parasite and lead to increased parasite loads, with IL-17 gene-deficient mice able to resolve the infection (Lopez Kostka *et al.*, 2009). Interestingly, leishmaniases caused by *L. braziliensis* and *Leishmania panamensis* are more effectively cleared in the presence of IL-17 and an underlying Th17 response (Castilho *et al.*, 2010; Vargas-Inchaustegui *et al.*, 2008). This trend also holds true for *L. donovani*, which elicits significant increases in IL-17 in infected peripheral blood mononuclear cells (Pitta *et al.*, 2009).

STAT3 is also implicated in the host immune response to VL due to activation by the immunosuppressive cytokine IL-10 (Biswas *et al.*, 2011), triggering the expression of IL-4 receptor alpha (IL-4Rα), with subsequent induction of arginase-1 (arg-1). Increased levels of arg-1 not only favour parasite intracellular growth and replication, but provide an anti-inflammatory environment that is associated with a non-healing phenotype (Biswas *et al.*, 2011). IL-10 is also known to contribute to susceptibility to CL, as it suppresses macrophage activation and prevents the expansion of Th1-type cells required for protective immunity (Cummings *et al.*, 2010). Studies using IL-10-deficient mice demonstrate that IL-10 contributes to susceptibility to *L. major*, *L. mexicana* and *L. panamensis* (Buxbaum and Scott, 2005; Castilho *et al.*, 2010; Thomas and Buxbaum, 2008). These findings together paint a rather complicated picture for the role of STAT3 in leishmaniasis, given the wide variety of activating cytokine ligands and distinct immune responses to various infective parasite species. With responses to IL-10 and IL-17, the STAT3 signalling pathway appears to balance the immune response between immunosuppression and inflammation. Future research will be needed to fully determine the role of STAT3 in leishmaniasis.

STAT4 and leishmaniasis

Numerous studies demonstrate that STAT4 is involved in the early development of an efficient Th1 immune response to *Leishmania*. STAT4 is present in the cell as an unphosphorylated, non-active monomer which is activated primarily by the cytokine IL-12. IL-23 also activates STAT4 and generally forms heterodimers with STAT3 (Paradowska-Gorycka *et al.*, 2010). When active, STAT4 increases the transcription of genes which are responsible for generation of IFN-γ and is an important intermediate in the development of a polarized Th1 CD4$^+$ T-cell response (Nishikomori *et al.*, 2002). The STAT4 signalling pathway is similar to signalling in other STATs and involves the binding of the cognate cytokine ligand IL-12 or IL-23 to its cell-surface receptor, which is preferentially expressed on a subset of T-cells. Conformational changes allow the binding of JAK2 and Tyk2 to the intracellular cytoplasmic tail. Transphosphorylation between the two kinases leads to phosphorylation of the receptor chains (IL-12Rβ2 for the IL-12 pathway and IL-23R for IL-23) by activated JAKs. Phosphorylated receptor chains serve as docking sites for subsequent STAT4 recruitment and activation by JAKs. Activated STAT4 molecules are released into the cytoplasm, form homo- or hetero-dimers and translocate to the nucleus where they bind to STAT4 binding sites in gene promoter regions and initiate transcription of target genes.

The importance of the IL-12-derived STAT4 activation pathway has been largely characterized by knockout and transgenic studies of the signal transduction molecules. The IL-12 cytokine itself has a complicated role in resistance to leishmaniasis due to the presence of an uncharacterized IL-12 independent pathway for IFN-γ production (Stamm *et al.*, 1999). The IL-12 dependent pathway has been shown to be crucial to resistance in C57BL/6 mice during *L. major* infection, but not essential for resistance to *L. mexicana* (Buxbaum *et al.*, 2002). Similar to *L. major*, IL-12 deficient C57BL/6 mice infected with *L. donovani* developed much higher parasite burdens when compared with wild type(WT) mice (Satoskar *et al.*, 2000), indicating differences in the mechanisms of resistance to *L. donovani* and *L. mexicana*. In susceptible BALB/c mice, use of recombinant IL-12 was sufficient to render them resistant to *L. major* (McDowell and Sacks, 1999) but transgenic expression of the IL-12Rβ2 gene was unable to confer resistance during *L. major* infection (Nishikomori *et al.*, 2001).

IL-23, another potential STAT4 activator, may contribute to the production of IFN-γ outside the IL-12 activation pathway and has been shown to be important in limiting the number of parasites in *L. donovani* infections in the absence of IL-12 signalling (Murray *et al.*, 2006). It is important to note that IL-23 was not itself sufficient to prevent infection in Murray's study, hinting that it serves only a secondary function during *L. donovani* infections. The effects of STAT4 mediated IL-23 activation remain to be fully investigated in other species of *Leishmania*.

In contrast with the cytokines involved with STAT4 activation, the role of STAT4 in leishmaniasis has been much more easily defined. STAT4 has been shown to be a crucial factor in the development of a Th1 immune response (Cummings *et al.*, 2010) and is essential for resistance to CL caused by *L. major* and *L. Mexicana* (Buxbaum *et al.*, 2002; Stamm *et al.*, 1999). Interestingly, while IL-12 is not essential in conferring resistance to *L. mexicana*, STAT4 is required for resistance against this parasite (Buxbaum *et al.*, 2002). This reinforces the idea that IL-12 independent pathways for STAT4 activation play a major role in the control of leishmaniasis in mouse models. Recent studies conducted by our laboratory have also shown that STAT4 is critical for immunity to VL caused by *L. donovani* (Oghumu *et al.*, 2013). Together, these studies demonstrate that STAT4 is indispensable for developing an effective response to *Leishmania* parasites.

STAT6 and leishmaniasis

STAT6 is a 94 kDa protein found in almost all cell types, but is highly expressed in T-cells, B-cells and mast cells (Chen and Jiang, 2013). In the cytoplasm, the protein is found in an inactive, unmodified form which requires activation by a similar pathway to other STATs. The initial steps of canonical STAT6 activation proceed by binding of either IL-13 or IL-4 to their respective receptors and recruitment of a common γ chain. Activation of receptor associated JAK kinases leads to phosphorylation of tyrosine residues on the cytoplasmic tail of the IL-4Rα chain, which serves as a docking site for STAT6. The activation of STAT6 follows the same pathway as other STATs, beginning with phosphorylation of STAT6 proteins, formation of dimers and translocation to the nucleus for transcription activation of IL-6 responsive genes (Hebenstreit *et al.*, 2006). Studies by many groups have shown that an alternative activation pathway for STAT6 can also occur by a variety of factors and in a number of cell types. This list of alternative activators includes IL-3 in IL-3 dependent cells (Quelle *et al.*, 1995), IL-15 in mast cells (Masuda *et al.*, 2000), platelet-derived growth factor (PDGF) in fibroblasts (Patel *et al.*, 1996), and IFN-α in B-cells (Fasler-Kan *et al.*, 1998). Such versatility in activation is accompanied by the formation of heterodimers with other STAT proteins, which is a common occurrence in various JAK/STAT signalling pathways (Adamson *et al.*, 2009; Delgoffe *et al.*, 2011; Murray, 2007).

As a molecule which responds to IL-13 and IL-4, STAT6 plays a major role in the development of a Th2-type T-cell response (Maier *et al.*, 2012). When activated, STAT6 not only contributes to the induction of Th2-associated cytokines (Asnagli *et al.*, 2002) through the up-regulation of GATA-binding protein 3 (GATA3), but it also prevents binding of other transcription factors such as NF-κB, thus leading to a heavily polarized Th2 response (Bennett *et al.*, 1997; Ouyang *et al.*, 1998). It is hypothesized that this inhibition of transcription is regulated epigenetically through chromatin modifications (Wei *et al.*, 2010) of sites involved in the development of a Th1 response. These activities make STAT6 a key player in the production of Th2-associated cytokines and in immune cell polarization towards a Th2 response.

It is not surprising that STAT6 is highly important in the host immune response to *Leishmania*. As discussed previously, a Th2 response is generally associated with susceptibility to *Leishmania*, accompanied by progressive, non-healing infection (Mougneau *et al.*, 2011). Moreover, IL-4 and IL-13 have been shown to contribute to the development of alternatively activated macrophages which is also associated with parasite survival and disease progression (Sacks and Noben-Trauth, 2002; Tripathi *et al.*, 2007).

Studies examining the role of STAT6 in *Leishmania* infection have utilized targeted knockouts of STAT6 as well as other mediators involved in the STAT6 signalling pathway. In a CL experimental mouse infection model using *L. mexicana* parasites, IL-4-deficient mice displayed increased resistance compared to WT mice, and this was associated with the increased production of Th1 cytokines and reduced production of Th2 cytokines (Satoskar *et al.*, 1995). A more recent study utilizing antibody mediated blockade of IL-4 in an *in vivo* model of *L. major* infection demonstrated that IL-4 inhibits the migration of Th1 cells to the infection site by attenuating Th1-associated chemokine receptor expression, thereby promoting parasite persistence (Lazarski *et al.*, 2013). In other studies using a number of knockout mouse models, STAT6 mediated IL-13 signalling was shown to play a major role in susceptibility to *L. major* infection (Dent *et al.*, 1999). STAT6-deficient mice have also been shown to mount a strong Th1 response to *L. mexicana*, showing its role in susceptibility to CL infection (Stamm *et al.*, 1998). Finally, in mice which are naturally susceptible to *L. major* infection characterized by a dominant Th2 response, genetic deletion of STAT6 results in a resistant phenotype associated with Th1 immune responses (Burgis and Gessner, 2007). It is evident that STAT6 and its activation pathway play a major role in determining the outcome of cutaneous forms of leishmaniasis.

These trends do not hold true for mouse models of infections by caused by *L. chagasi* and *L. donovani*, two species which cause VL in the New and Old World, respectively. IL-4 has been shown to play little or no role in susceptibility to *L. chagasi* or *L. donovani*, as IL-4-deficient mice display a similar, or even more susceptible phenotype than WT mice (Satoskar *et al.*, 1995; Wilson *et al.*, 1998). Indeed, some studies suggest a protective role for IL-4 in *L. donovani* infection and a requirement for this cytokine in effective drug therapy against VL (Alexander *et al.*, 2000; Satoskar *et al.*, 1995, 2003a,b). On the other hand, studies using hamster models of VL (which more closely resemble human disease) seem to suggest a role for STAT6 in the pathogenesis of *L. donovani* infection. A recent study showed that STAT6 is up-regulated in hamsters infected by *L. donovani*, causing alternative activation of macrophages, increased production of arg-1 and suppressed production of iNOS, conditions which favour parasite growth and replication (Osorio *et al.*, 2012). Interestingly, targeted knock-down of STAT6 via RNA interference (RNAi) suppressed arg-1induction and enhanced control of parasite growth (Osorio *et al.*, 2012). It is possible that other forms of *Leishmania* may similarly induce the expression of STAT6 in order to subvert the host immune response, although this remains to be seen. Nevertheless, the role of STAT6 in immunity to *Leishmania* as demonstrated by a number of studies suggests that the use of STAT6 inhibitors might be a viable approach in leishmaniasis therapy.

Conclusion

The various mechanisms developed by *Leishmania* parasites to evade host immune signalling pathways to establish infection in the human host include the manipulation of the JAK-STAT pathway (Abu-Dayyeh *et al.*, 2010; Contreras *et al.*, 2010; Gregory and Olivier, 2005; Matte and Descoteaux, 2010; McDowell and Sacks, 1999; Mol *et al.*, 2013; Naiyer *et al.*, 2013; Vargas-Inchaustegui *et al.*, 2008). This presents the possibility of exploiting STAT signalling through the use of antagonists, agonists or inhibitors as therapeutic tools in the management of leishmaniasis in humans. Several animal studies have demonstrated the potential of utilizing this tool in the treatment of *Leishmania* infection. For example, cystatin, a natural cysteine protease inhibitor has been successfully used to treat experimental murine VL caused by *L. donovani* (Das *et al.*, 2001), and has been shown to synergize with IFN-γ in activating the JAK/STAT signalling pathway. Specifically, cystatin increased IFN-γ-mediated activation of JAK1 and JAK2 as well as STAT1α phosphorylation, contributing to the leishmanicidal activity of macrophages via up-regulation of iNOS (Kar *et al.*, 2009). Two other effective anti-leishmanial drugs, sodium stibogluconate and miltefosine have been shown to suppress SHP-1 activation and restore macrophage responses to IFN-γ stimulation (Pathak and Yi, 2001; Wadhone *et al.*, 2009)

Inhibition of STATs that are associated with Th2 responses is another viable target for the development of anti-leishmanial drugs. Proof-of-concept of this approach can be seen in the use of targeted miRNA knock-down of STAT6 in *L. donovani*-infected cells (Osorio *et al.*, 2012). Whether or not this or other approaches that target the STAT signalling pathway are successful in the prevention or treatment of leishmaniasis remains to be seen.

References

Abu-Dayyeh, I., Hassani, K., Westra, E.R., Mottram, J.C., and Olivier, M. (2010). Comparative study of the ability of *Leishmania mexicana* promastigotes and amastigotes to alter macrophage signaling and functions. Infect. Immun. *78*, 2438–2445.

Adamson, A.S., Collins, K., Laurence, A., and O'Shea, J.J. (2009). The Current STATus of lymphocyte signaling: new roles for old players. Curr. Opin. Immunol. *21*, 161–166.

Aggarwal, B.B., Kunnumakkara, A.B., Harikumar, K.B., Gupta, S.R., Tharakan, S.T., Koca, C., Dey, S., and Sung, B. (2009). Signal transducer and activator of transcription-3, inflammation, and cancer: how intimate is the relationship? Ann. N. Y. Acad. Sci. *1171*, 59–76.

Alexander, J., and Bryson, K. (2005). T-helper (h)1/Th2 and *Leishmania*: paradox rather than paradigm. Immunol. Lett. *99*, 17–23.

Alexander, J., Carter, K.C., Al-Fasi, N., Satoskar, A., and Brombacher, F. (2000). Endogenous IL-4 is necessary for effective drug therapy against visceral leishmaniasis. Eur. J. Immunol. *30*, 2935–2943.

Asnagli, H., Afkarian, M., and Murphy, K.M. (2002). Cutting edge: identification of an alternative GATA-3 promoter directing tissue-specific gene expression in mouse and human. J. Immunol. *168*, 4268–4271.

Au-Yeung, N., Mandhana, R., and Horvath, C.M. (2013). Transcriptional regulation by STAT1 and STAT2 in the interferon JAK-STAT pathway. JAKSTAT *2*, e23931.

Barbi, J., Snider, H.M., Bhardwaj, N., Lezama-Davila, C.M., Durbin, J.E., and Satoskar, A.R. (2009). Signal transducer and activator of transcription 1 in T-cells plays an indispensable role in immunity to *Leishmania major* by mediating Th1 cell homing to the site of infection. FASEB J. *23*, 3990–3999.

Bates, P.A., and Rogers, M.E. (2004). New insights into the development biology and transmission mechanisms of *Leishmania*. Curr. Mol. Med. *4*, 601–609.

Behrmann, I., Smyczek, T., Heinrich, P.C., Schmitz-Van de Leur, H., Komyod, W., Giese, B., Müller-Newen, G., Haan, S., and Haan, C. (2004). Janus kinase (Jak) subcellular localization revisited: the exclusive membrane localization of endogenous janus kinase 1 by cytokine receptor interaction uncovers the jak·receptor complex to be equivalent to a receptor tyrosine kinase. J. Biol. Chem. *279*, 35486–35493.

Bennett, B.L., Cruz, R., Lacson, R.G., and Manning, A.M. (1997). Interleukin-4 suppression of tumor necrosis factor alpha-stimulated E-selectin gene transcription is mediated by STAT6 antagonism of NF-kappaB. J. Biol. Chem. *272*, 10212–10219.

Bhardwaj, N., Rosas, L.E., Lafuse, W.P., and Satoskar, A.R. (2005). *Leishmania* inhibits STAT1-mediated IFN-γamma signaling in macrophages: increased tyrosine phosphorylation of dominant negative STAT1beta by *Leishmania mexicana*. Int. J. Parasitol. *35*, 75–82.

Biswas, A., Bhattacharya, A., Kar, S., and Das, P.K. (2011). Expression of IL-10-triggered STAT3-dependent IL-4Ralpha is required for induction of arginase 1 in visceral leishmaniasis. Eur. J. Immunol. *41*, 992–1003.

Blanchette, J., Racette, N., Faure, R., Siminovitch, K.A., and Olivier, M. (1999). *Leishmania*-induced increases in activation of macrophage SHP-1 tyrosine phosphatase are associated with impaired IFN-γ-triggered JAK2 activation. Eur. J. Immunol. *29*, 3737–3744.

Boaventura, V.S., Santos, C.S., Cardoso, C.R., de Andrade, J., Dos Santos, W.L., Clarencio, J., Silva, J.S., Borges, V.M., Barral-Netto, M., Brodskyn, C.I., *et al.* (2010). Human mucosal leishmaniasis: neutrophils infiltrate areas of tissue damage that express high levels of Th17-related cytokines. Eur. J. Immunol. *40*, 2830–2836.

Bournazou, E., and Bromberg, J. (2013). Targeting the tumor microenvironment: JAK-STAT3 signaling. JAKSTAT *2*, e23828.

Brierley, M.M., and Fish, E.N. (2005). Stats: multifaceted regulators of transcription. J. Interferon Cytokine Res. *25*, 733–744.

Bromberg, J., and Darnell, J.E., Jr. (2000). The role of STATs in transcriptional control and their impact on cellular function. Oncogene *19*, 2468–2473.

Burgis, S., and Gessner, A. (2007). Unexpected phenotype of STAT6 heterozygous mice implies distinct STAT6 dosage requirements for different IL-4 functions. Int. Arch. Allergy Immunol. *143*, 263–268.

Buxbaum, L.U., and Scott, P. (2005). Interleukin 10- and Fcγ Receptor-Deficient Mice Resolve *Leishmania mexicana* Lesions. Infect. Immun. *73*, 2101–2108.

Buxbaum, L.U., Uzonna, J.E., Goldschmidt, M.H., and Scott, P. (2002). Control of New World cutaneous leishmaniasis is IL-12 independent but STAT4 dependent. Eur. J. Immunol. *32*, 3206–3215.

Camporeale, A., and Poli, V. (2012). IL-6, IL-17 and STAT3: a holy trinity in auto-immunity? Front. Biosci. (Landmark Ed) *17*, 2306–2326.

Castilho, T.M., Goldsmith-Pestana, K., Lozano, C., Valderrama, L., Saravia, N.G., and McMahon-Pratt, D. (2010). Murine model of chronic L. (Viannia) panamensis infection: role of IL-13 in disease. Eur. J. Immunol. *40*, 2816–2829.

Chen, H., and Jiang, Z. (2013). The essential adaptors of innate immune signaling. Protein Cell *4*, 27–39.

Chen, Z., and O'Shea, J.J. (2008). Th17 cells: a new fate for differentiating helper T-cells. Immunol. Res. *41*, 87–102.

Chowdhury, F.Z., and Farrar, J.D. (2013). STAT2: A shape-shifting anti-viral super STAT. JAKSTAT *2*, e23633.

Collison, L.W., Delgoffe, G.M., Guy, C.S., Vignali, K.M., Chaturvedi, V., Fairweather, D., Satoskar, A.R., Garcia, K.C., Hunter, C.A., Drake, C.G., *et al.* (2012). The composition and signaling of the IL-35 receptor are unconventional. Nat. Immunol. *13*, 290–299.

Contreras, I., Gomez, M.A., Nguyen, O., Shio, M.T., McMaster, R.W., and Olivier, M. (2010). *Leishmania*-induced inactivation of the macrophage transcription factor AP-1 is mediated by the parasite metalloprotease GP63. PLoS Pathog. *6*, e1001148.

Cummings, H.E., Tuladhar, R., and Satoskar, A.R. (2010). Cytokines and their STATs in cutaneous and visceral leishmaniasis. J. Biomed. Biotechnol. *2010*, 294389.

Das, L., Datta, N., Bandyopadhyay, S., and Das, P.K. (2001). Successful therapy of lethal murine visceral leishmaniasis with cystatin involves up-regulation of nitric oxide and a favorable T-cell response. J. Immunol. *166*, 4020–4028.

Delgoffe, G.M., Murray, P.J., and Vignali, D.A. (2011). Interpreting mixed signals: the cell's cytokine conundrum. Curr. Opin. Immunol. *23*, 632–638.

Dent, A.L., Doherty, T.M., Paul, W.E., Sher, A., and Staudt, L.M. (1999). BCL-6-deficient mice reveal an IL-4-independent, STAT6-dependent pathway that controls susceptibility to infection by *Leishmania major*. J. Immunol. *163*, 2098–2103.

Diefenbach, A., Schindler, H., Donhauser, N., Lorenz, E., Laskay, T., MacMicking, J., Rollinghoff, M., Gresser, I., and Bogdan, C. (1998). Type 1 interferon (IFNalpha/beta) and type 2 nitric oxide synthase regulate the innate immune response to a protozoan parasite. Immunity *8*, 77–87.

Dostalova, A., and Volf, P. (2012). *Leishmania* development in sand flies: parasite–vector interactions overview. Parasit. Vectors *5*, 276.

Fasler-Kan, E., Pansky, A., Wiederkehr, M., Battegay, M., and Heim, M.H. (1998). Interferon-alpha activates signal transducers and activators of transcription 5 and 6 in Daudi cells. Eur. J. Biochem. *254*, 514–519.

Forget, G., Siminovitch, K.A., Brochu, S., Rivest, S., Radzioch, D., and Olivier, M. (2001). Role of host phosphotyrosine phosphatase SHP-1 in the development of murine leishmaniasis. Eur. J. Immunol. *31*, 3185–3196.

Forget, G., Gregory, D.J., Whitcombe, L.A., and Olivier, M. (2006). Role of host protein tyrosine phosphatase SHP-1 in *Leishmania donovani*-induced inhibition of nitric oxide production. Infect. Immun. *74*, 6272–6279.

Gregory, D.J., and Olivier, M. (2005). Subversion of host cell signalling by the protozoan parasite *Leishmania*. Parasitology *130 Suppl*, S27–35.

Gupta, G., Oghumu, S., and Satoskar, A.R. (2013). Mechanisms of immune evasion in leishmaniasis. Adv. Appl. Microbiol. *82*, 155–184.

Harris, T.J., Grosso, J.F., Yen, H.R., Xin, H., Kortylewski, M., Albesiano, E., Hipkiss, E.L., Getnet, D., Goldberg, M.V., Maris, C.H., *et al.* (2007). Cutting edge: An *in vivo* requirement for STAT3 signaling in TH17 development and TH17-dependent autoimmunity. J. Immunol. *179*, 4313–4317.

Hebenstreit, D., Wirnsberger, G., Horejs-Hoeck, J., and Duschl, A. (2006). Signaling mechanisms, interaction partners, and target genes of STAT6. Cytokine Growth Factor Rev. *17*, 173–188.

Heinrich, P.C., Behrmann, I., Muller-Newen, G., Schaper, F., and Graeve, L. (1998). Interleukin-6-type cytokine signalling through the gp130/Jak/STAT pathway. Biochem. J. *334 (Pt 2)*, 297–314.

Herwaldt, B.L. (1999). Leishmaniasis. Lancet *354*, 1191–1199.

Hughes, A.L., and Piontkivska, H. (2003). Molecular phylogenetics of Trypanosomatidae: contrasting results from 18S rRNA and protein phylogenies. Kinetoplastid Biol. Dis. *2*, 15.

Imada, K., and Leonard, W.J. (2000). The Jak-STAT pathway. Mol. Immunol. *37*, 1–11.

Jacobson, N.G., Szabo, S.J., Weber-Nordt, R.M., Zhong, Z., Schreiber, R.D., Darnell, J.E., Jr., and Murphy, K.M. (1995). Interleukin 12 signaling in T-helper type 1 (Th1) cells involves tyrosine phosphorylation of signal transducer and activator of transcription (Stat)3 and Stat4. J. Exp. Med. *181*, 1755–1762.

Johnson, L.M., and Scott, P. (2007). STAT1 expression in dendritic cells, but not T-cells, is required for immunity to *Leishmania major*. J. Immunol. *178*, 7259–7266.

Kar, S., Ukil, A., and Das, P.K. (2009). Signaling events leading to the curative effect of cystatin on experimental visceral leishmaniasis: involvement of ERK1/2, NF-kappaB and JAK/STAT pathways. Eur. J. Immunol. *39*, 741–751.

Kima, P.E., and Soong, L. (2013). Interferon gamma in leishmaniasis. Front. Immunol. *4*, 156.

Kima, P.E., Ruddle, N.H., and McMahon-Pratt, D. (1997). Presentation via the class I pathway by *Leishmania amazonensis*-infected macrophages of an endogenous leishmanial antigen to CD8⁺ T-cells. J. Immunol. *159*, 1828–1834.

Kropf, P., Herath, S., Weber, V., Modolell, M., and Muller, I. (2003). Factors influencing *Leishmania major* infection in IL-4-deficient BALB/c mice. Parasite Immunol. *25*, 439–447.

Lazarski, C.A., Ford, J., Katzman, S.D., Rosenberg, A.F., and Fowell, D.J. (2013). IL-4 attenuates Th1-associated chemokine expression and Th1 trafficking to inflamed tissues and limits pathogen clearance. PLoS One *8*, e71949.

Lopez Kostka, S., Dinges, S., Griewank, K., Iwakura, Y., Udey, M.C., and von Stebut, E. (2009). IL-17 promotes progression of cutaneous leishmaniasis in susceptible mice. J. Immunol. *182*, 3039–3046.

McDowell, M.A., and Sacks, D.L. (1999). Inhibition of host cell signal transduction by *Leishmania*: observations relevant to the selective impairment of IL-12 responses. Curr. Opin. Microbiol. *2*, 438–443.

McGwire, B.S., and Satoskar, A.R. (2013). Leishmaniasis: clinical syndromes and treatment. QJM *107*, 7–14.

McMahon-Pratt, D., and Alexander, J. (2004). Does the *Leishmania major* paradigm of pathogenesis and protection hold for New World cutaneous leishmaniases or the visceral disease? Immunol. Rev. *201*, 206–224.

Maier, E., Duschl, A., and Horejs-Hoeck, J. (2012). STAT6-dependent and -independent mechanisms in Th2 polarization. Eur. J. Immunol. *42*, 2827–2833.

Masuda, A., Matsuguchi, T., Yamaki, K., Hayakawa, T., Kubo, M., LaRochelle, W.J., and Yoshikai, Y. (2000). Interleukin-15 induces rapid tyrosine phosphorylation of STAT6 and the expression of interleukin-4 in mouse mast cells. J. Biol. Chem. *275*, 29331–29337.

Matte, C., and Descoteaux, A. (2010). *Leishmania donovani* amastigotes impair gamma interferon-induced STAT1alpha nuclear translocation by blocking the interaction between STAT1alpha and importin-alpha5. Infect. Immun. *78*, 3736–3743.

Mol, M., Patole, M.S., and Singh, S. (2013). Immune signal transduction in leishmaniasis from natural to artificial systems: Role of feedback loop insertion. Biochim. Biophys. Acta *1840*, 71–79.

Mosser, D.M., and Edwards, J.P. (2008). Exploring the full spectrum of macrophage activation. Nat. Rev. Immunol. *8*, 958–969.

Mougneau, E., Bihl, F., and Glaichenhaus, N. (2011). Cell biology and immunology of *Leishmania*. Immunol. Rev. *240*, 286–296.

Murray, H.W., Tsai, C.W., Liu, J., and Ma, X. (2006). Responses to *Leishmania donovani* in mice deficient in interleukin-12 (IL-12), IL-12/IL-23, or IL-18. Infect. Immun. *74*, 4370–4374.

Murray, P.J. (2007). The JAK-STAT signaling pathway: input and output integration. J. Immunol. *178*, 2623–2629.

Naiyer, M.M., Saha, S., Hemke, V., Roy, S., Singh, S., Musti, K.V., and Saha, B. (2013). Identification and characterization of a human IL-10 receptor antagonist. Hum. Immunol. *74*, 28–31.

Najjar, I., and Fagard, R. (2010). STAT1 and pathogens, not a friendly relationship. Biochimie *92*, 425–444.

Nandan, D., and Reiner, N.E. (1995). Attenuation of gamma interferon-induced tyrosine phosphorylation in mononuclear phagocytes infected with *Leishmania donovani*: selective inhibition of signaling through Janus kinases and Stat1. Infect. Immun. *63*, 4495–4500.

Nandan, D., Lo, R., and Reiner, N.E. (1999). Activation of phosphotyrosine phosphatase activity attenuates mitogen-activated protein kinase signaling and inhibits c-FOS and nitric oxide synthase expression in macrophages infected with *Leishmania donovani*. Infect. Immun. *67*, 4055–4063.

Nandan, D., Yi, T., Lopez, M., Lai, C., and Reiner, N.E. (2002). *Leishmania* EF-1alpha activates the Src homology 2 domain containing tyrosine phosphatase SHP-1 leading to macrophage deactivation. J. Biol. Chem. *277*, 50190–50197.

Nishikomori, R., Gurunathan, S., Nishikomori, K., and Strober, W. (2001). BALB/c mice bearing a transgenic IL-12 receptor beta 2 gene exhibit a nonhealing phenotype to *Leishmania major* infection despite intact IL-12 signaling. J. Immunol. *166*, 6776–6783.

Nishikomori, R., Usui, T., Wu, C.Y., Morinobu, A., O'Shea, J.J., and Strober, W. (2002). Activated STAT4 has an essential role in Th1 differentiation and proliferation that is independent of its role in the maintenance of IL-12R beta 2 chain expression and signaling. J. Immunol. *169*, 4388–4398.

Nylen, S., and Gautam, S. (2010). Immunological perspectives of leishmaniasis. J. Glob. Infect. Dis. *2*, 135–146.

Oghumu, S., Gupta, G., Snider, H.M., Varikuti, S., Terrazas, C.A., Papenfuss, T.L., and Satoskar, A.R. (2013). STAT4 is critical for immunity but not for anti-leishmanial activity of antimonials in experimental visceral leishmaniasis. Eur. J. Immunol. *44*, 450–459.

Olivier, M., Gregory, D.J., and Forget, G. (2005). Subversion mechanisms by which *Leishmania* parasites can escape the host immune response: a signaling point of view. Clin. Microbiol. Rev. *18*, 293–305.

Osorio, E.Y., Zhao, W., Espitia, C., Saldarriaga, O., Hawel, L., Byus, C.V., Travi, B.L., and Melby, P.C. (2012). Progressive visceral leishmaniasis is driven by dominant parasite-induced STAT6 activation and STAT6-dependent host arginase 1 expression. PLoS Pathog. *8*, e1002417.

Ouyang, W., Ranganath, S.H., Weindel, K., Bhattacharya, D., Murphy, T.L., Sha, W.C., and Murphy, K.M. (1998). Inhibition of Th1 development mediated by GATA-3 through an IL-4-independent mechanism. Immunity *9*, 745–755.

Paradowska-Gorycka, A., Grzybowska-Kowalczyk, A., Wojtecka-Lukasik, E., and Maslinski, S. (2010). IL-23 in the pathogenesis of rheumatoid arthritis. Scand. J. Immunol. *71*, 134–145.

Park, C., Li, S., Cha, E., and Schindler, C. (2000). Immune response in Stat2 knockout mice. Immunity *13*, 795–804.

Patel, B.K., Wang, L.M., Lee, C.C., Taylor, W.G., Pierce, J.H., and LaRochelle, W.J. (1996). Stat6 and Jak1 are common elements in platelet-derived growth factor and interleukin-4 signal transduction pathways in NIH 3T3 fibroblasts. J. Biol. Chem. *271*, 22175–22182.

Pathak, M.K., and Yi, T. (2001). Sodium stibogluconate is a potent inhibitor of protein tyrosine phosphatases and augments cytokine responses in hemopoietic cell lines. J. Immunol. *167*, 3391–3397.

Perry, S.T., Buck, M.D., Lada, S.M., Schindler, C., and Shresta, S. (2011). STAT2 mediates innate immunity to Dengue virus in the absence of STAT1 via the type I interferon receptor. PLoS Pathog. *7*, e1001297.

Pitta, M.G., Romano, A., Cabantous, S., Henri, S., Hammad, A., Kouriba, B., Argiro, L., el Kheir, M., Bucheton, B., Mary, C., *et al.* (2009). IL-17 and IL-22 are associated with protection against human kala azar caused by *Leishmania donovani*. J. Clin. Invest. *119*, 2379–2387.

Quelle, F.W., Shimoda, K., Thierfelder, W., Fischer, C., Kim, A., Ruben, S.M., Cleveland, J.L., Pierce, J.H., Keegan, A.D., Nelms, K., *et al.* (1995). Cloning of murine Stat6 and human Stat6, Stat proteins that are tyrosine phosphorylated in responses to IL-4 and IL-3 but are not required for mitogenesis. Mol. Cell. Biol. *15*, 3336–3343.

Rauch, I., Muller, M., and Decker, T. (2013). The regulation of inflammation by interferons and their STATs. JAKSTAT *2*, e23820.

Ray, M., Gam, A.A., Boykins, R.A., and Kenney, R.T. (2000). Inhibition of interferon-gamma signaling by *Leishmania donovani*. J. Infect. Dis. *181*, 1121–1128.

Rebe, C., Vegran, F., Berger, H., and Ghiringhelli, F. (2013). STAT3 activation: A key factor in tumor immunoescape. JAKSTAT *2*, e23010.

Reithinger, R., Dujardin, J.C., Louzir, H., Pirmez, C., Alexander, B., and Brooker, S. (2007). Cutaneous leishmaniasis. Lancet Infect. Dis. *7*, 581–596.

Roberts, M.T. (2005). Current understandings on the immunology of leishmaniasis and recent developments in prevention and treatment. Br. Med. Bull. *75–76*, 115–130.

Rosas, L.E., Keiser, T., Pyles, R., Durbin, J., and Satoskar, A.R. (2003). Development of protective immunity against cutaneous leishmaniasis is dependent on STAT1-mediated IFN signaling pathway. Eur. J. Immunol. *33*, 1799–1805.

Rosas, L.E., Keiser, T., Barbi, J., Satoskar, A.A., Septer, A., Kaczmarek, J., Lezama-Davila, C.M., and Satoskar, A.R. (2005). Genetic background influences immune responses and disease outcome of cutaneous L. mexicana infection in mice. Int. Immunol. *17*, 1347–1357.

Rosas, L.E., Snider, H.M., Barbi, J., Satoskar, A.A., Lugo-Villarino, G., Keiser, T., Papenfuss, T., Durbin, J.E., Radzioch, D., Glimcher, L.H., *et al.* (2006). Cutting edge: STAT1 and T-bet play distinct roles in determining outcome of visceral leishmaniasis caused by *Leishmania donovani*. J. Immunol. *177*, 22–25.

Sacks, D., and Noben-Trauth, N. (2002). The immunology of susceptibility and resistance to *Leishmania major* in mice. Nat. Rev. Immunol. *2*, 845–858.

Sacks, D.L., and Perkins, P.V. (1985). Development of infective stage *Leishmania* promastigotes within phlebotomine sand flies. Am. J. Trop. Med. Hyg. *34*, 456–459.

Satoskar, A., Bluethmann, H., and Alexander, J. (1995). Disruption of the murine interleukin-4 gene inhibits disease progression during *Leishmania mexicana* infection but does not increase control of *Leishmania donovani* infection. Infect. Immun. *63*, 4894–4899.

Satoskar, A.R., Rodig, S., Telford, S.R., 3rd, Satoskar, A.A., Ghosh, S.K., von Lichtenberg, F., and David, J.R. (2000). IL-12 gene-deficient C57BL/6 mice are susceptible to *Leishmania donovani* but have diminished hepatic immunopathology. Eur. J. Immunol. *30*, 834–839.

Schonian, G., Kuhls, K., and Mauricio, I.L. (2011). Molecular approaches for a better understanding of the epidemiology and population genetics of *Leishmania*. Parasitology *138*, 405–425.

Spath, G.F., Schlesinger, P., Schreiber, R., and Beverley, S.M. (2009). A novel role for Stat1 in phagosome acidification and natural host resistance to intracellular infection by *Leishmania major*. PLoS Pathog. *5*, e1000381.

Stager, S., Alexander, J., Carter, K.C., Brombacher, F., and Kaye, P.M. (2003a). Both interleukin-4 (IL-4) and IL-4 receptor alpha signaling contribute to the development of hepatic granulomas with optimal anti-leishmanial activity. Infect. Immun. *71*, 4804–4807.

Stager, S., Alexander, J., Kirby, A.C., Botto, M., Rooijen, N.V., Smith, D.F., Brombacher, F., and Kaye, P.M. (2003b). Natural antibodies and complement are endogenous adjuvants for vaccine-induced CD8$^+$ T-cell responses. Nat. Med. *9*, 1287–1292.

Stamm, L.M., Raisanen-Sokolowski, A., Okano, M., Russell, M.E., David, J.R., and Satoskar, A.R. (1998). Mice with STAT6-targeted gene disruption develop a Th1 response and control cutaneous leishmaniasis. J. Immunol. *161*, 6180–6188.

Stamm, L.M., Satoskar, A.A., Ghosh, S.K., David, J.R., and Satoskar, A.R. (1999). STAT-4 mediated IL-12 signaling pathway is critical for the development of protective immunity in cutaneous leishmaniasis. Eur. J. Immunol. *29*, 2524–2529.

Steinman, R.M. (2007). Dendritic cells: understanding immunogenicity. Eur. J. Immunol. *37 Suppl 1*, S53–60.

Takeda, K., Kaisho, T., Yoshida, N., Takeda, J., Kishimoto, T., and Akira, S. (1998). Stat3 Activation Is Responsible for IL-6-Dependent T Cell Proliferation Through Preventing Apoptosis: Generation and Characterization of T Cell-Specific Stat3-Deficient Mice. J. Immunol. *161*, 4652–4660.

Terrazas, C.A., Terrazas, L.I., and Gomez-Garcia, L. (2010). Modulation of dendritic cell responses by parasites: a common strategy to survive. J. Biomed. Biotechnol. *2010*, 357106.

Thomas, B.N., and Buxbaum, L.U. (2008). FcgammaRIII mediates immunoglobulin G-induced interleukin-10 and is required for chronic *Leishmania mexicana* lesions. Infect. Immun. *76*, 623–631.

Tripathi, P., Singh, V., and Naik, S. (2007). Immune response to leishmania: paradox rather than paradigm. FEMS Immunol. Med. Microbiol. *51*, 229–242.

Vargas-Inchaustegui, D.A., Xin, L., and Soong, L. (2008). *Leishmania braziliensis* infection induces dendritic cell activation, ISG15 transcription, and the generation of protective immune responses. J. Immunol. *180*, 7537–7545.

Wadhone, P., Maiti, M., Agarwal, R., Kamat, V., Martin, S., and Saha, B. (2009). Miltefosine promotes IFN-γamma-dominated anti-leishmanial immune response. J. Immunol. *182*, 7146–7154.

Walters, L.L., Chaplin, G.L., Modi, G.B., and Tesh, R.B. (1989a). Ultrastructural biology of *Leishmania* (Viannia) panamensis (=*Leishmania braziliensis* panamensis) in *Lutzomyia gomezi* (Diptera: Psychodidae): a natural host–parasite association. Am. J. Trop. Med. Hyg. *40*, 19–39.

Walters, L.L., Modi, G.B., Chaplin, G.L., and Tesh, R.B. (1989b). Ultrastructural development of *Leishmania chagasi* in its vector, Lutzomyia longipalpis (Diptera: Psychodidae). Am. J. Trop. Med. Hyg. *41*, 295–317.

Wei, L., Vahedi, G., Sun, H.W., Watford, W.T., Takatori, H., Ramos, H.L., Takahashi, H., Liang, J., Gutierrez-Cruz, G., Zang, C., *et al.* (2010). Discrete roles of STAT4 and STAT6 transcription factors in tuning epigenetic modifications and transcription during T-helper cell differentiation. Immunity 32, 840–851.

Wilson, M.E., Young, B.M., Davidson, B.L., Mente, K.A., and McGowan, S.E. (1998). The importance of TGF-beta in murine visceral leishmaniasis. J. Immunol. *161*, 6148–6155.

Xin, L., Li, K., and Soong, L. (2008). Down-regulation of dendritic cell signaling pathways by *Leishmania amazonensis* amastigotes. Mol. Immunol. *45*, 3371–3382.

Leishmania Modulates Toll-like Receptor Signalling in Macrophages

8

Soumya kanti Ghosh, Kalavati M. Lalsare and Bhaskar Saha

Abstract

Toll-like receptors (TLRs) are essential pattern recognition receptors (PRR) of the innate immune system. Their sensitivity to different types of ligands make them appropriate immunological receptors against most type of pathogens. Macrophages, the immune cells that play host to the protozoan parasite *Leishmania,* express the entire TLR family of receptors along with other PRRs. TLRs play vital roles during *Leishmania* infection via regulation of proinflammatory and anti-inflammatory cytokines. *Leishmania* interacts with host cells and invade them through surface molecules such as lipophosphoglycan (LPG), phosphoglycans and other pathogen molecules. Being ligands to several TLR receptors, these molecules can regulate TLR signalling and iNOS expression along with oxidative bursts following infection. Leishmanial survival techniques exploit TLR–TLR cross-talk and TLR signalling mechanisms for survival. This brings changes in signalling pattern and cytokine secretions. This chapter deals with the impact of leishmanial infection on TLR signalling, a major component of innate immunity and a trigger factor for adaptive immune system.

Introduction

Every living being in nature owns a defence mechanism to avoid any external attack, be it plants through their thick bark and variable chemical secretions and exudates or arthropods with their chitin exoskeleton. Mammals have intricate immune systems to work as an army to defend entire body from external opportunistic pathogens, particularly, parasites. On the other hand, the co-evolution of parasites has enabled them to evade mammalian immune systems. The immune cells play vital roles in the recognition of pathogens, stimulation of the required signalling, the regulation of cytokine secretions and activation of pathogen killing simultaneously. The vector, sand fly injects *Leishmania* promastigotes into the epidermis where they encounter different immune cell types like Langerhans cells, dendritic cells, neutrophils and tissue macrophages. Macrophages are one of the major types of antigen-presenting cells with more than one receptor to recognize surface antigens of leishmanial parasite. TLR is one of the most vital receptor on antigen-presenting cells. Ligand–receptor interaction stimulates macrophage to initiate series of signalling responses which are necessary for leishmanial antigen presentation, cytokine secretion and parasite killing. Toll-like receptors regulate various effector functions including NO synthesis, reactive oxygen species generation, pro-inflammatory cytokine secretion followed by myeloid and lymphoid cell activation and anti-inflammatory cytokine secretion to prevent hyper-inflammation.

Toll-like receptors are the only pathogen recognizing receptors (PRRs) which are discovered early in comparison to other PRRs. Toll-like receptors are Type 1 transmembrane proteins which have an extracellular, a transmembrane and a cytoplasmic domains. Toll-like receptors were initially recognized in *Drosophila melanogaster* as a vital player for embryonic development. Later Toll receptors of *Drosophila melanogaster* and Toll-like receptors of human were observed as key regulators for host immune response (Lemaitre *et al.*, 2008). Human Toll-like receptors have ability of direct pathogen recognition, initiating signalling and recruitment of transcription factors, e.g. NF-κB transcription factor, AP1, Interferon regulating factors (IRFs) 3 and 7. TLRs recognize antigens through structural recognition sites on the TLR ectodomain. These specific structural motifs are called leucine-rich repeats. Leucine-rich repeats are 20 to 30 amino acid sequences which are rich in amino acid leucine. Every TLR has a different conserved sequence of LRR which is specific to the ligands they recognize.

Toll-like receptors are available on cell membrane as well as on the endosomal membrane of antigen-presenting cells (APC). The number of TLRs available on macrophage varies based upon the phylogenetic position of the animal, such as human or mice. Human immune cells express TLR 1–10 whereas in mice TLR1–13 are found but TLR10 is functionally not expressed. TLRs are localized at two different positions depending on the types of agonist they recognize; some are cell membrane localized and some localized on endosomal membrane. TLRs localized on cell membrane recognize the pathogen surface markers or antigens such as lipopolysaccharide, flagellin, lipophosphoglycan, diacyl and triacyl lipopeptide. Similarly TLRs localized in the endosomal membrane recognize nucleic acids, e.g. ssRNA, dsRNA, DNA, etc. (Medzhitov *et al.*, 2002). Protozoan parasites like *Leishmania* have several methods to counter-protect self from host defence mechanisms. *Leishmania* too uses Toll-like receptors for synthesis of anti-inflammatory cytokine for its survival. The pathways through which TLR perform their signalling and the ability of *Leishmania* to regulate this well established host receptor signalling for its own survival will be discussed in the chapter.

Other than TLR, there are other PRRs which too have same role as TLRs but with a variation in signalling mechanism. We will be discussing about these receptor mediated signalling pathways in context to *Leishmania* infection.

Innate immune receptors on antigen-presenting cell, macrophage

Receptors on macrophage

Macrophages are the major phagocytic cells which phagocytose a variety of pathogens such as viruses, bacteria and protozoa. This antigen-presenting cell internalizes pathogen via receptor mediated endocytosis. Macrophages possess various receptors on their surface and every receptor has a specific role in the recognition of the pathogen. Among all these receptors, some receptors present on the antigen-presenting cell play a vital role in the activation of adaptive immune response and also enable macrophages to be effector cells for the activation of innate immune system. Macrophages migrate through the lymphatic system to the various organs and tissues, where they serve multiple functions in different tissues (Roitt *et al.*, 2001). They are designated according to their tissue location as follows:

- Kupffer cells in the liver
- mesangial cells in the kidney
- alveolar cells in the lungs
- histocytes in the connective tissues
- microglial cells in the brain
- osteoclasts in bone.

Receptors involved in the endocytosis of pathogens by macrophages are as follows (Fig. 8.1):

1. complement receptor 1 (CR1), i.e. C3b receptor, CD35
2. complement receptor 3 (CR3), i.e. CD11b, MAC-1, fir C3bi
3. fibronectin receptor
4. mannose receptor
5. advanced glycosylation end product receptor
6. mannosyl-fucosyl receptors
7. NOD-like receptors
8. Toll-like receptors
9. phosphatidyl serine receptors
10. IgG receptors
    a. high-affinity FcγRI (CD64)
    b. low-affinity FcγRII (CD32)
    c. medium-affinity FcγRIII (CD16)
11. C-type Lectin receptor

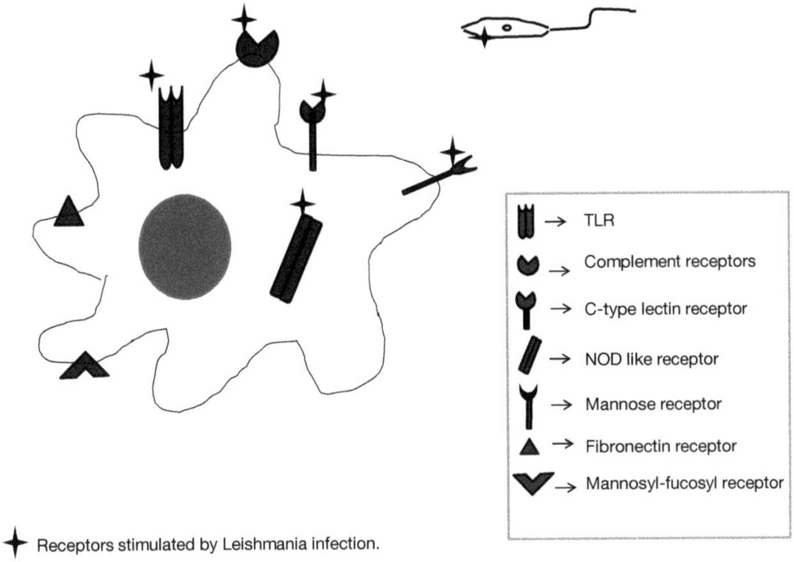

Figure 8.1 Host macrophage with the essential surface receptors. It comprises of *Leishmania* antigen recognizing receptors and non-recognizing receptors.

Role of receptors on cell surface

Some of the above-mentioned receptors are involved in the recognition and phagocytosis of *Leishmania* parasite. The process of *Leishmania* phagocytosis is of two types: zipper phagocytosis and coiling phagocytosis. During zipper phagocytosis, internalization of *Leishmania* is characterized by positive co-operativity, i.e. one receptor increases the expression of the other receptors on the surface of the surrounding membrane. The coiling phagocytosis is an asymmetrical way to phagocytose the *Leishmania*. CR1 and CR3 receptors are critical for both types of phagocytosis.

1 *Complement receptor 1 (CD35)*. This receptor binds to C3b complement protein and lipophosphoglycan (LPG), a major surface molecule of promastigotes of *Leishmania major* and helps in the internalization of promastigotes into the macrophages. This receptor is investigated by blocking the CR1 receptors on the macrophages, which significantly reduces the parasite load and oxidative burst in macrophages.
2 *Complement receptor 3 (MAC-1/CD11b)*. It is a heterodimeric receptor, which recognizes the iC3b, a hydrolysed product of C3b. It also binds to the LPG and gp63 of less virulent amastigotes. It has been reported that CR3 is the most potent receptor for the uptake of *Leishmania* promastigotes by macrophages as compared to the CR1receptor. *Leishmania* surface molecules such as lipophosphoglycan, proteophosphoglycan and parasite surface antigen 2 (PSA-2) have a leucine-rich repeat motif which is recognized by the CR3 and thus it facilitates the pathogen recognition, attachment and engulfment by macrophages (Mosser *et al.*, 1985).
3 *Complement receptor 4*. CR4 is a β-integrin family receptor present on the monocytes-derived macrophages, well studied in the phagocytosis of *Mycobacterium tuberculosis*. Previous studies shown that phagocytosis of bacteria, parasite by CR4 is opsonin dependent (C3bi) but further study by Zaffran *et al.* (1998) showed that CR4 recognizes the *Mycobacterium tuberculosis* in opsonin independent pathway and it directly recognize the tyrosine-phosphorylated bacterial proteins and gain the entry into the host cells (Zaffran *et al.*, 1998).
4 *Mannose receptors*. These receptors of macrophages are attached with the terminal mannose, fucose and N-acetyl glucosamine

of branched polysaccharides of *Leishmania* surface molecules. It does not recognize the amastigote form of *Leishmania*. Also, it does not activate the host immune response, like production of proinflammatory cytokines such as IL12, IL6 and TNFα. It plays a major role in the wound healing, tissue remodelling and hormones clearance during inflammatory response (Akilov *et al.*, 2007).

5 *Fibronectin receptor*. Fibronectin like protein gp63 or leishmanolysin is major protein present on the extracellular membrane of the *Leishmania* and like fibronectin, it binds with the fibronectin receptors of macrophages. Both complement and fibronectin receptors are the most efficient receptors for the phagocytosis of *Leishmania* by macrophages.

6 *Phosphatidyl serine receptors*. Phosphatidyl serine receptors are involved in the clearance and phagocytosis of inflammatory necrotic and apoptotic cells by macrophages. These receptors play a very important role in the prevention of tissue damage by necrotic cells and accumulation of apoptotic cells in the inflammatory area. Other receptors are also involved in phagocytosis of apoptotic cells including scavenger receptors, CD14, and lectin receptors.

7 *Lipopolysaccharide binding proteins*. One binding protein, CD14, a membrane phosphatidylinositol glycan-linked protein and scavenger receptors are binding proteins which have more affinity to lipopolysaccharides (LPS) of Gram-negative bacteria. Role of both receptors in the internalization of bacteria and other pathogenic organisms by macrophages is not known. LPS-binding proteins bind to *Mycobacterium tuberculosis* but phagocytosis may be mediated by other receptors. Scavenger receptors also bind to lipoteichoic acid of Gram-positive bacteria. CD14 recognizes the LAM (Lipoarabinomannan of *Mycobacterium tuberculosis* and triggers macrophages for IL-8 secretion. Toll-like receptor 4 (TLR4) even recognizes the LPS and induces the production of pro-inflammatory cytokines (Ernst *et al.*, 1998).

8 *IgG receptors*. IgG receptors bind to the Fc fragment of IgG antibody-coated microorganisms and facilitates the internalization of IgG-coated pathogen, activation and inactivation of immune response, cytokine production and development of T- and B-lymphoid cells. There are three different types of IgG receptors present on the leucocytes. First is the FcγRI (CD64), which is the high affinity receptor expressed on macrophages, dendritic cell and only binds with monomeric IgG (IgG1 and IgG3). Second one is FcγRII (CD32), which is the low affinity receptor found on PMN, macrophages, dendritic cells, mast cells and binds with complexed IgG2. Third one is the intermediate affinity receptor, FcγRII (CD16), which is expressed on NK cell, macrophages, platelet, mast cells and also binds with complex of IgG1 and IgG3 (Lynch *et al.*, 2000).

Comparison of other receptors against Toll-like receptor over versatility and efficacy of signalling on defence against infection

Pattern recognition receptor (PRR) is the initial responder against any foreign invasion which may be viral, bacterial, fungal or parasitic. There are different families of PRRs working together against infections. These include RLRs, NLRs, CLRs and TLRs, which are the different families of PRR (Table 8.1).

RIG-1-like receptor (RLR)

The RIG-1-like receptor (RLR) family comprises RIG-1, melanoma differentiation-associated gene-5 (MDA-5) and LGP2. RIG-1 and MDA5 are interferon (IFN)-secreting RNA helicases which have a central DEAD Box helicase with ATPase activity, CARD (N-terminal caspase recruitment domain) and a C-terminal regulatory domain. RLRs recognize viral dsRNA and interaction of dsRNA in c-terminal region triggers signalling via CARD–CARD interactions between helicase and adaptor protein IFN-β promoter stimulator-1 (IPS-1), which is followed by type I IFN production. Thus, RLRs recognize only dsRNA, among which RIG1 recognizes short dsRNA and 5′-triphosphate RNA whereas MDA-1 recognizes long dsRNA.

Table 8.1 Recognition of microbial component by variable pattern recognition receptors

	TLR	RLR	NLR	CLR
Bacteria				
Triacyl lipoprotein	+	–	–	–
Lipotechoic acid	+	–	–	–
Peptidoglycan	+	–	–	–
LPS	+	–	–	–
Flagellin	+	–	–	–
CpG DNA	+	–	–	–
Diamiopimelic acid	–	–	+	–
MDP	–	–	+	–
DNA	–	–	+	–
Virus				
DsRNA	+	+	–	–
ssRNA	+	–	–	–
CpG DNA	+	–	–	–
Envelope glycoprotein	+	–	–	–
DNA	–	–	+	–
Fungi				
Phospholipomannan	+	–	–	–
Zymosan	+	–	–	–
βGlycan/βGlucan	β-Glycan	–	–	β-Glucan
Mannan	+	–	–	–

NOD-like receptors (NLRs)

NOD-like receptors (NLRs) belongs to the innate immune receptor family and is known by centrally located NOD which induces oligomerization. NLR has a C-terminal LRR for ligand sensing and N-terminal CARD is responsible for initiation of signalling. NLR family has two members NOD1 and NOD2. NOD1 senses diaminopimelic acids, found in Gram-negative bacteria and NOD2 senses muramyl dipeptide, which is found on both Gram-negative and Gram-positive bacteria. Stimulation with any of the ligand or antigen results to proinflammatory cytokine secretions.

Member of inflammasome family NAcht LRR protein (NALP3) recognizes PAMPs through NLR. It recognizes viral and bacterial RNA. Among all types of pathogens, NLRs recognize specific Gram-negative and Gram-positive bacterial molecules with NALP3 recognizing viral and bacterial RNA.

C-type lectin receptors (CLRs)

CLRs are transmembrane receptor proteins which are known for their carbohydrate-binding domain. CLRs recognize carbohydrate molecules on bacteria, virus and fungi. Dectin-1 and dectin-2 comprise an immune receptor tyrosine-based activation motif (ITAM) coupled with CLRs which identify β-glycan of fungi. C-type lectin MINCLE on macrophage senses fungal infection and an endogenous protein, i.e. spliceosome associated protein 130 (SAP130) from necrotic host cells. CLR stimulation results in proinflammatory cytokine secretion. Thus, CLRs recognizes carbohydrate antigenic molecules on bacteria, virus and fungi (Takeuchi *et al.*, 2010; Mogensen *et al.*, 2009).

Toll-like receptors

Toll-like receptor is the most widespread innate immune receptor system. The human TLR system has 10 TLRs. Among these ten TLRs, TLR1, TLR2, TLR4, TLR5, TLR6 and TLR10 are on cell membrane, whereas TLR3, TLR7, TLR8 and TLR9 are located on endosomal membrane. TLRs have specificity to a wide spectrum of ligands:

1. TLR1–2 recognizes triacylated lipoproteins and peptidoglycan Gram-negative bacterial molecule.
2. TLR2–6 recognizes diacylated lipoproteins of Gram-positive bacterial molecule, even Gram-negative bacteria might possess diacylated lipoproteins.
3. TLR2 recognizes lipophosphoglycan a parasite surface molecule, zymosan and β-glycan which are fungal molecules.
4. TLR4 recognizes lipopolysaccharides a bacterial molecule and paclitaxel.
5. TLR5 recognizes flagellin present in bacteria and parasites.
6. TLR3 recognizes ds RNA specific to viral family like picornavirues, influenza virus, paramyxovirus along with some more viral types and TLR 3 even recognizes tRNA.
7. TLR7 recognizes ssRNA, imidazoquinolines and guanosine analogues.
8. TLR8 recognizes ssRNA, imidazoquinolines.
9. TLR9 recognizes CpG DNA, which can be observed in most of bacteria and parasites.

TLRs recognize molecules on pathogen surface as well as nucleic acids. TLRs pattern recognition capacity is much wider than the other PRRs, which have restricted ligand recognition ability such as RLRs have specificity to dsRNA of virus, NLRs have specificity to Gram-positive and Gram-negative bacterial molecules and CLRs recognize carbohydrate ligands on bacteria, virus and fungi. Collectively, TLRs are the most efficient and versatile innate immune receptors. A synergistic association of some of the PRRs with Toll-like receptors impacts proinflammatory cytokine and type I IFN secretions (Takeuchi *et al.*, 2010; Mogensen *et al.*, 2009).

Toll-like receptors: structure and its agonists

As a germline-encoded innate immune receptor, Toll-like receptors has been selected for evolutionary development to recognize pathogen-derived antigen. As earlier discussed it is a Type I transmembrane protein with N-terminal ligand recognition domain, a transmembrane helix and cytoplasmic C-terminal binding domain (Fig. 8.2). N-terminal ligand recognizing domain has multiple repeats of the amino acid leucine and thus it is called leucine-rich repeats region (LRR). LRR is structured or arranged in a specific pattern in different TLRs which enable the innate immune receptor to achieve specificity against ligands on the pathogen's surface or cytoplasm. Transmembrane domain contains a series of 20 uncharged and mostly hydrophobic residues. Transmembrane domain is essential in case of specific TLR for localization to endosomal membrane post infection. An endoplasmic reticulum transmembrane protein UNC93B interacts with TLR transmembrane domain to assist signalling

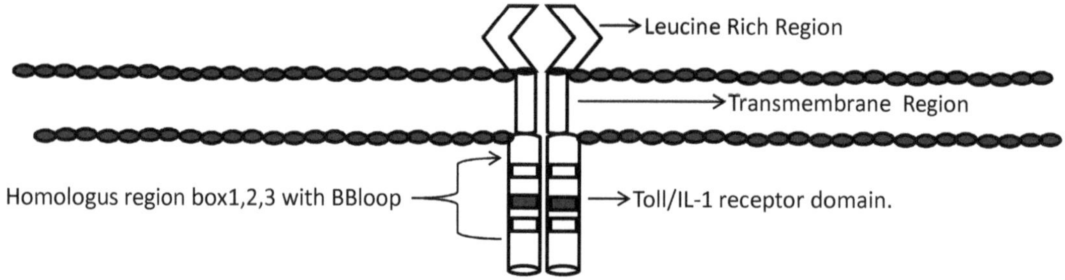

Figure 8.2 Structural details of Toll-like receptors. (1) An extracellular region with a leucine-rich domain. (2) A transmembrane region. (3) A cytoplasmic region with Toll/IL-1 receptor (TIR) domain. TIR region has three conserved homologous box regions, each with a BB loop as the connecting medium between them.

and endosomal localization. UNC93B interacts with the TLRs which recognize nucleic acid as ligands such as TLR3, TLR7 and TLR9. TLR9 is an endoplasmic reticulum localized protein, on stimulation by CpG its trans-membrane domain interacts with UNC93B which subsequently leads to TLR9 localization to endosome (Brinkmann *et al.*, 2007). This known observation has provided with a substantial outlook of the phenomenon of TLR3, -7 and -9 stimulation and translocation to endosomal membrane through trans-membrane domain. Cytoplasmic domain consists of TIR (Toll IL-1 receptor). They are called so because they share homology with signalling domain of the IL-1 receptor family. TIR has specificity to adaptor proteins such as MyD88 and TRIF. Both the adaptor proteins have a specific downstream signalling pattern. The effector transcription factor recruitment downstream of signalling and later cytokine synthesis even varies for both the adaptor proteins.

Leucine-rich repeats of ectodomain: the distinguishing factor of Toll-like receptors

Leucine rich repeats are the antigen recognizing and binding motifs on the extracellular domain of TLR. As the name suggests, the domain is rich in leucine. LRR is divided into a highly conserved segment (HCS) and variable segment (VS). The HCS part of the LRR consists of 11 residue motif LxxLxLxxNxL or some have 12 residue motif LxxLxLxxCxxL and Variable segment is characterized into seven classes of LRRs RI like, CC, Bacterial, SDS22-like, plant specific, Typical and PS. Typical, RI, CC and SD22-like are available in animals. An entire LRR domain is a single continuous structure with an arc shape. On the inner or concave side there are stacks of β-strands, forming parallel β-sheets with hydrophobic residues forming curve inside, whereas the outer side, i.e. the convex side, has a variety of secondary structures of α-helix, 3_{10}-helix, polyproline II helix and tandem arrangement of β-turn. Each time after ligand interaction two LRR ectodomain of TLRs dimerizes which is followed by signal transduction downstream (Brinkmann *et al.*, 2007; Matsushima *et al.*, 2007; Bell *et al.*, 2003).

Leucine-rich repeats of Toll-like receptors

TLRs have been grouped based on the LRR homology. They are grouped into six families: TLR 1/2/6 and 10, TLR3, TLR4, TLR5, TLR7/8 and 9 and TLR11/12 and 13. All the TLRs have significant features of extracellular domain ligand binding motifs which enabled them to characterize pathogen generated surface and intracellular molecules and identify antigens of their specificity. For example, TLR 1–2 heterodimer recognizes triacylated lipopeptides Pam3CSK whereas TLR2–6 heterodimer recognizes diacylated lipopeptide FSL/PamCysPamSK. This ability of different heterodimers to recognize the ligand lies in the modification of LRR. For example, TLR1–2 heterodimer LRR pocket is lined with hydrophobic residue which accommodates long lipid chains and thus it accommodates triacylated lipopeptide. Whereas for TLR2–6 heterodomain the TLR6 LRR pocket has side chains of two phenylalanine residues which blocks lipid binding channel, as the channel is less than half the size of TLR1. Thus, the structural confirmations provide specificity for diacylated over triacylated lipopeptide.

TLR4 agonist lipopolysaccharide (LPS) interacts with TLR4 LRR through MD2 co-receptor. The number of lipid chains in the lipid A portion is the deciding factor for virulence of LPS, i.e. LPS with five lipid chains is optimally virulent over six lipid chain LPS, which is 100-fold less virulent. MD-2 co-factor is considered to be a pocket where the lipid chain of LPS interacts through hydrophobic contact. Smaller lipid chain LPS through shifting position gets accommodated inside MD-2 pocket but at the same time LPS looses ability to procure ionic interaction with LRR9 of mTLR. Thus, the long lipid chain LPS shows optimal interaction with TLR4.

These examples provides substantial basic information regarding the way LRR of every TLR recognize ligands in their own unique manner and this brings specificity among TLR–ligand interaction (Takeuchi *et al.*, 2010).

TIR (Toll/IL-1 receptor) region of Toll-like receptor

TLR responds to pathogen attack on ligand–TLR interaction through ectodomain dimerization and subsequently stimulates downstream signalling by TIR dimerization. TIR is an essential 200 amino acid conserved structural region of Toll-like receptor. TIR region has conserved region, mostly found are BBloop, with helix αA, strand βB and aromatic side chain of (F/Y) DA motif. BB loop connects five stranded β sheet and surrounding five helical α-helix. BB loop is even found in adjacent to the conserved box 1 and 2 of TIR region. Mutation in these positions significantly reduces the signalling of the TLR receptor. On stimulation the dimerization of TIR region of TLR is recognized by the TIR binding domain on adaptor proteins such as MyD88, TRIF, TIRAP and TRAM which initiates downstream signalling and is followed by cytokine secretions. TIR dimerization is substantially dependent on BB loop along with contributions from the DD-loop and αC helix. Other than regulation of TIR-TIR dimer formation, BB-loop is also involved in interaction with adaptor protein during signal transduction. These conserved regions of TIR enable downstream signalling through interaction with TIR domain of adaptor molecules and regulate the type of cytokine secretion and any mutation in these conserved region reduce signalling tremendously (Akira *et al.*, 2004).

Toll-like receptors and their agonists

Toll-like receptors have a significant role in the innate immune system as a pathogen recognizing receptor. TLRs have acquired the significance owing to their ability to recognize a variety of antigenic molecules of pathogen in simultaneous with capability to regulate two different signalling pathways which decides the type of cytokine to be synthesized. Each Toll-like receptor has specific ligand for interaction. Each ligand generates any one recognized adaptor mediated signalling, i.e. either MyD88- or TRIF-dependent signalling. Pathogens exploit this nature of signalling and on contrary to the normal innate immune response against infection it express certain regulating proteins and antigenic molecules which directs antigen-presenting cells to synthesize anti-inflammatory cytokines to nullify effects of proinflammatory cytokine on infection.

Every TLR has specificity to ligands due to characteristic structural signature regions on ligands which are complementary to the TLR specific leucine-rich repeats.

- *TLR1/2/6*. TLR1 and TLR6 cannot form a homodimer but individually form a heterodimer with TLR2. This gives TLR1–2 heterodimer and TLR6–2 heterodimer. TLR1–2 heterodimer recognizes triacylated lipoproteins such as Pam_3CSK, peptidoglycan and TLR2–6 heterodimer recognizes diacylated lipoproteins such as PamCysPamSK or FSL. While discussing differences between TLR1 and TLR6, it was clear that TLR1 accommodates long lipid chains and thus have gained specificity to triacylated lipoproteins which allows Pam_3CSK and peptidoglycans as ideal ligands for TLR1–2 heterodimer. On the contrary, owing to the presence of two phenylalanine residue on the side chain in TLR6 LRR region blocks long-chain binding.
- *TLR2–2*. TLR2–2 homodimer recognizes lipophosphoglycan (LPG), which is a major cell surface molecule for *Leishmania* parasite. Lipophosphoglycan is a long chain of phosphodisaccharide mannose and galactose domains with a hexaglycosyl unit and lipid moiety and has been found to be involved in several functions to support parasite survival.
- *TLR4*. Most essential ligand for TLR4 is lipopolysaccharides (LPS). LPS binds to LPS binding protein which is subsequently recognized by CD14 a glycosylphosphatidylinositol molecule. CD14 is mostly observed on monocytes, macrophage and neutrophils. CD14 next interact with MD2 to accommodate LPS inside MD2 for further interaction with the TLR4 LRR region.
- *TLR5*. Flagellin, the important constituent of flagella, is the vital and only agonist of TLR5. Flagellin is an evolutionary conserved pathogen-associated molecular pattern and is recognized by conserved LRR sequence of receptor TLR5.
- *TLR10*. It is the one Toll receptor not available in mouse antigen-presenting cells but is exclusively available on human antigen-presenting cells. TLR10 dimerises with TLR2 and recognizes triacylated lipopeptides. Profilin has been reported as an agonist for TLR–10.

- *TLR9*. TLR9 is localized to endosomal membrane and being a cytoplasmic Toll receptor it recognizes nucleic acids. The unmethylated CpG is detected by TLR9.
- *TLR7 and TLR8*. These both recognize ssRNA along with imidazoquinolines; in contrast, TLR3 recognizes dsDNA (Takeuchi *et al.*, 2010) (webresource 1).

Agonists against Toll-like receptors presented by *Leishmania major* and *Leishmania donovani* on initiation of infection in antigen-presenting cells.

With infection of one type, i.e. parasite or bacterial species, specific TLRs get stimulated instead of all TLRs participating in the effector downstream signalling to subvert pathogen survival. Similarly, *Leishmania* parasite also stimulates specific TLR signalling. The selection of TLRs to be stimulated on infection commonly depends on the type of agonist presented by the parasite.

TLR2 and MyD88 adaptor protein on *Leishmania major* infection

Direct activation of TLR2 via leishmanial surface molecule has been observed. One of the most abundant *L. major* surface molecule, i.e. lipophosphoglycan (LPG) is observed to stimulate and up-regulate TLR2 mediated signalling on antigen-presenting cells such as macrophage as well as in NK cells which is followed by enhanced TNF-α and IFN-γ synthesis. Species and strain specific structural differences can be observed in LPG phosphoglycan length and the type of saccharides contributing as its building block which impacts the intensity of infection of the parasite. Lipophosphoglycan is a long chain of phosphorylated saccharide repeating units linked with glycosylphosphatidylinositol. Different parasites have variable saccharide groups forming phosphoglycan unit of LPG. *Leishmania major* have galactose and mannose as phosphorylated glycan backbone for LPG. Efficient infection can be established by *Leishmania major* only at promastigote stage, as at promastigote stage *Leishmania* attains metacyclic lipophosphoglycan which is having multiple phosphorylated disaccharides with multiple branching of β1–3Gal-transferase and β1–3Ara-transferase side chains and active LPG to stimulate TLR 2 dependent signalling. LPG-mediated TLR2 stimulation through MyD88-dependent manner engage in proinflammatory cytokine secretion such as IL-12, IFN-γ and TNF-α. LPG stimulation too trigger nitric oxide and reactive oxygen species production. *Leishmania major* LPG is even observed to induce production of suppressors against cytokine signalling proteins such as SOCS-1 and SOCS-3. Expression of such suppressors diminish proinflammatory cytokine secretion through abrogating TLR4 signalling, which concludes the role of LPG in proinflammatory response as well as in cytokine suppression (deVeer *et al.*, 2003; Oliver *et al.*, 2005).

The dual role of lipophosphoglycan establishes its position as a regulator of innate immune response during *Leishmania major* infection. These properties of lipophosphoglycan enable it to be the most essential antigen on the *Leishmania major* surface.

Flagellin and TLR5 receptor

Leishmania parasite has flagella as its motility factor. Flagella functions as locomotion and sensory organelle. Flagella assists in chemotactic movement of the parasite which is merely a response to the environmental chemical state. Flagella comprises basic structural features which are distinguished into three fundamental structures. First is basal body which depends on the proton motive force to work as motor, second is torsion hook and finally the filament which is composed of subunits of protein flagellin. Flagellin is the antigenic factor which elicits TLR5-mediated signalling for proinflammatory cytokine secretions. The N-terminal and C-terminal chains of flagellin are packed as α-helical structures which are localized centrally and are conserved structures. The flagellin variable region is a β-sheet structure. It is the monomer of flagellin that induces TLR5 signalling, not the flagellar structure. The stimulating antigenic moiety of flagellin is compact in flagellum but is exposed as monomer. The LRR region on the TLR5 extracellular domain is observed to identify a single monomer. This hints that two TLR5 monomer units form TLR5 homodimer. Flagellin monomer-mediated TLR5 signalling activates the NF-κB through MAPK pathways which are followed by encoding immune mediators. As the signalling pathway is

MyD88 mediated; thus, the transcription factor playing the prominent role here would be NF-κB (Didilaurent *et al.*, 2004).

Leishmanial cytidine phosphate guanosone (CpG) and TLR9

Almost all foreign pathogens have unmethylated CpG which allows the host innate immune system to identify pathogen invasion. Innate immune receptors have a vital role to play, Toll-like receptor is one such innate immune receptor family which has developed specificity towards such unmethylated CpG. TLR9 on stimulation signals for proinflammatory cytokine synthesis to supersede pathogen infection development. On *Leishmania* infection too TLR9 senses unmethylated CpG and via MyD88-dependent pathway signals for NF-κB nuclear translocation and induction of proinflammatory cytokine synthesis.

Types of Toll-like receptor-mediated signalling

Toll-like receptor signalling induces synthesis of variable cytokines through activation of several transcription factors, e.g. NF-κB, AP1 and IRFs. Every TLR has a specific pattern of signalling which enable them to further synthesis variable pro- inflammatory or anti-inflammatory cytokine through the above mentioned transcription factors. This differential signalling patterns obtained, explicitly depends on adaptor proteins which binds with TIR region.

MyD88-dependent pathways

MyD88-dependent pathway is followed by mostly all TLRs except endosomal membrane localized TLR3 and translocalized TLR4 to endosomal membrane. MyD88 signalling induces synthesis of proinflammatory cytokines IL-6 and IL-12p40. TIRAP a co-adaptor molecule with MyD88 is vital as connecting link between TLR and MyD88. The adaptor molecules have a death domain at the N-terminal end and TIR domain at the C-terminal end. TIR domain interacts with TLR and death domain interacts with IL-1 receptor-associated kinase (IRAK). Out of four IRAKs (IRAK-1, IRAK-2, IRAK-4 and IRAK-M), IRAK-4 interacts with the death domain of MyD88 and thus gets activated simultaneously with IRAK-1. IRAKs later dissociates from TIR region and interacts with TNFR-associated factors 6 (TRAF6); thus, IRAK-1 phosphorylates TRAF 6. TRAF6 acts as E3 ubiquitin protein ligase which in complex with E2 ubiquitin-conjugating enzyme complex (ubc13 and Uev1A) catalyses formation of lysine63 linked polyubiquitin chain on TRAF6 itself. TGF-β activated kinase-1(TAK-1) with TAK binding protein 1 (TAB1, TAB2 and TAB3) gets activated by unconjugated polyubiquitin K63 chain and phosphorylates I-κB kinase (IKK-β) and MAP kinase kinase 6. TAK1 subsequently activates the MAP kinases c-jun-N-terminal kinase (JNK), P^{38}, which downstream regulates AP1 and target cytokine gene. The IKK complex IKKα, IKKβ and NF-κB essential modulator (NEMO) phosphorylates IκBa an NF-κB inhibitor. Phosphorylated I-κB undergoes ubiquitin-proteosome degradation, which allows NF-κB translocation to the nucleus and which is followed by proinflammatory cytokine secretion (Fig. 8.3A).

TRIF-dependent pathway

The presence of another pathway was revealed when MyD88-deficient mice retained signalling on LPS stimulation. Although the inflammatory cytokine synthesis was absent but DC maturation, IFNβ, interferon inducible genes and IRF3 transcription factor activations were observed in MyD88-deficient mice.

TIR-domain containing adaptor protein inducing IFN-β (TRIF) is an adaptor protein other than MyD88. TRIF is even known as TICAM-1. TLR3 and TLR4 are the only TLRs which follow TRIF dependent pathway (Fig. 8.4). TRIF have co-adaptor molecule TRAM which connects TLR with TRIF. Unlike TLR4, stimulated TLR3 signals through TRIF but in the absence of TRAM protein. TRAM adaptor with GOLD domain (TAG) is a negative regulator of the TRIF dependent pathway. TRIF interacts with TRAF3 and TRAF6 via TRAF binding motifs which is N-terminal localized. TRAF-3 activates two IKK-mediated kinases first TANK-binding kinase (TBK1) and IKK-i/IKK-ε. TBK1 and IKK-i phosphorylates IRF3 and IRF7, IRF3:IRF7 dimers translocate to the nucleus and stimulates type 1 IFN and interferon inducible genes (Fig.

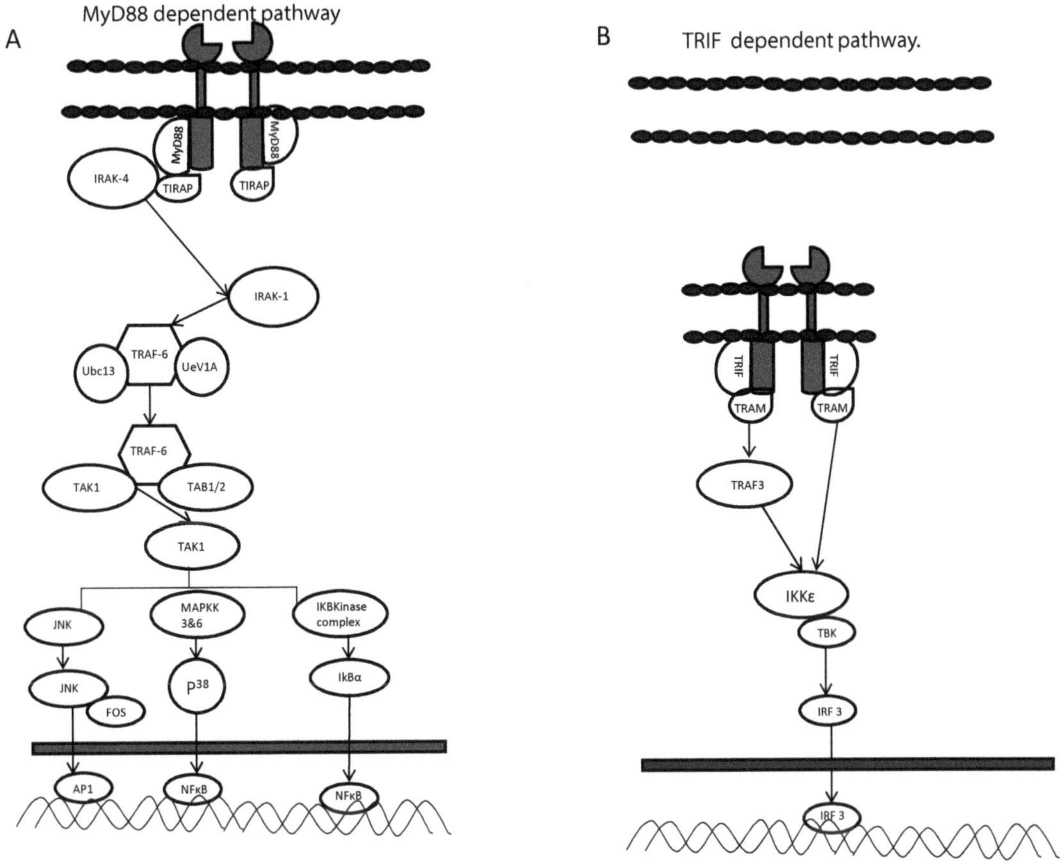

Figure 8.3 TLR signals through two pathways: (A) MyD88-dependent pathway which signals through the adaptor protein MyD88 with a co-adaptor molecule TIRAP. MyD88 pathway has important mediator downstream signalling proteins IRAK-1 and 4, TRAF6, TAK/TAB. (B) TRIF-dependent pathway which signals through adaptor protein TRIF and co-adaptor molecule TRAM. TRIF pathway has important mediators signalling proteins TRAF6 and TBK. Abbreviations: MyD88, myleoid differentiation primary response 88; TIRAP, TIR domain containing adaptor protein; IRAK, IL-1 receptor associated kinase; TRAF, tumour necrosis factor receptor associated factor; TAK, TGF-β activated kinase; TAB, TAK1 binding protein; TRIF, TIR domain containing adaptor inducing protein; TRAM, translocating chain associated membrane; IKK, IκB kinase; TBK, TANK-binding kinase.

8.3B). IKK-i also phosphorylates STAT-1 which facilitates induction of set of IFN inducible gene (Lee, *et al* 2007; Akira *et al.*, 2006).

Transcription factors involved in Toll-like receptor signalling

There are specific transcription factors which are activated via Toll-like receptor signalling. These are certain vital transcription factors which upon inhibition impairs innate immune response, and like TLRs they are even regulated by other pattern recognition receptors (Table 8.2).

- *NF-κB*. Nuclear factor kappa-light-chain-enhancer of activated B-cells is a protein complex which regulates DNA transcription. It is present in all the cell types and it responds to cellular stimuli. Along with the property to play major role in immune response to infection, aberrant regulation of NF-κB lead to cancer, inflammation, autoimmune disease.
- *AP1*. This is a transcription factor with heterodimeric protein with FOS or JUN or ATF7. It recognizes and binds to enhancer heptamer motif 5′-TGA[CG]TCA-3′. AP1 transcription

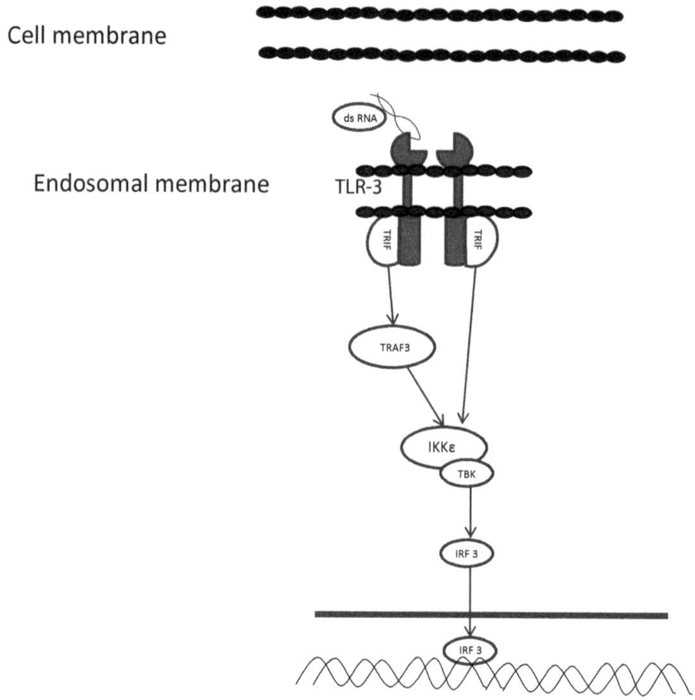

Figure 8.4 TLR3 is a cytoplasmic TLR on the endosomal membrane. It recognizes dsRNA, follows TRIF-dependent pathway and initiates type I interferon synthesis through IRF3 transcription factor.

Table 8.2 List of TLR and the transcription factors they activate

	NF-κB	AP1	IRF FAMILY
TLR	TLR1, 2, 4, 5, 6, 7, 8 and 9	TLR3, 7 and 9	TLR3, 4, 7, 8 and 9

factor induces cytokine synthesis via regulating cytokine gene expression on infection.

- *IRF (interferon regulating factor) family*. These are proteins that regulate transcription of the interferon cytokine family. The major IRFs that participate in TLR-mediated signalling are IRF-3 and IRF-7 (Lee *et al.*, 2007; Akira *et al.*, 2006; O'Neill *et al.*, 2013).

Interaction of *Leishmania* and macrophages

Leishmania in direct interaction with macrophages

Promastigotes and amastigotes are the two developmental phases of *Leishmania* parasite which possesses different types of surface molecules for the attachment and entry into their respective host cell. The major surface molecule present on the parasite is lipophosphoglycan (LPG) which mask the entire surface area of *Leishmania* parasite. LPG in metacyclic form is present on the surface of promastigotes but not on the amastigotes. LPG plays important role in the parasite-host cell interaction and gain direct entry into the macrophages (major phagocytic cell). LPG has a highly conserved tripartite structure and possesses lysoalkylphosphatidylinositol lipid anchor, phosphorylated hexasaccharide glycan core and phosphoglycan (PG). Among all the glycol conjugate proteins such as proteophosphoglycan (PPG), gp63 metalloproteinase, glycophosphatidylinositol lipid (GIPL) and lipophosphoglycan (LPG) present on the *Leishmania* parasite, only LPG and GP63 metalloproteinase are implicated in the host–parasite interaction and survival in the intracellular and extracellular harsh environment in host cells. Complex glycol conjugate LPG and PPG present in *Leishmania* inhibits the action of mid gut hydrolytic enzyme and secretion of protease enzyme.

Role of lipophosphoglycan in *Leishmania* invasion and survival in host macrophage

During infection *Leishmania* parasite requires lipophosphoglycan (LPG) to attain adhesion to the surface of the host macrophage. For example, LPG interacts with mannose fucose receptor of macrophage. Mannan-binding protein binds to the mannose residue of the LPG structure. This process enables the formation of C3 convertase simultaneously with the generation of c3b. This compliment protein helps leishmanial adhesion to the macrophage surface. C-reactive protein receptor even triggers phagocytosis of *Leishmania donovani* metacyclic promastigotes via interaction with parasite LPG. Post phagocytosis LPG inhibits its fusion between parasitophorous vacuole and lysosomes which avoid parasite killing mechanism by being protected from acidic condition and hydrolytic enzymes which in consequence support parasite proliferation. This mechanism has been confirmed by infecting macrophage with LPG-deficient *Leishmania major* and *Leishmania donovani*. Although other factors helped in parasite survival but rapid fusion between parasitophorous vacuole and lysosome was observed. LPG has been found as an essential molecule to protect *Leishmania* during differentiation to amastigote from promastigote although it does not have any significant role in the development of amastigote form.

Leishmania escapes into macrophage through neutrophils

Neutrophils are the first cell recruited at the site of the *Leishmania* infection. The lifespan of neutrophils is short and apoptosis of neutrophils releases various types of components. These components damage the surrounding tissue and to minimize the harmful effect of these components, macrophages and dendritic cells phagocytose the neutrophils. Early recruitment of neutrophil is driven by various events like wound repair, tissue damage by sandfly bite or by infection. *Leishmania* infection is a vector-based infection, which induces the production of IL-8 and neutrophils chemotactic factor by *Leishmania* parasite.

The potential roles of neutrophils in parasite infection includes primary immune response against the *Leishmania* such as secretion of peptides, cytokines, chemokines and expression of inducible nitric oxide synthase which is required for the parasite killing.

Polymorphonuclear neutrophil granulocytes (PMNs) are the first leucocyte which infiltrates to the site of *Leishmania* infection and engulf the parasite. Previously, it has been shown that neutrophils encounter *Leishmania* parasite, but it is neither killed nor multiplied within the phagolysosome of neutrophils. It is observed that neutrophils harbouring parasite secrete high levels of MIP-1 and β cytokines which are required for the recruitment of major phagocytic cells such as macrophages and dendritic cells. Macrophage internalizes the infected neutrophils and secretes anti-inflammatory cytokines such as TGF-β. The *Leishmania* parasite takes advantage of these apoptotic neutrophils for their silent entry into the macrophages. This phenomenon is called as Trojan Horse effect. In this mechanism *Leishmania* use PMN as a temporary host cell for their unrecognized entry into the macrophages. Infected neutrophils gets engulfed into the macrophages and there parasite survives and multiply.

PMN are not a primary host cell for parasite infection because of their very short life span as they readily undergo apoptosis within 6–12 hours. But after *Leishmania* infection, apoptotic death program of neutrophils delay up to 42 hours to 2 days. *Leishmania* parasite hide inside the neutrophils and evades the killing mechanism of oxidative burst, degranulation of phagolysosomal compartment, nitric oxide-dependent killing (Ritter *et al.*, 2009; Zanderbergen *et al.*, 2004; Flavia *et al.*, 2012).

Inhibition of spontaneous apoptosis of neutrophils by *Leishmania major* infection

Leishmaniasis is a vector born disease caused by intracellular parasite of the genus *Leishmania*, transmitted to the host by infected female sand fly during the blood meal. After biting by sand fly, the local inflammatory processes are initiated followed by the vasodilation of blood vessel, secretion of chemokines, extravasations of number of immune cells. Primary cell that swarms to the site of tissue damage is polymorphonuclear neutrophil granulocytes (PMN) which provides early line of host defence against the *Leishmania* parasite. Neutrophils infected with *Leishmania major* promastigotes are found to undergo reduced apopstotic state which is verified by various apoptotic assay such as TUNEL assay, annexin V staining, morphological assay (Ger van Zandbergen *et al.*, 2002). This study further confirms that *Leishmania* infection to neutrophils show antiapoptotic effect. Antiapoptotic effect of *Leishmania* has been observed in live parasite infection, this infers modulation in the programmed cell death by reducing the expression of caspase-3 activity and other proapoptotic molecules. Inhibitory role of *Leishmania* is investigated by measuring the antiapoptotic cytokines, IL-8 and GM-CSF which get secreted after the infection of *Leishmania major* with polymorphonuclear cells. This study has revealed that viable parasites are amenable to protect the neutrophils from programmed cell death. It has been already reported that *Leishmania* promastigotes do not completely halts the programmed cell death of neutrophils, but only play a major role in delaying neutrophil apoptosis. *In vivo* studies have also shown that *Leishmania* infection prolongs the survival of PMNs (Aga *et al.*, 2002).

Neutrophil–parasite interaction

Neutrophils are the predominant cells localized at the *Leishmania*-infected or tissue damaged state by the intra peritoneal injection. These infiltrated neutrophil cells rapidly phagocytose *Leishmania* parasite by two mechanisms: Opsonin-dependent (complement-dependent), a protective mechanism and opsonin-independent pathway, a non-protective mechanism. Neutrophils have both the protective and non-protective role against *Leishmania* infection. The protective role of neutrophils through opsonin-dependent pathway is observed to kill the promastigotes and amastigotes form of *Leishmania donovani* within one hour by the mechanism of oxidative burst. *Leishmania* loaded PMN escape the local immune defence mechanism of host cell including the Th1 response, anti-leishmanial function, NET-dependent parasite killing, oxidative burst.

In an *in vitro* model during early phase of *Leishmania* infection, neutrophils phagocytose the parasite in the presence of fresh human serum. Rate of phagocytosis is observed to be high in the presence of the activated serum in comparison to heat inactivated serum which supports that complement component are even involved in the engulfment of *Leishmania*. The best-studied complement components are C3bi and C3, attached to the complement receptor 3 (CR3) present on the major phagocytic cells and immunoglobulins which bind through the Fc receptor present on phagocytic cells and other components which are also involved are mannan-binding lectin. All these receptors are involved in receptor-mediated phagocytosis or opsonin-dependent phagocytosis. Some other receptors present on sentinel cells recognize the specific structure of pathogen and phagocytose the parasite. This type of phagocytosis is opsonin independent and non-protective, which invade the innate and adaptive immune response of the host (Laufs *et al.*, 2002; Guimarães-Costa *et al.*, 2009).

Evasion of anti-leishmanial mechanism of neutrophils by *Leishmania* infection

Upon the infection of promastigotes and amastigotes form of *Leishmania*, the neutrophils play a major role in the first line of host defence against the parasite. Neutrophils have a surprising mechanism to trap the micro-organisms in the extracellular environment using the neutrophil extracellular trap (NET), this is also called as NETosis. Formation of NET by the dead neutrophil consists of secreted DNA molecules and granular proteins which forms the network. The present study of (Guimarães-Costa *et al.*, 2009) shows that the formation of NET is induced by the PMA stimulated and *Leishmania*-infected neutrophils. NET formation provides the suitable environment to internalize the *Leishmania* by polymorphonuclear neutrophil granulocytes and NET formation thus suppresses the immune response against the *Leishmania* infection(Guimarães-Costa *et al.*, 2009).

Toll-like receptor signalling upon stimulation with *Leishmania* surface antigen

All the TLRs upon stimulation with respective ligands initiate downstream signalling which include chain phosphorylation of the downstream proteins and ubiquitination (Fig. 8.5). On *Leishmania* infection few of the TLRs attains stimulation with respect to the availability of antigenic molecules on parasite surface. *Leishmania* uses these TLR signalling pathways to elicit immune response and maintain survival.

Toll-like receptor 4

Toll-like receptor 4 is a cell surface receptor containing 12 leucine-rich repeats which recognize and bind to agonist lipopolysaccharide. TLR 4 follows MyD88-dependent pathway and thus the adaptor protein binding to TIR of TLR4 is MyD88 protein (Fig. 8.6A). TLR4 too have one more co-receptor protein associated to it, i.e. CD14. CD14 can bind LPS only in the presence of LBP (lipopolysaccharide-binding protein). A chain of association of LPS, LBP and CD14 stimulate TLR 4 and thus activate the downstream signalling.

As earlier mentioned that TLRs always convey message to downstream proteins with two types of adaptor proteins, i.e. MyD88 or TRIF. Cell membrane localized TLR4 interacts with adaptor protein MyD88 via its TIR intracellular domain. TLR4 use TIRAP to recruit MyD88 adaptor protein. TIRAP is another adaptor protein which serves as a bridge between TLR4 and MyD88 on LPS stimulation. The subsequent downstream signalling proteins are IRAK4, IRAK1, TRAF6 and TBK1 respectively. Upon stimulation, the TIR domain homodimer interact with TIRAP and subsequently activate MyD88 adaptor moiety. MyD88 adaptor protein associate with downstream protein kinase group known as IRAK4 and IRAK1 (IL-1R associated kinase) and thus leading to sequential activation of both. TNF receptor associated factor 6 (TRAF6) gets activated through activated IRAK-1. Post activation of TRAF6, Ubc13 and UeV1A ubiquitinates

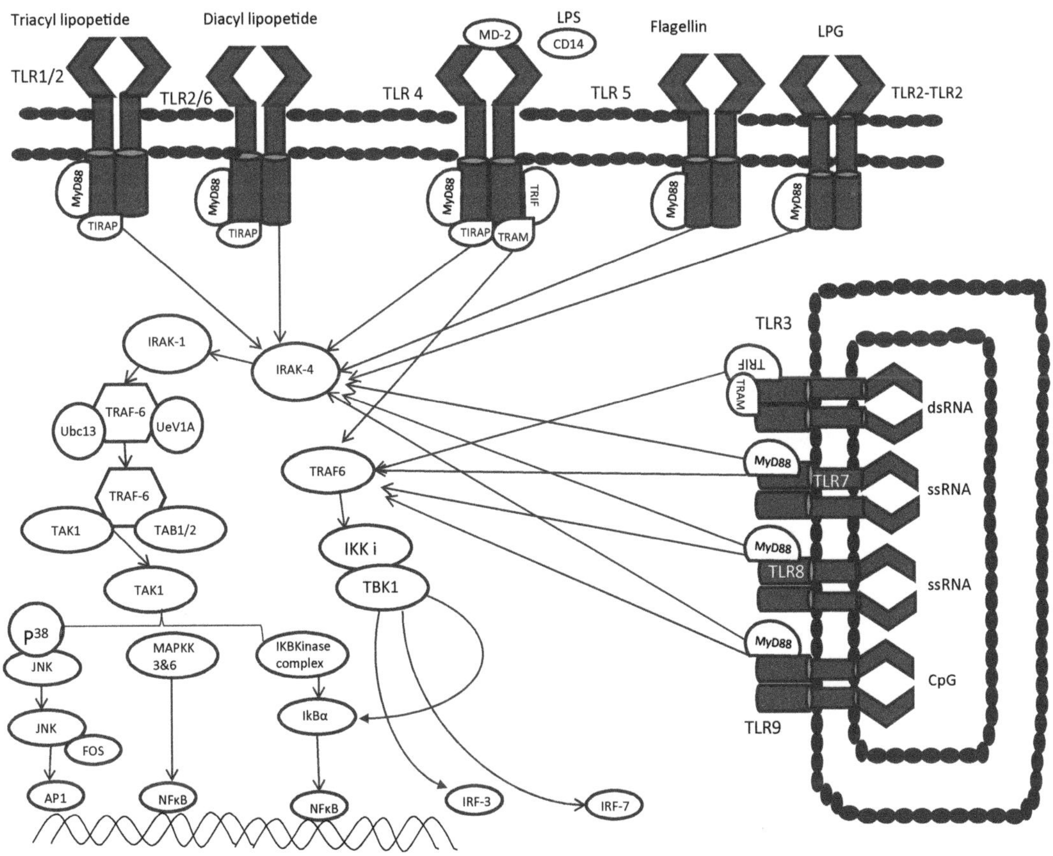

Figure 8.5 Cumulative representation of TLR signalling on stimulation.

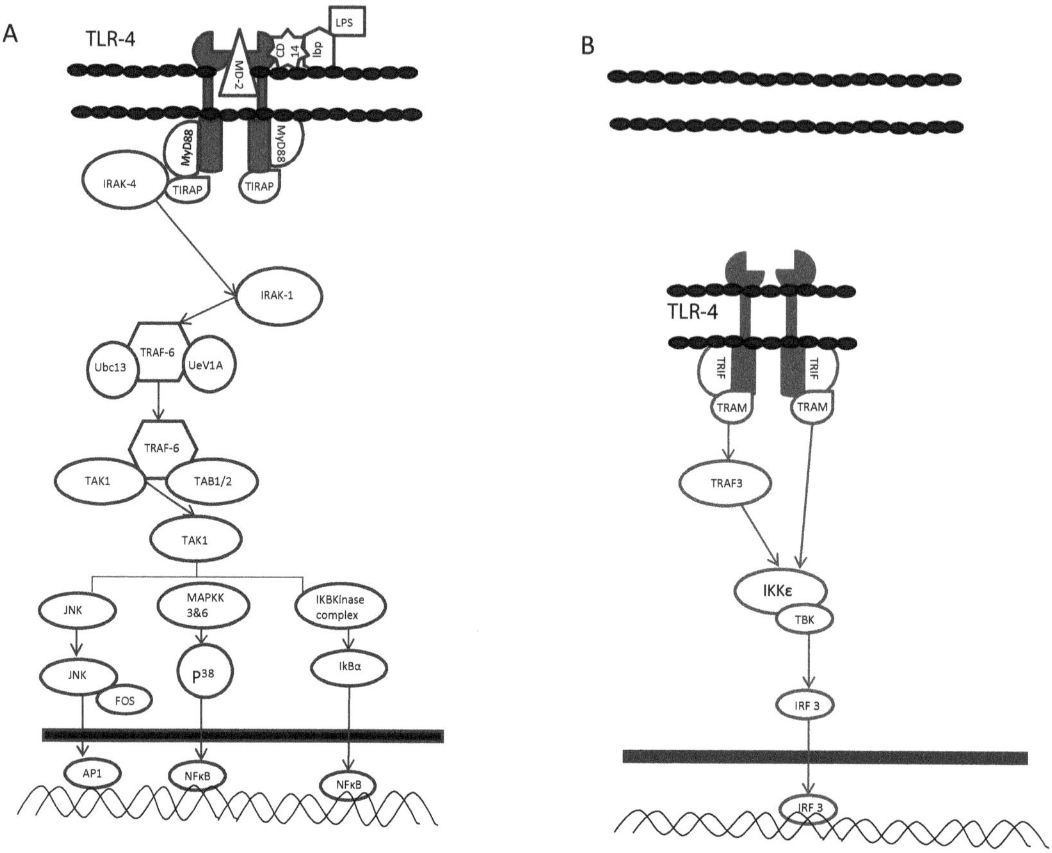

Figure 8.6 (A) TLR4 gets stimulated by the ligand lipopolysaccharide via another host surface proteins CD14 and MD-2. It follows the MyD88-dependent pathway. (B) Post stimulation, TLR4 even translocate to the endosomal membrane. It follows the TRIF-dependent pathway post translocating to endosomal membrane.

TRAF6 and I-κB kinase γ. Ubiquitinated TRAF6 associates with TAK1 and TAB1/2 complex and activates TAK1 (a MAPKKK family protein). TAK1 subsequently activates MAPKinases, i.e. c-jun-N-terminal kinase (JNK) and p38 which in downstream regulates AP1, NF-κB activation and thus monitors pro-inflammatory cytokine synthesis (Lee *et al.*, 2007; Ritter *et al.*, 2009).

TLR 4 on stimulation even localizes to the cytoplasm and proceeds with TRIF dependent signalling pathway (Fig. 8.6B). Through this pathway TLR4 regulates type I IFN cytokine family response. The change is observed in the choice of adaptor protein, i.e. TRIF protein instead of MyD88. Translocating chain-associated membrane (TRAM) is the co-adaptor protein for TRIF. The TRIF-dependent pathway is followed by TLR4 and TLR3, of which only TLR4 has TRAM as co-adaptor molecule for TRIF. The hypothesis and experimental evidence underlying the transitions between the cell membrane location and early endosome localization of TLR4 are interesting. It has been observed that upon TLR4 stimulation MyD88+TIRAP-dependent signalling is initiated at PtdIns(4,5)P_2, which is followed by a drop in PtdIns(4,5)P_2 concentration from invaginating membrane, which results in the release of TIRAP–MyD88 complex. This release helps to engage TRIF–TRAM complex to associate with TIR of TLR4.

Why is TLR4 required to translocate to the early endosome to signal IFN-β synthesis? TRAF3, a downstream protein, is one of the key factors in IFN-β synthesis but, as it is not available on cell membrane, TLR4 must be translocated to the early endosomal membrane, where TRAM recruits TRIF at TIR of TLR4. TRIF–TRAM complex upon activation recruits TRAF3 activation. TRAF3 activates IKKε and TBK complex downstream and thus further activates interferon regulating factor-3 (IRF-3), which localizes in the nucleus and regulates the synthesis of IFN-β.

TLR 4 signalling results in extensive secretion of proinflammatory cytokines, e.g. IL-12, IFN-alpha. These cytokines are responsible for the Th1 response which helps in leishmanicidal activity. Lipopolysaccharide content on the *Leishmania* surface is very low and the role of proinflammatory cytokine is non-significant in leishmanicidal activity. This observation is with reference to the activity of lipophosphoglycan as a repressor of proinflammatory cytokine secretion via TLR4 signalling through the action of suppressor proteins SOCS-1 and SOCS-3 (Fig. 8.7). At the same time TLR4 signalling activate NF-κB, which translocate to nucleus as a result NF-κB transcribe mRNA for iNOS. iNOS has leishmanicidal activity as it synthesizes NO radical, which inhibits leishmanial survival (deVeer *et al.*, 2003).

Toll-like receptors 1, 2 and 6

TLR2, a cell membrane localized TLR, has 19 leucine-rich repeats. TLR2 has a wide variety of ligand recognition ability. TLR2 have acquired this ability by forming heterodimer with TLR1 and TLR6 respectively. TLR1–TLR2 heterodimer recognizes *triacylated lipoproteins,* whereas TLR6–2 heterodimer recognizes *diacylated lipoproteins.* TLR2 homodimers recognize lipophosphoglycan and peptidoglycan. TLR2

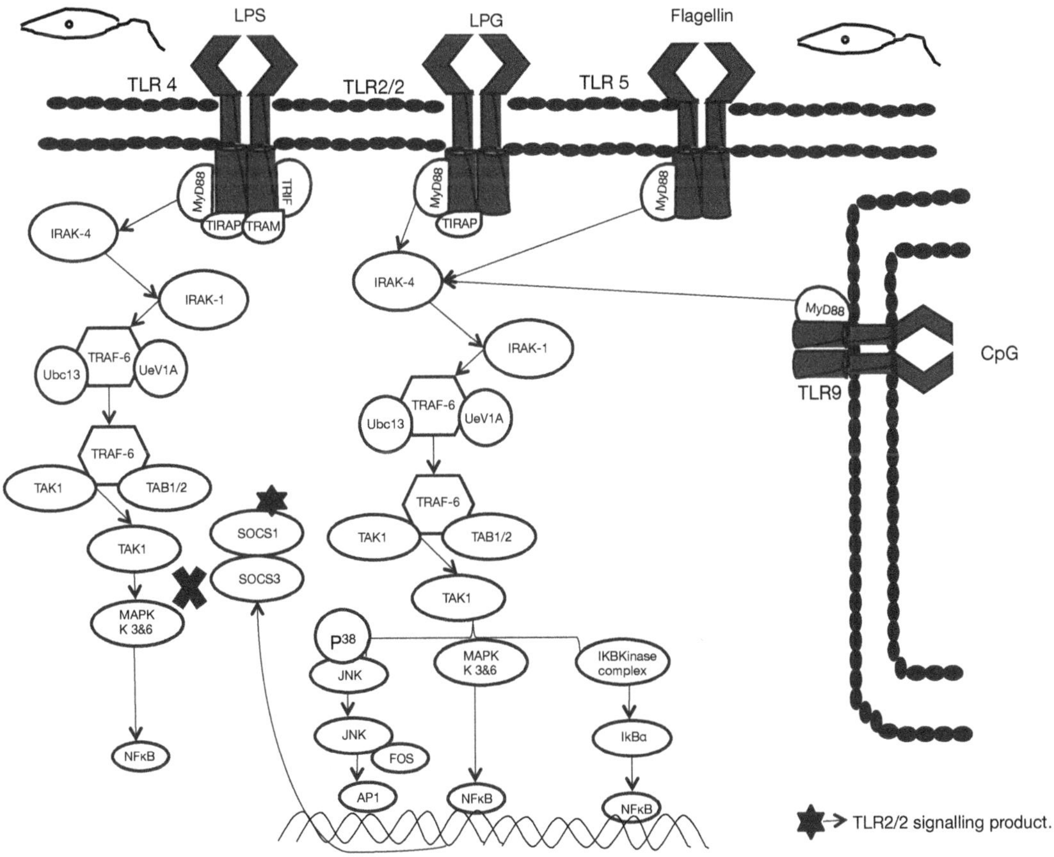

Figure 8.7 TLR–TLR crosstalk between TLR2 and TLR4 on TLR2 stimulation with lipophosphoglycan (LPG). LPG-mediated signalling stimulates synthesis of SOCS1 and SOCS3 which inhibits TLR4-mediated signalling. *TLR2,5&9 follows MyD88-dependent pathways, here TLR5 and TLR9 signalling are not influencing TLR2 signalling.

follows MyD88-dependent pathway and thus adaptor protein MyD88 binds to the TIR of TLR. TLR2 too uses TIRAP to recruit MyD88 adaptor protein (Fig. 8.8). The subsequent downstream signalling proteins involved are IRAK4, IRAK1, TRAF6 and TBK1 respectively. Upon stimulation the TIR domain interact with TIRAP which bridge TIR and MyD88 interaction and subsequently activate MyD88 adaptor moiety. MyD88 adaptor protein associate with downstream protein kinase group known as IRAK4 and IRAK1 (IL-1R-associated kinase) and both are activated. TNF receptor associated factor 6 (TRAF6) gets activated through IRAK-1. Post activation of TRAF6, Ubc13 and UeV1A ubiquitinates TRAF6 and I-κB kinase γ. Ubiquitinated TRAF6 associates with TAK1 and TAB1/2 complex and activates TAK1 (an MAPKKK family protein). TAK1 subsequently activate MAPKinases, i.e. c-jun-N-terminal kinase (JNK) and p38 which in downstream regulates AP1, NF-κB activation and thus monitors pro-inflammatory cytokine synthesis.

The signalling pattern is observed in all the three type of stimulation, i.e. on TLR2–2 homodimer, TLR2–1 and TLR2–6 heterodimer stimulation. Lipophosphoglycan is an agonist of TLR2–2 homodimer. As lipophosphoglycan is the major surface molecule of virulent *Leishmania* parasite, it has major role to play with the first line of invasion of parasite. Promastigotes of *Leishmania major* and *Leishmania donovani* have metacyclic lipophosphoglycan and bear potential to infect host cells. TLR2 homodimer on stimulation with ligand LPG signal through MyD88-dependent pathway but its response

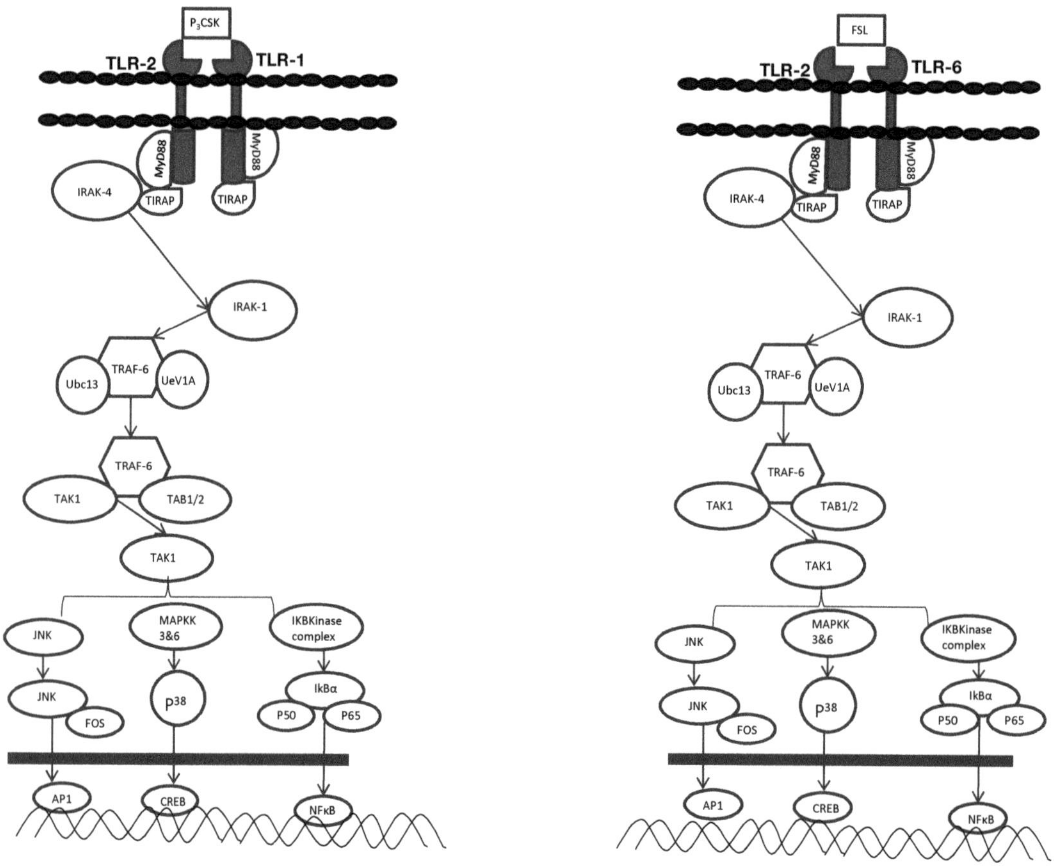

Figure 8.8 TLR2 form heterodimers with TLR-1 and TLR-6. TLR1–TLR2 heterodimer recognizes triacyl lipopeptides whereas TLR6–TLR2 heterodimer recognizes diacyl lipopetides. Both follow the MyD88-dependent pathway.

leads to synthesis of anti-inflammatory cytokine such as IL-10, IL-4. The same observation is made from TLR1–TLR2 heterodimer signalling upon stimulation with agonist Pam_3CSK and PGN. On other hand, TLR2–TLR6 heterodimer signalling provides host protection through anti-leishmanial response. Diacylated lipoprotein agonist stimulates TLR2–TLR6 heterodimer which is followed by MyD88-dependent signalling and proinflammatory cytokine secretions (West *et al.*, 2006).

Leishmania donovani infection regulates TLR2 signalling via monitoring the recruitment of MAP kinase family protein downstream. During the infection synthesis of IL12p40 is suppressed and instead increase in synthesis of IL-10 is observed. Parasite modulates the MAP kinase family protein recruitment, where MAPK p38 phosphorylation is suppressed and ERK gets activated. This is followed by suppression of IL-12 synthesis and an increase in the IL-10 synthesis which through Th2 immune response down-regulates IL-12 synthesis. The entire process helps in disease susceptibility.

Toll-like receptor 5

TLR5 is a cell membrane-localized protein homodimer with 15 leucine-rich repeat. Agonist for TLR5 is flagellin protein, which is available in flagella of bacteria and parasites. In case of *Leishmania* infection TLR5 plays vital role as PRR because flagellin is one of the PAMPs present on *Leishmania* parasite. TLR5 too upon stimulation signals its downstream proteins through MyD88 adaptor protein (Fig. 8.9). The signalling intermediates for TLR5 signalling are similar as already described for TLR4. Transcription factor NF-κB is activated on signalling and mediates proinflammatory cytokine synthesis. Proinflammatory cytokines cause Th0 phase T-cells to differentiate to Th1 type cells, which synthesize several cytokines to mitigate the severity of infection.

Toll-like receptor 9

If we focus on Toll-like receptors present in the cytoplasm, i.e. on the membrane of endosomes, the TLR which has a vital role in regulating immune response via self or cross-talk in *Leishmania* infection is TLR9. TLR9 is a 26 leucine-rich repeat containing receptor. It recognizes CpG as its agonist. CpG is a short form of cytosine–phosphate–guanine where cytosine is followed by guanine with a phosphate link. CpG agonist for TLR9 is an unmethylated form. Unmethylated CpG of *Leishmania* parasite acts as ligand for TLR9 which leads to MyD88-dependent signalling (Fig. 8.10). Through MyD88 adaptor protein TLR9 activates IRF (interferon regulating factor) 3, 7 as well as NF-κB transcription factors. TLR9 also signals through MyD88-dependent pathway but without TIRAP as co-adaptor protein. The subsequent downstream proteins are IRAK4, IRAK1, TRAF6 and TBK1 respectively. Upon stimulation the TIR domain activates MyD88 adaptor moiety. MyD88 adaptor protein associates with downstream protein kinase group known as IRAK4 and IRAK1 (IL-1R-associated kinase) and thus leading to activation of both. TNF receptor associated factor 6 (TRAF6) gets activated through activated IRAK-1. Post activation of TRAF6, Ubc13 and UeV1A ubiquitinates TRAF6 and I-κB kinase γ. Ubiquitinated TRAF6 associate with TAK1 and TAB1/2 complex and activate TAK1 (an MAPKKK family protein). TAK1 subsequently activates MAPKinases, i.e. c-jun-N-terminal kinase (JNK) and p38 which in downstream regulates AP1, NF-κB activation and thus monitors pro-inflammatory cytokine synthesis.

Even TRAF3 mediated signalling is observed which is most common for TLR9,7,3 and cytoplasm localized TLR4 when it follows TRIF dependent pathway. TLR9 too activates TRAF3 via MyD88 as TIR domain adaptor protein. TRAF3 activation facilitates complex formation of IRAK-1, TBK-1 and IKKε finally activating transcription factors IRF-3/IRF-7 leading to type 1 IFN cytokine synthesis.

The role of TLR9 is well defined as a source for IL-12/IL-23p40 synthesis. During infection the response of TLR9 varies with respect to cell type. On stimulation with agonist during infection in dendritic cells TLR9 shows robust signalling which is associated with synthesis of proinflammatory cytokine secretion, i.e. IL-12/IL-23p38. In contrast to DCs, the ability of macrophages and monocytes to respond to TLR9 agonist during infection is low. Thus, during *Leishmania* infection due to activated TLR2 immunomodulatory

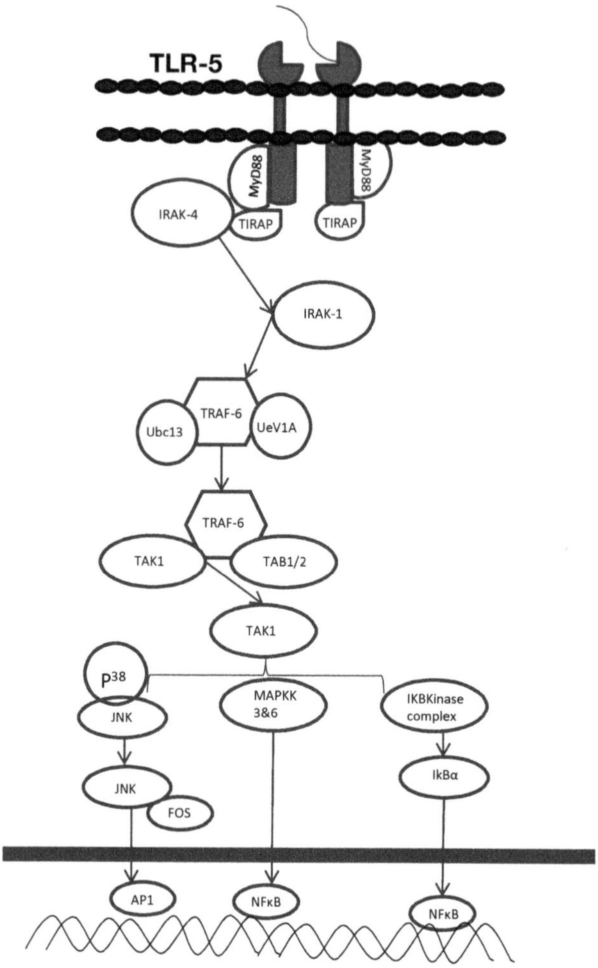

Figure 8.9 Flagellin protein is the agonist for TLR5. TLR5 follows the MyD88-dependent pathway for downstream signalling.

effect, low TLR9 signalling along with inefficient pro-inflammatory cytokine response is observed among macrophages, monocytes and even in DCs. This response of macrophage on infection implies decrease in TLR9 signalling in macrophage on *Leishmania* infection.

The cumulative effect of *Leishmania* infection on Toll-like receptors and intra-pattern recognition receptor signalling

Toll-like receptors share the stage as PRR with other PRRs such as CARD helicases, NOD-like receptors and C-type lectin type receptors. Some of these PRRs synergistically operate with TLR on initiation of infection to subvert pathogen survival. On leishmanial infection some of these PRRs partner with TLR to regulate the synthesis of proinflammatory cytokines. SHP-1 is a phosphotyrosine phosphatase, a cytoplasmic protein, even have been found to interfere with TLR signalling on *Leishmania donovani* infection.

Toll-like receptors and other pathogen recognition receptors

Toll-like receptors and CARD helicase

CARD (caspase activation and recruitment domain) are the interaction motifs and are involved in the process of inflammation and apoptosis. A CARD domain such as RIG-1 and MDA5

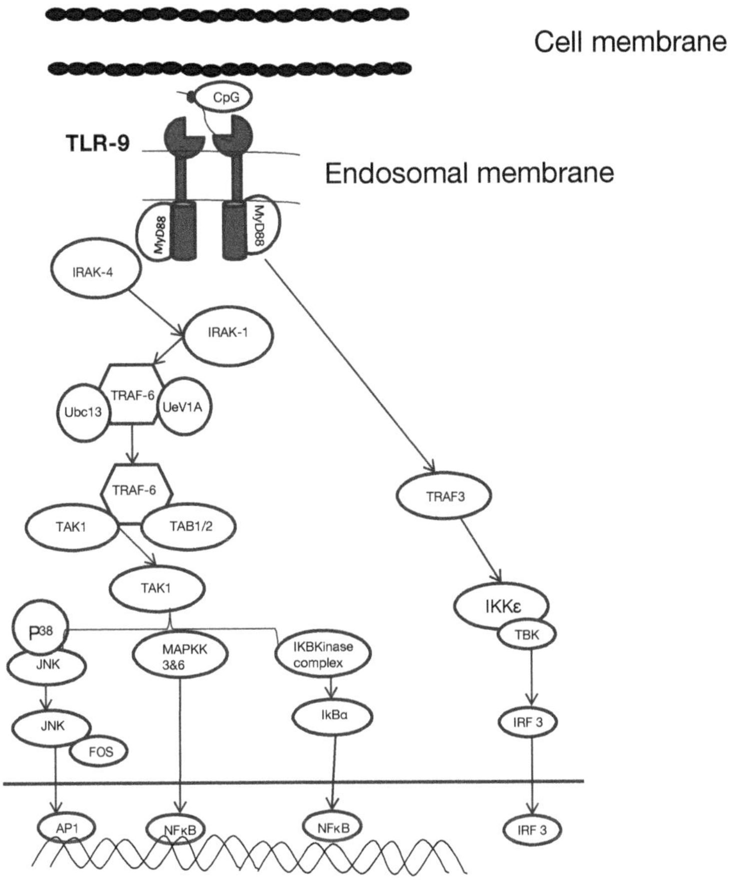

Figure 8.10 TLR9 is an endosomal membrane localized TLR, which gets stimulated by the unmethylated CpG of the pathogen. It follows the MyD88-dependent pathway. It even signals through TRAF3 protein to stimulate interferon protein synthesis.

along with TLR 3, 4, 7, 8 and 9 activates IRF3 and/or IRF7 to induce type 1 interferon. Cross-talk of CARD with Toll-like receptor has a synergistic as well as antagonistic impact on CARD-mediated signalling. First, CARD helicase domains RIG-1 and MDA5 both are type 1 IFN inducible. TLR 3, 4, 7, 8 and 9 through activated IRF 3 and 7 induce type 1 IFN secretion. Stimulation of such TLRs should synergistically up-regulate REG-1 and MDA5 response. On the other hand, as an example of the antagonistic impact A20 a ubiquitin-modifying enzyme terminates Toll-like receptor-induced activity of transcription factors NFκB and inhibit synthesis of proinflammatory cytokine IL-8 in antigen-presenting cells. However, significant regulation of RIG-1 signalling by A20 is not clear (Lee *et al.*, 2007).

Toll-like receptor and SHP1

Leishmanial infection not only stimulates TLR directly but also elicits several other vital proteins which via subsequent signalling through TLR intermediate proteins govern transcription factor activation and cytokine synthesis.

SHP1 (SRC homology 2 domain phosphotyrosine phosphatases) is a protein tyrosine phosphatase (PTP) that has been found to be associated with a conserved motif present on multiple kinase proteins. It binds to conserved KTIM protein and simultaneously regulate IL-1 receptor associated kinase 1 (IRAK-1) and other MAP kinase family protein. KTIM is an immunoreceptor tyrosine-based inhibitory receptor (ITAM)-like motif. KTIM is kinase tyrosine-based inhibitory receptor. This motif allows interaction between IRAK-1 and SHP-1 followed by inhibition of IRAK-1 activity. This mechanism even regulates TLR downstream signalling, keeping IRAK-1 as a mediator of leishmanial infection.

Leishmania donovani and SHP1

SHP1 is one of the prominent targets during leishmanial survival through *Leishmania* surface protease GP63. GP63 selectively cleaves intracytosolic PTPs. A good example is SHP-1, which is activated by GP63 activity. IFN-γ synthesis during infection initiates a response signalling via kinase proteins, e.g. JAK-STAT and MAPK family proteins (ERK). These kinase proteins on phosphorylation induce iNOS synthesis and thus result in production of NO, which is lethal to *Leishmania* parasite. GP63 protease enzyme activates SHP-1 which inhibits phosphorylation of JAK2 and ERK1/2 resulting inhibition of I NOS synthesis simultaneously inhibits NO production (Kaye *et al.*, 2011; Shio *et al.*, 2012).

Changes in cytokine synthesis upon synergy within Toll-like receptor stimulations

On stimulation of each Toll-like receptor, the type of cytokines to be synthesized along with fold increase is completely governed by single ligand stimulation and combination of ligand stimulation. However, the combinatorial ligand stimulation holds vital significance in modulating the level of induction of pro- or anti-inflammatory cytokine. In addition to ligand combination, adaptor protein selection too features significance in regulating the intensity of cytokine synthesis.

On pair wise stimulation of TLR in combinations of MyD88-MyD88-dependent and MyD88-TRIF dependent pathway (i.e. TLR 5 with TLR 2 and TLR 5 with TLR 3 or TLR4) fold increase or decrease in the synthesis of cytokines IL-10, IL-6 and IL-12p40 were observed. Among the combinations of MyD88-MyD88 and MyD88-TRIF pathways the fold increase in cytokine synthesis is observed in collectively activated MyD88-dependent and TRIF dependent manner. Pro-inflammatory cytokines such as IL-6 and IL-12p40 showed a rise in secretion with combined TLR stimulations. Thus, the response of a specific class of cytokines might differ upon combinatorial signalling. Similar results have been obtained in both human moDC (monocyte derived dendritic cells) and mouse moDC. These observations indicate that the combination of MyD88 and TRIF dependent pathways play role in regulating fold increase of cytokine synthesis. TLR-dependent synergistic cytokine secretion can be obtained through fold activation and cooperation of IRF, NF-κB, MAPK, PI-3K and STAT signalling pathways.

Leishmanial infection too can elicit synergistic cytokine synthesis if TLR with MyD88 and TRIF adaptor protein can simultaneously be stimulated. Leishmanial surface molecules provide with opportunity to elicit MyD88 signalling through agonist such as LPG (lipophosphoglycan), flagellin along with TRIF signalling with agonist such as (LPS) lipopolysaccharide. Host cells such as macrophage may use combinatorial TLR stimulation to subvert parasite survival with much greater innate immune response (Suet *et al.*, 2013 Makela *et al.*, 2009; Bohnenkamp *et al.*, 2007).

Toll-like receptor–Toll-like receptor cross-talk

During infection, pathogens with more than one surface molecule along with variable form of nucleic acids as antigenic molecules do tend to stimulate multiple pattern recognition receptors. Toll-like receptors attain specificity to many of the antigenic molecules and thus more than one TLR is stimulated. Inter-TLR regulations which can even be mentioned as TLR–TLR cross-talk regulates the balance between proinflammatory and anti-inflammatory cytokine secretions. TLRs on stimulation in combination have been observed to regulate the expression level and cytokine secretion of each other. This signalling development during parasite infection too plays a significant role to subvert infection and vice versa where parasite uses it as a significant mechanism to escape host innate immune response. Similarly, during *Leishmania* infection it stimulates more than one TLR and inter-TLR cross-talk is expected.

TLR 9 plays a very crucial role in parasite control and host survival on infection as it has been observed to regulate IL12 and thus control

Th1 cell differentiation with indirect regulation of type I IFN. The behaviour of TLR9 is even a cell type dependent, the above mentioned pattern of TLR9 signalling is observed in dendritic cells. In monocytes/macrophages TLR9 even regulate IL12 but decreased capability of monocyte/macrophage to respond to TLR9 agonist have lowered the IL12/IL-23p40 expression level. TLR2 play immunomodulatory role where it has dual role first, to regulate IL12/IL-23p40 by dendritic cells and second to elevate TNFα production by macrophage. It has been verified that TLR2 do not regulate TLR9 expression but still directly interfere with the capacity of dendritic cells to synthesize IL-12/IL-23p40. In such scenario the most appropriate reasons behind might be manipulation in TLR9 signalling pathway, transcription factor activation by TLR2 immunomodulatory effect. Thus, TLR2 and TLR9 combinatorial signalling on infection resulted to antagonistic effect which modulates innate immune response. Similar observation might be even expected in leishmanial infection (Fig. 8.11).

Another group of TLRs, i.e. TLR7/8 or TLR8 alone, are vital immunoregulators. Magnitude of IL12p70 expression has been low in general with individual TLR stimulation but with TLR7/8 or TLR8 alone stimulation expression of IL12p70 is found elevated in human PBMC. But with TLR9 agonist stimulation poor induction of IL-12p70 is obtained with poor augmentation of cytokine induction in combination with other TLRs.

Leishmania infection stimulates multiple TLR signalling due to the presence of many TLR-specific ligands on its surface as well as in cytoplasm. This allows simultaneous stimulation of more than one TLR, which indicates that the parasite might use the TLR–TLR cross-talk mechanism to escape host response such as lipophosphoglycan-stimulated down-regulation of proinflammatory cytokine secretion and inhibition of TLR4-mediated signalling through activating suppressor protein SOCS-1 and SOCS-3 (Gravina *et al.*, 2013; Holley *et al.*, 2012; Ghosh *et al.*, 2007).

Conclusion and future trends

Leishmania parasite, like other pathogens, uses host regulatory mechanisms for its own survival. TLR2-stimulated signalling through vital *Leishmania* surface molecule lipophosphoglycan plays role as an immunomodulator and has been observed to regulate TLR4 signalling pathways along with proinflammatory cytokine secretion. The regulatory signals might elevate parasite survival or host parasite killing efficiency due to macrophage activation. Still *Leishmania* parasite for its survival disables several host defence signalling to overcome host parasite killing efficiency. Toll-like receptors (TLR) play vital role

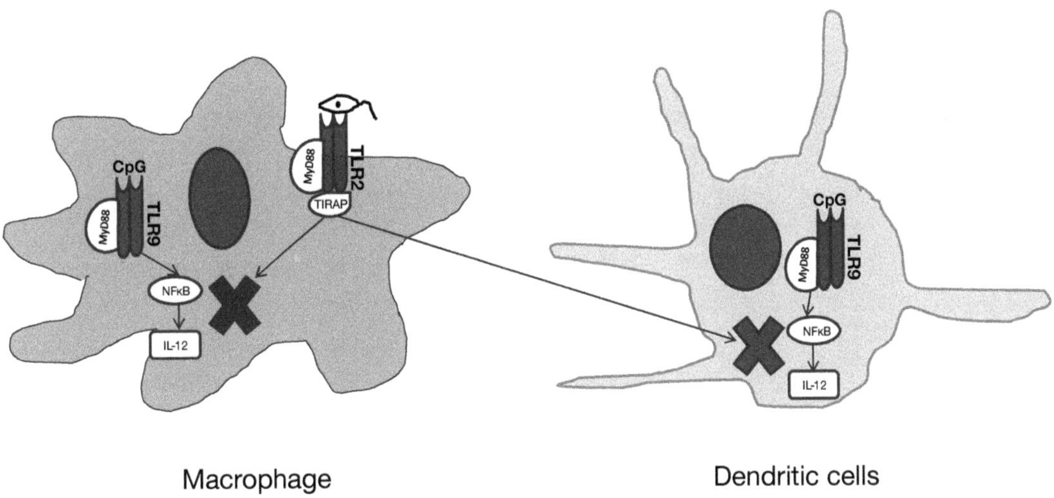

Figure 8.11 TLR9 and TLR2 crosstalk: LPG-mediated TLR2 signalling regulating TLR9 signalling in macrophage as well as in dendritic cells.

as the source signalling stimulator which parasite exploits. Different types of TLRs on infection undergo stimulation and signal to recover the host immune response. TLR 3 has been observed to stimulate proinflammatory cytokine synthesis on *Leishmania donovani* infection. Although *Leishmania* is not reported to be the carrier of dsRNA, still TLR3 stimulation on *Leishmania* infection to increase proinflammatory cytokine synthesis raises speculations and questions. These types of signalling need to be extensively studied and verified regarding the types of *Leishmania* generated ligands elicit the TLRs.

Toll-like receptor signalling elicits vital immune responses, be it proinflammatory cytokine secretions or type 1 interferon secretions, and indirectly supports differentiation of Th0 cells to Th1 cells. These properties show it to be one of the major regulatory cell surface receptors. Antigen molecules or agonists of TLRs can be used for vaccine development. Further studies on TLR signalling of $CD8^+$ cell in the event of *Leishmania* infection and the analysis of possibilities of cellular cross-talk between lymphoid cells and antigen-presenting cells will provide a new horizon of understanding of relations between innate and adaptive immune system responses.

Web resources

Toll-like receptor review. (2012). Invivogen: http://www.invivogen.com/review-tlr

References

Aga, E., Katschinski, D.M., Zandbergen, G.V., Laufs, H., Hansen, B., Muller, K., Solbach, W., and Laskay, T. (2002). Inhibition of spontaneous apoptosis of neutrophil granulocytes by the intracellular parasite *Leishmania major*. J. Immunol. *169*, 898–905.

Akilov, O.E., Kausuboski, R.E., Carter, C.R., and McDowel, M.A. (2007). The role of mannose receptor during experimental Leishmaniasis. J. Leukoc. Biol. *81*, 1188–1196.

Akira, S., and Kawai, T. (2006). TLR signaling. Cell Growth Diff. *76*, 816–825.

Akira, S., and Takeda, K. (2004). Toll-like receptor signaling. Nat. Rev. *4*, 499–510.

Bell, J.K., Mullen, G.E., Leifer, C.A., Mazzoni, A., Davies, D.R., and Segal, D.M., (2003). Leucine-rich repeats and pathogen recognition in Toll-like receptor. Trends Immunol. *24*, 528–533.

Blackwell, J.M., Ezekowitz, R.A.B., Roberts, M.B., Channon, J.Y., Sim, R.B., and Gordon, S. (1985). Macrophage complement and lectin-like receptors bind *Leishmania* in the absence of serum. J. Exp. MEO. *162*, 324–331.

Bohnenkamp, H.R., Papazisis, K.T., Burchell, J.M., and Taylor-Papadimitriou, J. (2007). Synergism of Toll-like receptor-induce interleukin-12p70 secretion by monocyte-derived dendritic cells is mediated through p38 MAPK and lowers the threshold of T-helper cell type 1 responses. Cell. Immunol. *247*, 72–84.

Brinkmann, M.M., Spooner, E., Hobe, K., Beutler, B., Ploegh, L.H., and Kim, Y.K. (2007). The interaction between the ER membrane protein UNC93B and TLR3,7 and 9 is crucial for TLR signaling. J. Cell Biol. *177(2)*, 265–275.

DeVeer, M.J., Curtis, J.M., Balsdwin, T.M., DiDonato, J.A., Sexton, A., McConville, M.J., Handman, E., and Schofield, L. (2003). MyD88 is essential for clearance of *Leishmania major* Possible role for lipophosphoglycan and Toll-like receptor2 signaling. Eur. J.Immunol. *33*, 2822–2831.

Didilaurent, A., Ferrero, I., Otten, L.A., Berteand, D., Reinhardt, M., Carlsen, H., Blomhoff, R., Akira, S., Kraehenbuhl, J.P., and Sirard, J.C. (2004). Flagellin promotes myeloid differentiation factor88-dependent development of Th2-type response. J. Immunol. *172*, 6922–6930.

Ernst, J.D. (1998). Macrophage Receptors for Mycobacterium tuberculosis. Infect. Immun. *66*, 1277–1281.

Flavia, L., Riberio, G., and Sacks, D. (2012). The influence of early neutrophil–*Leishmania* interactions on the host immune reponse to infection. Front. Cell. Infect. Microbiol. *2*, 1–8.

Ghosh, T.K., Mickelson, D.J., Solberg, J.C., Lipson, K.E., Inglefield, J.R., and Alkan, S.S. (2007). TLR–TLR cross-talk in human PBMC resulting in synergistic and antagonistic regulation of type-1 and 2 interferons, IL-12 and TNK-α. Int. Immunopharmacol. *21*, 1111–1121.

Gravina, H.D., Antonelli, L., Gazzinelli, R.T., and Ropert, C. (2013). Differential use of TLR2 and TLR9 in the regulation of immune responses during the infection with Trypanosoma cruzi. PLoS One. *8*, 1–10.

Guimarães-Costa, A.B., Nascimento, T.C.M., Froment, G.S., Soares, R.P.P., Morgado, F.N., Conceição-Silva, F., and Saraiva, E.M. (2009). *Leishmania amazonensis* promastigotes induce and are killed by neutrophil extracellular trap. PNAS. *106*, 6748–6753.

Holley, M.M., Zhang, Y., Lehrmann, E., Wood, W.H., Becker, K.G., and Kielian, T. (2012). Toll-like receptor 2 (TLR2)–TLR9 cross-talk dictates IL-12 family cytokine production in microglia. Glia. *60*, 29–42.

Kaye, P., and Scot, P. (2011). Leishmaniasis: complexity at the host–pathogen interface. Nat. Rev. *9*, 604–614.

Laufs, H., Muller, K., Fleischer, J., Reiling, N., Jahnke, N., Jensenius, C.J., Solbach, W., and Laskay, T. (2002). Intracellular survival of *Leishmania major* in neutrophil granulocytes after uptake in the absence of heat-labile serum factors. Infect. Immun. *70*, 826–835.

Lee, M.S., and Kim, Y.J., (2007). Signaling pathways downstream of pattern-recognition receptors and their cross-talk. Annu. Rev. Biochem. *76*, 447–480.

Lemaitre, B., and Leulier, F. (2008). Toll-like receptors-taking an evolutionary approach. Nat. Rev. *9*, 165–178.

Lynch, G.R. (2000). Regulatory roles for FcgRIII (CD16) and FcgRII (CD32) in the development of T- and B-lineage lymphoid cells. J. Leukoc. Biol. *67*, 279–284.

Makela, S.M., Strengell, M., Pietila, T.E., Osterlund, P., and Julkinen, I. (2009). Multiple signaling pathways contribute to synergistic TLR ligand-dependent cytokine gene expression in human monocyte-derived macrophages and dendritic cells. J. Leukoc. Biol. *85*, 664–672.

Matsushima, N., Tanaka, T., Enkhbayar, P., Mikami, T., Taga, M., Yamada, K., and Kuroki, Y. (2007). Comparative sequence analysis of leucine-rich repeats (LRRs) within vertebrate Toll-like receptors. BMC Genomics *8*, 124–144.

Medzhitov, R., and Barton, G.M. (2002). Toll-like receptors and their ligands. Curr. Top. Microbiol. *270*, 81–92.

Mendoza, A.V., Carrero, J.C., and Sosa, M.R. (2013). Parasitic infections: a role for C-type lectins receptors. Biomed. Res. Int. *2013*, 1–11.

Mogensen. T.H. (2009). Pathogen recognition and inflammatory signaling in innate immune defences. Clin. Microbiol. Rev. *22*, 240–262.

Mosser, D.M., and Edelson, P.J. (1985). The mouse macrophage receptor for c3bi (cr3) is a major mechanism in the phagocytosis of leishmania promastigotes. J. Immunol. *135*, 2785–2789.

Oliver, M., Gregory, D.J., and Forget, G. (2005). Subversion mechanism by which *Leishmania* parasite can escape the host immune response: a signaling point of view. Clin. Microbiol. Rev. *18*, 293–305.

O'Neill, L.A.J., Golenbock, D., and Bowie, A.G., (2013). The history of Toll-like receptors-redefining innate immunity. Nat. Rev. Immunol. *13*, 453–460.

Penke, L.R., Sudan, R., Sathishkumar, S., and Saha, B. (2012). Prostaglandin E2 receptors have differential effects on *Leishmania major* infection. Parasite Immunol. 35, 51–54.

Peter, C., Aebischer, T., Stierhof, Y.D., Fuchs, M., and Overath, P. (1995). The role of macrophage receptors in adhesion and uptake of *Leishmania mexicana* amastigotes. J. Cell Sci. *108*, 3715–3724.

Rebecca, A.G., and Belosevic, M. (1993). Comparison of receptors required for entry of *Leishmania major* amastigotes into macrophages. Infect. Immun. *61*, 1553–1558.

Ritter, U., Frischknecht, F., and Zandbergen, G.V. (2009). Are neutrophils important host cells for *Leishmania* parasites? Cell. *25*, 505–510.

Roitt, M.I., Brostoff,J., and Male,D.K. (2001). Immunology. 6th edition. (Elsevier health sciences), pp. 3, 6, 15–17, 106, 147–161.

Shio, M.T., Hassani, K., Isnard, A., Ralph, B., Contrearus, I., Gomez, M.A., Abu- Dayyeh, I., and Oliver, M. (2012). Host cell signaling and *Leishmania* mechanisms of evasion. J. Trop. Med. *2012*, 1–14.

Suet Ting Tan, R., Lin, B., Liu, Q., Kellogg-Tucker, L., Ho, B., Leung, B.P., and Ling Ding, J. (2013). The synergy in cytokine production through MyD88-TRIF pathways is co-ordinated with ERK phosphorylation in macrophages. Immunol. Cell Biol. *91*, 377–387.

Takeuchi, O., and Akira, S. (2010). Pattern recognition receptors and inflammation. Cell *140*, 805–816.

Thon, V., Vlkova, M., Freiber, G.T., Lit, Z.J., and Lo, K.J. (2005). The expression of Fc gamma receptors on leukocytes and clinical course of common variable immunodeficiency (CVID). Scr. Med. *78(6)*, 315–322.

West, P.A., Koblansky, A.A., and Ghosh, S. (2006). Recognition and signaling by Toll-like receptors. Annu. Rev. Cell Dev. Biol. 22, 409–437.

Zaffran, Y., Zhang, L., and Ellner, J.J. (1998). Role of CR4 in Mycobaterium tuberculosis – human macrophages binding and signal transduction in the absence of serum. Infect. Immun. *66*, 4541–4544.

Zanderbergen, G.V., Klinger, M., Mueller, A., Dannenberg, S., Gebert, A., Solbach, W., and Laskay, T. (2004). Cutting edge: neutrophil granulocyte serves as a vector for *Leishmania* into macrophages. J. Immunol. *173*, 6521–6525.

Role of T-cells in *Leishmania* Infection

9

Chiranjib Pal and Sunil Martin

Abstract

T-cells play a critical role in the immunoregulation of anti-leishmanial immune response. In leishmaniasis, Th1-related cytokine production seems to be crucial for host control of parasite burden and clinical cure. Visceral and diffuse cutaneous leishmaniasis are characterized by negative skin test for parasite antigens and failure to produce Th1 cytokines, whereas tegumentary leishmaniasis is characterized by a positive skin test and the ability of peripheral blood mononuclear cells (PBMCs) to produce IL-12 and IFN-γ (Th1 cytokine). In fact, IL-4 and IL-13 (Th2 cytokines) promote disease progression in cutaneous leishmaniasis, whereas IL-4 seems to enhance protective type-1 responses in visceral leishmaniasis. Thus, immune response to intracellular parasites shifted from the Th1/Th2 paradigm of resistance/susceptibility to a more biologically relevant regulatory and counter-regulatory network of diverse T-cell subsets. Moreover, the presence of antigen-specific regulatory T-cell subsets may provide an environment that contributes to the balance between Th1 and Th2 cells. Finally, the involvement of $CD8^+$ T-cells has been described, but the modality of their role and function in this kind of infection has not been expanded so far. Recently discovered Th17, Th9 and Tfh subsets and related cytokines have been reported to perform diverse functions in the course of leishmaniasis as a whole. *Leishmania* is one of the first microbes credited to employ T-reg cells for immunoevasion. IL-10-producing $Foxp3^+$ and $Foxp3^-$ T-reg subtypes perpetuate immunosuppression whereas inflammatory $CD8^+$ T-cells counter the immunosuppression, forcing the parasite to metastasize. A detailed understanding of these immune-regulatory checkpoints may expose novel drug targets, which may help in developing therapies against diseases with similar molecular aetiology.

Introduction

During a pathogenic encounter, the antigen-presenting cells of the innate defence are activated, and phagocytose and destroy the pathogen. Moreover, the danger signals generated initiate an adaptive T-cell and B-cell response which eventually secretes series of co-surface proteins and cytokines to limit the infection and palliate the inflammation-induced immunopathology. It would appear that these cytokines are secreted by groups of reciprocally regulating specialized T-helper cell subsets broadly categorized as Th1 and Th2 subsets with unique cytokine and transcription factor signatures (Bluestone *et al.*, 2009). This hypothesis temporarily satisfied the quest for a T-cell subset that mediates tissue damage and B-cell help for antibody generation. Nonetheless, experimental evidence and expanding world of cytokines would initiate a search for the novel IL-17-secreting T-helper cells. Th-17 cells were thus implied to be instigating tissue damage during autoimmunity and offering protection against fungal infections with unique functions at the mucosal surfaces (Jin *et al.*, 2008). Th17 cells are known to be generated *in vitro* by the combined engagement of TGF-β and IL-6. Such *in vitro* cytokine combination system gives way to the identification of IL-9-producing Th9 cells. Originally described as a Th2 cytokine in leishmaniasis, IL-9 also contributes to the tissue damage in EAE model (Kaplan, 2013). The notion that helper T-cells were regulated by professional suppressor T-cell subsets died out and eventually shaped in the form of T-reg cells. Incidentally, *L. major* is the first microbe reported to employ T-reg cells to sabotage host immune response. Parallel studies on the proinflammatory and anti-inflammatory CD8$^+$ T-cell subsets overlaid to the existing picture of anti-leishmanial T-cell response significantly contributed to the growing network of key concepts in the T-cell subsets diversity.

The widening horizons of T-helper subsets

Th1/Th2-balancing resistance and susceptibility

The Th1/Th2 balance hypothesis emerged in the late 1980s, originating from studies in mice of two subtypes of T-helper cells differing in cytokine secretion and functions (Mosmann *et al.*, 1986). It has been proposed that the Th1 and Th2 cells were 'important regulators of the class of immune response.' The concept consequently was smeared to human immunity (Mosmann *et al.*, 1989), and a decade after the original discovery the effects of Th1 and Th2 in disease became a main research aim ((Mosmann *et al.*, 1996; Abbas *et al.*, 1996). Presently much of the literature raises the Th1/Th2 balance concept to the level of epitome. Although Th1 and Th2 cells are now virtually ascribed with the responsibility for coordinating the immune system, critical investigators are finding discrepancies in the hypothesis (Singh et el., 1999). A hallmark of potent immunity against intracellular pathogens is the development of an optimal T-cell response that exhibits low apoptosis and rapid proliferative potential.

In the first instance protective immunity against all species is by consensus recognized as being Th1 dependent. However, as the causative agents of both Old World and New World cutaneous leishmaniasis as well as visceral leishmaniasis diverged in evolutionary terms 40–80 million years ago (McMahon-Pratt and Alexander, 2004) they have had significant time to develop different mechanisms to survive within the mammalian host. As a consequence these various species have provided excellent tools to dissect different pathways of subverting the development of protective Th1 responses. However, part of their evaluation of the 'Th1/Th2' paradigm results from the identification of further CD4$^+$ T-cell populations that can significantly influence disease outcomes. Such populations include CD4$^+$ T-cell regulatory populations, as well as further CD4$^+$ T-helper populations Th17, Th9, and T follicular helper (fh) cells (Bettelli *et al.*, 2007; Korn *et al.*, 2009; Jäeger and Kuchroo, 2010; Crotty, 2011; Peterson, 2012). There is also increasing evidence of plasticity in function of different CD4$^+$ T-cell populations that while adding to the perceived complexity of host–pathogen interactions may also clarify previous apparently anomalous reports.

In visceral leishmaniasis (Fig. 9.1), the determining factors have not yet been completely identified, but a *Leishmania* specific cellular immune response seems to play a fundamental role in the control of infection. We know that resistance to disease in VL is determined by an appropriate Th1 response, whereas susceptibility is mediated by a Th2 response, characterized by the stimulation of high titres of antibody production (Awasthi *et al.*, 2004). In VL, susceptibility is associated with an inability to produce a macrophage-stimulating cytokine profile (Th1 profile), including IFN-γ and IL-12; instead, production of immunosuppressive cytokines, such as IL-10 (Th2 profile) is observed (Awasthi *et al.*, 2004).

Earlier reports on *Leishmania* infection suggested that immune response in human CL and VL were associated with an interaction of Th1/Th2 cytokines at both cellular as well as transcriptional levels and reduced lymphoproliferative responses (Mansueto *et al.*, 2012). As the studies progressed, IL-10 emerged as the most persuasive factor for VL pathogenesis as evidenced from recent reports (Saha *et al.*, 2007; Nylen *et al.*, 2007; Mondal *et al.*, 2010). A close look at the critical findings in the field of immunology of human VL suggests important roles of proinflammatory and anti-inflammatory cytokines in disease protection and pathogenesis. Few of the very first reports on VL immunopathogenesis revealed up-regulation of IL-4 (Kemp *et al.*, 1993; Mary *et al.*, 1999) and IL-10 (Holaday *et al.*, 1993) and loss of function

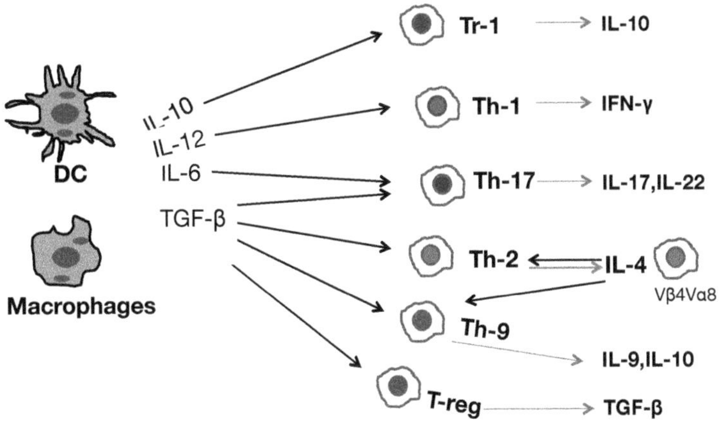

Figure 9.1 CD4+ T-cell subsets involved in the adaptive immune response during *Leishmania* infection - Depending on the immune microenvironment the interaction *Leishmania* infected Macrophages or Dendritic Cells and CD4+ T-cells, results in the generation of cytokines such as IL6, IL-12, IL-10 and TGF-β. Il-12 drives Th1 response and IFN-γ production from Th1 cells induces classical macrophage activation, nitric oxide production, and parasite killing. IL-10 from DC/Macrophages drives the IL-10 producing Tr1 cells and facilitates Th-2 differentiation along with IL-4. TGF-β drives the generation of T-reg cells and together with IL-6 influences the expansion of Th17 that causes neutrophil entry in to lesion which is often associated with severe pathology. TGF-β along with IL4 from Vβ4Vα8 CD4+ T-cells could potentiate Th9 development. Type-2 cytokines from Th2/Th9 populations induce alternative macrophage activation, increased arginase expression, and parasite expansion.

of IFN-γ (Holaday *et al.*, 1993; Ghalib *et al.*, 1993, 1995; Bacellar *et al.*, 1996) as the indication of active disease. In fact IL-4 was initially regarded as a marker for active disease (Mary *et al.*, 1999).

However, later observations exhibited its mixed response towards VL. Whereas some reports indicated lower IL-4 levels in nodal and portal areas of *in situ* liver lesions as well as in serum during disease (Ansari *et al.*, 2006) and its corresponding enhancement after cure (Hailu *et al.*, 2005), others suggested enhancement of IL-4+ neutrophils in active VL (Peruhype-Magalhaes *et al.*, 2005). Investigations using different samples, such as tissue sections (Ghosh *et al.*, 2006), plasma, whole blood, peripheral blood mononuclear cells (PBMC), mRNA and splenic aspirate cells (Bhattacharya and Ali, 2013), have confirmed that IL-10 is the principal contributor to VL pathology. Results from flow cytometric studies of cytokines also support this finding (Lagler *et al.*, 2003; Peruhype-Magalhaes *et al.*, 2005). PKDL, a disease manifestation of VL, is also characterized by enhanced IL-4 (Ismail *et al.*, 1999; Ansari *et al.*, 2006) as well as IL-10 expression and perhaps the subsequent masking effect exerted on IFN-γ by IL-10 is apparent from a generally heightened response of this cytokine. Like IFN-γ, IL-12 is also responsible for a protective response.

Recently, it has been shown that TLR7-mediated activation of IRF-5 is essential for the development of Th1 responses to *L. donovani* in the spleen during chronic infection. It could be further added that IRF-5 deficiency leads to the incapacity to control *L. donovani* infection in the liver and to the formation of smaller granulomas (Paun *et al.*, 2011). Granulomas in $Irf5_{-/-}$ mice are characterized by an increased IL-4 and IL-10 response and concomitant low iNOS expression (Paun *et al.*, 2011).

During human VL infection, highly polarized Th2 cell type responses do not appear to develop; instead there is an initial Th1-Th2 cell response. Although one study noted that IFN-γ levels in the sera from Indian patients with active VL were not elevated above those in healthy controls (Sundar *et al.*, 1997), increased levels of IFN-γ have been detected in other patients with active disease (Karp *et al.*, 1993; Kenney *et al.*, 1998). Increased productions of IL-4 and IL-10 have been detected following infection (Karp *et al.*, 1993; Carvalho *et al.*, 1994; Sundar *et al.*, 1997; Kenney *et al.*, 1998). However, the elevated levels of IL-10 that are observed after infection are markedly decreased with the resolution of disease (Karp *et al.*, 1993; Kenney *et al.*, 1998). IL-10 is a potent inhibitor of macrophage activation and can block pathways leading to the production of reactive nitrogen intermediates.

IFN-γ production from T-cells is also inhibited by IL-10, via its ability to inhibit accessory cell function rather than by a direct effect on the T-cells (Fiorentino *et al.*, 1991). Since elevated levels of IL-10 correlate with active VL, it has been suggested that the cellular immune depression that is associated with active disease may be due to the potent inhibitory effects of IL-10 rather than a lack of IFN-γ+ (Holaday *et al.*, 1993; deMedeiros *et al.*, 1998). Recently, it has been revealed that IL-27 produced by macrophages induce IL- 21 from T-cell sources leading to the expansion of Ag-specific, IL-10-producing T-cells (Ansari *et al.*, 2011).

Recent studies caught attention to another lymphokine, IL-15, for its role against infectious diseases (Woldmann *et al.*, 1999) and its ability to enhance both Th1 responses, by increasing IFN-γ production from NK and T-cells (Lauwerys *et al.*, 2000), and Th2 responses, by augmenting IL-5 and IL-13 production (Loza *et al.*, 2002). In particular, it has been demonstrated that endogenous IL-15 plays a role in the suppression of Th2 cytokines in acute VL patients, even though it does not enhance the production of Th1 cytokines (Milano *et al.*, 2002), indicating a potential protective role of IL-15 against leishmaniasis caused by the indirect effect on Th1, due to the restriction of Th2 proliferation (Milano *et al.*, 2002). Furthermore, activation of macrophage cultures, *in vitro*, by IL- 15 determines a significant anti-leishmanial activity, comparable to IFN-γ-induced one (D'Agostino *et al.*, 2004). IL-15 primes the intracellular killing of *Leishmania* in macrophages which is linked with increase in synthesis of IL-12 (D'Agostino *et al.*, 2004). Altogether, these data indicate IL-15 could, directly or indirectly, have role as activator of leishmanicidal activity, by inducing IL-12 production.

The role of IL-4 in VL is controversial. However, studies in mice suggested IL-4 may be protective in some circumstances, and can promote resistance, rather than susceptibility, to intracellular pathogens: in effect, IL-4$^{-/-}$ and IL-4Rα$^{-/-}$ animals are more susceptible to disease than their wild-type counterparts (Satoskar *et al.*, 1995). IL-4 and IL-4Rα signalling is essential not only for optimal clearance of *L. donovani* from liver and limiting infection in spleen, following primary infection, but also for effective T-cell-dependent chemotherapy, and for vaccine-induced resistance (Stager *et al.*, 2003). As evidence, in absence of IL-4, type-1 responses and IFN-γ production fail to be maintained following chemotherapy or to be induced by vaccination (Satoskar *et al.*, 1995; Stager *et al.*, 2003).

In cutaneous leishmaniasis (Fig. 9.2), the traditional counter-regulatory roles for Th1 and Th2 cells and interaction of their signatory cytokines IFN-γ and IL-4 remain subjects to important debate. For example, the archetypal Th2 cytokines IL-4 and IL-13 need not essentially counter-regulate a type-1 response as initially proposed, but can also in certain disease models or experimental conditions drive, facilitate, or promote a Th1 response (Alexander *et al.*, 2000; Biedermann *et al.*, 2001; Stager *et al.*, 2003a,b; Murray *et al.*, 2006; McFarlane *et al.*, 2011). Furthermore, Th2 responses can also be induced independently without influence of signatory cytokine IL-4 (Mohrs *et al.*, 2000). IL-4/IL-13 mediated Th1 activities include inducing macrophage and dendritic cell IL-12 production (Hochrein *et al.*, 2000; McDonald *et al.*, 2004), enhancing IFN-γ production (Noble and Kemeny, 1995), or synergizing with IFN-γ for enhanced anti-microbial activity (Bogdan *et al.*, 1991; Lean *et al.*, 2003). These studies emphasize the pleiotropic activities of IL-4 and IL-13. However, some contradictory reports suggested that IL-13 might act independently from IL-4, and the effects of IL-13 and IL-4 might be additive (Matthews *et al.*, 2000). Thus, many apparently conflicting reports on IL-4/IL-13 influences during infections with different species or strains of *Leishmania* may result in large part from the hierarchy of importance of different target cell/IL-4 and/or IL-13 interactions with in the overall global network of IL-4/IL-13 activities in an individual host model system.

Th2 cells were identified at the same time as

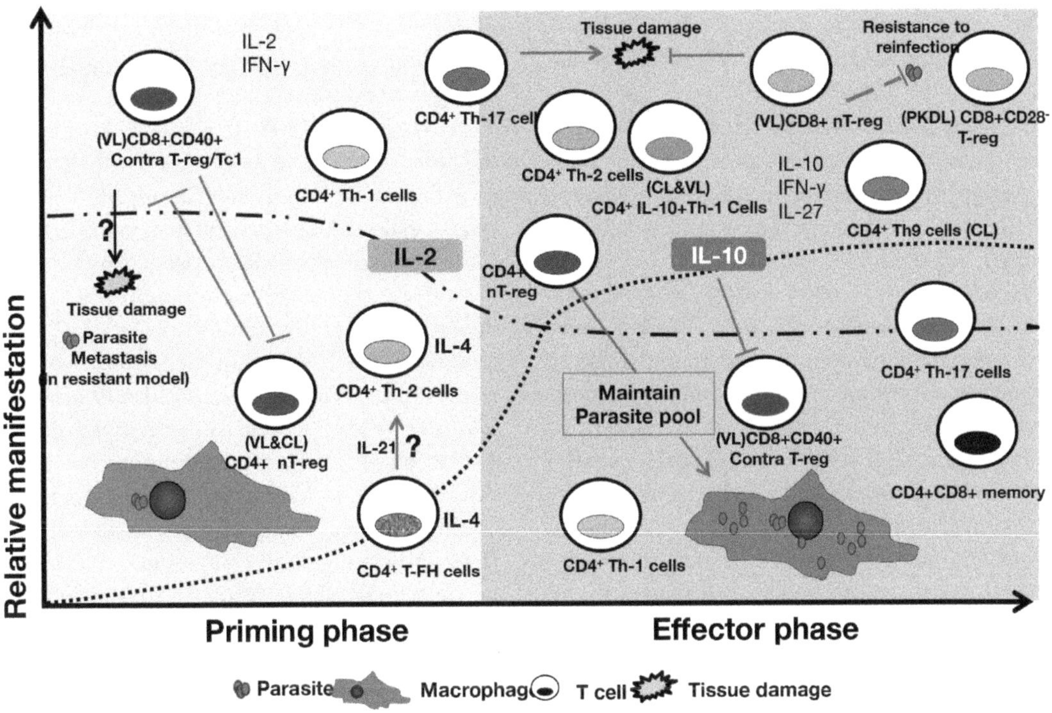

Figure 9.2 Role of T-cell in antileishmanial immune response emerging picture. Leishmania infection can be broadly categorized in to early priming phase and later effector phase based on the nature of the immune response. During the early phase as the parasite attempts to establish inside the host macrophages the type I immune response such as and Tc1 are activated as the IL-2 production goes up keeping the parasite number checked. IL-4 producing Th-2 cells are slowly increased as the infection progresses which are relatively rapid in susceptible models. Though a direct role of IL-21 in directing the course of infection is not yet established, IL-21 seems to support Th2 response. The presence of inflammatory CD8$^+$ T-cells forces the parasite migrates apart from inflicting severe tissue damage in certain models. As the immune response advances to effector phase anti-inflammatory mechanisms are triggered with an increase in parasite load and IL-10 production from the Th1 cells. IL-9 producing Th-9 cells are also a source of IL-10. Also joining to the IL-10 producing subsets are CD8$^+$ T-reg cells, CD8$^+$CD28- subsets especially in the settings of PKDL. The anti-inflammatory axis of the immune system suppresses the type I response by apoptosis (IL-10 induces cell death in contra-T-reg cells) or conversion (IL-10 production from Th1). On the other hand, T-reg cells are known to impart resistance to reinfection by a contrivance not yet completely deciphered.

Th1 cells and characterized as producing IL-4, IL-5, IL-9 and IL-13. They were associated primarily with immunity to helminths and the pathology associated with inflammatory airways diseases. Initially thought to be IL-4 dependent in their generation via STAT 6 activation and GATA 3 induction it is now recognized that GATA 3 is the key regulator of Th2 differentiation and can be induced independently of IL-4/IL-13 and STAT 6 activation (Ouyang *et al.*, 1998, 2000; Brewer *et al.*, 1999; Mohrs *et al.*, 2000). Early studies in susceptible BALB/c mice demonstrated that a single T-cell epitope derived from the parasite LACK antigen (*Leishmania* homologue of receptors for activated C kinase) induced rapid IL-4 production by Vβ4Vα8CD4$^+$ T-cells which apparently rendered T-cells unresponsive to IL-12 and correlated with lesion development (Launois *et al.*, 1995). Thus, a particularly strong case for the predominant role of IL-4 and a Th2 response in non-healing cutaneous disease was established. Nevertheless, other studies on resistant C57BL/6 mice noted that this strain also produced IL-4 early during infection and yet developed an unimpaired Th1 biased response (Scott *et al.*, 1996).

IL-4 has been shown for a long time to play a predominant role in directing the functional differentiation of *Leishmania* reactive CD4$^+$

Th2 cells observed in susceptible BALB/c mice post infection with *L. major*. However, the use of IL-4R$\alpha^{-/-}$ BALB/c mice has generated controversial results. Indeed, IL-4R$\alpha^{-/-}$ BALB/c mice infected with *L. major* either contained infection or still developed progressive disease. These observations suggest that another cytokine than IL-4 might be involved in susceptibility to infection with *L. major*. Results showing that IL-10$^{-/-}$ and IL-4R$\alpha^{-/-}$IL-10$^{-/-}$ BALB/c mice limit parasite replication demonstrate that IL-10 signalling contribute to susceptibility to *L. major* infection (Kopf *et al.*, 1996; Noben-Trauth *et al.*, 1996; Mohrs *et al.*, 2000). This raised the question whether other cytokines could be regulating Th1 development. As IL-4 and IL-13 share a common signalling pathway through the IL-4 receptor α (IL-4Rα) chain and as IL-4R$\alpha^{-/-}$ mice were found to be more resistant than IL-4$^{-/-}$ mice, a compensatory disease exacerbating role for IL-13 has been suggested in the absence of IL-4 (Noben-Trauth *et al.*, 1999). While arguments remain over whether IL-4 and IL-13 play a significant role in *L. major* disease progression in BALB/c mice, IL-10 indisputably does promote parasite persistence in resistant mice and multiplication in susceptible mice (Belkaid *et al.*, 2001; Noben-Trauth *et al.*, 2003). IL-10 could be generated via either Fcγ-mediated uptake of antibody-coated amastigotes (Kane and Mosser, 2001), or produced by Th2 cells, or CD4^{+} CD25+ (Belkaid *et al.*, 2001, 2002a; Sacks and Anderson, 2004) or CD4^{+}CD25^{-}FoxP3^{+} T regulatory cells (Lages *et al.*, 2008) or indeed CD4^{+}CD25^{-}FoxP3^{-} Th1 population (Anderson *et al.*, 2007). The paramount role of IL-4 in promoting disease progression is, however, dependent on mouse strain and site of infection. Thus, the growth of both *L. mexicana* and *L. amazonensis* is primarily IL-4 dependent in rump infections in BALB/c and C57BL/6 mice (McMahon-Pratt and Alexander, 2004). By contrast the growth of these parasites in the footpad of C57BL/6 mice, though not BALB/c mice, is IL-4 independent (Jones *et al.*, 2000; McMahon-Pratt and Alexander, 2004) and the contribution of IL-10 to this activity may also depend on whether *L. amazonensis* or *L. mexicana* is studied (Padigel *et al.*, 2003a; Buxbaum and Scott, 2005). Overall, there is no evidence whatsoever that Th2 cytokines can be beneficial during infection with New World cutaneous leishmaniasis and often they play the classic counter-protective role.

IL-4, unlike IL-13, IL-10 and TGF- β, which are only partially involved, seems to make the greatest contribution in non-healing cutaneous *L. mexicana* infection (Alexander and Bryson, 2005). The cathepsin-like cysteine peptidase (CPB) has been identified as the likely *L. mexicana* virulence factor inducing IL-4 production (Denise *et al.*, 2003). The role of IL-4 in subverting development of Th1 response has been clearly identified in *L. mexicana* infection by several studies; contrariwise similar studies, on the closely related parasite *L. amazonensis*, have no evidence, bordering on secondary role IL-4 in non-healing *L. amazonensis* infection (Ueda-Nakamura, 2002; Cameron *et al.*, 2004). In this setting, ability of CPB to proteolytically degrade the NF-κB family of signalling proteins might be directly responsible of Th1 responses inhibition (Ueda-Nakamura, 2002; Cameron *et al.*, 2004). Thus, IL-4-independent mechanisms might prevail (Ueda-Nakamura, 2002; Cameron *et al.*, 2004). As a matter of fact, appropriate CL models show IL-4 and IL-13 (Th2 cytokines) can promote disease progression, whereas IL-4 role in VL is not well defined, as its ability to enhance protective type-1 response.

The exact pathogenic mechanism of ML has not been established till date. While it has been suggested that ML may represent a polar hypersensitivity reaction to *Leishmania* infection (Azulay *et al.*, 1995), other studies have reported a combination of type 1 and type 2 cytokines in lesions with up-regulation of mRNA for IFN-γ, TNF-γ, IL-4 and IL-10 (Caceres-Dittmar *et al.*, 1993; Pirmez *et al.*, 1993). Reports have been shown that lymphocytes from ML patients produce high levels of IFN-γ and TNF-α in response to antigenic stimulation (Carvalho *et al.*, 1985; Ribeiro-de-Jesus *et al.*, 1998). However, ML patients have a decreased ability to produce and to respond to IL-10 after stimulation with *Leishmania* antigen suggests that the response to this major immunoregulatory cytokine is impaired in ML patients. This could represent a major mechanism whereby immune responses are deregulated, leading to this severely disfiguring form of human disease (Bacellar *et al.*, 2002).

The major source of *Leishmania* antigen-induced IFN-γ and TNF-α in ML patients is CD4+ Th1 cells (Bacellar *et al.*, 2002). The pathways underlying the improper production of and response to IL-10 during ML remain hard to pin down. One possibility is that the concentrations of IFN-γ in ML patients down-regulate the expression of IL-10R in lymphocytes and consequently prevent IL-10 modulatory effects. The IL-10-specific ligand, IL-10R (Liu *et al.*, 1994), is structurally related to interferon receptors (Ho *et al.*, 1993) and might be negatively influenced by high levels of this Th1-type cytokine. Another possibility is that the increased secretion of other proinflammatory cytokines observed in ML patients could also down-regulate IL-10R as shown in acute exanthematic psoriatic epidermis in which the decrease in expression of the IL-10 receptor has been reported to be influenced by the proinflammatory cytokine IL-8 (Michel *et al.*, 1997).

The involvement of major cytokines in *Leishmania* infection is summarized in Table 9.1.

Th17 and related responses

In 2000, a novel cytokine chain, p19, was discovered by a computational sequence screen looking for homologues of the IL-6 subfamily of proteins (Oppmann *et al.*, 2000). This p19 chain forms heterodimers with the p40 chain of IL-12, and this novel cytokine was named IL-23; thus, all approaches that targeted the p40 chain of IL-12 would affect both IL-12 and IL-23. By creating IL-23p19 knockout (KO) mice and comparing them with IL-12p35-deficient mice, Cua and colleagues (Cua *et al.*, 2003) at the DNAX Research Institute demonstrated that IL-23 and not IL-12 was crucial for the induction of EAE. In the follow-up study (Langrish *et al.*, 2005.), they showed that IL-23 expands/generates IL-17-producing T-cells that are capable of inducing EAE when adoptively transferred into naive wild-type mice. These IL-17-producing T-cells were dramatically reduced in the central nervous system (CNS) of IL-23p19-deficient mice. On the basis of this and other studies, investigators proposed that IL-17-producing T-cells are a distinct T-helper cell subset, which was named Th17 cells (Langrish *et al.*, 2005; Park *et al.*, 2005; Harrington *et al.*, 2005). However, as IL-23 receptor is not expressed on naive T-cells, and, IL-23 was not able to generate *de novo* IL-17-producing T-cells from sorted naive T-cells (Bettelli *et al.*, 2006), the question arises whether IL-23 indeed is the differentiation factor of Th17 cells, and, if not, then what are the factors needed for the differentiation of Th17 cells.

Th17 cell lineage development is induced by cytokines such as IL-6, IL-23 and TGF-β (Ouyang *et al.*, 2009) which in turn activate the STAT 3 pathway (Marwaha *et al.*, 2012) leading to the expression of the transcription factor retinoic-acid orphan related receptors (ROR-α and

Table 9.1 CD4+T-cells and cytokines involved in different forms of leishmaniasis

	Healing		Non-healing	
Disease	CD4+ T-cells	Cytokines	CD4+ T-cells	Cytokines
Cutaneous leishmaniasis	*L. major:* Th1	IL-12, IL-1α, IL-18	*L. major:* excessive Th2 response and/or defective APC function and/or defective Th1 response	IL-4 (?) IL-13, IL-10, TGF-β
	L. mexicana: Th1	IL-23, IL-27, IFN- γ	*L. mexicana*: excessive Th2 response and/ or defective APC function and/or defective Th1 response	IL-4, IL-13 (?), IL-10 (?)
	L. amazonensis: Th1	IFN- γ, GM-CSF, IL-10, TGF- β	*L. amazonensis:* excessive Th2 response and/or defective APC function and/or defective Th1 response	IL-4 (?)
Visceral leishmaniasis	Experimental: Th1, Th2 (?)	IL-12, IL-4 (?), IFN- γ _IL-17	Experimental: Defective Th1 response (?). Excessive Th2 response (?)	IL-4 (?), IL-10, TGF-β (?)
	Human: Th1, Th2 (?)	IL-2, IFN- γ, IL-12, IL-15, IL-17, IL-4 (?)	Human: Defective Th1 response (?). Excessive Th2 response (?)	IL-4 (?), IL-10, TGF-β (?)

ROR-γt) (Alexander *et al.*, 2012). Th17 cells are the key producers of IL-17A and IL-17F, in addition to other cytokines such as IL-21 and IL-22 (Bending *et al.*, 2011). IL-17 performs an important function in neutrophil development and since neutrophils are regarded as an important effector during *Leishmania* infection, it would be possible that Th17 cells might have a significant role in this disease (Charmoy *et al.*, 2010). However, the exact function they perform in this disease depends largely on the parasite species and the genetic background of the host.

Pitta *et al.* (2009) first reported the involvement of Th17 cells and its importance in *L. donovani* infections. They showed that naturally resistant subjects with enhanced IL-17 and IL-22 responses react more rapidly to *L. donovani*, attracting strong effectors of innate immunity and recruiting Th1 cells to tissues. These cells would consecutively enhance the microbicidal activity of phagocytes. Thus, Th17 and Th1 cells may play complementary roles in protection against *L. donovani*, with both being required for complete protection. Interestingly, Th1 cells may play the important role of down-regulating the Th17 response after the infection is controlled, thereby preventing tissue damage resulting from the out-of-control expansion of the Th17 cell population during Th1–Th17 association. Moreover, they showed that the IL-22 response induced by *L. donovani* was more dependent on IL-6 than on IL-23, and conversely the IL-17 response was more dependent on IL-23 than on IL-6. However, other studies have also reported different cytokine requirements for IL-22 and IL-17 production, with IL-22 production dependent on IL-6 but not on TGF-β, and with IL-17 production dependent on TGF-β and IL-6 (Zheng *et al.*, 2007). IL-23 is required to maintain the Th17 response (Veldhoen *et al.*, 2006; Yen *et al.*, 2006). The lower level of production of IL-6 and IL-23 in cultures from subjects with VL patients may have contributed to the weaker IL-17 and IL-22 responses of these subjects. Their observations suggest a defect rendering T-cells less responsive to key regulators, such as IL-1β and IL-6 that are crucial for Th17 induction.

A study on American CL revealed that IL-17 is expressed in the peripheral blood and tissue of patients with Leishmaniasis, and there was an association between the number of cells expressing IL-17 and the intensity of the inflammatory infiltration. These data, together with the observation of a correlation between IL-17 and TNF-α, point to the participation of IL-17 in the pathogenesis of leishmaniasis (Bacellar *et al.*, 2009). Recently, the involvement of IL-17 in PKDL pathogenesis and its association with TNF-α and NO production have also been demonstrated (Katara *et al.*, 2012).

However, conflicting report in *L. major*-infected BALB/c mice revealed that DC-derived IL-23, in addition to IL-1α and IL-12p80, contributes to disease susceptibility *in vivo*. IL-23-mediated IL-17 induction appears to be accountable for persisting and unfavourable infiltration of *Leishmania* lesions with inflammatory neutrophils as earlier findings have demonstrated that disease progression in BALB/c mice is linked with an excess of neutrophil response (Lopez *et al.*, 2009).

Th17 cell development and IL-17 production are down-regulated by IL-27/WSX-1 signalling (Fitzgerald *et al.*, 2007). Prior to the characterization of Th17 cells increased *L. major* growth in IL-27R$^{-/-}$ (WSX-1$^{-/-}$) C57 BL/6 mice in the healing C57BL/6 model was attributed to a diminished Th1 response (Artis *et al.*, 2004). Subsequently, it has been shown that IL-27R$^{-/-}$ (WSX-1$^{-/-}$) C57 BL/6 mice infected with a virulent strain of *L. major*, that induced non-healing disease in these normally resistant mice, had enhanced Th17 development and increased pathology though no difference in parasite numbers compared with wild-type controls (Anderson *et al.*, 2009). However, as the IL-27/WSX-1 signalling pathway has also been associated with promoting Th1 cell development by inhibiting IL-4 production (Artis *et al.*, 2004) and also in enhancing IL-10 production from CD4$^+$ T-cells (Anderson *et al.*, 2009) this can often make interpretation of effects utilizing IL-27R$^{-/-}$ (WSX$^{-/-}$) mice difficult.

Finally, the uncertainty surrounding the role of IL-17 is emphasized in studies involving the *L. (viannia)* subgenus. Healing in a murine model of *L. braziliensis* infection (Vargas-Inchaustegui *et al.*, 2008) as well as resolution of *L. panamensis* infections in IL-13$^{-/-}$ mice (Castilho *et al.*, 2010) is associated with elevated IL-17 as well as IFN-γ levels. Overall the role of Th17 cells in leishmaniasis remains unclear, and both disease-promoting and protective responses have been attributed to their influence. Undoubtedly, the infecting species as well as the genetic background of the host play a role in Th17 cell activities.

Th-9 – susceptibility factor?

Very little is known on the role of IL-9 in leishmaniasis. Reports indicate that IL-9 is induced by *L. major* infection and transiently expressed during the first days post infection. Expression correlated with the expansion of Ag-specific Th2 cells and purified CD4$^+$ T-cells produced IL-9 during polyclonal or Ag-specific re-stimulation (Gessner *et al.*, 1993). No further information is available, with a potential role of IL-9 in leishmaniasis still unexplored.

Th9 cells are often considered as the subset of Th2 cells, generated by IL-4, but requiring the additional influence of TGF-β to be reprogrammed, as well as IL-2 (Jäeger and Kuchroo, 2010; Nowak and Noelle, 2010). Conversely, expansion of the Th9 cell population is inhibited by IFN-γ. Th9 cells express the transcription factor PU.1 but do not express T-bet, GATA 3, ROR-γt, or Foxp3. Th9 cells produce both IL-9 and IL-10 although IL-10 expression is not regulated by PU.1. Th9 cells are considered as the main source of IL-9. However, this cytokine can also be produced by Th17 and T-reg cells. Th9 cells and IL-9 production have been also associated with Mast cell recruitment and expansion, the pathology associated with atopic allergy and immunity to nematode infection (Faulkner *et al.*, 1998).

Interestingly, it has been observed that IL-9 synthesis was only induced in susceptible BALB/c but not in resistant C57BL/6 or DBA mice (Gessner *et al.*, 1993; Nashed *et al.*, 2000). Higher levels of IL-9 were observed in susceptible BALB/c mice and the ability of CD4$^+$ T-cells to produce IL-9 persisted as infection progressed. The early studies of disease promoting activity of IL-9 was confirmed in BALB/c mice rendered deficient in IL-9 by vaccination with a recombinant IL-9/Freund's adjuvant complex (Arendse *et al.*, 2005). The depletion of IL-9, *in vivo*, in BALB/c mice results more resistant to *L. major* infection and a switch in the Th2/Th1 balance in favour of a protective type-1 response has been observed. This resulted in a switch from a classically to an alternatively activated macrophage phenotype as characterized by increased microbicidal activity associated with higher NO production and reduced arginase-1 expression.

Follicular helper CD4$^+$ T-cells (T-fh)

Tfh cells were initially proposed as a separate lineage based on their failure to express Th1/Th2/Th17 cytokines and lineage-specific transcription factors (King *et al.*, 2008; Vogelzang *et al.*, 2008; Nurieva *et al.*, 2008). However, recent reports showed that Tfh cells developed *in vivo* during Th1, Th2 or Th17 immune responses may and express IFN-γ, IL-4 or IL-17a (Jinfang *et al.*, 2010). Therefore, Tfh cells seem to be heterogeneous and to have a close relationship to Th1, Th2 or Th17 cells. Although Tfh cells express high levels of Bcl-6 and Bcl-6 is critical for Tfh cell development (Jinfang *et al.*, 2010), the involvement of other lineage-specific transcription factors, including T-bet and GATA-3, in such cytokine-secreting Tfh cells has not been determined. Tfh cells may represent a particular state of Th1, Th2 or Th17 cells, particularly those Th1/Th2/Th17 cells that migrate to the B-cell follicle. However, the alternative possibility is that naive CD4$^+$ T-cells differentiate into Tfh cells on certain forms of stimulation and then they acquire IL-4, IFN-γ or IL-17-producing capacity. Interestingly, in the gut, Tfh cells mainly originate from T-regs (Jinfang *et al.*, 2010). Data from *Leishmania* infection in IL-21R KO mice suggests that IL-21 is known to sustain Th-2 functions than directly regulating the course of infection (Frohlich *et al.*, 2007). Not much is known regarding the role of Tfh cells in *Leishmania* infection. However, a recent study argued for the role of 'B helper' follicular T-cell lineages in *L. major* progression. This study claimed that the bulk of IL-4 produced in the draining lymph nodes of susceptible mice is produced not, as expected, by Th2, but rather by these follicular CD4$^+$ T-cell lineages (Reinhardt *et al.*, 2009).

CD8$^+$ T-cells response

The available data suggest that CD8$^+$ T-cells are involved in leishmanicidal activity, and that the latter is modulated by IFN-γ and IL-2 secretion (Tsagozis *et al.*, 2005). A variety of CD8$^+$ T-cells effector mechanisms, involving direct cytotoxic activity, via perforin/granzyme pathway, and apoptosis induction, via Fas–FasL interaction, have been highlighted in several reports (Bertho *et al.*, 2000; Tsagozis *et al.*, 2003, 2005). Furthermore, CD8$^+$ lymphocytes secrete various cytokines (especially IFN-γ) and chemokines [especially ones belonging to C-C group, i.e. Regulated upon Activation, Normal T-cell Expressed, and presumably Secreted (RANTES)], which might have key role in acquired cell-mediated immunity against parasite, i.e. by promoting CD4$^+$ Th1 cell development (Tsagozis *et al.*, 2003). The protective effect of challenging the Immune BALB/c mice may be ascribed to the generation of IFN-γ from CD8$^+$ T-cells (Muller *et al.*, 1993; Stefani *et al.*, 1994). Studies on the *Leishmania* infection in MHC class I and II-deficient mice of resistant background and β_2-microglobulin mice questioned the protective role of CD8$^+$ T-cells during *Leishmania* infections (Erb *et al.*, 1996).

Protective phenotype associated with IFN-γ production from CD8$^+$ T-cells have been observed when immune BALB/c mice rechallenged with live parasites (Muller *et al.*, 1993; Stefani *et al.*, 1994). MHC class I-deficient mice were showing the curative phenotype (Locksley *et al.*, 1993) ruling out the protective role CD8$^+$ T-cells during primary infections whereas MHC class II-deficient mice of resistant background which were resistant for *L. major* infection otherwise showed fatal disease compared with wild type (Erb *et al.*, 1996). In agreement with these observations, another study showed that CD8- or β_2-microglobulin-deficient mice are resistant to *Leishmania* infection and Th1 immunity in these mice is maintained long-term.

It has also been shown that low parasite dose and inoculation into a dermal site during the parasite transmission played a decisive role in disease protection of C57BL/6 mice associated with CD8$^+$ T-cell accumulations in the skin whereas CD8$^{-/-}$ mice abrogated the protection (Belkaid *et al.*, 2002). Further, β_2-microglobulin$^{-/-}$ and perforin$^{-/-}$ mice primed with a leishmanial antigen were not able to control a challenge infection after vaccination (Colmenares *et al.*, 2003), suggesting the role of CD8$^+$ T-cell in leishmaniasis. Taken together, these observations suggest that the role of CD8$^+$ T-cells is still not defined in *Leishmania* infection. It may perhaps be possible that the requirement of CD4$^+$ T-cells helps in the generation of long-lasting memory against *Leishmania* parasite by CD8$^+$ T-cells depending upon the early priming of CD4$^+$ T-cells and the leishmanial antigen.

T-reg cells as master regulators of immunosuppression

Immunoregulation in the adoptive immune system is more specialized and compartmentalized involving T-reg cells (regulatory T-cells) which would block the exaggerated activation of the helper T-cells subsets and thus is an attractive target for the microbes to shield the immune pressure. There was an explosion of literature in the past decade about the heterogeneity and function of T-reg cells in the antipathogenic immune response.

T-reg cells classification

Regulatory T-cells are master regulators of dominant tolerance in the periphery (Sakaguchi *et al.*, 2013). Generally T-reg cells maintain a high receptivity for IL-2 (CD25hi), a low receptivity for IL-7 (CD127) and expression of Foxp3 – member of the forkhead/winged-helix family of transcriptional factor. Foxp3 mutation causes a heritable genetic disease in humans [immune dysregulation, polyendocrinopathy, enteropathy X-linked (IPEX) syndrome] and in mouse ('scurfy') which is associated with T-reg dysfunction (Bennett *et al.*, 2001). T-reg cells are poised to suppress the proliferation and function of the hyperactive immune cells to maintain homeostasis through diverse mechanisms (Pandiyan *et al.*, 2011; Wilke *et al.* 2010). In contrast pathogens and tumours employ T-reg cells to evade effective immune response directed against them (Dasgupta and Saxena, 2012). Growing research in this field opened up more questions (about the identity, origin, suppressive functions, stability, mobility and metabolism of T-reg cells) than answers. Three major T-reg subsets which are extensively studied in the parasitic diseases are natural T-reg cells (nT-regs), induced T-regs (iT-reg) and CD8$^+$ T-reg cells.

Natural T-reg cells (nT-regs)

All T-cell lineages including the naive T-reg cells with the phenotype of CD4$^+$CD25$^+$ CD127^{low}FOXp3$^+$ stems in the thymus and constitute 5–10% of the peripheral CD4$^+$ T-cells in mice and human (Pacholczyk and Kern, 2008). The role of this unique population was identified by the classic day 3 thymectomy-induced autoimmunity in the neonatal mouse (Sakaguchi *et al.*, 1995). Day 7 thymectomized mice did not develop autoimmune disease, suggesting the homing of suppressor cells on day 3 from the thymus. nT-reg cells are suggested to be generated due to the defective positive selection in the thymus as T-cells with intermediate TCR affinity to self-antigens develops and exit the thymus as nT-regs (Lee *et al.*, 2011). nT-reg cells proliferate in response to foreign antigen and maintain a stable pool in the periphery (Fisson *et al.*, 2003).

Induced T-reg cells (iT-regs)

Antigen specific activation of the emigrated naive T-cells in the periphery induces their differentiation into both 'activated naive' and 'converted T-reg cells' subsets functionally identical to the nT-regs. nT-regs cells have been shown to further differentiate into effector T-reg cells and terminal effector nT-regs cells with unique cell surface marker expression (Sakaguchi *et al.*, 2010). Also contributing to the effector TReg cell pool are converted T-reg-like cells, which are derived from naive T-cells. The induced T-reg cells originate from CD4$^+$CD25$^-$Foxp3 progenitors under sub-optimal stimulation. Tr1cells (CD4$^+$IL-10$^+$) and Th3 cells (CD4$^+$LAP$^+$) are some of the examples of converted T-reg cells especially in gut environment (Chatenoud, 2011).

CD8$^+$ T-regs

The existence of CD8$^+$ T-reg cells were originally proposed by Gershon and Kondo in the early 1980s (Cantor *et al.*, 1976). Accumulating evidence indicate the role played by CD8$^+$ T-reg cells in autoimmune diseases and transplantation tolerance. CD8$^+$ CD25$^+$, CD8$^+$ CD122$^+$, CD8$^+$ CD45RC^{low} induces non-specific immunosuppression whereas CD8$^+$ CD28$^-$, CD8$^+$ CD75s+, CD8$^+$ CD45RC^{hi} TC1 and TCR peptide-specific CD8$\alpha\alpha$ T-reg cells suppress the immune response antigen specifically (Tang *et al.*, 2005). Qa-1-restricted CD8$^+$ T-regs are also described which recognizes Qa-1 in mice (HLA-E in human). CD8$^+$ T-reg cells were suggested to hamper anti-tumour immune response and erode protective immunity to chronic viral infections (Hu *et al.*, 2013; Muthu Raja *et al.*, 2012). It is suggested that CD8$^+$ T-reg cell represents the diversity among the T-reg diaspora and may protect the host from extensive tissue damage by the inflammatory CTL responses.

Human T-reg cells and mouse T-reg cells

Mouse and human T-reg cells are fundamentally similar in function. However, there are considerable differences between them in the aspects of molecular regulation and signalling, phenotype and mechanisms of suppression. In murine studies CD4$^+$CD25$^+$FOXP3$^+$ phenotype almost invariably represent the T-reg cells whereas in human this population can be more heterogeneous (memory subset, activation phenotype, naive T-reg cells and memory T-reg cells) (Baecher-Allan *et al.*, 2005; Taams *et al.*, 2001). Naive T-reg cells and memory T-reg cells can be classified using CD45RA and CD45RO in human and by CD45RB in mouse (Sakaguchi *et al.*, 2010). Unlike in mouse, human T-reg cells express two splice variants of Foxp3, namely isoform 1 and isoform 2 (in which exon 2 is spliced out) (Allan *et al.*, 2005). Furthermore, human T-reg cells are developed in the 'Hassall's corpuscle' inside the thymic medulla wherein TSLP (thymic stromal lymphopoietin) activated the immature CD11c$^+$ DC to support the budding nT-reg cells (Watanabe *et al.*, 2004, 2005).

Mechanisms of T-reg mediated immunosuppression

T-reg cells regulate the immune response primarily by four different mechanisms:

1 Inhibitory cytokines (membrane bound TGF-β, TGF- β, IL-10 and IL-35);

2. Cytolysis of the effector cells (Granzyme A, Granzyme B and perforin);
3. Metabolic interruption (IL-2 depletion, cAMP release and blocking of TCR signalling in effector cells);
4. Generating immunosuppressive APC (inhibiting the DC maturation and function by inhibitory receptors, expansion of MDSCs) (Vignali, 2012).

Evolution of the concept of T-regs in leishmaniasis – a time line

Parasites exploit the host machinery for its own survival without killing them (McGwire and Kulkarni, 2010). A crucial strategy to survive in the host is to evade the detection by the host immune system by inducing immunosuppression.

Early recognition phase

The idea that *Leishmania* causes chronic immunosuppression perpetuated by a specialized group of cells known as suppressor T-cells (Ts) was known as early as the 1980s (Howard *et al.*, 1980; Nussenzweig, 1982). Howard *et al* reported clearance of the parasite in the thymectomized mice and suppression of DTH by the T-cell enriched fraction (as few as 10^6 cells) in the uninfected mice (Liew *et al.*, 1982). Efforts were made to identify the cellularity and mechanisms of suppression of Ts cells. These cells were narrated to be adherent cells of Thy-1+, Lyt-1+2-, and I-J- phenotype mediating suppression through secretory factors (Perez, 1983).The immunodepression upon PHA stimulation of the splenocytes were strongly co-related with the progressive diminishing of IL-2 activity 2–8 weeks post infection (Reiner and Finke, 1983). Deletion of precursors of T-cells by cyclophosphamide and non-lethal irradiation was suggested to resolve the infection in non-curing murine models (Ulczak and Blackwell, 1983).

Reformation phase

As early as the 1990, Hviid *et al.* (1990) observed the bifurcation of a T-cell population into $CD25^-$ and $CD25^+$ cells with a lower cellular volume in *Leishmania donovani*-infected human PBMCS. Further down the line *L. panamensis* infection was also found to augment the expression of IL-2 receptor CD25 (Isaza *et al.*, 1996). Fabienne Tacchini-Cottier group formally addressed the connection between leishmaniasis and $CD4^+CD25^+$ regulatory T-cells (Aseffa *et al.*, 2002). It was observed that early depletion of regulatory T-cells by antiCD25 antibodies resulted in bursting of IL-4 production and influences the course of the disease. Further evidence for the Th1 and Th2 regulation of regulatory T-cells came from the studies on the susceptible and resistance model of *L. major* and colitis (Xu *et al.*, 2003). Thus, *Leishmania* become the first microbe to be well described for its direct role in actively engaging T-reg cells for modulating the immunosuppression (Kaye and Scott, 2011).

In a murine model of *L. major* infection the regulatory T-cells were revealed to play a quintessential role in maintaining the parasite population after healing offering immunity to reinfection (concomitant immunity) (Belkaid *et al.*, 2002; Mendez *et al.*, 2004). By then it became increasingly clear that T-reg cells suppress Th1 cells in resistant model (C57BL/7) and Th2 cells in susceptible BALB/c model. In fact depletion of T-reg cells or genetically knocking down IL-10 establishes immunity to non-healing infection despite Th1 polarization, suggesting that T-reg cells directly influence the immunity and tolerance (Anderson *et al.*, 2005).

The focus was soon shifted to the cellular behaviour of T-reg cells. A seminal paper from Yasmine Belkaid's group showed that the alpha (E) beta (7) integrin CD103 is required for the local retainment of T-reg cells at the site of infection and that blocking or deleting this molecule may help clearing the pathogen. With this study CD103 was also recognized as a T-reg marker and joined in the litany of the growing T-reg cell markers (Suffia *et al.*, 2005). Till then $CD4^+CD25^+Foxp3^+$ cells expressing GITR was considered to be regulatory T-cells. The first conclusive report for the involvement of IL-10 secreting $CD4^+CD25^+$ T-reg cells in the visceral leishmaniasis (*L. amazonensis*) came from research group from UT galavastin (Ji *et al.*, 2005). Antibody mediated blocking of IL-10R or CD25 could enhance the resistance to reinfection by eliminating the T-reg cells providing proof of the principle that deletion of T-reg-related functions can enhance the vaccine efficacy (Stober *et al.*, 2005; Tabbara *et al.*, 2005).

Not surprisingly TGF-β is also indicated to be a key player in orchestrating immunosuppression in parasitic diseases (Buzoni-Gatel *et al.*, 2001; Gantt *et al.*, 2003). A subset of $CD4^+CD25^+$ T-cells expressing TGF-β was shown to expand in response to *L. guyanensis* infection. Healthy patients with no history of *Leishmania* infection (Kariminia *et al.*, 2005). Since TGF-b blockade restores the *L. guyanensis* induced immunosuppression but not anti CD3-induced immunosuppression, targeting TGF-b could be one of the attractive strategies to enhance the ant parasitic immune response.

By then, there was explosion of interest in the antigen specificity of T-reg cells and factors that control the induction and proliferation of T-reg cells. In *L. major* infection antigen experienced T-cells were found to be responsible for the parasite persistence than the naive T-cells (Grey *et al.*, 2006). The studies addressing the foreign antigen specificity of natural T-reg cells (nT-reg) accumulated at the site of infection clearly demonstrated antigen-specific proliferation of nT-reg cells at the site of infection. *Leishmania*-infected DCs expanded $CD4^+CD25^{hi}$ nT-reg cells but not by the infected MHC-II KO DCs (Suffia *et al.*, 2006). In fact, evidence from the clinical data of *L. amazonensis* suggests the persistence and accumulation of T-reg cells (Campanelli *et al.*, 2006). Although at that point of time, the markers of nT-reg cells were disputed with activation markers especially in human T-reg cells. These experimental data provide important insights to the cellular mechanisms of T-reg proliferation (Peters and Sacks, 2006). This recruitment of T-reg cells to the site of infection was mediated by chemokines MIP-1α, MIP-1β, or RANTES as T-reg cells lacking CCR-5 fails to reach out to the site of infection.

The growing idea of the phenotype of T-reg cells that regulate *Leishmania* infection by then was that of $CD4^+CD25^{hi}CD103^+CCR5^+$ $IL\text{-}10^+$ cells. However, in three separate reports $CD4^+CD25^-FOXP3^-IFN\text{-}\gamma^+$ T-cells were demonstrated to be the major source of IL-10 in both murine *L. major* infection and human patients of *L. donovani* (Anderson *et al.*, 2007; Nagase *et al.*, 2007; Nylen *et al.*, 2007). IL-2 was also shown to cross regulate IL-12 responsiveness as IL-2 neutralization restores the IL-12 induced proliferation of $CD4^+$ effector cells (Ramer-Tait *et al.*, 2011). Furthermore, IL-12 was suggested to be essential for the expansion of $CD4^+IL\text{-}10^+$ populations and IL-6 inhibits this population (Stager *et al.*, 2006). Taken together these reports gave strong basis for the primacy of IL-10-producing Th1 cells in orchestrating immunosuppression in leishmaniasis.

CD25 is the alpha chain of IL-2 receptor and therefore an enhanced receptivity of IL-2 may have a functional role in regulating the biology of T-reg cells. During the priming phase of *Leishmania donovani* infection IL-2 was revealed to induce IL-10 and the generation of T-reg cells. Hence IL-2 neutralization provides proof of the principle for phase specific immunotherapy (Bodas *et al.*, 2006). Interestingly the killed parasites are known to expand the nT-reg cells and abrogate infection induced immunity in healed mouse (Okwor *et al.*, 2009). Thus, the role of both nT-reg cells and IL-10-producing Th1 cells were emerging as a possible cause of chronic immunosuppression in leishmaniasis (Rodrigues *et al.*, 2009). In fact IL-10 was indicated to determine the efficacy of chemotherapy against visceral leishmaniasis.

IL-27 is another cytokine that is known to have pleiotropic effect in regulating IL-10 and IL-17 production from $CD4^+$T-cells during *L. major* infection. IL-27R blockade was found to reduce the IL-10 and IFN-γ production from $CD4^+$ cells but enhances the IL-4 and IL-17 production (Anderson *et al.*, 2009).

IL-10 production from $CD8^+$ T-cells were reported as a key contributing factor of immunosuppression in *L. chagasi* infection (Holaday *et al.*, 1993). Ganguly *et al.* (2010) described an increased peripheral $CD8^+$ $CD28^-$,$CD3^+$ $IL\text{-}10^+$ and $CD3^+Foxp3^+$ lymphocytes in Indian patients with PKDL (post-kala-azar dermal leishmaniasis), which were and regressed with treatment. Furthermore these cells were anergic to mitogenic signals indicating a new player in the stage of immunosuppression in leishmaniasis. Thus, the PKDL study exposed $CD8^+$ $CD28^-$ as yet another player in the *Leishmania* induced complex immunosuppressive mechanism. In canine leishmaniasis there was an increase in $CD3^+CD8^+$ T-cell numbers without an apparent increase in the nT-reg cell population (Cortese *et al.*, 2013). Therefore further studies systematically addressing the role of $CD8^+$ T-cell subsets which are suppressive will expand our understanding of *Leishmania*-induced immunosuppression.

Emerging picture (Fig. 9.2)

Inflammatory responses often fuel the pathogenesis of cutaneous leishmaniasis. T-reg cells are known to inhibit inflammation and hasten the wound healing process. Intriguingly, in chronic dermal leishmaniasis (CDL) patients, impaired Foxp3 functions are associated with the pathogenesis. Furthermore, healing and resolution of the dermal lesions were associated with *Leishmania* specific $CD25^{hi}CD127^-$ $CD4^+$ T-cell with regulatory functions (Rodriguez-Pinto *et al.*, 2012). T-reg cells are also shown to interfere with immunity against reinfection in *L. braziliensis* (Falcão *et al.*, 2012). Apart from hampering the tissue damage, IL-10 produced by these T-reg cells preserves *Leishmania* in the host. This population of the parasites can not only impart resistance to reinfection but also create a long term reservoir for its alternate host – *sandfly* (Yazdanbakhsh and Sacks, 2010). In the light of this study it is reasonable to speculate that T-regs can play a dual role in the pathogenesis of Leishmaniasis.

In the context of increasing incidence of VL in the immunocompromised patients, the study of the kinetics of leishmaniasis in the immunocompromised mouse models revealed the role of $CD4^+Foxp3^+$ cells in the persistence of the parasites in the liver (Tiwananthagorn *et al.*, 2012). Further studies in this model could reveal important insights about the organ specific immune regulation in VL.

More recently there were attempts to kinetically understand the mechanisms of generation of IL-10-producing $CD4^+$T-cells. Dendritic cell (DC) expressed key co-stimulatory molecule CD40 was found to be essential for the generation of IL-10-producing T-reg cells during the course of *L. donovani* infection. DC expressing lower level CD40 elicited T-reg cells than CD40-deficient T-reg cells. A higher CD40 expression led to the generation of contra-T-reg cells (Martin *et al.*, 2010a). Along the same line tracking of antigen-specific T-cells using major histocompatibility complex class II (pMHCII) tetramers during the course of *L. major* infection indicated that IL-10-producing Th-1 cells are developed during an early antigenic stimulation (Pagan *et al.*, 2013). Intriguingly, using conditional cell-specific depletion and adoptive transfer experiments demonstrated that CD11chiDCs producing IL-10 and IL-27 induces IL-10 producing Th-1 cells (Owens *et al.*, 2012).

Contra-regulation

The concept of contrasuppression was originally described by Gershon as a mechanism that counter regulate the functions of T-cells. However, the advances in molecular biology failed to support T-cells as IJ moiety was not found in the H2 complex (Simpson, 2008). Along with this finding the model of contrasuppression also faded away. The regulatory activity that disables the suppressive functions of T-reg cells or enables the targets of suppression (immune cells) can be named as contra-regulation. The mechanism of contrasuppression was suggested to be the engagement of various cytokine and co stimulatory receptors such as IL2, IL6, IL-7, IL-21,GITR, OX40, TIM-1, TLR2, TLR8 and blocking of PD-1 CD40 and CD137 (Lehner, 2008). More recently, the possibility of having a cell population to check the misdirected activity of T-regs had been explored (Hill *et al.*, 2008).

A T-cell lineage with pathogenic properties known as Th17 cells which produce IL-17 is gaining recognition for its distinct contra T-reg functions (Bettelli *et al.*, 2006). $CD4^+CD44^{hi}$ cells had been implied to have 'contra-conversion' activity which prevents the TGF-β induction of Foxp3 in the naive $CD4^+$ T-cells (Hill *et al.*, 2008). During the course of an experimental visceral Leishmaniasis, $CD8^+$ CD40+ cells with contra-regulatory functions against T-reg cells develops during the priming face which will be apoptosed by IL-10-producing T-reg cells during the effector phase (Martin *et al.*, 2010b). There are increasing evidence for the existence of a reciprocity between vaccine-induced IFN-γ-producing $CD8^+$ T-cells and IL-10-producing $CD4^+$ cells in *L. donovani* infection (Basu *et al.*, 2007). Nonetheless, it is reported that CD8$^+$ T-cells may perpetuate severe pathology and metastasis of the parasite (Novais *et al.*, 2013).

Another line of research which may throw light into the contra-regulatory mechanism of T-reg generation is the study of infection kinetics in Galactin-3KO mice (Fermino *et al.*, 2013). Galactin-3-deficient T-reg cells were found to have an increase in the suppressive functions and IL-10 production. More molecules like Galactin-3 with counter-regulatory functions may expand the spectrum of therapeutic choices.

T-reg biology: lesson learned from leishmaniasis

Leishmaniasis has expanded our understanding of the biology of T-reg cells. In fact, *Leishmania* was one of the first microbes known to expand nT-reg cells. Listed below are some of the major concepts revealed about the functions of T-reg cells from the infection models of *Leishmania:*

- Active role of T-reg cells in immunosuppression during infection (Belkaid *et al.*, 2002).
- T-reg cells not only suppress Th1 cells but also suppress Th2 cells (Aseffa *et al.*, 2002).
- T-reg cell accumulate and proliferate at the site of the microbial infection (Suffia *et al.*, 2006)
- T-reg cells as necessary evil for the homeostasis- T-reg cells not only prevent productive immune response against the microbe but also establish resistance to reinfection and prevent tissue damage due to immunopathology (Mendez *et al.*, 2004).
- Discovery of CD103, CCR-5 as a marker of T-reg cells (Suffia *et al.*, 2005; Yurchenko *et al.*, 2006).
- IL-10 as the archetypical immunosuppressive cytokine produced from nT-reg cells and Th-1 cells (Anderson *et al.*, 2007; Nylen *et al.*, 2007)
- The concept of regulatory T-cells as a functional state as opposed to fixed phenotype (Martin *et al.*, 2010a)
- IL-2, IL27 as key factors for the survival of T-reg cells and therefore the phase specific immunotherapy (Anderson *et al.*, 2009; Bodas *et al.*, 2006).
- Heterogeneity of regulatory T-cell phenotypes such as CD8$^+$ CD25$^+$ and CD8$^+$ CD28$^-$ T-reg cells (Bourreau *et al.*, 2007; Saha *et al.*, 2007).
- CD8$^+$ CD40$^+$ T-cells as contra-regulatory T-cells (Martin *et al.*, 2010b)
- Role of Gal-3 in T-reg-mediated immunosuppression (Fermino *et al.*, 2013).

T-reg targeting in anti-leishmanial immunotherapy

The insights gained by decades of research on the role of T-reg cells in leishmaniasis expanded the horizon of possibilities to target T-reg cells to cure parasitic diseases. But most of these trials are attempted in the preclinical mouse models. There is a desperate clinical demand for inventive therapeutics against leishmaniasis.

Antibody-mediated targeting of T-reg functions

Blocking the IL-10 signalling by anti-IL10R antibodies might be the first of its kind to be tested in the preclinical models of *L. major* infection (Belkaid, 2001). AntiCD25 antibody was used to deplete T-reg cells (Aseffa 2002). Although this method is effective it is possible that this mode of treatments can exclude the effector T-cells since many activated effector cells and memory cells express CD25 – the IL-2R. Same applied to the strategies targeting IL-2 directly because of the quintessential nature of the cytokine, required by most of the T-cell subsets (Bodas *et al.*, 2002). Anti-FR4 (folate receptor abundantly expressed in T-reg cells) was found to reduce the parasite load in *L. donovani* infection and was more effective than antiCD25 treatment..

Interrupting the mobility and retainment of T-reg cells

CCR-5 neutralizing antibodies were found to completely block the migration of T-reg cells to the site of infection and hence can be a useful strategy to curb the infection (Ekaterina Yurchenko, 2006). Maraviroc (UK-427857, Pfizer) is an FDA approved drug which is used to block the entry of HIV which antagonizes CCR-5 (Abel, 2009). Hence this drug may be of choice for treating the HIV–*Leishmania* co-infection. It can also be argued that CCR-5 is also essential for the homing of the effecter cells to the site of inflammation. In fact, in a clinical trial of GvHD, Maraviroc was shown to block the homing of the inflammatory T-cells to the liver and the gut but not to the skin opening (Reshef, 2012). Maraviroc may thus represent a potential treatment option for cutaneous leishmaniasis rather than visceral leishmaniasis.

Adoptive immunotherapy therapy

Adoptive transfer of *Leishmania* antigen pulsed DCs were found to tip the balance of immune response in favour of the host and accelerate parasite elimination. The immune restoration is associated with increase in the $CD4^{+}IL\text{-}2^{+}IFN\text{-}\gamma^{+}$ cells and reduction in the IL-10-producing T-reg cells (Martin *et al.*, 2010a; Schwarz *et al.*, 2013). This is especially evident when DCs are stimulated with TLR ligands (LPS) or antiCD40 before adoptively transferring them to the infected animals in the preclinical models. For e.g. DC vaccine pulsed with *L. infantum* histones +CpG elicited protective Th-1 response with reduced IL-10 production (Carrion *et al.*, 2007). In fact CD40 sufficient DCs induced host protective $CD8^{+}$ T-cells secreting IFN-g and granzyme. Adoptive transfer of these $CD8^{+}$ T-cells alone was found to reduce the parasite burden in the infected hosts (Martin *et al.*, 2010b).

Although there are no formal clinical trials directly targeting T-reg cells and its functionaries such as IL-10, some of the commercially available drugs which are immunomodulatory drugs (Leish-110f®) are known to increase the IL-12/IL-10 ratio (Probst *et al.*, 1997). Furthermore, T-reg cell proliferation and functions are an important monitor of the therapeutic prognosis. Immunotherapeutics that directly aim to modulate T-reg cells may revolutionize current anti-parasite therapies.

Unanswered questions and future perspectives

1 The immune system is exquisitely complex and therefore, in order to have a realistic understanding of the cellular and molecular dynamics of immune cells during the course of infection, robust and multiplex assays need to be employed. The power of CyTOF mass cytometry and RNAseq can be utilized to unearth unexpected and exciting cellular and molecular targets that fine tune and modulate the regulatory T-cell biology. Novel genetic approaches such as CRISPR can be employed to delete the genes of interest *in vitro* in the human systems.
2 Detailed molecular mechanisms of the generation and function of IL-10-producing $CD4^{+}$ T-cells during *Leishmania* infection with special focus on the surface markers and transcription factors.
3 Is there a unique marker for the subset of $CD4^{+}CD25^{-}FOXP3^{-}IL\text{-}10^{+}IFNg^{+}$ T-cells which can be therapeutically targeted?
4 Can we target CD49b and LAG-3 to defuse IL-10-producing Tr1 cells (Gagliani *et al.*, 2013)?
5 Since IL-10+Th-1 cells are known to be induced in the early TCR stimulation, an exquisite investigation in to the nature of the early antigen and the molecular elements required to better understand the early events during an infection may reveal novel early therapeutic intervention points.
6 The mechanistic insights into the role of T-reg cells in imparting resistance to reinfection conferred by *Leishmania*.
7 Because most of the T-cell subsets are known to express a constellation of innate receptors like TLRs and NLRs, these factors ought to be considered while tracing the biochemical dynamics of T-reg generation.
8 Combination therapy against leishmaniasis with the clinical objective of parasite clearance and least tissue damage by phase specific combination therapy. Although IL-10, IL-2, CCR-5, CD103, CD40 and FR-4 are experimented in the preclinical settings, a comprehensive protocol to arrive at the sweet spot of combination, balancing host protection and parasite clearance, is lacking.
9 *Leishmania* is proposed to modulate the metabolic machinery of the T-reg cells by targeting mTOR in the innate immune cells to change the cytokine milieu towards an anti-inflammatory spin ensuring its survival

inside the host (Martin *et al.*, 2012). Mouse models in which mTOR signalling machinery is genetically altered can be an effective approach to experimentally address this conundrum. *Leishmania* model will also give unique insights to the metabolic regulation of the immune system as well.

10 Understanding the detailed molecular mechanisms of *Leishmania*-induced immunosuppression can enlighten novel targets of signaling molecules which can be exploited to develop drugs against various autoimmune diseases and allergies.

11 Utilizing humanized mice as preclinical models of infection can accelerate the translation of the insights gained through the previous data.

Concluding remarks

Leishmania is classically the best known model to study the Th1–Th2 paradigm. However, this concept is increasingly challenged with comprehensive studies using the advancing molecular and cellular tools. Recently Th-17 cells are shown to impact the immunopathology whereas Th-9 cells are inferred to contribute to the susceptibility to infection. The rediscovery of the $CD4^+CD25^+Foxp3^+$ nT-reg cells initiated series of research formally establishing the active role of T-reg cells in the *Leishmania* induced immunosuppression. Current evidence from the preclinical mouse models and clinical samples indicate that *Leishmania* employs IL-10-secreting nT-reg cells and Th1 cells to regulate the immune system. Ironically nT-reg cells support sub-clinical loads of parasites in the healed individuals which may prevent reinfection possibly by maintaining the memory T-cell pool. Furthermore, owing to its natural role to block the exaggerated inflammation, T-regs prevents the immunopathology and tissue damage during the course of infection as per the studies in cutaneous and dermal leishmaniasis. $CD8^+$ T-cells such as $CD8^+$ T-reg cells and contra T-reg cells modulate the course of infection especially in visceral leishmaniasis. Intriguingly, $CD8^+$ cytotoxic cells can perpetuate host damage by imparting severe immunopathology and metastasis of the parasite. Leishmaniasis is thus an illustrious model to dissect the kinetics of the detailed cellular and molecular mechanisms of organ specific immunity. The identification of additional T-cell subset network with explicit molecular details of the pathways of generation will not only enlighten additional drug targets in other parasitic diseases but also benefits any disease model in which T-reg cells plays a central role.

References

Abel, S., Back, D.J., and Vourvahis, M. (2009). Maraviroc: pharmacokinetics and drug interactions. Antivir. Ther. *14*, 607–618.

Alexander, J., and Bryson, K. (2005). T helper (h)1/Th2 and *Leishmania*: paradox rather than paradigm. Immunol. Lett. *99*, 17–23.

Allan, S.E., Passerini, L., Bacchetta, R., Crellin, N., Dai, M., Orban, P.C., Ziegler, S.F., Roncarolo, M.G., and Levings, M.K. (2005). The role of 2 FOXP3 isoforms in the generation of human $CD4^+$ Tregs. J. Clin. Invest. *115*, 3276–3284.

Anderson, C.F., Mendez, S., and Sacks, D.L. (2005). Nonhealing infection despite Th1 polarization produced by a strain of *Leishmania major* in C57BL/6 mice. J. Immunol. *174*, 2934–2941.

Anderson, C.F., Oukka, M., Kuchroo, V.J., and Sacks, D. (2007). CD4(+)CD25(-)Foxp3(-) Th1 cells are the source of IL-10-mediated immune suppression in chronic cutaneous leishmaniasis. J. Exp. Med. *204*, 285–297.

Anderson, C.F., Stumhofer, J.S., Hunter, C.A., and Sacks, D. (2009). IL-27 regulates IL-10 and IL-17 from $CD4^+$ cells in nonhealing *Leishmania major* infection. J. Immunol. *183*, 4619–4627.

Aseffa, A., Gumy, A., Launois, P., MacDonald, H.R., Louis, J.A., and Tacchini-Cottier, F. (2002). The early IL-4 response to *Leishmania major* and the resulting Th2 cell maturation steering progressive disease in BALB/c mice are subject to the control of regulatory $CD4^+CD25+$ T-cells. J. Immunol. *169*, 3232–3241.

Baecher-Allan, C., Wolf, E., and Hafler, D.A. (2005). Functional analysis of highly defined, FACS-isolated populations of human regulatory $CD4^+$ CD25+ T-cells. Clin. Immunol. *115*, 10–18.

Basu, R., Bhaumik, S., Haldar, A.K., Naskar, K., De, T., Dana, S.K., Walden, P., and Roy, S. (2007). Hybrid cell vaccination resolves *Leishmania donovani* infection by eliciting a strong $CD8^+$ cytotoxic T-lymphocyte response with concomitant suppression of interleukin-10 (IL-10) but not IL-4 or IL-13. Infect. Immun. 75, 5956–5966.

Belkaid, Y., Hoffmann, K.F., Mendez, S., Kamhawi, S., Udey, M.C., Wynn, T.A., and Sacks, D.L. (2001). The role of interleukin (IL)-10 in the persistence of *Leishmania major* in the skin after healing and the therapeutic potential of anti-IL-10 receptor antibody for sterile cure. J. Exp. Med. *194*, 1497–1506.

Belkaid, Y., Piccirillo, C.A., Mendez, S., Shevach, E.M., and Sacks, D.L. (2002). CD4+CD25+ regulatory T-cells control *Leishmania major* persistence and immunity. Nature *420*, 502–507.

Bertho, A.L., Santiago, M.A., Da-Cruz, A.M., and Coutinho, S.G. (2000). Detection of early apoptosis and cell death in T CD4+ and CD8+ cells from lesions of patients with localized cutaneous leishmaniasis. Braz. J. Med. Biol. Res. *33*, 317–328.

Bettelli, E., Carrier, Y., Gao, W., Korn, T., Strom, T.B., Oukka, M., Weiner, H.L., and Kuchroo, V.K. (2006). Reciprocal developmental pathways for the generation of pathogenic effector TH17 and regulatory T-cells. Nature *441*, 235–238.

Bluestone, J.A., Mackay, C.R., O'Shea, J.J., and Stockinger, B. (2009). The functional plasticity of T-cell subsets. Nat. Rev. Immunol. *9*, 811–816.

Bodas, M., Jain, N., Awasthi, A., Martin, S., Penke Loka, R.K., Dandekar, D., Mitra, D., and Saha, B. (2006). Inhibition of IL-2 induced IL-10 production as a principle of phase-specific immunotherapy. J. Immunol. *177*, 4636–4643.

Bourreau, E., Ronet, C., Couppie, P., Sainte-Marie, D., Tacchini-Cottier, F., and Launois, P. (2007). IL-10 producing CD8+ T-cells in human infection with *Leishmania guyanensis*. Microbes Infect. *9*, 1034–1041.

Buzoni-Gatel, D., Debbabi, H., Mennechet, F.J., Martin, V., Lepage, A.C., Schwartzman, J.D., and Kasper, L.H. (2001). Murine ileitis after intracellular parasite infection is controlled by TGF-β-producing intraepithelial lymphocytes. Gastroenterology *120*, 914–924.

Campanelli, A.P., Roselino, A.M., Cavassani, K.A., Pereira, M.S., Mortara, R.A., Brodskyn, C.I., Goncalves, H.S., Belkaid, Y., Barral-Netto, M., Barral, A., *et al.* (2006). CD4+CD25+ T-cells in skin lesions of patients with cutaneous leishmaniasis exhibit phenotypic and functional characteristics of natural regulatory T-cells. J. Infect. Dis. *193*, 1313–1322.

Cantor, H., Shen, F.W., and Boyse, E.A. (1976). Separation of helper T-cells from suppressor T-cells expressing different Ly components. II. Activation by antigen: after immunization, antigen-specific suppressor and helper activities are mediated by distinct T-cell subclasses. J. Exp. Med. *143*, 1391–1340.

Carrion, J., Nieto, A., Soto, M., and Alonso, C. (2007). Adoptive transfer of dendritic cells pulsed with *Leishmania infantum* nucleosomal histones confers protection against cutaneous leishmaniosis in BALB/c mice. Microbes Infect. *9*, 735–743.

Cortese, L., Annunziatella, M., Palatucci, A.T., Rubino, V., Piantedosi, D., Di Loria, A., Ruggiero, G., Ciaramella, P., and Terrazzano, G. (2013). Regulatory T-cells, Cytotoxic T lymphocytes and a T1 cytokine profile in dogs naturally infected by *Leishmania infantum*. Res. Vet. Sci. *95*, 942–949.

Falcão, S.C., de Moura, T.R., Clarencio, J., Brodskyn, C., Barral, A., and de Oliveira, C.I. (2012). The presence of Tregs does not preclude immunity to reinfection with *Leishmania braziliensis*. Int. J. Parasitol. *42*, 771–780.

Fermino, M.L., Dias, F.C., Lopes, C.D., Souza, M.A., Cruz, A.K., Liu, F.T., Chammas, R., Roque-Barreira, M.C., Rabinovich, G.A., and Bernardes, E.S. (2013). Galectin-3 negatively regulates the frequency and function of CD4(+) CD25(+) Foxp3(+) regulatory T-cells and influences the course of *Leishmania major* infection. Eur. J. Immunol. *43*, 1806–1817.

Frohlich, A., Marsland, B.J., Sonderegger, I., Kurrer, M., Hodge, M.R., Harris, N.L., and Kopf, M. (2007). IL-21 receptor signaling is integral to the development of Th2 effector responses *in vivo*. Blood *109*, 2023–2031.

Gagliani, N., Magnani, C.F., Huber, S., Gianolini, M.E., Pala, M., Licona-Limon, P., Guo, B., Herbert, D.R., Bulfone, A., Trentini, F., *et al.* (2013). Coexpression of CD49b and LAG-3 identifies human and mouse T regulatory type 1 cells. Nat. Med. *19*, 739–746.

Ganguly, S., Mukhopadhyay, D., Das, N.K., Chaduvula, M., Sadhu, S., Chatterjee, U., Rahman, M., Goswami, R.P., Guha, S.K., Modak, D., *et al.* (2010). Enhanced lesional Foxp3 expression and peripheral anergic lymphocytes indicate a role for regulatory T-cells in Indian post-kala-azar dermal leishmaniasis. J. Invest. Dermatol. *130*, 1013–1022.

Gantt, K.R., Schultz-Cherry, S., Rodriguez, N., Jeronimo, S.M., Nascimento, E.T., Goldman, T.L., Recker, T.J., Miller, M.A., and Wilson, M.E. (2003). Activation of TGF-β by *Leishmania chagasi*: importance for parasite survival in macrophages. J. Immunol. *170*, 2613–2620.

Ghalib, H.W., Piuvezam, M.R., Skeiky, Y.A., Siddiq, M., Hashim, F.A., el-Hassan, A.M., Russo, D.M., and Reed, S.G. (1993). Interleukin 10 production correlates with pathology in human *Leishmania donovani* infections. J. Clin. Invest. *92*, 324–329.

Ghalib, H.W., Whittle, J.A., Kibin, M., Hashim, F.A., el-Hassan, A.M., Grabstein, K.H., Trinchieri, G., and Reed, S.G. (1995). IL-12 enhances Th1-type responses in human *Leishmania donavani* infections. J. Immunol. *154*, 4623–4629.

Gray, P.M., Reiner, S.L., Smith, D.F., Kaye, P.M., and Scott, P. (2006). Antigen-experienced T-cells limit the priming of naive T-cells during infection with *Leishmania major*. J. Immunol. *177*, 925–933.

Hill, J.A., Hall, J.A., Sun, C.M., Cai, Q., Ghyselinck, N., Chambon, P., Belkaid, Y., Mathis, D., and Benoist, C. (2008). Retinoic acid enhances Foxp3 induction indirectly by relieving inhibition from CD4+CD44hi cells. Immunity *29*, 758–770.

Holaday, B.J., Pompeu, M.M., Jeronimo, S., Texeira, M.J., Sousa Ade, A., Vasconcelos, A.W., Pearson, R.D., Abrams, J.S., and Locksley, R.M. (1993). Potential role for interleukin-10 in the immunosuppression associated with kala azar. J. Clin. Invest. *92*, 2626–2632.

Howard, J.G., Hale, C., and Liew, F.Y. (1980). Immunological regulation of experimental cutaneous leishmaniasis. III. Nature and significance of specific suppression of cell-mediated immunity in mice highly susceptible to *Leishmania tropica*. J. Exp. Med. *152*, 594–607.

Hu, Z., Zhang, W., and Usherwood, E.J. (2013). Regulatory CD8+ T-cells associated with erosion of immune surveillance in persistent virus infection suppress *in vitro* and have a reversible proliferative defect. J. Immunol. *191*, 312–322.

Hviid, L., Sorensen, A.L., Kharazmi, A., and Theander, T.G. (1990). Functional and phenotypic changes in human lymphocytes after coincubation with *Leishmania donovani in vitro*. Infect. Immun. *58*, 3163–3167.

Isaza, D.M., Restrepo, M., Restrepo, R., Caceres-Dittmar, G., and Tapia, F.J. (1996). Immunocytochemical and histopathologic characterization of lesions from patients with localized cutaneous leishmaniasis caused by *Leishmania panamensis*. Am. J. Trop. Med. Hyg. *55*, 365–369.

Ji, J., Masterson, J., Sun, J., and Soong, L. (2005). CD4+CD25+ regulatory T-cells restrain pathogenic responses during *Leishmania amazonensis* infection. J. Immunol. *174*, 7147–7153.

Jin, D., Zhang, L., Zheng, J., and Zhao, Y. (2008). The inflammatory Th 17 subset in immunity against self and non-self antigens. Autoimmunity *41*, 154–162.

Kaplan, M.H. (2013). Th9 cells: differentiation and disease. Immunol. Rev. *252*, 104–115.

Kariminia, A., Bourreau, E., Pascalis, H., Couppie, P., Sainte-Marie, D., Tacchini-Cottier, F., and Launois, P. (2005). Transforming growth factor beta 1 production by CD4+ CD25+ regulatory T-cells in peripheral blood mononuclear cells from healthy subjects stimulated with *Leishmania guyanensis*. Infect. Immun. *73*, 5908–5914.

Kaye, P., and Scott, P. (2011). Leishmaniasis: complexity at the host–pathogen interface. Nat. Rev. Microbiol. *9*, 604–615.

Lehner, T. (2008). Special regulatory T-cell review: The resurgence of the concept of contrasuppression in immunoregulation. Immunology *123*, 40–44.

Liew, F.Y., Hale, C., and Howard, J.G. (1982). Immunologic regulation of experimental cutaneous leishmaniasis. V. Characterization of effector and specific suppressor T-cells. J. Immunol. *128*, 1917–1922.

McGwire, B.S., and Kulkarni, M.M. (2010). Interactions of antimicrobial peptides with *Leishmania* and trypanosomes and their functional role in host parasitism. Exp. Parasitol. *126*, 397–405.

Mansueto, P., Pepe, I., Seidita, A., Scozzari, F., Vitale, G., Arcoleo, F., Elvira, I., Cillari, E., Rini, G.B., Napoli, N., *et al.* (2012). Significance of persistence of antibodies against *Leishmania infantum* in Sicilian patients affected by acute visceral leishmaniasis. Clin. Exp. Med. *12*, 127–132.

Martin, S., Agarwal, R., Murugaiyan, G., and Saha, B. (2010a). CD40 expression levels modulate regulatory T-cells in *Leishmania donovani* infection. J. Immunol. *185*, 551–559.

Martin, S., Pahari, S., Sudan, R., and Saha, B. (2010b). CD40 signaling in CD8+ CD40+ T-cells turns on contra-T regulatory cell functions. J. Immunol. *184*, 5510–5518.

Mendez, S., Reckling, S.K., Piccirillo, C.A., Sacks, D., and Belkaid, Y. (2004). Role for CD4(+) CD25(+) regulatory T-cells in reactivation of persistent leishmaniasis and control of concomitant immunity. J. Exp. Med. *200*, 201–210.

Muthu Raja, K.R., Kubiczkova, L., Rihova, L., Piskacek, M., Vsianska, P., Hezova, R., Pour, L., and Hajek, R. (2012). Functionally suppressive CD8 T regulatory cells are increased in patients with multiple myeloma: a cause for immune impairment. PLoS One 7, e49446.

Nagase, H., Jones, K.M., Anderson, C.F., and Noben-Trauth, N. (2007). Despite increased CD4+Foxp3+ cells within the infection site, BALB/c IL-4 receptor-deficient mice reveal CD4+Foxp3-negative T-cells as a source of IL-10 in *Leishmania major* susceptibility. J. Immunol. *179*, 2435–2444.

Novais, F.O., Carvalho, L.P., Graff, J.W., Beiting, D.P., Ruthel, G., Roos, D.S., Betts, M.R., Goldschmidt, M.H., Wilson, M.E., de Oliveira, C.I., *et al.* (2013). Cytotoxic T-cells mediate pathology and metastasis in cutaneous leishmaniasis. PLoS Pathog. *9*, e1003504.

Nussenzweig, R.S. (1982). Parasitic disease as a cause of immunosuppression. N. Engl. J. Med. *306*, 423–424.

Nylen, S., Maurya, R., Eidsmo, L., Manandhar, K.D., Sundar, S., and Sacks, D. (2007). Splenic accumulation of IL-10 mRNA in T-cells distinct from CD4+CD25+ (Foxp3) regulatory T-cells in human visceral leishmaniasis. J. Exp. Med. *204*, 805–817.

Okwor, I., Liu, D., Beverley, S.M., and Uzonna, J.E. (2009). Inoculation of killed *Leishmania major* into immune mice rapidly disrupts immunity to a secondary challenge via IL-10-mediated process. Proc. Natl. Acad. Sci. U.S.A. *106*, 13951–13956.

Owens, B.M., Beattie, L., Moore, J.W., Brown, N., Mann, J.L., Dalton, J.E., Maroof, A., and Kaye, P.M. (2012). IL-10-producing Th1 cells and disease progression are regulated by distinct CD11c(+) cell populations during visceral leishmaniasis. PLoS Pathog. *8*, e1002827.

Pagan, A.J., Peters, N.C., Debrabant, A., Ribeiro-Gomes, F., Pepper, M., Karp, C.L., Jenkins, M.K., and Sacks, D.L. (2013). Tracking antigen-specific CD4+ T-cells throughout the course of chronic *Leishmania major* infection in resistant mice. Eur. J. Immunol. *43*, 427–438.

Perez, H. (1983). Factors influencing the host response to *Leishmania mexicana*. Ciba Found. Symp. *99*, 157–173.

Peters, N., and Sacks, D. (2006). Immune privilege in sites of chronic infection: *Leishmania* and regulatory T-cells. Immunol. Rev. *213*, 159–179.

Probst, P., Skeiky, Y.A., Steeves, M., Gervassi, A., Grabstein, K.H., and Reed, S.G. (1997). A *Leishmania* protein that modulates interleukin (IL)–12, IL-10 and tumor necrosis factor-alpha production and expression of B7–1 in human monocyte-derived antigen-presenting cells. Eur. J. Immunol. *27*, 2634–2642.

Ramer-Tait, A.E., Petersen, C.A., and Jones, D.E. (2011). IL-2 limits IL-12 enhanced lymphocyte proliferation during *Leishmania amazonensis* infection. Cell. Immunol. *270*, 32–39.

Reiner, N.E., and Finke, J.H. (1983). Interleukin 2 deficiency in murine Leishmaniasis donovani and its relationship to depressed spleen cell responses to phytohemagglutinin. J. Immunol. *131*, 1487–1491.

Reshef, R., Luger, S.M., Hexner, E.O., Loren, A.W., Frey, N.V., Nasta, S.D., Goldstein, S.C., Stadtmauer, E.A., Smith, J., Bailey, S., *et al.* (2012). Blockade of lymphocyte chemotaxis in visceral graft-versus-host disease. N. Engl. J. Med. 367, 135–145.

Rodrigues, O.R., Marques, C., Soares-Clemente, M., Ferronha, M.H., and Santos-Gomes, G.M. (2009). Identification of regulatory T-cells during experimental *Leishmania infantum* infection. Immunobiology *214*, 101–111.

Rodriguez-Pinto, D., Navas, A., Blanco, V.M., Ramirez, L., Garcerant, D., Cruz, A., Craft, N., and Saravia, N.G. (2012). Regulatory T-cells in the pathogenesis and healing of chronic human dermal leishmaniasis caused by *Leishmania* (Viannia) species. PLoS Negl. Trop. Dis. *6*, e1627.

Saha, S., Mondal, S., Ravindran, R., Bhowmick, S., Modak, D., Mallick, S., Rahman, M., Kar, S., Goswami, R., Guha, S.K., *et al.* (2007). IL-10- and TGF-beta-mediated susceptibility in kala-azar and post-kala-azar dermal leishmaniasis: the significance of amphotericin B in the control of *Leishmania donovani* infection in India. J. Immunol. *179*, 5592–5603.

Sakaguchi, S., Miyara, M., Costantino, C.M., and Hafler, D.A. (2010). FOXP3+ regulatory T-cells in the human immune system. Nat. Rev. Immunol. *10*, 490–500.

Schwarz, T., Remer, K.A., Nahrendorf, W., Masic, A., Siewe, L., Muller, W., Roers, A., and Moll, H. (2013). T-cell-derived IL-10 determines leishmaniasis disease outcome and is suppressed by a dendritic cell based vaccine. PLoS Pathog. *9*, e1003476.

Simpson, E. (2008). Special regulatory T-cell review: Regulation of immune responses – examining the role of T-cells. Immunology *123*, 13–16.

Stager, S., Maroof, A., Zubairi, S., Sanos, S.L., Kopf, M., and Kaye, P.M. (2006). Distinct roles for IL-6 and IL-12p40 in mediating protection against *Leishmania donovani* and the expansion of IL-10+ $CD4^+$ T-cells. Eur. J. Immunol. *36*, 1764–1771.

Stober, C.B., Lange, U.G., Roberts, M.T., Alcami, A., and Blackwell, J.M. (2005). IL-10 from regulatory T-cells determines vaccine efficacy in murine *Leishmania major* infection. J. Immunol. *175*, 2517–2524.

Suffia, I., Reckling, S.K., Salay, G., and Belkaid, Y. (2005). A role for CD103 in the retention of $CD4^+CD25+$ Treg and control of *Leishmania major* infection. J. Immunol. *174*, 5444–5455.

Suffia, I.J., Reckling, S.K., Piccirillo, C.A., Goldszmid, R.S., and Belkaid, Y. (2006). Infected site-restricted Foxp3+ natural regulatory T-cells are specific for microbial antigens. J. Exp. Med. *203*, 777–788.

Taams, L.S., Smith, J., Rustin, M.H., Salmon, M., Poulter, L.W., and Akbar, A.N. (2001). Human anergic/suppressive CD4(+)CD25(+) T-cells: a highly differentiated and apoptosis-prone population. Eur J Immunol *31*, 1122–1131.

Tabbara, K.S., Peters, N.C., Afrin, F., Mendez, S., Bertholet, S., Belkaid, Y., and Sacks, D.L. (2005). Conditions influencing the efficacy of vaccination with live organisms against *Leishmania major* infection. Infect. Immun. *73*, 4714–4722.

Tang, X.L., Smith, T.R., and Kumar, V. (2005). Specific control of immunity by regulatory CD8 T-cells. Cell. Mol. Immunol. *2*, 11–19.

Tsagozis, P., Karagouni, E., and Dotsika, E. (2003). CD8(+) T cells with parasite-specific cytotoxic activity and a Tc1 profile of cytokine and chemokine secretion develop in experimental visceral leishmaniasis. Parasite Immunol. *25*, 569–579.

Tsagozis, P., Karagouni, E., and Dotsika, E. (2005). Function of CD8+ T lymphocytes in a self-curing mouse model of visceral leishmaniasis. Parasitol. Int. *54*, 139–146.

Tiwananthagorn, S., Iwabuchi, K., Ato, M., Sakurai, T., Kato, H., and Katakura, K. (2012). Involvement of CD4(+) Foxp3(+) regulatory T-cells in persistence of *Leishmania donovani* in the liver of alymphoplastic aly/aly mice. PLoS Negl. Trop. Dis. *6*, e1798.

Ulczak, O.M., and Blackwell, J.M. (1983). Immunoregulation of genetically controlled acquired responses to *Leishmania donovani* infection in mice: the effects of parasite dose, cyclophosphamide and sublethal irradiation. Parasite Immunol. *5*, 449–463.

Vignali, D.A. (2012). Mechanisms of T(reg) Suppression: Still a Long Way to Go. Front. Immunol. *3*, 191.

Watanabe, N., Hanabuchi, S., Soumelis, V., Yuan, W., Ho, S., de Waal Malefyt, R., and Liu, Y.J. (2004). Human thymic stromal lymphopoietin promotes dendritic cell-mediated $CD4^+$ T-cell homeostatic expansion. Nat. Immunol. 5, 426–434.

Watanabe, N., Wang, Y.H., Lee, H.K., Ito, T., Cao, W., and Liu, Y.J. (2005). Hassall's corpuscles instruct dendritic cells to induce $CD4^+CD25+$ regulatory T-cells in human thymus. Nature *436*, 1181–1185.

Xu, D., Liu, H., Komai-Koma, M., Campbell, C., McSharry, C., Alexander, J., and Liew, F.Y. (2003). $CD4^+CD25+$ regulatory T-cells suppress differentiation and functions of Th1 and Th2 cells, *Leishmania major* infection, and colitis in mice. J. Immunol. *170*, 394–399.

Yazdanbakhsh, M., and Sacks, D.L. (2010). Why does immunity to parasites take so long to develop? Nat. Rev. Immunol. *10*, 80–81.

Yurchenko, E., Tritt, M., Hay, V., Shevach, E.M., Belkaid, Y., and Piccirillo, C.A. (2006). CCR5-dependent homing of naturally occurring $CD4^+$ regulatory T-cells to sites of *Leishmania major* infection favors pathogen persistence. J. Exp. Med. *203*, 2451–2460.

Vaccine Biology of *Leishmania* Infection

10

Abdus Sabur and Nahid Ali

Abstract

Leishmaniasis, a vector-borne parasitic disease, imposes devastating impact on global health. A proficient preventive measure against the disease is lacking. But a number of promising vaccine candidates for leishmaniasis have been reported. This has been achieved by efforts put forward in understanding the immuno-biology associated with leishmaniasis, breakthroughs in search of antigenic components and advances in delivery system and adjuvant biology. The initial part of the chapter deals with the limitations and challenges in vaccine development against leishmaniasis pertaining to issues like less understood host–parasite interaction during *Leishmania* infection and immunological requirements for protection. In the subsequent sections major advances in search of antigenic components and vaccine formulations are depicted. This section highlights various approaches and the developmental status of experimental vaccines for leishmaniasis. The limitations associated with each vaccine category are also pointed. An update of some impact making vaccines for canine leishmaniasis is briefed. The therapeutic potential of some of the vaccine candidates are also discussed. Finally, the role played by various adjuvants and delivery system in vaccine development against leishmaniasis is detailed. In a nutshell, the chapter sheds light on the developmental history, with recent updates and future prospects for a successful vaccine for leishmaniasis.

Introduction

Leishmaniasis is the third most significant vector-borne parasitic disease with respect to morbidity, mortality and worldwide prevalence (Hotez *et al.*, 2004; Stockdale and Newton, 2013; Mathers *et al.*, 2007). The disease is endemic in at least 98 countries with 2 million new cases occurring each year and 350 million people are at risk of developing the disease (Alvar *et al.*, 2012; Who, 2010). Leishmaniasis takes a yearly toll of around 20,000 to 40,000 human lives and affects the socioeconomic wellbeing of millions of people worldwide (Alvar *et al.*, 2012). The disease is a group of manifestations caused by about 20 species of protozoan parasite *Leishmania* and transmitted by about 30 species of sandfly vector of the genus *Lutzomyia* and *Phlebotomus* in human and some selected mammals (http://www.who.int/tdr/publications/documents/swg_leish.pdf). Depending upon the parasite strain, sandfly vector and geographical location, transmissions of the disease may be zoonotic, anthroponotic or both (Desjeux, 1992).

Clinically, leishmaniasis is manifested in three forms (http://www.who.int/leishmaniasis/disease_epidemiology/en/). Most commonly, it presents as a self-healing, localized, to non-healing, diffuse, ulcerative form on skin and hence is known as cutaneous leishmaniasis (CL). *Leishmania major* and *Leishmania tropica* are the most common causative agents of CL. In some parts of Latin America infections due to *Leishmania panamensis*, *Leishmania guyanensis* and *Leishmania braziliensis* cause progressive ulcers of mucous membrane most often destroying and deforming the oral and nasal cavity. This disease form is known as mucocutaneous leishmaniasis (ML). Species like *Leishmania infantum* and *Leishmania donovani* affect the visceral organs such as liver and spleen and cause the most deadly form of leishmaniasis called visceral leishmaniasis (VL) or kala azar (Kaye and Scott, 2011). Although the clnical manifestations of leishmaniasis are dependent on parasite strain, immunological status of host and type of vector, there are several reports of cross-infection.

The *Leishmania* parasite completes its life cycle in two stages, the promastigote form which develops in the sandfly vector, and the amastigote form which develops in the macrophage of mammalian host. Following inoculation of infective promastigotes under the skin during bloodmeal of an infected female sandfly, there is a very complex interplay of host–pathogen interaction and immune response (Kaye and Scott, 2011). The outcome of this interplay decides the fate of the parasite to be cleared or established in the host. Furthermore, this also determines whether the parasite established within will cause a symptomatic disease or remain asymptomatic (Murray *et al.*, 2005). Immunosuppressed individuals are therefore more prone to this disease, and one of the major challenges to combat its increasing epidemiology is co-infection with HIV (Andreani *et al.*, 2012).

Other challenges in the control of leishmaniasis include zoonotic transmission of the disease. Except for the Indian Subcontinent and Central Africa, the disease transmission is largely zoonotic where infected dogs serve as the major reservoir (Palatnik-de-Sousa, 2012). Therefore control of canine leishmaniasis is equally important as human leishmaniasis. Change in habitat of sandfly due to deforestation, global warming, along with population migration is a major setback in the control of leishmaniasis (WHO, 2010). Besides, lack of suitable, fast and reliable diagnostic methods to detect asymptomatic carriers, often results in blood transfusion from infected to non-infected individuals (Dey and Singh, 2006). Additionally, immigration to and emigration from endemic foci has greatly affected the containment of the disease (Mott *et al.*, 1995).

Control measures of the disease are primarily focused on prevention of sandfly bites, controlling the population of sandflies and managing animal reservoirs (Stockdale and Newton, 2013). Prevention of sandfly bites by using insecticide-treated nets and insect repellents, although quite effective, is not convenient, cost-effective or user-friendly for the poorest sections of the society. Controlling the sandfly population size by spraying insecticides such as DDT initially had a major impact in decreasing the incidence of leishmaniasis, but the disease has since returned as a result of the development of tolerance to DDT (Chappuis *et al.*, 2007; Dinesh *et al.*, 2010; Sharma and Singh, 2008). Managing animal reservoirs by dog culling is an unethical and impractical approach. Although use of permethrin-impregnated dog collars has been found to be quite effective in control of canine leishmaniasis, this approach is suitable for controlling leishmaniasis in pet dogs only (Murray *et al.*, 2005; Stockdale and Newton, 2013).

Therapeutic interventions against the disease are met with few options for drugs that too with several limitations. Pentavalent antimonials are first choice of drugs for leishmaniasis except in Indian subcontinent, where almost 65% of patients show resistance to this group of drugs (Stauch *et al.*, 2012). Besides drug resistance, antimonials are toxic and need long course of treatment through painful intramuscular or intravenous injections. Alternatively, amphotericin B (AmB) is an antifungal drug commonly used to treat leishmaniasis but this is hepato- and nephrotoxic. Liposomal formulations of AmB such as AmBisome is less toxic but costly. Miltefosine is the only oral drug for leishmaniasis, but it is teratogenic and prone to resistance (Ejazi and Ali, 2013). Hence, one of the best ways to control the disease would be a cost-effective preventative vaccine.

However, there is no vaccine for any form of leishmaniasis in humans. One of the major hurdles in vaccine development for leishmaniasis is the poorly understood mechanism of host–parasite interactions and the complex immunobiology associated with *Leishmania* infection. The initial part of this chapter deals with a brief summary of the present understanding of the immunology of *Leishmania* infection and the future directions of queries and prospects in perspective for a successful vaccine against leishmaniasis. This is followed by updates of developmental history in this field, the challenges ahead and some possible measures to overcome these hurdles. Finally, the role of adjuvants and delivery systems in vaccine development is detailed.

Immunology of *Leishmania* infection

Leishmania infection, like most infectious diseases, is an outcome of evasion and modulation of both innate and adaptive arm of immune system (Soong, 2012). Approximately 100–100,000 metacyclic promastigotes are inoculated under the skin of mammalian host by a single bite of an infected sandfly (Kimblin *et al.*, 2008). The promastigotes along with the salivary proteins and associated tissue injury initiate a cascade of immunological response. Despite such responses the parasites in many cases do succeed to establish a safe niche inside the host. The following sections would discuss the art of survival by the parasite and the arm of parasite clearance by the host.

Innate immune response

Innate immunity is the first line of defence and comprises a non-specific response to infectious agents without generation of memory. The key players of innate immunity are the phagocytes, complement system and natural killer (NK) cells assisted by TLR synergy.

Neutrophils

Recruitment (early and late)

Tissue injury associated with sandfly bite along with its inoculum of salivary peptides and microbiota triggers secretion of a series of alarmins and chemoattractants which results in the recruitment of cellular infiltrates at the bite site. In experimental studies involving cutaneous leishmaniasis in C57BL/6 mice, neutrophils are observed to infiltrate within 30 minutes to the site of infection (Peters *et al.*, 2008). Although neutrophils are recruited even in case of sterile bite of sand fly, it has been shown that the promastigotes produce a neutrophil chemotactic factor and induce IL-8 secretion which supplements the chemoattraction of neutrophils to the bite site (Muller *et al.*, 2001; van Zandbergen *et al.*, 2002). Experimental infection of mice with *L. major*, results in activation of C3 and up-regulation of CXCL1, a functional homologue of IL-8 in mice in the dermal tissues which drives the early neutrophil influx (Jacobs *et al.*, 2005; Hol *et al.*, 2010). A subsequent influx of neutrophils is observed a week after the infection and this is driven by myeloid receptor proteins (MRPs) especially MRP-8 and MRP-14 secreted by the early neutrophils from the bite site (Contreras *et al.*, 2013). Secondary recruitment of neutrophils is also mediated by IL-17 secretion by Th17 cells (Gonzalez-Lombana *et al.*, 2013). Massive secondary neutrophils recruitment is associated with exuberated cutaneous and mucocutaneous lesions and hence late neutrophils are more pathological than protective in function for CL and ML.

Neutrophils as Trojan horse in *Leishmania* infection

The early neutrophils rapidly engulf the metacyclic promastigotes by complement and opsonin-dependent phagocytosis (Laufs *et al.*, 2002). For efficient killing of promastigotes fusion of tertiary granules with the phagolysosome resulting in acidification and oxidative burst is necessary. But the surface marker LPG of the parasite interferes with the fusion and hence leads to parasite survival (Panaro *et al.*, 1996). Besides, there are no reports of transformation from promastigote to amastigote inside the neutrophils. Since neutrophils are short-lived granulocytes they undergo apoptosis or necrosis soon after *Leishmania* uptake. The surviving parasites in the apoptotic/necrotic bodies are internalized by macrophages without generation of a prominent immune response. Therefore neutrophils act as the Trojan horse model of *Leishmania* infection (Ritter *et al.*, 2009). However, the role of neutrophils as Trojan horse is contradicted by the finding that there is no evident change of parasite load in macrophages and dendritic cells even in neutrophil depleted experimental mice models (Peters *et al.*, 2008; Ritter *et al.*, 2009).

Neutrophils as scavengers in *Leishmania* infection

Neutrophils secrete a series of molecules which activates the macrophages for efficient killing of the parasites. Notably among them are tumour necrosis factors (TNFs) and MRPs which catalyses reactive oxygen species (ROS) production in macrophages (Contreras *et al.*, 2013). Role of neutrophils in *Leishmania* infection as evaluated by studies on neutrophil-depleted mice models was rendered inconclusive (Ribeiro-Gomes and Sacks, 2012). In another aspect, neutrophil extracellular traps (NETs) comprising of DNA, histone and microbicidal granules are formed upon stimuli of microbial infection. NET formation upon *Leishmania* infection is triggered by LPG, and the protein contents like histones (H2A/H2B) of NETs are found to reduce the pathogenesis by killing of *L. amajonesis* promastigotes (Guimarães-Costa *et al.*, 2009; Abi Abdallah and Denkers, 2012). The results suggest that neutrophils are involved in both protection and pathology of the diseases.

Macrophages

Recruitment and phagocytosis

Macrophages are the primary host of *Leishmania* parasites in the mammalian body. Dermal resident macrophages are the cells to phagocytose promastigotes after neutrophils and they become the most dominant parasitized cells within 24 hours of *Leishmania* infection (Peters *et al.*, 2008). Monocytes infiltrate at the site of sandfly bite and are the primary phagocytes along with resident macrophages of the skin that takes up the parasite in free form, or that those are trapped in apoptotic bodies of neutrophils or released from the neutrophils. The parasites reside in the parasitophorous vacuole formed by phagocytosis. Phagocytosis of promastigotes by macrophages is mostly CR1, CR3 and sometimes FcγR mediated (Da Silva *et al.*, 1989; Ueno and Wilson, 2012). Hence, parasites that are either opsonized with C3 or IgG are more efficiently phagocytosed. Furthermore, once inside the phagosomal compartment, promastigotes are transformed to amastigotes, the more resistant form to the harsh acidic and proteolytic environment of the phagolysosome (Antoine, 1995). The fate of the parasites depends on the type of immune response developed and signalling cascade generated following infection as discussed in the following sections.

Macrophages as host for the parasite

The promastigote surface glycolipid lipophosphoglycan (LPG) interferes with the phagolysosomal maturation by disruption of phagosomal lipid microdomains and accumulation of F actin at the periphery of phagosomes (Vinet *et al.*, 2009; Moradin and Descoteaux, 2012; Lodge and Descoteaux, 2005; Dermine *et al.*, 2000). This allows time for susceptible promastigotes to transform into the more resistant amastigote form. In addition, gp63 a metalloprotease on the promastigote surface degrades the proteases of the phagolysosome, and thus prevents its own degradation (Kulkarni *et al.*, 2006). The impairment of phagosome results in the reduction of ROS production which is necessary for efficient killing of both amastigotes and promastigotes. The intrusive assembly of NADPH oxidase and proton ATPase on the altered phagosome makes the parasite find a ROS free and less acidic niche inside the parasitophorous vacuole (Vinet *et al.*, 2009). Moreover reactive oxygen burst is a result of IL-12, IFN-γ induced activation of JAK-STAT pathway which leads to up-regulation of inducible nitric oxide synthase (*i*NOS). On the contrary, production of immunosuppressive cytokines like IL-10, IL-4 and IL-13 leads to suppression of *i*NOS and up-regulation of arginase production (Gordon, 2003). Both *i*NOS and arginase act on a common substrate L-arginine but, whereas *i*NOS catalyses NO production, arginase action leads to production of polyamines that serve as nutrients for the parasites (da Silva and Floeter-Winter, 2014). Impairment of phagosomal organization, degradation of phagolysosomal proteins, induction of IL-10 production by host IgG-opsonized parasite, down-regulation of IL-12 and up-regulation of LPG driven Th2 cytokine IL-4 productions, etc. are some of the strategies employed by the parasites to establish a safe niche in the macrophages (Isnard *et al.*, 2012; Shadab and Ali, 2011; Moradin and Descoteaux, 2012).

Macrophages as graveyard of *Leishmania*

Macrophages when suitably stimulated can bring about oxidative burst of NO, and reactive oxygen intermediates in the phagolysosome and it is the most effective way of parasite clearance from the host. Therefore determining the underlying mechanism and signalling cascade for oxidative burst in macrophages is the first step towards vaccine development. Three signalling cascades have been under investigation for *Leishmania* infection. They are namely JAK2-STAT1 dependent pathway, protein kinase C-dependent pathway and the mitogen activated protein kinase 1-mediated pathway. These signalling cascades determine the fate of the parasite inside macrophage (Shadab and Ali, 2011). However, *Leishmania* has devised mechanisms to modulate these signal cascades for its survival. Various surface markers of *Leishmania* induce various phosphatases like SHP1, PP2A, etc. that modulate the signalling cascades so as to reduce nitric oxide and TNF- α suitable for parasite survival (Kar *et al.*, 2010; Blanchette *et al.*, 1999; Dey *et al.*, 2007). But when a macrophage is suitably activated, the otherwise host can turn to be the graveyard of *Leishmania* parasite. Another mode of parasite clearance is the induction of apoptosis through Fas–FasL interaction of infected macrophages and NK cells (Chakour *et al.*, 2009).

Dendritic cells

Dendritic cells are a heterogeneous group of bone marrow-derived leucocytes that are specialized in phagocytosis, antigen processing and presentation to stimulate T-cell subsets. During *Leishmania* infection, the DCs following parasite capture and maturation migrate to the draining lymph nodes to present the antigen to T-cells (Banchereau *et al.*, 1998; Steinman, 2007). Antigen presentations through MHC II activate $CD4^+$ T-cells, whereas through MHC I the $CD8^+$ T-cells are activated. Besides antigen presentation through MHC II and MHC I, dendritic cells can secrete cytokines, such as IL-12, that can trigger Th1 immune response resulting in IFN-γ mediated oxidative burst in cells harbouring the parasites (Leon *et al.*, 2007; von Stebut *et al.*, 1998). DCs can also be activated by TLR-mediated response which recognizes pathogen associated molecular patterns (PAMPs) of leishmanial surface proteins. Hence DCs play a very important role in both healing of *Leishmania* infection as well as imparting protective immunity.

However, *Leishmania* has evolved to subvert the protective function of DCs by altering antigen presentation and impairing signal transduction in the dendritic cells. Like macrophages, activation of DCs is suppressed by degradation of intracellular signalling proteins and down-regulation of production of proinflammatory cytokines and their receptors (Xin *et al.*, 2008). In BALB/c mice *L. major* has been shown to induce up-regulation of IL-4 receptors in Langerhans cells which promotes Th2 response (Moll *et al.*, 2002). Activation and expansion of Th2 cells leads to disease progression in cutaneous leishmaniasis. Additionally, Langerhans cells, a subset of skin DCs, have shown to induce expansion of regulatory T-cells (Kautz-Neu *et al.*, 2011). Such DCs with impaired antigen presentation might play a pathogenic role in disease exacerbation and may be the cause of visceralization in case of visceral leishmaniasis. Hence dendritic cells play a pivotal role in coordination and direction of both innate and adaptive immune response and determine the fate of infection.

Natural killer (NK) cells

Natural killer cells are an important cellular arm of innate immunity. During *Leishmania* infection, NK cells infiltrate at the infection site from the draining lymph nodes. CXCL10 is the primary chemoattractant (Muller *et al.*, 2001). PAMPs on the surface of *Leishmania* are recognized by the TLRs on NK cells and their activation leads to secretion of various proinflammatory cytokines like IFN-γ and TNF-α (Becker *et al.*, 2003; Scharton and Scott, 1993). Hence, NK cells are important mediator for inducing Th1 response, obligatory for protection against leishmaniasis (Scharton and Scott, 1993). However, experimental studies show that mice deficient in NK cells can also be cured of *Leishmania* infection (Satoskar *et al.*, 1999). This finding can be explained by the fact that IL-12 production by cells other than NK cells primes the protective Th1 response in these mice (Wakil *et al.*, 1998). Thus, NK cells are supplements in the defensive arm of the host against leishmaniasis.

Complement system

The complement system comprises a group of serum proteins that play a vital role in the non-cellular arm of innate immunity. The complement system works by either degrading the pathogen or assisting in phagocytosis. During *Leishmania* infection, complements recognize the parasite and bind to surface markers on the pathogen, especially LPG, which helps in CR1- and

C3R-mediated uptake by macrophages (da Silva *et al.*, 1988; Mosser and Brittingham, 1997). But the parasite surface metalloprotease gp63 cleaves C3 to iC3b which deactivates the lytic component and helps in silent entry to the host macrophages (Brittingham *et al.*, 1995). Hence *Leishmania* has evolved to use the foe as friend in its establishment in the host cell.

Toll-like receptor activation and evasion

Toll-like receptors (TLRs) are vital arms of innate immune cells and these pattern recognition receptors (PRRs) are expressed on macrophage, dendritic cells, NK cells, etc. These molecules recognize pathogen-associated molecular patterns (PAMPs) and trigger a series of downstream signalling events which result in the production of proinflammatory cytokines, chemokines, and antimicrobial peptides (Kawai and Akira, 2010). Out of the 13 different types of TLRs discovered so far, *Leishmania* infection results in up-regulation of a few, namely TLR1, TLR2, TLR3, TLR4 and TLR9 (Faria *et al.*, 2012). The adaptor molecule of most TLR signalling is MyD88 and knockout MyD88$^{-/-}$ BALB/c mice were found to be more susceptible to *L. major* infection (Debus *et al.*, 2003). Studies with *L. panamensis* showed TLR dependent increase in TNF-α mediated macrophage activation, whereas impaired TLR signal pathway failed to substantially activate the macrophages (Gallego *et al.*, 2011). TLR9 which recognizes bacterial DNA is the most significant of all TLRs for protection against leishmaniasis and it has been found that TLR9 knockout mice as more susceptible to *L. major* infection (Liese *et al.*, 2007). While other TLRs like TLR2 and TLR4 are also important in imparting significant protection against leishmaniasis, agonist of such TLRs are hotspots in the adjuvant biology of *Leishmania* vaccines (Raman *et al.*, 2012; Singh *et al.*, 2012).

Adaptive immune system

Adaptive or acquired immunity is the most potent and effective means of protection from pathogenic infection. Most importantly the response generated is specific for a given pathogen or its types (Cooper and Herrin, 2010). However, the onset of adaptive immune response in individuals exposed to pathogenic agent for the first time requires time and hence pathogens in many cases succeed to establish acute infection. Interestingly, the memory generated with the first infection makes subsequent responses quick and efficient (Weng *et al.*, 2012). Generation of memory has hence been the baseline of vaccine development. For chronic diseases like leishmaniasis even if the incubation period is long enough for adaptive immune response to initiate, modulation and suppression of the host immune response are prominent strategies taken by the parasite for its successful infection (Kaye and Scott, 2011). Therefore apart from activation of appropriate arm immune response, prevention of immune modulation in favour of the pathogen is requisite for protection from such diseases. Thus, understanding the immunobiology of adaptive immune response associated with leishmaniasis is precondition in vaccine development against the disease.

Th1 response

Since *Leishmania* is an obligate intracellular parasite it can evade the humoral antiparasitic components of the host, and cell mediated immune response is therefore more efficient for prevention and cure from *Leishmania* infection. An indispensable arm of cell mediated response is represented by CD4$^+$ Th1 cells. This group of T-helper cells is distinguished by their ability to produce pro-inflammatory cytokines like IFN-γ, TNF-α and IL-2 (Romagnani, 1999). Cytokines released by Th1 cells exert anti-leishmanial activity by either inducing oxidative burst in the parasitized cell or by stimulating it to undergo apoptosis. Notably IFN-γ activates the JAK-STAT pathway to induce nitric oxide production while TNF-α incites apoptotic signals in the infected cells (Shadab and Ali, 2011). Therefore activation of Th1 cells leads to anti-leishmanial activity. In fact patients with leishmaniasis have been observed to elicit marked Th1 response during the healing stage. As evident from its protective role, search for vaccine candidates against leishmaniasis begins with immunogenic components that can trigger Th1 response. Differentiation and activation of Th1 cells during antigen presentation by APCs through MHC II to naive CD4$^+$ T-cells is mediated by a key cytokine IL-12. Hence IL-12 plays a significant role in driving immune response to a protective Th1 type (Tacchini-Cottier *et al.*, 2012). Although Th1 response is associated with protection, overindulgence of inflammatory Th1 cytokine response with little Th2 or T-reg is detrimental for cutaneous and mucocutaneous disease. Since Th1 cytokines are proinflammatory in action excess Th1 response results in exuberated cutaneous lesions. Indeed mucocutaneous leishmaniasis is caused more due to inflammatory and delayed type hypersensitive (DTH) response than the parasitic load (Gaze *et al.*, 2006; Ronet *et al.*, 2011). Therefore optimal balance of Th1 response for different types of leishmaniasis needs to be standardized for vaccination strategies.

Th2 response

T-helper type 2 (Th2) cells are a group of CD4$^+$ T-cells characterized by their ability to produce cytokines like IL-4, IL-5, IL-6, IL-10 and IL-13 that induce humoral response. Differentiation of Th2 cells from naive T-cells is mediated under the influence of cytokines like IL-4 and IL-2 during antigen presentation through MHC II by APCs. Besides its role as an autocrine in self-activation of Th2 cells, IL-4 induces B-cell to generate antibody response. During active *Leishmania* infection there is undeniably profound anti-leishmanial antibody titre in sera of patients, indicating that a dominant Th2 response is generated but with no evident protective role. Furthermore, Th2 response acts antagonistically to protective Th1 response and up-regulates immunosuppressive cytokine IL-10 production by B-cells resulting in chronic progression of the leishmaniasis. This has been the basis of formation of Th1 and Th2 dichotomy which explains that progression of CL is due to dominant Th2 response while Th1 response ensures healing (Scott *et al.*, 1988). Hence suppression of Th2 response is a key strategy in developing protective immune response against CL. However, the Th1/Th2 dichotomy cannot be generalized and other immunological factors govern the outcome of leishmaniasis (Alexander and Brombacher, 2012). This is evident from experimental VL studies where despite substantial IL-4 production, cure could be achieved with a dominant Th1 response and suppression of IL-10 (Cillari *et al.*, 1995). In this case a small degree of IL-4 production was found to be beneficial for developing protective immunity against VL. Hence the exact role of Th2 response in leishmaniasis outcome and concurrent effect on protective immunity needs to be investigated in details for all forms leishmaniasis before it can be successfully exploited in vaccine development.

Cytotoxic T-cell response

CD8$^+$ T-cells are an important group of lymphocytes that recognize MHC I-specific peptides via TCRs expressed on their surface. The antigenic peptides associated with MHC I are usually derived from endogenous proteins. However, exogenous peptides can also be presented on MHC I via internalization of antigenic components followed by proteasome or TAP dependent antigen processing. The method by which exogenous peptides are expressed on MHC I molecule and presented to CD8$^+$ T-cells for their activation is known as cross presentation (Bertholet *et al.*, 2006). Cross presentation of leishmanial antigens is evident from activation of CD8$^+$ cells during *Leishmania* infection (Ruiz and Becker, 2007). *Leishmania* derived antigens are processed either by proteasome machinery in the cytosol or by proteasome independent pathways in the phagosomal compartment to be expressed on MHC I molecules (Kima *et al.*, 1997; Bertholet *et al.*, 2006). Activated CD8$^+$ T-cells are functionally categorized in two types. In the first category CD8$^+$ T-cells differentiate into CTLs that exert cytotoxic activity on the pathogen by releasing perforins, granzyme A/B or granulysin but do not produce IFN-γ (Trapani and Smyth, 2002). The other type of activated CD8$^+$ T-cells actively produces inflammatory cytokine IFN-γ. During *Leishmania* infection although CTLs have not been directly reported to exert cytotoxic effect on the parasites, they are believed to induce cellular lysis of parasitized cells. But noticeably, release of IFN-γ is the most effective anti-leishmanial protective function of activated CD8$^+$ T-cells (Stefani *et al.*, 1994). Apparently various experimental studies on cutaneous leishmaniasis show that while IFN-γ-producing CD8$^+$ T-cells impart protection, the CTLs were associated with exuberated lesions and pathology of CL (Faria *et al.*, 2009; Belkaid *et al.*, 2002; Stager and Rafati,

2012). Moreover, the memory generated with CD8$^+$ T-cells imparts long-lasting protection. On the other side some CD8$^+$ T-cells acquire regulatory function by their ability to produce IL-10 along with IFN-γ. Although IL-10 production is beneficial for containing the disease symptoms of ML, this has been associated with chronic VL progression and subsequent development of PKDL (Gaze *et al.*, 2006; Ganguly *et al.*, 2008; Bourreau *et al.*, 2007). Hence, the role of CD8$^+$ T-cells in leishmaniasis needs to be clearly defined and targeted accordingly in vaccine development.

Role of T-regs

Uncontrolled inflammatory response can cause detrimental effects such as tissue injury, hypersensitive reactions that in many cases may prove fatal. A group of CD4$^+$ cells known as regulatory T-cells (T-regs) counteract inflammatory response by secreting immunosuppressive cytokines IL-10 and TGF-β. T-regs are broadly classified as natural and inducible. While all natural T-regs (nT-regs) are FoxP3$^+$ CD25^{high} and develop in the thymus, inducible T-regs can be either FoxP3$^+$ or FoxP3$^-$ and they are spawned in the periphery. The characteristic cytokine IL-10 secreted by T-regs negatively affects the protective immune response required for *Leishmania* clearance (Anderson *et al.*, 2008). IL-10 modulates nitric oxide production by inducing arginase1 in place of *i*NOS. This is evident from observation that peripheral blood mononuclear cells (PBMCs) of VL patients are less active to IFN-γ response while IL-10 blockade results in their reactivation (Carvalho *et al.*, 1994). Hence despite high IFN-γ, parasite clearance is greatly affected due to the negative effect of IL-10 regulation. Evidently activation of T-regs is associated with non-curative and relapsing VL and development of PKDL (Saha *et al.*, 2007). Therefore evading or reducing T-reg activation is a necessary challenge for vaccine development against VL. Additionally it has been observed that mice defective in IL-10 production results in development of sterile immunity to *L. major* infection (Belkaid *et al.*, 2001). But, interestingly, no infection derived immunity was generated in these mice models. This lead to speculation that despite hampering protective immune response, IL-10 can generate long term immunity by allowing parasite survival in low numbers. Moreover, optimal activation of T-regs helps to condense the disease symptoms in ML patients (Faria *et al.*, 2005). Therefore optimization of T-reg activation and methods to regulate T-reg population for different forms of leishmaniasis should be addressed separately for their application in vaccine developmental strategies.

Role of Th17 cellular response

Th17 cells are a group of CD4$^+$ T-cells that secrete characteristic cytokines such as IL-17A, IL-17F along with IL-21 and IL-22 (Sundrud and Trivigno, 2013). Differentiation of naive CD4$^+$ T-cells to Th17 subtype is induced under the influence of IL-6, IL-23, IL-1β and TGF-β. The key cytokine IL-17, secreted by Th17 cells, is a known neutrophil chemoattractant. Studies with the experimental CL mice model have pointed to the role of IL-17 in exacerbating neutrophil-associated CL pathological severity (Gonzalez-Lombana *et al.*, 2013; Lopez Kostka *et al.*, 2009). Role of IL-17 on pathology of ML is contradictory, while findings from experimental ML animal models suggest protective function. On the contrary studies on ML patients indicate pathological association of IL-17. IL-27 is an inhibitor of Th17 cell differentiation (Rousseau *et al.*, 2010; Anderson *et al.*, 2009). In experimental knockout IL-27R mice, *L. donovani* challenge is associated with liver pathology, suggesting that Th17 might play a protective role against disease pathology of VL (Rosas *et al.*, 2006). However, due to lack of extensive studies on this recently discovered cell lineage, role of Th17 response is leishmaniasis is not completely clear.

Challenges in vaccine development against leishmaniasis

Development of a suitable vaccine for leishmaniasis is a difficult goal. The major obstacles, limitations and challenges for a successful vaccine against the disease are discussed below.

1 Understanding the appropriate immunological requirements for vaccine development against different form of leishmaniasis.

2 Identification of components that can generate such responses.
3 Determination of reliable and efficient ways and methods to evaluate vaccine efficacy.
4 Development of suitable animal models for preclinical evaluation of vaccine efficacy.
5 Generation of long term memory for protection and ways to effectively evaluate memory cell production during vaccination.
6 Discovery of suitable adjuvants and delivery system to stimulate protective immune response.
7 To bring the cost of production within an affordable limit for poor countries.
8 To guarantee uncompromised safety with good quality profile of vaccine formulation.
9 Development of vaccine formulation that qualifies for needle free immunization.

Rationale of vaccine development against leishmaniasis

Historically cutaneous leishmaniasis was observed to impart almost lifelong immunity to reinfection. On this basis leishmanization was a regular practice in the Middle East countries. Furthermore, infections with other forms of leishmaniasis also impart substantial resilience to subsequent infection. Besides, only a small proportion of people infected with *Leishmania* parasites develop full clinical symptoms of the disease while a majority of them are either asymptomatic or self-curing. These findings strongly suggest feasibility of a successful vaccine against leishmaniasis. Therefore search for vaccine against leishmaniasis gained a tremendous momentum in the past decades giving rise to a big list of potential vaccine candidates against leishmaniasis.

General methods to evaluate vaccine potentials against leishmaniasis

Any formulation to be approved by the FDA as licensed/registered vaccine needs successful evaluation at preclinical and clinical level. Preclinical studies are conducted on experimental animal models starting with small mammals and subsequently proceeding to large animals like dogs and primates. For leishmaniasis the common experimental models used are Golden Syrian hamsters and BALB/c mice owing to their susceptibility to the disease. Immune responses generated in these models following immunization are considered as a hint of immunogenicity in the human counterpart. But this consideration is not a true reflection of human immunity. A recent and more appropriate alternative is the use of humanized nude/SCID mice. The protective efficacy is assayed by parasite count in spleen, liver and bone marrow monocytes for VL, and dermal macrophages for CL. The techniques used for parasite count are generally limiting dilution assay (LDA) and Leishman–Donovan unit (LDU). Recent approaches for parasite count include techniques like parasite specific RT-PCR. Toxicity as well as duration of protection is determined by the time-dependent survival of experimentally infected animals. DTH response generated in immunized mice models is evaluated by time dependent footpad swelling in the experimental animals injected with the antigens at that site.

PBMCs isolated from human (patient, cured and/or healthy individuals) blood are stimulated with vaccine formulation and assayed for cell proliferation, cytokine response and parasite clearance.

Once in clinical trial, safety and immunogenicity studies are conducted in phases I and II. Determination of immunogenicity of a particular vaccine candidate is generally evaluated by immunodiagnostic methods of leishmaniasis such as Montenegro skin test (MST), leishmanin skin test (LST) or by direct agglutination test. While the first confirms generation of DTH response to leishmanial antigens, the latter decide *Leishmania* specific antibody production in the human subject. Cytokine assay of blood sera of immunized individuals or their PBMCs upon antigen stimulation are also evaluated for immunogenicity of vaccine formulation. But as far as the protective efficacy of the vaccine is considered, it is determined by the incidence of leishmaniasis in immunized population as compared to the control group.

Antigenic components of *Leishmania* vaccine

Efforts put forward in development of a successful vaccine for leishmaniasis have witnessed several novel approaches which resulted in a long list of promising vaccine formulations. These range from live cellular forms to dead acellular forms, from protein fractions to recombinant subunit proteins, and from naked DNA to virosomal forms. Depending upon the physiochemical nature of the constituent vaccine, they have been categorized into the following types.

Live virulent parasites

The most primitive but effective vaccination strategy for CL is leishmanization which involves inoculation of live virulent *L. major* promastigotes from culture, or amastigotes from pus of infected CL lesions to a healthy subject. Leishmanization successfully prevented undesired CL infection in immunologically healthy subjects. But with onset of HIV epidemics and quality control hindrance, leishmanization had to be abandoned in all countries except Uzbekistan, where it is still practised. Among other live virulent parasites used in vaccination strategies is the naturally attenuated *L. donovani* that has lost its visceralization capability and cause cutaneous leishmaniasis in Sri Lanka. Such an attenuated parasite isolated from Sri Lanka conferred protective immunity to BALB/c mice against visceral leishmaniasis (McCall *et al.*, 2013). Killed but metabolically active parasites generated by physico-chemical means like low-dose UV irradiation, amotosalen treatment, or by γ-irradiation are partially virulent parasites that have lost their replicative power (Das and Ali, 2012). Such parasites have been effectively established to impart protection in experimental models of various forms of leishmaniasis (Bruhn *et al.*, 2012; Alexander, 1982; Datta *et al.*, 2012). In another approach, live non-pathogenic strains like *Leishmania tarentolae* was used for vaccination and found to be suitably protective against experimental *L. donovani* challenge in BALB/c mice (Breton *et al.*, 2005). A major advantage of using live non-genetically modified parasite for vaccination is that they mimic natural infection. Consequently, apart from protective immunity a long term protective memory is generated. But due to lack of homogeneity (quality control) and undesired pathological association in immunocompromised individuals (safety), use of live virulent parasite is less encouraging. Besides, the chance of relapse with changing immunological status of individuals makes the use of virulent parasites as vaccine candidate more objectionable.

First generation vaccine

In an attempt to overcome the safety let-down of live virulent parasites, killed *Leishmania* parasites grouped as first generation vaccine candidates were introduced. Autoclaved *L. amazonensis* isolates were used as the vaccine candidate by Mayrink *et al.* (1979) and evaluated for prophylaxis against CL and ML in Brazil and then in Ecuador and Colombia. Mayrink's vaccine failed to impart significant protective immunity in vaccinated groups (Velez *et al.*, 2005; Armijos *et al.*, 2004). Following Mayrink's approach Convit *et al.* (2003) introduced autoclaved *Leishmania mexicana* supplemented with BCG in Venezuela (Castes *et al.*, 1994). Autoclaved *L. mexicana* + BCG was found to be insignificant in imparting prophylaxis to CL and ML of the new world. A similar approach for prevention of CL and VL was undertaken in Iran using autoclaved *L. major* (ALM) + BCG. Multiple doses programme was initiated to enhance the immunogenicity of this vaccine formulation from Razi Seum Institute, Iran. The efficacy of two dose vaccine formulation was approximately 43% in LST converted individuals in Sudan (Khalil *et al.*, 2000; Mutiso *et al.*, 2013). However, in order to overcome the requirement for multiple doses of ALM + BCG for significant LST conversion, the vaccine formulation was supplemented with another adjuvant in the form of aluminium hydroxide (alum) (Khalil *et al.*, 2006). Alum ALM + BCG was found to impart 69.3% protection to canine leishmaniasis (Mohebali *et al.*, 2004). But as evident from human trials, despite high immunogenicity, this vaccine formulation was found to be poor in prophylactic efficacy against CL in Iran and VL in Sudan. In a nutshell the various efficacy trials with first generation vaccines, despite good safety value, were meta-analysed to be inadequately protective to various forms of leishmaniasis (Noazin *et al.*, 2009). But efforts to enhance the prophylactic potential of first generation vaccine against leishmaniasis are being undertaken by using different adjuvants. Notably IL-12 adjuvanted with ALM could induce Th1 response but did not protect vervet monkeys from CL, (Gicheru *et al.*, 2001). Among other important adjuvants, monophosphoryl lipid A (MPLA), when used along with formalin-killed *Leishmania* promastigotes, could significantly protect BALB/c mice from *L. major* and *L. donovani* infection. But adverse reactions were observed in these mice models. Hence supplements of a suitable adjuvant which is safe for human use can redefine the preventative value of first generation vaccine against leishmaniasis. Despite failure of field trials for a prophylactic first generation vaccine for leishmaniasis, their therapeutic potential is very promising (Okwor and Uzonna, 2009). The success of first generation vaccine as immuno-chemotherapeutic agents will be detailed in the topic pertaining to therapeutic vaccine for leishmaniasis. However, it is worth mentioning that different batch-to-batch preparations of killed parasite leads to different quality of vaccine formulation. Hence quality control is a major issue associated with first generation vaccine candidates.

Second-generation vaccines

Second-generation vaccine constitutes of protein fractions or defined subunit polypeptides. In viewpoint of the failure of first generation vaccine candidates for prophylaxis against leishmaniasis, a more defined approach was undertaken by using peptidal fractions or defined proteins or their cocktails as antigenic components. These antigenic components are primarily of the parasitic origin. But in some approaches the antigenic peptides represent proteins derived from the sandfly saliva. The potentials of second generation prophylactic vaccine against leishmaniasis are detailed below.

Peptide fractions

A few noteworthy developments with protein fractions as vaccine for leishmaniasis is summarized. The fucose mannose ligand (FML) consists of a mixture of glycoproteins derived from surface of *L. donovani*. The major antigenic component of FML is gp36. FML adjuvanted with deacylated saponin derived from *Quillaja saponaria* was found to be safe, immunogenic and protective to experimental VL models of BALB/c mice and hamsters. The FML saponin formulation, also known as Leishmune imparted 76–80% protection to canine visceral leishmaniasis (CVL) in field trial and is the first vaccine to be licensed for prophylaxis against CVL (Dantas-Torres, 2006; Palatnik-de-Sousa, 2012). However, Leishmune is still to enter clinical trial for leishmaniasis in human. Leishmune is also a superior immuno-chemotherapeutic agent in canine leishmaniasis imparting almost sterile cure with amphotericin B and allopurinol (Borja-Cabrera *et al.*, 2010). Among other noteworthy vaccination approaches with peptide fractions is the total membrane antigen of *L. donovani* (LAg). LAg entrapped in liposomes was shown to impart protective immunity to *L. donovani* challenge in BALB/c mice (Ali and Afrin, 1997). Cationic LAg entrapped liposomal formulation induced robust DTH response and *L. donovani* specific IgG2a antibody response (Afrin and Ali, 1997; Mazumdar *et al.*, 2004). Additionally in different studies using solubilized *Leishmania* antigens (SLA) alone or with CpG was shown to be efficiently protective both in liposomal and non-liposomal formulation against experimental VL and CL (Shargh *et al.*, 2012; Heravi Shargh *et al.*, 2012; Golali *et al.*, 2012). SLA from *L. donovani* entrapped in cationic liposome could induce a strong Th1 and IgG2a antibody response in BALB/c mice and hamster for efficient protection from *L. donovani* (Bhowmick *et al.*, 2007). In another finding, *L. major* culture-derived soluble exogenous antigens (SEAgs) comprising lipophosphoglycan and gp46/M2/PS-2 as secreted antigens, was found to be significantly protective to murine CL model (Tonui *et al.*, 2004). SEAgs used alone or with adjuvants like alum, recombinant IL-12, or montanide ISA 720 were shown to induce T-cell proliferation with mixed Th1 and Th2 response, and significantly reduced lesion size and parasite burden in experimental CL mice model. In another advancement, *L. donovani* promastigote soluble antigens (sLAg) entrapped liposomes comprising lipids derived from *Escherichia coli* elicited parasite-specific activation of $CD8^+$ and $CD4^+$ T-cell immune response and protected experimental BALB/c mice and hamster from *L. donovani* infection (Sharma *et al.*, 2006). Despite good prophylactic efficacy of various protein fractions in experimental leishmaniasis, a major drawback of these vaccine candidates is difficulty in purification steps for large scale production and to maintain homogeneity.

Defined proteins

Basically two major approaches are implied for search of proteins to be considered as vaccine candidates. In the first approach, the proteins that are associated with virulence or important physiological function of the parasite are targeted so as to preclude *Leishmania* pathogenesis. In the other approach proteins that are associated with transmission of the parasite are selected so that *Leishmania* transmission could be effectively blocked. Supported by the availability of the whole genome sequence of several *Leishmania* species, over a hundred formulations and at least 34 antigenic peptides have been reported as vaccine candidates for leishmaniasis (Costa *et al.*, 2011). Defined peptides evaluated in experimental vaccination against leishmaniasis have been listed in Table 10.1.

Lipophosphoglycan (LPG)

Surface glycocalyx of *Leishmania* promastigotes have been identified as the first defined antigenic components for leishmaniasis. LPG is important in *Leishmania* virulence by hampering phagosomal maturation forming a barrier in hydrolytic digestion of promastigotes in the phagolysosome. However, LPG acts as *Leishmania* PAMPs recognized by NK cells leading to up-regulation of TLR2. On this basis LPG are promising vaccine candidates that up-regulate IFN-γ and TNF-α production to exert anti-leishmanial activity. Studies with LPG immunized mice infected with *L. major* were shown to be less infective to sandfly indicating that they can be a good transmission blocking vaccine candidate (Tonui *et al.*, 2001).

Leishmanolysin (gp63)

gp63 is a zinc-based metalloprotease that is expressed in all species of *Leishmania* and is an important virulence factor associated with degradation of phagosomal proteins and complement proteins during *Leishmania* infection. Native gp63 was found to be protective against experimental CL and VL. However, recombinant gp63 (rgp63) expressed in *Escherichia coli* and adjuvanted with BCG could not induce peripheral blood leucocytes of vervet monkeys to produce IFN-γ and failed to protect these animals from *L. major* challenge. On the contrary rgp63 expressed in BCG could impart protection in experimental mice models of CL. In other studies with rgp63 used alone or along with CpG ODN in cationic liposomal formulation could effectively protect BALB/c mice from CL as well as VL (Jaafari *et al.*, 2006, 2007; Bhowmick *et al.*, 2008). Monophosphoryl lipid A-trehalose dicorynomycolate (MPL-TDM) used as an adjuvant to gp63 entrapped in DSPC liposome was found to be a superior formulation in imparting robust long term protective immunity to BALB/c mice against *L. donovani* challenge (Mazumder *et al.*, 2011a). In another studies by the same group it was found that using DNA vaccine of GP63 for priming and protein gp63 for boosting (heterologous prime boosting approach), a long term effective protection could be established in experimental VL models (Mazumder *et al.*, 2011b).

Ribosomal proteins

A vaccine using *L. infantum* ribosomal proteins and saponin as adjuvant was shown to induce a protective response in BALB/c mice challenged with *L. chagasi* and *L. amazonensis* (Chavez-Fumagalli *et al.*, 2010). Immunization with extracts of ribosomal proteins adjuvanted with CpG ODN could significantly reduce parasite burden and lesion size, and resistance to reinfection by *L. major* in BALB/c mice model (Ramirez *et al.*, 2010). These studies indicate that ribosomal proteins, along with a suitable adjuvant, can be a promising vaccine for leishmaniasis.

Sandfly salivary proteins

During bloodmeal of infected sandfly, a mixture of salivary proteins are co-inoculated with *Leishmania* parasite. These salivary components play a vital role in immunomodulation favouring establishment of the parasites in the host cell. Consequently, a low dose of 100–1000 *Leishmania* organisms is sufficient for natural infection, in

Table 10.1 Subunit protein vaccine candidates for experimental leishmaniasis

Antigen	Vaccine formulation	Animal model	Challenge infection	Protective efficacy	Disease	Reference
Lipophosphoglycan	LPG	BALB/c mice	*L. major*	Low transmission in sandflies	CL	Tonui *et al.* (2001)
Leishmanolysin (gp63)	gp63	BALB/c CBA mice	*L. mexicana*	Protection	CL	Russell and Alexander (1988)
gp63	rgp63 + MPL-TDM + cationic liposome	BALB/c mice	*L. donovani*	Protection	VL	Mazumder *et al.* (2011)
Thiol-specific antioxidant (TSA)	TSA + IL-12	Mice BALB/c	*L. major*	Partial protection	CL	Webb *et al.* (1998)
L. major homologue to eukaryotic stress-inducible protein (LmSTI1)	LmSTI1	Mice BALB/c	*L. major*	Protection	CL	Scott *et al.* (1988)
Leishmania elongation initiation factor (LeIF)	rLeIF	BALB/c mice	*L. major*	Partial protection	CL	Skeiky *et al.* (1998)
Hydrophilic acylated surface protein B family (HASPB)	rHASPB1	Mice BALB/c	*L. donovani*	Protection	VL	Stager *et al.* (2000)
Sterol 24-c-methyltransferase (SMT)	SMT + MPL-SE	C57BL/6 mice	*L. infantum*	Protection	VL	Goto *et al.* (2007)
Kinetoplastid membrane protein 11 (KMP11)	rKMP11 + PLGA nanoparticle	BALB/c mice	*L. braziliensis*	Partial protection	CL	Santos *et al.* (2012)
Methionine aminopeptidase 45 (Ld p45)	rLdp45	Hamster	*L. donovani*	Protection	VL	Gupta *et al.* (2012)
Leishmania homologue of receptors for activated C-kinase	rLACK (truncated) + IL12	BALB/c mice	*L. major*	Protection	CL	Mougneau *et al.* (1995)
Elongation-factor 2 (eIF2)	reIF2	Hamster	*L. donovani*	Protection	VL	Kushawaha *et al.* (2011)
Protein disulfide isomerase (PDI)	rLdPDI	Hamster	*L. donovani*	Th1 cytokine response from PBMCs	VL	Kushawaha *et al.* (2012)
Open reading frame F (ORFF)	rORFF + CpG-ODN	BALB/c mice	*L. donovani*	Protection	VL	Tewary *et al.* (2004)
γ-Glutamyl cysteine synthetase protein (LdγGCS)	r LdγGCS + NIV	BALB/c	*L. donovani*	Protection	VL	Henriquez *et al.* (2010)
Recombinant protein similar to ribosomal protein P0 (F14)	rF14 + MPL	Hamster	*L. donovani*	Protection	VL	Bhardwaj *et al.* (2009)
Maxadilan (MAX)	Synthetic MAX + Freund's adjuvant	CBA and BALB/c mice	*L. major*	Protection	CL	Morris *et al.* (2001)
Virulence factor A2	A2	BALB/c, C57BL/6 mice	*L. donovani*	Protection	VL	Ghosh *et al.* (2001)

contrast to the high doses (100,000–10,000,000 parasites) required to cause disease in experimental animals in the absence of salivary proteins (Belkaid *et al.*, 1998). Hence, following the rationale that an anti-salivary protein response would significantly reduce parasite transmission, the *Leishmania* transmission blocking efficacy of various salivary protein extracts of sandfly have been investigated in experimental animals and dogs. Out of the 35 proteins identified through reverse antigen screening, maxadilan is an important constituent (Valenzuela *et al.*, 2004). Synthetically prepared maxadilan of *Lutzomyia longipalis* could effectively protect CBA mice from *L. major* infection (Morris *et al.*, 2001). Notably, recombinant LJM11 from *L. longipalis* saliva is found to impart protection to BALB/c mice against *L. major* infection (Gomes *et al.*, 2012a). Similarly LJM19, 11-kDa salivary protein imparted protective immunity to hamsters challenged with *L. infantum* (Gomes *et al.*, 2008). In another study LJM17 and LJL143 two immunogenic components of salivary proteins from *L. longipalis* could significantly induce Th1 and Th2 responses in dogs and effectively clear parasites from infected macrophages *in vitro* (Collin *et al.*, 2009). However, efficacy of salivary proteins has been questioned with the observation that people in endemic areas are frequently exposed to sterile sandfly bite and hence to the salivary proteins, which should otherwise have been sufficient to contain the disease.

Other important defined vaccine candidates include heat shock proteins (HSP70 and HSP83), thiol-specific antioxidant, *L. major* homologue to eukaryotic stress-inducible protein (LmSTI1), histone proteins, *Leishmania* homologue of receptors for activated C kinase (LACK), Cysteine proteinases (CPA, CPB and CPC), etc., used alone or in defined cocktails were shown to prevent experimental leishmaniasis in various animal models.

Fusion proteins

A vaccine protective against all forms of leishmaniasis and immunogenic enough to impart robust protection requires multiantigenic components. This has been attributed by the success of purified protein fractions in experimental leishmaniasis. Therefore, defined cocktail of antigens and/or recombinant fusion protein has emerged as potent vaccine candidates for leishmaniasis. Notably, fusion of three antigens comprising *L. major* stress-inducible protein 1 (LmSTI1), thiol specific antioxidant (TSA) and *Leishmania* elongation initiation factors (LeIF) was produced as Leish111f. This recombinant polyprotein, in a stable emulsion of adjuvant MPL, was the first second generation vaccine to be registered for human trial against CL, VL and ML in Colombia, India and Peru respectively. Leish111F + MPL-SE was found to be safe and immunogenic as evident from phase I and phase II results (Chakravarty *et al.*, 2011; Velez *et al.*, 2009). Phase III trials with Leish-F1 + MPL-SE, the upgraded version of Leish111f+ MPL-SE, have been initiated and the results of efficacy tests are expected in the near future. Produced by the same group at IDRI, Seattle, WA, USA, a fusion of two proteins named Leish-F3 used with a more homogeneous synthetic analogue of MPL, GLA-SE is undergoing phase I trial in India and the USA against VL (http://www.idri.org/press-2-22-12.php). In another development, KSAC, a triple fusion polyprotein with MPL-SE, was found to be cross-protective against human and canine VL (Goto *et al.*, 2011; Gomes *et al.*, 2012b).

Despite the success of fusion polyproteins in experimental leishmaniasis, with promising results in clinical trials, a major disadvantage of using recombinant fusion proteins is their instability and requirements for cold storage and transport. In addition, durability of protection from protein-based vaccine needs to be enhanced for better cost-effective prophylactic control of leishmaniasis in poor nations.

Third generation vaccine

DNA encoding genes for one or more antigens has been used in vaccination studies against leishmaniasis. Basically, third-generation vaccines includes two major approaches. In the naked DNA approach prokaryotic or eukaryotic plasmids encoding antigenic gene/s are used for vaccination. In the other type known as virosomal approach antigenic genes are inserted in the genome of viruses like adenovirus, vaccinia virus or influenza virus and these genetically modified viruses are employed for vaccination. The principal approaches with third-generation vaccines is considered in the following sections.

Naked vaccine encoding single antigen

Immunization of naked DNA containing sequence of N-terminal domain of proteophosphoglycan of

L. donovani was found to be protective to *L. donovani* challenge in hamster by increased production of IFN-γ, TNF-α, and IL-12, and down-regulation of TGF-β, IL-4 and IL-10. Additionally, intramuscular DNA vaccination with gene encoding γ-glutamyl cysteine synthetase protected BALB/c mice against *L. donovani* challenge. Out of the many naked DNA vaccination employed, LACK gene has been most frequently tested against various forms of leishmaniasis. Plasmids containing LACK gene has been found to be protective to both genetically resistant and susceptible mice from infection with *L. major* (Gurunathan *et al.*, 1997). Additionally, administration of IL-12 or IL-22 enhanced the protective performance of LACK gene encoding DNA against *L. major* infection (Hezarjaribi *et al.*, 2013). Intranasal immunization of LACK containing plasmids has been found to be protective to *L. amazonensis* as well as *L. chagasi* infection in experimental BALB/c mice (Pinto *et al.*, 2004; Gomes *et al.*, 2007; De Oliveira Gomes *et al.*, 2012). In a similar approach DNA encoding KMP11 gene with exogenous administration of IL-12 protected BALB/c mice from *L. major* as well as *L. donovani* infection. In a recent advance, DNA encoding *Leishmania* haemoglobin receptor (HbR) could induce sterile immunity to *L. donovani* infection in hamster and BALB/c mice for long duration (Guha *et al.*, 2013b). In a major advancement plasmid based vaccine called LEISHDNAVAX expressed in MIDGE vector is presently undergoing preclinical evaluation.

Multiantigenic DNA vaccine

In order to enhance immunogenicity and cross-protectivity, multiantigenic DNA vaccination studies were employed. Cocktail DNA vaccine encoding LACK gene as well as gene corresponding to thiol-specific antioxidant (TSA) was found to be protective against *Leishmania* challenge. Cocktail of plasmid encoding KMP11 gene along with salivary peptides LJM19 gene was protective to experimental *L. chagasi* challenge in hamsters (da Silva *et al.*, 2011). Naked DNA vaccine which includes genes of open reading frame F (ORFF) along with ubiquitin protected BALB/c mice against both antimony resistant as well as susceptible *L. major*. Other studies using cocktail DNA vaccine encoding CPA, CPB and CPC delivered through cationic solid lipid nanoparticles induced protective immunity to BALB/c mice against *L. major* infection (Doroud *et al.*, 2011). Multiantigenic strategies can be considered superior in terms of efficacy, durability and cross-protectiveness. However, better effectiveness of multiantigenic DNA vaccine was contradicted by studies using a multiantigenic DNA vaccine encoding KMP11, TRYP, LACK and gp63 which failed to protect dogs against *L. infantum* (Rodriguez-Cortes *et al.*, 2007).

Heterologous prime boost approach (DNA/recombinant virus, virus/DNA)

In a novel heterologous prime boost (HPB) approach where KMP11 naked DNA was used for priming, and vaccinia expressing the same protein was used as booster, imparted superior protection to hamsters against VL (Guha *et al.*, 2013). In another study, BALB/c mice were primed with pcDNA3 containing gene encoding tryparedoxin perodixase (TRYP) and boosted with vaccinia virus expressing the same protein. This heterologous vaccination approach imparted protective immunity to *L. major* challenge in BALB/c (Stober *et al.*, 2007). The same vaccination approach was also found to be safe and immunogenic to dogs (Carson *et al.*, 2009). In a major advance with virus mediated vaccination programme, a safety and immunogenicity study using adenovirus encoding hydrophilic acylated surface protein B (HASPB) and KMP11 (ChAd63-KH) against VL is being conducted in the United Kingdom (http://apps.who.int/trialsearch/Trial.aspx?TrialID=ISRCTN07766359).

The experimental DNA vaccines for leishmaniasis show potential as effective prophylactic agents. But possible outcome of long term adverse effects like autoimmune disorder, cancer and immune tolerance, makes the use of DNA vaccine a big concern for human vaccination purpose.

Genetically modified live vaccine

In consideration of the complexity of the host–pathogen interaction in leishmaniasis, the best prophylactic measure should be a mimic of natural infection. Hence genetically modified *Leishmania* parasites have been explored as a vaccine candidate in experimental leishmaniasis. Notably *L. tarentolae* were transformed to express A2 protein and in another study with trigene fusion protein A2-CPA-CPB (-CTE). These recombinant parasites were found to be safe and protect BALB/c mice from *L. infantum* infection (Mizbani *et al.*, 2009; Saljoughian *et al.*, 2013). Among other genetically modified, dihydrofolate reductase thymidylate synthase (dhfr-ts) knockout *Leishmania* was found to protective to *L. major* and *L. amazonensis* challenge in experimental mice (Veras *et al.*, 1999). Similarly, serine protease deletion of *L. mexicana* was found to be protective in various animal models of *L. major* infection. Centrin (Ld cen$^{-/-}$) and p27 (Ld p27$^{-/-}$) knockout *L. donovani* in separate studies were found to be protective to homologous as well as heterologous challenge in experimental leishmaniasis (Selvapandiyan *et al.*, 2009; Dey *et al.*, 2013; Fiuzav, 2013). Other knockout parasites include phosphomannomutase (PMM)-deficient *L. major* and HSP70 II null mutant *L. infantum* which were found to protect mice from experimental CL (Kedzierski *et al.*, 2008; Carrion *et al.*, 2011). These studies indicate the cross protective efficacy of genetically modified parasites as a vaccine candidate for leishmaniasis. However, gain of virulence by mutational event can cause devastating impact on human health. Therefore, focus of live vaccination has included non-leishmanial microorganism as vaccine candidate. Genetically engineered *Lactococcus lacti* expressing LACK and IL-12 when administered orally was found to trigger anti-leishmanial immune responses and protect BALB/c mice from *L. major* infection (Hugentobler *et al.*, 2012). The approach is a leap forward in development of an oral vaccine for leishmaniasis.

Dendritic cell-based vaccine

Different route of immunization in experimental animals with vaccine formulation resulted in different immunological outcome (Bhowmick *et al.*, 2009). This has been primarily attributed to the role of various APC subsets in disease outcome and immunity. Consequently, a new approach in which dendritic cells are pulsed with antigens *ex vivo* and then administered as vaccine, has emerged. Indeed, naive mice immunized with dendritic cells pulsed with *L. donovani* soluble antigens (SLDA) imparted partial protection against *L. donovani* challenge (Ahuja *et al.*, 1999). Complete resilience was observed with pulsed DCs engineered for expression of IL-12 production (Ahuja *et al.*, 1999). Similar results with DC based vaccine with other antigens have been reported, highlighting the feasibility of DC-based vaccine. In hybrid cell vaccination approach involving APCs, hybridization of allogeneic macrophage with KMP11 pulsed macrophages was carried out. The resulting hybrid macrophage used as vaccine candidate induced profound antigen specific CD8^{+} T-cell response (Basu *et al.*, 2007). Apart from prophylactic function, DC-based vaccines are promising therapeutic candidates. However, owing to the cost of maintenance and the arduous preparation protocol, the use of DCs as vaccine candidates is less appealing.

Vaccine for canine leishmaniasis

Control of canine leishmaniasis is important as the animal serves as the major reservoir for transmission of leishmaniasis in zoonotic foci. Till date field trials for at least seven vaccine candidates for canine leishmaniasis have been conducted. The first trial using Convit's vaccine formulation (killed *L. braziliensis* + BCG) no significant protection could be observed. In another field study using first generation vaccine, autoclaved *L. major* + Alum protected dogs against natural infection of *L. infantum* to an efficacy of 69.3%. Dunan *et al.* (1989) used 67–94 kDa protein fraction of *L. infantum* but this lyophilized semipurified protein fraction was found to be ineffective in phase III trial in France. *L. donovani* FML along with acylated quillazza saponin was found to be 76–80% protective in field trials in Brazil. This vaccine is licensed as Leishmune. LiESAP is a 54-kDa excreted protein of *L. infantum* adjuvanted with MDP and imparted 92% protection in field trial in France (Palatnik-de-Sousa, 2012; Lemesre *et al.*, 2007). A similar vaccine composed of LiESAp + QA21 (adjuvant derived from *Quillazza saponaria*) is licensed as Cani-Leish after undisclosed field trial result in Europe. Recombinant A2 protein derived from *L. donovani* is licensed as a vaccine for canine leishmaniasis as Leish-Tec. Hence, at present there are three registered vaccines for canine leishmaniasis (Palatnik-de-Sousa, 2012).

Therapeutic vaccines

Many anti-*Leishmania* drugs are immunomodulatory and hence aiming to exploit cross-talk of immune modulation for various drugs in favour of parasite clearance, use of therapeutic vaccine was introduced (Ali and Bhattacharya, 2013). Despite poor prophylactic efficacy, first generation vaccine candidates were a big success as therapeutic agent. Indeed Mayrink's vaccine (killed *L. amazonensis*) was found to reduce the dose of antimonials. It is now registered as an adjunct to antimony therapy in Brazil (Machado-Pinto *et al.*, 2002). With similar findings Convit's vaccine with killed *L. mexicana* is used as immunotherapeutic agent to treat CL in initial stage, failing which administration of antimonials is initiated (Convit *et al.*, 2003). Additionally alum precipitated ALM+BCG in clinical trial as immunochemotherapeutic agent for PKDL in Sudan gave promising initial results (Musa *et al.*, 2008). Leish111f + MPL-SE with antimonials showed promising therapeutic efficacy in the treatment of canine VL (Trigo *et al.*, 2010).

Adjuvants and delivery system

It is evident from the various preclinical and clinical studies that adjuvants and delivery systems play a vital role in inducing, directing and enhancing protective immunity of various antigenic components against leishmaniasis. Major roles played by adjuvants and delivery systems are stimulation of a particular set of immune response or targeting antigens to a defined cellular or tissue foci or sustaining slow and persistent release of antigens. Roles of various adjuvants and delivery system used in vaccination studies for leishmaniasis are discussed below.

Aluminium salts (alum)

Aluminium hydroxide is the first adjuvant approved by FDA for human vaccination. The salt is known to induce humoral immunity. Therefore, in respect to cellular immune response required for protection from *Leishmania* infection, alum has been used as a carrier of other adjuvants like IL-12, BCG, GM-CSF, CpG, and MPL (Raman *et al.*, 2012). Indeed alum precipitated ALM+BCG was shown to impart better DTH response and has been successfully employed in chemotherapeutic agent in PKDL treatment (Musa *et al.*, 2008).

Interleukin-12 (IL-12)

Interleukin-12 is a proinflammatory cytokine which drives Th1 response. Recombinant IL-12 used with SLA and LACK induced profound protective immune response against *L. major* infection but memory generated is short lived. Suggesting sustained IL-12 is required for long term protection. IL-12 has also been found to induce protective immune response to DNA vaccine encoding LACK gene and KMP11 gene. However, owing to the high toxicity of IL-12, the adjuvant is not suitable for human use (Raman *et al.*, 2012).

Toll-like receptor agonists

Among notable adjuvants used in vaccine formulations against leishmaniasis, TLR4 agonists monophosphoryl lipid (MPL) and glucopyranosyl lipid A (GLA) are approved for human administration. Various clinical trials using MPL-SE with Leish F1, the adjuvant MPL has been shown to be safe and immunogenic. However, in order to maintain a homogenous immunogenicity a synthetic analogue of MPL called GLA-SE has been developed. Safety and immunogenicity of GLA-SE is under way. Among other TLR agonists, imiquimod, a TLR7/8 agonist, has been approved for topical administration in humans. TLR9 agonists, such as CpG ODNs, and other oligonucleotides have been shown to be very effective adjuvant to various experimental formulations of vaccine against leishmaniasis. However, CpG has not yet been used in any trial of leishmaniasis vaccine (Raman *et al.*, 2012).

Saponins

Saponins are plant-based amphipathic glycosides known for their immunostimulatory properties (Tadokoro *et al.*, 1996). These molecules trigger a humoral response. Saponins used as adjuvant to FML significantly protected dogs from CVL in field trials, and the first licensed vaccine against CVL comprised acylated saponin adjuvant. However, because of toxicity concerns, saponins have not been used in any human vaccine trials.

Liposomes

Liposomes are phospholipid-based artificial vesicles used commonly for delivery of drugs and vaccine components. In addition, they are good adjuvants for sustained release of entrapped component and can be manipulated for targeted delivery. Various liposome based vaccine formulation have been tried and tested for different types of leishmaniasis. However, positively charged liposomes were found to be more effective than neutral or negatively charged liposomes in experimental leishmaniasis vaccine (Bhowmick *et al.*, 2007). Evidently cationic liposomes encapsulating different antigens have been found to trigger polarized Th1 response for protection from various forms of experimental leishmaniasis (Badiee *et al.*, 2013). Furthermore, cationic liposomes entrapped LAg was found to impart better protection compared to BCG + LAg and MPL-TDM + LAg against experimental VL (Ravindran *et al.*, 2010). Although liposomal formulations of therapeutic agent like amphotericin B have been approved for human use, no cationic liposome-based vaccination trials has been conducted so far for leishmaniasis.

Other important adjuvants and delivery systems used in experimental leishmaniasis include GM-CSF, ISCOMS, solid lipid nanoparticles, PLGA nanoparticles, etc. (Badiee *et al.*, 2013; Raman *et al.*, 2012).

Summary

Leishmaniasis is a group of disease manifestations with devastating impact on human health. Control of the disease is daunting due to lack of proper chemotherapy. Preventative methods like reservoir and vector control, or personal measures are either not adequate or practical. Hence a preventative vaccine for leishmaniasis is required.

Leishmania is an opportunistic parasite that primarily resides in phagocytes of mammalian host. Innate arm of host immunity is subverted by the parasites and henceforth exploited for disease establishment. However, macrophage, the primary host cell, when activated to induce reactive oxidative burst, can lead to parasite clearance. This could be achieved by activation of inflammatory cytokine-producing Th1 and CD8$^+$ T-cells. Furthermore immune response that acts antagonistically to proinflammatory response involving Th2 and T-regs are detrimental for protection against leishmaniasis. However, the exact role played by adaptive immune response is not fully understood and varies with different forms of leishmaniasis.

Vaccination with live parasites was practised in many parts of the word for cutaneous leishmaniasis but had to be abandoned due to safety concern. First generation vaccines were found to be poor in prophylaxis. Second and third generation vaccines were promising in experimental models but only a few vaccine candidates are being clinically evaluated. Other approaches in vaccine search include genetically modified parasites and dendritic cell-based vaccines.

Therapeutic vaccines have been registered for human use, and more are anticipated in the near future. Three canine vaccines are in the market and more efficient vaccines are being developed.

Adjuvants and delivery systems play an important role in elicitation of suitable and sustained immune response. TLR agonists, cationic liposomes, saponins and lipid nanoparticles are promising adjuvants in the development of vaccines against leishmaniasis.

Although a preventative leishmanial vaccine for human use is still lacking, promising candidates are anticipated in the near future.

Acknowledgements

I sincerely thank all my lab members especially Sarfaraz Ahmed Ejazi, Md Asad, Md Shadad, Mithun Maji, Smriti Mondal, Pradyot Bhattacharya, Somsubhra Thakur Chowdhury and Sadatul Islam for their valuable guidance and insightful suggestions.

Reference

Abi Abdallah, D.S., and Denkers, E.Y. (2012). Neutrophils cast extracellular traps in response to protozoan parasites. Front. Immunol. 3, 382.

Afrin, F., and Ali, N. (1997). Adjuvanticity and protective immunity elicited by *Leishmania donovani* antigens encapsulated in positively charged liposomes. Infect. Immun. *65*, 2371–2377.

Ahuja, S.S., Reddick, R.L., Sato, N., Montalbo, E., Kostecki, V., Zhao, W., Dolan, M.J., Melby, P.C., and Ahuja, S.K. (1999). Dendritic cell (DC)-based anti-infective strategies: DCs engineered to secrete IL-12 are a potent vaccine in a murine model of an intracellular infection. J. Immunol. *163*, 3890–3897.

Alexander, J. (1982). A radioattenuated *Leishmania major* vaccine markedly increases the resistance of CBA mice to subsequent infection with *Leishmania mexicana* mexicana. Trans. R. Soc. Trop. Med. Hyg. *76*, 646–649.

Alexander, J., and Brombacher, F. (2012). T-helper1/t helper2 cells and resistance/susceptibility to leishmania infection: is this paradigm still relevant? Front. Immunol. *3*, 80.

Ali, N., and Afrin, F. (1997). Protection of mice against visceral leishmaniasis by immunization with promastigote antigen incorporated in liposomes. J. Parasitol. *83*, 70–75.

Ali, N., and Bhattacharya, P. (2013). Translating immune cell cross-talk into a treatment opportunity for visceral leishmaniasis. Immunotherapy *5*, 1025–1027.

Alvar, J., Velez, I.D., Bern, C., Herrero, M., Desjeux, P., Cano, J., Jannin, J., and den Boer, M. (2012). Leishmaniasis worldwide and global estimates of its incidence. PLoS One *7*, e35671.

Anderson, C.F., Lira, R., Kamhawi, S., Belkaid, Y., Wynn, T.A., and Sacks, D. (2008). IL-10 and TGF-beta control the establishment of persistent and transmissible infections produced by *Leishmania tropica* in C57BL/6 mice. J. Immunol. *180*, 4090–4097.

Anderson, C.F., Stumhofer, J.S., Hunter, C.A., and Sacks, D. (2009). IL-27 regulates IL-10 and IL-17 from CD4$^+$ cells in nonhealing *Leishmania major* infection. J. Immunol. *183*, 4619–4627.

Andreani, G., Lodge, R., Richard, D., and Tremblay, M.J. (2012). Mechanisms of interaction between protozoan parasites and HIV. Curr. Opin. HIV AIDS *7*, 276–282.

Antoine, J.C. (1995). [Biology of macrophages–*Leishmania* interactions]. Pathol. Biol. (Paris) *43*, 215–223.

Armijos, R.X., Weigel, M.M., Calvopina, M., Hidalgo, A., Cevallos, W., and Correa, J. (2004). Safety, immunogenecity, and efficacy of an autoclaved *Leishmania amazonensis* vaccine plus BCG adjuvant against New World cutaneous leishmaniasis. Vaccine *22*, 1320–1326.

Badiee, A., Heravi Shargh, V., Khamesipour, A., and Jaafari, M.R. (2013). Micro/nanoparticle adjuvants for anti-leishmanial vaccines: present and future trends. Vaccine *31*, 735–749.

Banchereau, J., and Steinman, R.M. (1998). Dendritic cells and the control of immunity. Nature *392*, 245–252.

Basu, R., Bhaumik, S., Haldar, A.K., Naskar, K., De, T., Dana, S.K., Walden, P., and Roy, S. (2007). Hybrid cell vaccination resolves *Leishmania donovani* infection by eliciting a strong CD8$^+$ cytotoxic T-lymphocyte response with concomitant suppression of interleukin-10 (IL-10) but not IL-4 or IL-13. Infect. Immun. *75*, 5956–5966.

Becker, I., Salaiza, N., Aguirre, M., Delgado, J., Carrillo-Carrasco, N., Kobeh, L.G., Ruiz, A., Cervantes, R., Torres, A.P., Cabrera, N., *et al.* (2003). *Leishmania* lipophosphoglycan (LPG) activates NK cells through Toll-like receptor-2. Mol. Biochem. Parasitol. *130*, 65–74.

Belkaid, Y., Kamhawi, S., Modi, G., Valenzuela, J., Noben-Trauth, N., Rowton, E., Ribeiro, J., and Sacks, D.L. (1998). Development of a natural model of cutaneous leishmaniasis: powerful effects of vector saliva and saliva preexposure on the long-term outcome of *Leishmania major* infection in the mouse ear dermis. J. Exp. Med. *188*, 1941–1953.

Belkaid, Y., Hoffmann, K.F., Mendez, S., Kamhawi, S., Udey, M.C., Wynn, T.A., and Sacks, D.L. (2001). The role of interleukin (IL)-10 in the persistence of *Leishmania major* in the skin after healing and the therapeutic potential of anti-IL-10 receptor antibody for sterile cure. J. Exp. Med. *194*, 1497–1506.

Belkaid, Y., Von Stebut, E., Mendez, S., Lira, R., Caler, E., Bertholet, S., Udey, M.C., and Sacks, D. (2002). CD8$^+$ T-cells are required for primary immunity in C57BL/6 mice following low-dose, intradermal challenge with *Leishmania major*. J. Immunol. *168*, 3992–4000.

Bertholet, S., Goldszmid, R., Morrot, A., Debrabant, A., Afrin, F., Collazo-Custodio, C., Houde, M., Desjardins, M., Sher, A., and Sacks, D. (2006). *Leishmania* antigens are presented to CD8$^+$ T-cells by a transporter associated with antigen processing-independent pathway *in vitro* and *in vivo*. J. Immunol. *177*, 3525–3533.

Bhardwaj, S., Vasishta, R.K., and Arora, S.K. (2009). Vaccination with a novel recombinant *Leishmania* antigen plus MPL provides partial protection against *L. donovani* challenge in experimental model of visceral leishmaniasis. Exp. Parasitol. *121*, 29–37.

Bhowmick, S., Ravindran, R., and Ali, N. (2007). Leishmanial antigens in liposomes promote protective immunity and provide immunotherapy against visceral leishmaniasis via polarized Th1 response. Vaccine *25*, 6544–6556.

Bhowmick, S., Ravindran, R., and Ali, N. (2008). gp63 in stable cationic liposomes confers sustained vaccine immunity to susceptible BALB/c mice infected with *Leishmania donovani*. Infect. Immun. *76*, 1003–1015.

Bhowmick, S., Mazumdar, T., and Ali, N. (2009). Vaccination route that induces transforming growth factor beta production fails to elicit protective immunity against *Leishmania donovani* infection. Infect. Immun. *77*, 1514–1523.

Blanchette, J., Racette, N., Faure, R., Siminovitch, K.A., and Olivier, M. (1999). *Leishmania*-induced increases in activation of macrophage SHP-1 tyrosine phosphatase are associated with impaired IFN-γamma-triggered JAK2 activation. Eur. J. Immunol. *29*, 3737–3744.

Borja-Cabrera, G.P., Santos, F.N., Santos, F.B., Trivellato, F.A., Kawasaki, J.K., Costa, A.C., Castro, T., Nogueira, F.S., Moreira, M.A., Luvizotto, M.C., *et al.* (2010). Immunotherapy with the saponin enriched-Leishmune vaccine versus immunochemotherapy in dogs with natural canine visceral leishmaniasis. Vaccine *28*, 597–603.

Bourreau, E., Ronet, C., Couppie, P., Sainte-Marie, D., Tacchini-Cottier, F., and Launois, P. (2007). IL-10 producing CD8+ T-cells in human infection with *Leishmania guyanensis*. Microbes Infect. *9*, 1034–1041.

Breton, M., Tremblay, M.J., Ouellette, M., and Papadopoulou, B. (2005). Live nonpathogenic parasitic vector as a candidate vaccine against visceral leishmaniasis. Infect. Immun. *73*, 6372–6382.

Brittingham, A., Morrison, C.J., McMaster, W.R., McGwire, B.S., Chang, K.P., and Mosser, D.M. (1995). Role of the *Leishmania* surface protease gp63 in complement fixation, cell adhesion, and resistance to complement-mediated lysis. J. Immunol. *155*, 3102–3111.

Bruhn, K.W., Birnbaum, R., Haskell, J., Vanchinathan, V., Greger, S., Narayan, R., Chang, P.L., Tran, T.A., Hickerson, S.M., Beverley, S.M., *et al.* (2012). Killed but metabolically active *Leishmania infantum* as a novel whole-cell vaccine for visceral leishmaniasis. Clin. Vaccine Immunol. *19*, 490–498.

Carrion, J., Folgueira, C., Soto, M., Fresno, M., and Requena, J.M. (2011). *Leishmania infantum* HSP70-II null mutant as candidate vaccine against leishmaniasis: a preliminary evaluation. Parasit. Vectors *4*, 150.

Carson, C., Antoniou, M., Ruiz-Arguello, M.B., Alcami, A., Christodoulou, V., Messaritakis, I., Blackwell, J.M., and Courtenay, O. (2009). A prime/boost DNA/Modified vaccinia virus Ankara vaccine expressing recombinant *Leishmania* DNA encoding TRYP is safe and immunogenic in outbred dogs, the reservoir of zoonotic visceral leishmaniasis. Vaccine *27*, 1080–1086.

Carvalho, E.M., Bacellar, O., Brownell, C., Regis, T., Coffman, R.L., and Reed, S.G. (1994). Restoration of IFN-γamma production and lymphocyte proliferation in visceral leishmaniasis. J. Immunol. *152*, 5949–5956.

Castes, M., Blackwell, J., Trujillo, D., Formica, S., Cabrera, M., Zorrilla, G., Rodas, A., Castellanos, P.L., and Convit, J. (1994). Immune response in healthy volunteers vaccinated with killed leishmanial promastigotes plus BCG. I: Skin-test reactivity, T-cell proliferation and interferon-gamma production. Vaccine *12*, 1041–1051.

Chakour, R., Allenbach, C., Desgranges, F., Charmoy, M., Mauel, J., Garcia, I., Launois, P., Louis, J., and Tacchini-Cottier, F. (2009). A new function of the Fas-FasL pathway in macrophage activation. J. Leukoc. Biol. *86*, 81–90.

Chakravarty, J., Kumar, S., Trivedi, S., Rai, V.K., Singh, A., Ashman, J.A., Laughlin, E.M., Coler, R.N., Kahn, S.J., Beckmann, A.M., *et al.* (2011). A clinical trial to evaluate the safety and immunogenicity of the LEISH-F1+MPL-SE vaccine for use in the prevention of visceral leishmaniasis. Vaccine *29*, 3531–3537.

Chappuis, F., Sundar, S., Hailu, A., Ghalib, H., Rijal, S., Peeling, R.W., Alvar, J., and Boelaert, M. (2007). Visceral leishmaniasis: what are the needs for diagnosis, treatment and control? Nat. Rev. Microbiol. *5*, 873–882.

Chavez-Fumagalli, M.A., Costa, M.A., Oliveira, D.M., Ramirez, L., Costa, L.E., Duarte, M.C., Martins, V.T., Oliveira, J.S., Olortegi, C.C., Bonay, P., *et al.* (2010). Vaccination with the *Leishmania infantum* ribosomal proteins induces protection in BALB/c mice against *Leishmania chagasi* and *Leishmania amazonensis* challenge. Microbes Infect. *12*, 967–977.

Cillari, E., Vitale, G., Arcoleo, F., D'Agostino, P., Mocciaro, C., Gambino, G., Malta, R., Stassi, G., Giordano, C., Milano, S., *et al.* (1995). In vivo and *in vitro* cytokine profiles and mononuclear cell subsets in Sicilian patients with active visceral leishmaniasis. Cytokine *7*, 740–745.

Collin, N., Gomes, R., Teixeira, C., Cheng, L., Laughinghouse, A., Ward, J.M., Elnaiem, D.E., Fischer, L., Valenzuela, J.G., and Kamhawi, S. (2009). Sand fly salivary proteins induce strong cellular immunity in a natural reservoir of visceral leishmaniasis with adverse consequences for *Leishmania*. PLoS Pathog. *5*, e1000441.

Contreras, I., Shio, M.T., Cesaro, A., Tessier, P.A., and Olivier, M. (2013). Impact of Neutrophil-Secreted Myeloid Related Proteins 8 and 14 (MRP 8/14) on Leishmaniasis Progression. PLoS Negl. Trop. Dis. *7*, e2461.

Convit, J., Ulrich, M., Zerpa, O., Borges, R., Aranzazu, N., Valera, M., Villarroel, H., Zapata, Z., and Tomedes, I. (2003). Immunotherapy of american cutaneous leishmaniasis in Venezuela during the period 1990–99. Trans. R. Soc. Trop. Med. Hyg. *97*, 469–472.

Cooper, M.D., and Herrin, B.R. (2010). How did our complex immune system evolve? Nat. Rev. Immunol. *10*, 2–3.

Costa, C.H., Peters, N.C., Maruyama, S.R., de Brito, E.C., Jr., and Santos, I.K. (2011). Vaccines for the leishmaniases: proposals for a research agenda. PLoS Negl. Trop. Dis. *5*, e943.

Dantas-Torres, F. (2006). Leishmune vaccine: the newest tool for prevention and control of canine visceral leishmaniosis and its potential as a transmission-blocking vaccine. Vet. Parasitol. *141*, 1–8.

Das, A., and Ali, N. (2012). Vaccine prospects of killed but metabolically active *Leishmania* against visceral leishmaniasis. Expert Rev. Vaccines *11*, 783–785.

Da Silva, R.P., Hall, B.F., Joiner, K.A., and Sacks, D.L. (1989). CR1, the C3b receptor, mediates binding of infective *Leishmania major* metacyclic promastigotes to human macrophages. J. Immunol. *143*, 617–622.

Datta, S., Manna, M., Khanra, S., Ghosh, M., Bhar, R., Chakraborty, A., and Roy, S. (2012). Therapeutic immunization with radio-attenuated *Leishmania* parasites through i.m. route revealed protection against the experimental murine visceral leishmaniasis. Parasitol Res *111*, 361–369.

Debus, A., Glasner, J., Rollinghoff, M., and Gessner, A. (2003). High levels of susceptibility and T-helper 2 response in MyD88-deficient mice infected with *Leishmania major* are interleukin-4 dependent. Infect. Immun. *71*, 7215–7218.

De Oliveira Gomes, D.C., Schwedersky, R.P., Barbosa De-Melo, L.D., Da Silva Costa Souza, B.L., De Matos Guedes, H.L., Lopes, U.G., and Rossi-Bergmann, B. (2012). Peripheral expression of LACK-mRNA induced by intranasal vaccination with

PCI-NEO-LACK defines the protection duration against murine visceral leishmaniasis. Parasitology *139*, 1562–1569.

Dermine, J.F., Scianimanico, S., Prive, C., Descoteaux, A., and Desjardins, M. (2000). *Leishmania* promastigotes require lipophosphoglycan to actively modulate the fusion properties of phagosomes at an early step of phagocytosis. Cell. Microbiol. *2*, 115–126.

Desjeux, P. (1992). Human leishmaniases: epidemiology and public health aspects. World Health Stat. Q. *45*, 267–275.

Dey, A., and Singh, S. (2006). Transfusion transmitted leishmaniasis: a case report and review of literature. Indian J. Med. Microbiol. *24*, 165–170.

Dey, R., Majumder, N., Bhattacharjee, S., Majumdar, S.B., Banerjee, R., Ganguly, S., Das, P., and Majumdar, S. (2007). *Leishmania donovani*-induced ceramide as the key mediator of Akt dephosphorylation in murine macrophages: role of protein kinase Czeta and phosphatase. Infect. Immun. *75*, 2136–2142.

Dey, R., Dagur, P.K., Selvapandiyan, A., McCoy, J.P., Salotra, P., Duncan, R., and Nakhasi, H.L. (2013). Live attenuated *Leishmania donovani* p27 gene knockout parasites are nonpathogenic and elicit long-term protective immunity in BALB/c mice. J. Immunol. *190*, 2138–2149.

Dinesh, D.S., Das, M.L., Picado, A., Roy, L., Rijal, S., Singh, S.P., Das, P., Boelaert, M., and Coosemans, M. (2010). Insecticide susceptibility of Phlebotomus argentipes in visceral leishmaniasis endemic districts in India and Nepal. PLoS Negl. Trop. Dis. *4*, e859.

Doroud, D., Zahedifard, F., Vatanara, A., Najafabadi, A.R., Taslimi, Y., Vahabpour, R., Torkashvand, F., Vaziri, B., and Rafati, S. (2011). Delivery of a cocktail DNA vaccine encoding cysteine proteinases type I, II and III with solid lipid nanoparticles potentiate protective immunity against *Leishmania major* infection. J. Control. Release *153*, 154–162.

Dunan, S., Frommel, D., Monjour, L., Ogunkolade, B.W., Cruz, A., and Quilici, M. (1989). Vaccination trial against canine visceral leishmaniasis. Phocean Veterinary Study Group on Visceral Leishmaniasis. Parasite Immunol. *11*, 397–402.

Ejazi, S.A., and Ali, N. (2013). Developments in diagnosis and treatment of visceral leishmaniasis during the last decade and future prospects. Expert Rev. Anti Infect. Ther. *11*, 79–98.

Faria, D.R., Gollob, K.J., Barbosa, J., Jr., Schriefer, A., Machado, P.R., Lessa, H., Carvalho, L.P., Romano-Silva, M.A., de Jesus, A.R., Carvalho, E.M., *et al.* (2005). Decreased *in situ* expression of interleukin-10 receptor is correlated with the exacerbated inflammatory and cytotoxic responses observed in mucosal leishmaniasis. Infect. Immun. *73*, 7853–7859.

Faria, D.R., Souza, P.E., Duraes, F.V., Carvalho, E.M., Gollob, K.J., Machado, P.R., and Dutra, W.O. (2009). Recruitment of CD8(+) T-cells expressing granzyme A is associated with lesion progression in human cutaneous leishmaniasis. Parasite Immunol. *31*, 432–439.

Faria, M.S., Reis, F.C., and Lima, A.P. (2012). Toll-like receptors in leishmania infections: guardians or promoters? J. Parasitol. Res. *2012*, 930257.

Fiuza, J.A., Santiago Hda, C., Selvapandiyan, A., Gannavaram, S., Ricci, N.D., Bueno, L.L., Bartholomeu, D.C., Correa-Oliveira, R., Nakhasi, H.L., and Fujiwara, R.T. (2013). Induction of immunogenicity by live attenuated *Leishmania donovani* centrin deleted parasites in dogs. Vaccine *31*, 1785–1792.

Gallego, C., Golenbock, D., Gomez, M.A., and Saravia, N.G. (2011). Toll-like receptors participate in macrophage activation and intracellular control of *Leishmania* (Viannia) panamensis. Infect. Immun. *79*, 2871–2879.

Ganguly, S., Das, N.K., Panja, M., Pal, S., Modak, D., Rahaman, M., Mallik, S., Guha, S.K., Pramanik, N., Goswami, R., *et al.* (2008). Increased levels of interleukin-10 and IgG3 are hallmarks of Indian post-kala-azar dermal leishmaniasis. J. Infect. Dis. *197*, 1762–1771.

Gaze, S.T., Dutra, W.O., Lessa, M., Lessa, H., Guimaraes, L.H., Jesus, A.R., Carvalho, L.P., Machado, P., Carvalho, E.M., and Gollob, K.J. (2006). Mucosal leishmaniasis patients display an activated inflammatory T-cell phenotype associated with a nonbalanced monocyte population. Scand. J. Immunol. *63*, 70–78.

Ghosh, A., Zhang, W.W., and Matlashewski, G. (2001). Immunization with A2 protein results in a mixed Th1/Th2 and a humoral response which protects mice against *Leishmania donovani* infections. Vaccine *20*, 59–66.

Gicheru, M.M., Olobo, J.O., Anjili, C.O., Orago, A.S., Modabber, F., and Scott, P. (2001). Vervet monkeys vaccinated with killed *Leishmania major* parasites and interleukin-12 develop a type 1 immune response but are not protected against challenge infection. Infect. Immun. *69*, 245–251.

Golali, E., Jaafari, M.R., Khamesipour, A., Abbasi, A., Saberi, Z., and Badiee, A. (2012). Comparison of *in vivo* Adjuvanticity of Liposomal PO CpG ODN with Liposomal PS CpG ODN: Soluble *Leishmania* Antigens as a Model. Iran. J. Basic. Med. Sci. *15*, 1032–1045.

Gomes, D.C., Pinto, E.F., de Melo, L.D., Lima, W.P., Larraga, V., Lopes, U.G., and Rossi-Bergmann, B. (2007). Intranasal delivery of naked DNA encoding the LACK antigen leads to protective immunity against visceral leishmaniasis in mice. Vaccine *25*, 2168–2172.

Gomes, R., Teixeira, C., Teixeira, M.J., Oliveira, F., Menezes, M.J., Silva, C., de Oliveira, C.I., Miranda, J.C., Elnaiem, D.E., Kamhawi, S., *et al.* (2008). Immunity to a salivary protein of a sand fly vector protects against the fatal outcome of visceral leishmaniasis in a hamster model. Proc. Natl. Acad. Sci. U.S.A. *105*, 7845–7850.

Gomes, R., Oliveira, F., Teixeira, C., Meneses, C., Gilmore, D.C., Elnaiem, D.E., Kamhawi, S., and Valenzuela, J.G. (2012a). Immunity to sand fly salivary protein LJM11 modulates host response to vector-transmitted leishmania conferring ulcer-free protection. J. Invest. Dermatol. *132*, 2735–2743.

Gomes, R., Teixeira, C., Oliveira, F., Lawyer, P.G., Elnaiem, D.E., Meneses, C., Goto, Y., Bhatia, A., Howard, R.F.,

Reed, S.G., *et al.* (2012b). KSAC, a defined *Leishmania* antigen, plus adjuvant protects against the virulence of *L. major* transmitted by its natural vector Phlebotomus duboscqi. PLoS Negl. Trop. Dis. *6*, e1610.

Gonzalez-Lombana, C., Gimblet, C., Bacellar, O., Oliveira, W.W., Passos, S., Carvalho, L.P., Goldschmidt, M., Carvalho, E.M., and Scott, P. (2013). IL-17 mediates immunopathology in the absence of IL-10 following *Leishmania major* infection. PLoS Pathog 9, e1003243.

Gordon, S. (2003). Alternative activation of macrophages. Nat. Rev. Immunol. 3, 23–35.

Goto, Y., Bogatzki, L.Y., Bertholet, S., Coler, R.N., and Reed, S.G. (2007). Protective immunization against visceral leishmaniasis using *Leishmania* sterol 24-c-methyltransferase formulated in adjuvant. Vaccine 25, 7450–7458.

Goto, Y., Bhatia, A., Raman, V.S., Liang, H., Mohamath, R., Picone, A.F., Vidal, S.E., Vedvick, T.S., Howard, R.F., and Reed, S.G. (2011). KSAC, the first defined polyprotein vaccine candidate for visceral leishmaniasis. Clin. Vaccine Immunol. *18*, 1118–1124.

Guha, R., Das, S., Ghosh, J., Naskar, K., Mandala, A., Sundar, S., Dujardin, J.C., and Roy, S. (2013a). Heterologous priming-boosting with DNA and vaccinia virus expressing kinetoplastid membrane protein-11 induces potent cellular immune response and confers protection against infection with antimony resistant and sensitive strains of *Leishmania (Leishmania) donovani*. Vaccine *31*, 1905–1915.

Guha, R., Gupta, D., Rastogi, R., Vikram, R., Krishnamurthy, G., Bimal, S., Roy, S., and Mukhopadhyay, A. (2013b). Vaccination with leishmania hemoglobin receptor-encoding DNA protects against visceral leishmaniasis. Sci. Transl. Med. *5*, 202ra121.

Guimarães-Costa, A.B., Nascimento, M.T., Froment, G.S., Soares, R.P., Morgado, F.N., Conceição-Silva, F., and Saraiva, E.M. (2009). *Leishmania amazonensis* promastigotes induce and are killed by neutrophil extracellular traps. Proc. Natl. Acad. Sci. U.S.A. *106*, 6748–6753.

Gupta, R., Kushawaha, P.K., Tripathi, C.D., Sundar, S., and Dube, A. (2012). A novel recombinant *Leishmania donovani* p45, a partial coding region of methionine aminopeptidase, generates protective immunity by inducing a Th1 stimulatory response against experimental visceral leishmaniasis. Int. J. Parasitol. *42*, 429–435.

Gurunathan, S., Sacks, D.L., Brown, D.R., Reiner, S.L., Charest, H., Glaichenhaus, N., and Seder, R.A. (1997). Vaccination with DNA encoding the immunodominant LACK parasite antigen confers protective immunity to mice infected with *Leishmania major*. J. Exp. Med. *186*, 1137–1147.

Henriquez, F.L., Campbell, S.A., Roberts, C.W., Mullen, A.B., Burchmore, R., and Carter, K.C. (2010). Vaccination with recombinant *Leishmania donovani* gamma-glutamylcysteine synthetase fusion protein protects against *L. donovani* infection. J. Parasitol. *96*, 929–936.

Heravi Shargh, V., Jaafari, M.R., Khamesipour, A., Jalali, S.A., Firouzmand, H., Abbasi, A., and Badiee, A. (2012). Cationic liposomes containing soluble *Leishmania* antigens (SLA) plus CpG ODNs induce protection against murine model of leishmaniasis. Parasitol. Res. *111*, 105–114.

Hezarjaribi, H.Z., Ghaffarifar, F., Dalimi, A., Sharifi, Z., and Jorjani, O. (2013). Effect of IL-22 on DNA vaccine encoding LACK gene of *Leishmania major* in BALB/c mice. Exp. Parasitol. *134*, 341–348.

Hol, J., Wilhelmsen, L., and Haraldsen, G. (2010). The murine IL-8 homologues KC, MIP-2, and LIX are found in endothelial cytoplasmic granules but not in Weibel-Palade bodies. J. Leukoc. Biol. *87*, 501–508.

Hotez, P.J., Remme, J.H., Buss, P., Alleyne, G., Morel, C., and Breman, J.G. (2004). Combating tropical infectious diseases: report of the Disease Control Priorities in Developing Countries Project. Clin. Infect. Dis. *38*, 871–878.

Hugentobler, F., Di Roberto, R.B., Gillard, J., and Cousineau, B. (2012). Oral immunization using live Lactococcus lactis co-expressing LACK and IL-12 protects BALB/c mice against *Leishmania major* infection. Vaccine *30*, 5726–5732.

Isnard, A., Shio, M.T., and Olivier, M. (2012). Impact of *Leishmania* metalloprotease GP63 on macrophage signaling. Front. Cell. Infect. Microbiol. 2, 72.

Jaafari, M.R., Ghafarian, A., Farrokh-Gisour, A., Samiei, A., Kheiri, M.T., Mahboudi, F., Barkhordari, F., Khamesipour, A., and McMaster, W.R. (2006). Immune response and protection assay of recombinant major surface glycoprotein of *Leishmania* (rgp63) reconstituted with liposomes in BALB/c mice. Vaccine *24*, 5708–5717.

Jaafari, M.R., Badiee, A., Khamesipour, A., Samiei, A., Soroush, D., Kheiri, M.T., Barkhordari, F., McMaster, W.R., and Mahboudi, F. (2007). The role of CpG ODN in enhancement of immune response and protection in BALB/c mice immunized with recombinant major surface glycoprotein of *Leishmania* (rgp63) encapsulated in cationic liposome. Vaccine *25*, 6107–6117.

Jacobs, T., Andra, J., Gaworski, I., Graefe, S., Mellenthin, K., Kromer, M., Halter, R., Borlak, J., and Clos, J. (2005). Complement C3 is required for the progression of cutaneous lesions and neutrophil attraction in *Leishmania major* infection. Med. Microbiol. Immunol. *194*, 143–149.

Kar, S., Ukil, A., Sharma, G., and Das, P.K. (2010). MAPK-directed phosphatases preferentially regulate pro- and anti-inflammatory cytokines in experimental visceral leishmaniasis: involvement of distinct protein kinase C isoforms. J. Leukoc. Biol. *88*, 9–20.

Kautz-Neu, K., Noordegraaf, M., Dinges, S., Bennett, C.L., John, D., Clausen, B.E., and von Stebut, E. (2011). Langerhans cells are negative regulators of the anti-*Leishmania* response. J. Exp. Med. *208*, 885–891.

Kawai, T., and Akira, S. (2010). The role of pattern-recognition receptors in innate immunity: update on Toll-like receptors. Nat. Immunol. *11*, 373–384.

Kaye, P., and Scott, P. (2011). Leishmaniasis: complexity at the host–pathogen interface. Nat. Rev. Microbiol. *9*, 604–615.

Kedzierski, L. (2010). Leishmaniasis Vaccine: Where are We Today? J. Glob. Infect. Dis. 2, 177–185.

Kedzierski, L., Curtis, J.M., Doherty, P.C., Handman, E., and Kedzierska, K. (2008). Decreased IL-10 and IL-13 production and increased CD44hi T-cell recruitment contribute to *Leishmania major* immunity induced by non-persistent parasites. Eur. J. Immunol. *38*, 3090–3100.

Khalil, E.A., El Hassan, A.M., Zijlstra, E.E., Mukhtar, M.M., Ghalib, H.W., Musa, B., Ibrahim, M.E., Kamil, A.A., Elsheikh, M., Babiker, A., *et al.* (2000). Autoclaved *Leishmania major* vaccine for prevention of visceral leishmaniasis: a randomised, double-blind, BCG-controlled trial in Sudan. Lancet *356*, 1565–1569.

Khalil, E.A., Musa, A.M., Modabber, F., and El-Hassan, A.M. (2006). Safety and immunogenicity of a candidate vaccine for visceral leishmaniasis (Alum-precipitated autoclaved *Leishmania major* + BCG) in children: an extended phase II study. Ann. Trop. Paediatr. *26*, 357–361.

Kima, P.E., Ruddle, N.H., and McMahon-Pratt, D. (1997). Presentation via the class I pathway by *Leishmania amazonensis*-infected macrophages of an endogenous leishmanial antigen to CD8+ T-cells. J. Immunol. *159*, 1828–1834.

Kimblin, N., Peters, N., Debrabant, A., Secundino, N., Egen, J., Lawyer, P., Fay, M.P., Kamhawi, S., and Sacks, D. (2008). Quantification of the infectious dose of *Leishmania major* transmitted to the skin by single sand flies. Proc. Natl. Acad. Sci. U.S.A. *105*, 10125–10130.

Kulkarni, M.M., McMaster, W.R., Kamysz, E., Kamysz, W., Engman, D.M., and McGwire, B.S. (2006). The major surface-metalloprotease of the parasitic protozoan, *Leishmania*, protects against antimicrobial peptide-induced apoptotic killing. Mol. Microbiol. *62*, 1484–1497.

Kushawaha, P.K., Gupta, R., Sundar, S., Sahasrabuddhe, A.A., and Dube, A. (2011). Elongation factor-2, a Th1 stimulatory protein of *Leishmania donovani*, generates strong IFN-γamma and IL-12 response in cured *Leishmania*-infected patients/hamsters and protects hamsters against *Leishmania* challenge. J. Immunol. *187*, 6417–6427.

Kushawaha, P.K., Gupta, R., Tripathi, C.D., Sundar, S., and Dube, A. (2012). Evaluation of *Leishmania donovani* protein disulfide isomerase as a potential immunogenic protein/vaccine candidate against visceral Leishmaniasis. PLoS One *7*, e35670.

Laufs, H., Muller, K., Fleischer, J., Reiling, N., Jahnke, N., Jensenius, J.C., Solbach, W., and Laskay, T. (2002). Intracellular survival of *Leishmania major* in neutrophil granulocytes after uptake in the absence of heat-labile serum factors. Infect. Immun. *70*, 826–835.

Lemesre, J.L., Holzmuller, P., Goncalves, R.B., Bourdoiseau, G., Hugnet, C., Cavaleyra, M., and Papierok, G. (2007). Long-lasting protection against canine visceral leishmaniasis using the LiESAp-MDP vaccine in endemic areas of France: double-blind randomised efficacy field trial. Vaccine *25*, 4223–4234.

Leon, B., Lopez-Bravo, M., and Ardavin, C. (2007). Monocyte-derived dendritic cells formed at the infection site control the induction of protective T-helper 1 responses against *Leishmania*. Immunity *26*, 519–531.

Liese, J., Schleicher, U., and Bogdan, C. (2007). TLR9 signaling is essential for the innate NK cell response in murine cutaneous leishmaniasis. Eur. J. Immunol. *37*, 3424–3434.

Lodge, R., and Descoteaux, A. (2005). *Leishmania donovani* promastigotes induce periphagosomal F-actin accumulation through retention of the GTPase Cdc42. Cell Microbiol. *7*, 1647–1658.

Lopez Kostka, S., Dinges, S., Griewank, K., Iwakura, Y., Udey, M.C., and von Stebut, E. (2009). IL-17 promotes progression of cutaneous leishmaniasis in susceptible mice. J. Immunol. *182*, 3039–3046.

McCall, L.I., Zhang, W.W., Ranasinghe, S., and Matlashewski, G. (2013). Leishmanization revisited: immunization with a naturally attenuated cutaneous *Leishmania donovani* isolate from Sri Lanka protects against visceral leishmaniasis. Vaccine *31*, 1420–1425.

Machado-Pinto, J., Pinto, J., da Costa, C.A., Genaro, O., Marques, M.J., Modabber, F., and Mayrink, W. (2002). Immunochemotherapy for cutaneous leishmaniasis: a controlled trial using killed *Leishmania* (*Leishmania*) *amazonensis* vaccine plus antimonial. Int. J. Dermatol. *41*, 73–78.

Mathers, C.D., Ezzati, M., and Lopez, A.D. (2007). Measuring the burden of neglected tropical diseases: the global burden of disease framework. PLoS Negl. Trop. Dis. *1*, e114.

Mayrink, W., da Costa, C.A., Magalhaes, P.A., Melo, M.N., Dias, M., Lima, A.O., Michalick, M.S., and Williams, P. (1979). A field trial of a vaccine against American dermal leishmaniasis. Trans. R. Soc. Trop. Med. Hyg. *73*, 385–387.

Mazumder, S., Maji, M., and Ali, N. (2011a). Potentiating effects of MPL on DSPC bearing cationic liposomes promote recombinant GP63 vaccine efficacy: high immunogenicity and protection. PLoS Negl. Trop. Dis. *5*, e1429.

Mazumder, S., Maji, M., Das, A., and Ali, N. (2011b). Potency, efficacy and durability of DNA/DNA, DNA/protein and protein/protein based vaccination using gp63 against *Leishmania donovani* in BALB/c mice. PLoS One *6*, e14644.

Mazumdar, T., Anam, K., and Ali, N. (2004). A mixed Th1/Th2 response elicited by a liposomal formulation of *Leishmania* vaccine instructs Th1 responses and resistance to *Leishmania donovani* in susceptible BALB/c mice. Vaccine *22*, 1162–1171.

Mizbani, A., Taheri, T., Zahedifard, F., Taslimi, Y., Azizi, H., Azadmanesh, K., Papadopoulou, B., and Rafati, S. (2009). Recombinant *Leishmania* tarentolae expressing the A2 virulence gene as a novel candidate vaccine against visceral leishmaniasis. Vaccine *28*, 53–62.

Mohebali, M., Khamesipour, A., Mobedi, I., Zarei, Z., and Hashemi-Fesharki, R. (2004). Double-blind randomized efficacy field trial of alum precipitated autoclaved *Leishmania major* vaccine mixed with BCG against canine visceral leishmaniasis in Meshkin-Shahr district, I.R. Iran. Vaccine *22*, 4097–4100.

Moll, H., Scharner, A., and Kampgen, E. (2002). Increased interleukin 4 (IL-4) receptor expression and IL-4-induced decrease in IL-12 production by Langerhans cells infected with *Leishmania major*. Infect. Immun. *70*, 1627–1630.

Moradin, N., and Descoteaux, A. (2012). *Leishmania* promastigotes: building a safe niche within macrophages. Front. Cell. Infect. Microbiol. *2*, 121.

Morris, R.V., Shoemaker, C.B., David, J.R., Lanzaro, G.C., and Titus, R.G. (2001). Sandfly maxadilan exacerbates infection with *Leishmania major* and vaccinating against it protects against L. major infection. J. Immunol. *167*, 5226–5230.

Mosser, D.M., and Brittingham, A. (1997). *Leishmania*, macrophages and complement: a tale of subversion and exploitation. Parasitology *115 Suppl*, S9–23.

Mott, K.E., Nuttall, I., Desjeux, P., and Cattand, P. (1995). New geographical approaches to control of some parasitic zoonoses. Bull. World Health Organ. *73*, 247–257.

Mougneau, E., Altare, F., Wakil, A.E., Zheng, S., Coppola, T., Wang, Z.E., Waldmann, R., Locksley, R.M., and Glaichenhaus, N. (1995). Expression cloning of a protective *Leishmania* antigen. Science *268*, 563–566.

Muller, K., van Zandbergen, G., Hansen, B., Laufs, H., Jahnke, N., Solbach, W., and Laskay, T. (2001). Chemokines, natural killer cells and granulocytes in the early course of *Leishmania major* infection in mice. Med. Microbiol. Immunol. *190*, 73–76.

Murray, H.W., Berman, J.D., Davies, C.R., and Saravia, N.G. (2005). Advances in leishmaniasis. Lancet *366*, 1561–1577.

Musa, A.M., Khalil, E.A., Mahgoub, F.A., Elgawi, S.H., Modabber, F., Elkadaru, A.E., Aboud, M.H., Noazin, S., Ghalib, H.W., and El-Hassan, A.M. (2008). Immunochemotherapy of persistent post-kala-azar dermal leishmaniasis: a novel approach to treatment. Trans. R. Soc. Trop. Med. Hyg. *102*, 58–63.

Mutiso, J.M., Macharia, J.C., Kiio, M.N., Ichagichu, J.M., Rikoi, H., and Gicheru, M.M. (2013). Development of *Leishmania* vaccines: predicting the future from past and present experience. J. Biomed. Res. 27, 85–102.

Noazin, S., Khamesipour, A., Moulton, L.H., Tanner, M., Nasseri, K., Modabber, F., Sharifi, I., Khalil, E.A., Bernal, I.D., Antunes, C.M., *et al.* (2009). Efficacy of killed whole-parasite vaccines in the prevention of leishmaniasis: a meta-analysis. Vaccine *27*, 4747–4753.

Okwor, I., and Uzonna, J.E. (2009). Immunotherapy as a strategy for treatment of leishmaniasis: a review of the literature. Immunotherapy *1*, 765–776.

Palatnik-de-Sousa, C.B. (2012). Vaccines for canine leishmaniasis. Front. Immunol. *3*, 69.

Panaro, M.A., Puccini, V., Faliero, S.M., Marzio, R., Marangi, A., Lisi, S., and Brandonisio, O. (1996). *Leishmania donovani* lipophosphoglycan (LPG) inhibits respiratory burst and chemotaxis of dog phagocytes. New Microbiol. *19*, 107–112.

Peters, N.C., Egen, J.G., Secundino, N., Debrabant, A., Kimblin, N., Kamhawi, S., Lawyer, P., Fay, M.P., Germain, R.N., and Sacks, D. (2008). In vivo imaging reveals an essential role for neutrophils in leishmaniasis transmitted by sand flies. Science *321*, 970–974.

Pinto, E.F., Pinheiro, R.O., Rayol, A., Larraga, V., and Rossi-Bergmann, B. (2004). Intranasal vaccination against cutaneous leishmaniasis with a particulated leishmanial antigen or DNA encoding LACK. Infect. Immun. *72*, 4521–4527.

Raman, V.S., Duthie, M.S., Fox, C.B., Matlashewski, G., and Reed, S.G. (2012). Adjuvants for *Leishmania* vaccines: from models to clinical application. Front. Immunol. *3*, 144.

Ramirez, L., Iborra, S., Cortes, J., Bonay, P., Alonso, C., Barral-Netto, M., and Soto, M. (2010). BALB/c mice vaccinated with *Leishmania major* ribosomal proteins extracts combined with CpG oligodeoxynucleotides become resistant to disease caused by a secondary parasite challenge. J. Biomed. Biotechnol. *2010*, 181690.

Ravindran, R., Bhowmick, S., Das, A., and Ali, N. (2010). Comparison of BCG, MPL and cationic liposome adjuvant systems in leishmanial antigen vaccine formulations against murine visceral leishmaniasis. BMC Microbiol. *10*, 181.

Ribeiro-Gomes, F.L., and Sacks, D. (2012). The influence of early neutrophil–*Leishmania* interactions on the host immune response to infection. Front. Cell. Infect. Microbiol. *2*, 59.

Ritter, U., Frischknecht, F., and van Zandbergen, G. (2009). Are neutrophils important host cells for *Leishmania* parasites? Trends Parasitol. *25*, 505–510.

Rodriguez-Cortes, A., Ojeda, A., Lopez-Fuertes, L., Timon, M., Altet, L., Solano-Gallego, L., Sanchez-Robert, E., Francino, O., and Alberola, J. (2007). Vaccination with plasmid DNA encoding KMPII, TRYP, LACK and GP63 does not protect dogs against *Leishmania infantum* experimental challenge. Vaccine *25*, 7962–7971.

Romagnani, S. (1999). Th1/Th2 cells. Inflamm. Bowel. Dis. *5*, 285–294.

Ronet, C., Beverley, S.M., and Fasel, N. (2011). Mucocutaneous leishmaniasis in the New World: the ultimate subversion. Virulence *2*, 547–552.

Rosas, L.E., Satoskar, A.A., Roth, K.M., Keiser, T.L., Barbi, J., Hunter, C., de Sauvage, F.J., and Satoskar, A.R. (2006). Interleukin-27R (WSX-1/T-cell cytokine receptor) gene-deficient mice display enhanced resistance to *Leishmania donovani* infection but develop severe liver immunopathology. Am. J. Pathol. *168*, 158–169.

Rousseau, F., Basset, L., Froger, J., Dinguirard, N., Chevalier, S., and Gascan, H. (2010). IL-27 structural analysis demonstrates similarities with ciliary neurotrophic factor (CNTF) and leads to the identification of antagonistic variants. Proc. Natl. Acad. Sci. U.S.A. *107*, 19420–19425.

Ruiz, J.H., and Becker, I. (2007). CD8 cytotoxic T-cells in cutaneous leishmaniasis. Parasite Immunol. *29*, 671–678.

Russell, D.G., and Alexander, J. (1988). Effective immunization against cutaneous leishmaniasis with defined membrane antigens reconstituted into liposomes. J. Immunol. *140*, 1274–1279.

Saha, S., Mondal, S., Ravindran, R., Bhowmick, S., Modak, D., Mallick, S., Rahman, M., Kar, S., Goswami, R., Guha, S.K., *et al.* (2007). IL-10- and TGF-beta-mediated susceptibility in kala-azar and post-kala-azar dermal leishmaniasis: the significance of amphotericin

B in the control of *Leishmania donovani* infection in India. J. Immunol. *179*, 5592–5603.

Saljoughian, N., Taheri, T., Zahedifard, F., Taslimi, Y., Doustdari, F., Bolhassani, A., Doroud, D., Azizi, H., Heidari, K., Vasei, M., *et al.* (2013). Development of novel prime-boost strategies based on a tri-gene fusion recombinant L. tarentolae vaccine against experimental murine visceral leishmaniasis. PLoS Negl. Trop. Dis. 7, e2174.

Santos, D.M., Carneiro, M.W., de Moura, T.R., Fukutani, K., Clarencio, J., Soto, M., Espuelas, S., Brodskyn, C., Barral, A., Barral-Netto, M., *et al.* (2012). Towards development of novel immunization strategies against leishmaniasis using PLGA nanoparticles loaded with kinetoplastid membrane protein-11. Int. J. Nanomedicine. 7, 2115–2127.

Satoskar, A.R., Stamm, L.M., Zhang, X., Satoskar, A.A., Okano, M., Terhorst, C., David, J.R., and Wang, B. (1999). Mice lacking NK cells develop an efficient Th1 response and control cutaneous *Leishmania major* infection. J. Immunol. *162*, 6747–6754.

Scharton, T.M., and Scott, P. (1993). Natural killer cells are a source of interferon gamma that drives differentiation of CD4$^+$ T-cell subsets and induces early resistance to *Leishmania major* in mice. J. Exp. Med. *178*, 567–577.

Scott, P., Natovitz, P., Coffman, R.L., Pearce, E., and Sher, A. (1988). Immunoregulation of cutaneous leishmaniasis. T-cell lines that transfer protective immunity or exacerbation belong to different T-helper subsets and respond to distinct parasite antigens. J. Exp. Med. *168*, 1675–1684.

Selvapandiyan, A., Dey, R., Nylen, S., Duncan, R., Sacks, D., and Nakhasi, H.L. (2009). Intracellular replication-deficient *Leishmania donovani* induces long lasting protective immunity against visceral leishmaniasis. J. Immunol. *183*, 1813–1820.

Shadab, M., and Ali, N. (2011). Evasion of Host Defence by *Leishmania donovani*: Subversion of Signaling Pathways. Mol. Biol. Int. *2011*, 343961.

Shargh, V.H., Jaafari, M.R., Khamesipour, A., Jaafari, I., Jalali, S.A., Abbasi, A., and Badiee, A. (2012). Liposomal SLA co-incorporated with PO CpG ODNs or PS CpG ODNs induce the same protection against the murine model of leishmaniasis. Vaccine *30*, 3957–3964.

Sharma, S.K., Dube, A., Nadeem, A., Khan, S., Saleem, I., Garg, R., and Mohammad, O. (2006). Non PC liposome entrapped promastigote antigens elicit parasite specific CD8$^+$ and CD4$^+$ T-cell immune response and protect hamsters against visceral leishmaniasis. Vaccine *24*, 1800–1810.

Sharma, U., and Singh, S. (2008). Insect vectors of *Leishmania*: distribution, physiology and their control. J Vector Borne Dis *45*, 255–272.

da Silva, M.F., and Floeter-Winter, L.M. (2014). Arginase in leishmania. Subcell. Biochem. *74*, 103–117.

da Silva, R.A., Tavares, N.M., Costa, D., Pitombo, M., Barbosa, L., Fukutani, K., Miranda, J.C., de Oliveira, C.I., Valenzuela, J.G., Barral, A., *et al.* (2011). DNA vaccination with KMP11 and Lutzomyia longipalpis salivary protein protects hamsters against visceral leishmaniasis. Acta Trop. *120*, 185–190.

da Silva, R.P., Hall, B.F., Joiner, K.A., and Sacks, D.L. (1988). CR1 mediates binding of L. major metacyclic promastigotes to human macrophages. Mem. Inst. Oswaldo Cruz *83 Suppl 1*, 459–463.

Singh, R.K., Srivastava, A., and Singh, N. (2012). Toll-like receptor signaling: a perspective to develop vaccine against leishmaniasis. Microbiol. Res. *167*, 445–451.

Sheiky, Y.A., Kennedy, M., Kaufman, D., Borges, M.M., Guderian, J.A., Scholler, J.K., Ovendale, P.J., Picha, K.S., Morrissey, P.J., Grabstein, K.H., *et al.* (1998). LeIF: a recombinant *Leishmania* protein that induces an IL-12-mediated Th1 cytokine profile. J. Immunol. *161*, 6171–6179.

Soong, L. (2012). Subversion and Utilization of Host Innate Defense by *Leishmania amazonensis*. Front. Immunol. 3, 58.

Stager, S., and Rafati, S. (2012). CD8(+) T-cells in leishmania infections: friends or foes? Front. Immunol. *3*, 5.

Stager, S., Smith, D.F., and Kaye, P.M. (2000). Immunization with a recombinant stage-regulated surface protein from *Leishmania donovani* induces protection against visceral leishmaniasis. J. Immunol. *165*, 7064–7071.

Stauch, A., Duerr, H.P., Dujardin, J.C., Vanaerschot, M., Sundar, S., and Eichner, M. (2012). Treatment of visceral leishmaniasis: model-based analyses on the spread of antimony-resistant *L. donovani* in Bihar, India. PLoS Negl. Trop. Dis. *6*, e1973.

von Stebut, E., Belkaid, Y., Jakob, T., Sacks, D.L., and Udey, M.C. (1998). Uptake of *Leishmania major* amastigotes results in activation and interleukin 12 release from murine skin-derived dendritic cells: implications for the initiation of anti-*Leishmania* immunity. J. Exp. Med. *188*, 1547–1552.

Stefani, M.M., Muller, I., and Louis, J.A. (1994). *Leishmania major*-specific CD8$^+$ T-cells are inducers and targets of nitric oxide produced by parasitized macrophages. Eur. J. Immunol. *24*, 746–752.

Steinman, R.M. (2007). Lasker Basic Medical Research Award. Dendritic cells: versatile controllers of the immune system. Nat. Med. *13*, 1155–1159.

Stober, C.B., Lange, U.G., Roberts, M.T., Alcami, A., and Blackwell, J.M. (2007). Heterologous priming-boosting with DNA and modified vaccinia virus Ankara expressing tryparedoxin peroxidase promotes long-term memory against *Leishmania major* in susceptible BALB/c Mice. Infect. Immun. *75*, 852–860.

Stockdale, L., and Newton, R. (2013). A review of preventative methods against human leishmaniasis infection. PLoS Negl. Trop. Dis. *7*, e2278.

Sundrud, M.S., and Trivigno, C. (2013). Identity crisis of Th17 cells: Many forms, many functions, many questions. Semin. Immunol.

Tacchini-Cottier, F., Weinkopff, T., and Launois, P. (2012). Does T-helper differentiation correlate with resistance or susceptibility to infection with L. major? Some insights from the murine model. Front. Immunol. 3, 32.

Tadokoro, C.E., Macedo, M.S., and Abrahamsohn, I.A. (1996). Saponin adjuvant primes for a dominant interleukin-10 production to ovalbumin and to

Trypanosoma cruzi antigen. Immunology *89*, 368–374.

Tewary, P., Sukumaran, B., Saxena, S., and Madhubala, R. (2004). Immunostimulatory oligodeoxynucleotides are potent enhancers of protective immunity in mice immunized with recombinant ORFF leishmanial antigen. Vaccine *22*, 3053–3060.

Tonui, W.K., Mbati, P.A., Anjili, C.O., Orago, A.S., Turco, S.J., Githure, J.I., and Koech, D.K. (2001). Transmission blocking vaccine studies in leishmaniasis: I. Lipophosphoglycan is a promising transmission blocking vaccine molecule against cutaneous leishmaniasis. East Afr. Med. J. *78*, 84–89.

Tonui, W.K., Mejia, J.S., Hochberg, L., Mbow, M.L., Ryan, J.R., Chan, A.S., Martin, S.K., and Titus, R.G. (2004). Immunization with *Leishmania major* exogenous antigens protects susceptible BALB/c mice against challenge infection with L. major. Infect. Immun. *72*, 5654–5661.

Trapani, J.A., and Smyth, M.J. (2002). Functional significance of the perforin/granzyme cell death pathway. Nat Rev Immunol 2, 735–747.

Trigo, J., Abbehusen, M., Netto, E.M., Nakatani, M., Pedral-Sampaio, G., de Jesus, R.S., Goto, Y., Guderian, J., Howard, R.F., and Reed, S.G. (2010). Treatment of canine visceral leishmaniasis by the vaccine Leish-111f+MPL-SE. Vaccine *28*, 3333–3340.

Ueno, N., and Wilson, M.E. (2012). Receptor-mediated phagocytosis of *Leishmania*: implications for intracellular survival. Trends Parasitol. *28*, 335–344.

Valenzuela, J.G., Garfield, M., Rowton, E.D., and Pham, V.M. (2004). Identification of the most abundant secreted proteins from the salivary glands of the sand fly Lutzomyia longipalpis, vector of *Leishmania chagasi*. J. Exp. Biol. *207*, 3717–3729.

Velez, I.D., Gilchrist, K., Arbelaez, M.P., Rojas, C.A., Puerta, J.A., Antunes, C.M., Zicker, F., and Modabber, F. (2005). Failure of a killed *Leishmania amazonensis* vaccine against American cutaneous leishmaniasis in Colombia. Trans. R. Soc. Trop. Med. Hyg. *99*, 593–598.

Velez, I.D., Gilchrist, K., Martinez, S., Ramirez-Pineda, J.R., Ashman, J.A., Alves, F.P., Coler, R.N., Bogatzki, L.Y., Kahn, S.J., Beckmann, A.M., *et al.* (2009). Safety and immunogenicity of a defined vaccine for the prevention of cutaneous leishmaniasis. Vaccine *28*, 329–337.

Veras, P., Brodskyn, C., Balestieri, F., Freitas, L., Ramos, A., Queiroz, A., Barral, A., Beverley, S., and Barral-Netto, M. (1999). A dhfr-ts- *Leishmania major* knockout mutant cross-protects against *Leishmania amazonensis*. Mem. Inst. Oswaldo Cruz *94*, 491–496.

Vinet, A.F., Fukuda, M., Turco, S.J., and Descoteaux, A. (2009). The *Leishmania donovani* lipophosphoglycan excludes the vesicular proton-ATPase from phagosomes by impairing the recruitment of synaptotagmin V. PLoS Pathog. *5*, e1000628.

Wakil, A.E., Wang, Z.E., Ryan, J.C., Fowell, D.J., and Locksley, R.M. (1998). Interferon gamma derived from CD4(+) T-cells is sufficient to mediate T-helper cell type 1 development. J. Exp. Med. *188*, 1651–1656.

Webb, J.R., Campos-Neto, A., Ovendale, P.J., Martin, T.I., Stromberg, E.J., Badaro, R., and Reed, S.G. (1998). Human and murine immune responses to a novel *Leishmania major* recombinant protein encoded by members of a multicopy gene family. Infect. Immun. *66*, 3279–3289.

Weng, N.P., Araki, Y., and Subedi, K. (2012). The molecular basis of the memory T-cell response: differential gene expression and its epigenetic regulation. Nat. Rev. Immunol. *12*, 306–315.

WHO (2010). Control of the Leishmaniases. Report of a meeting of the WHO Expert Committee on the Control of Leishmaniases, Geneva, 22–26 March 2010. *949*, 1–186.

Xin, L., Li, K., and Soong, L. (2008). Down-regulation of dendritic cell signaling pathways by *Leishmania amazonensis* amastigotes. Mol Immunol *45*, 3371–3382.

van Zandbergen, G., Hermann, N., Laufs, H., Solbach, W., and Laskay, T. (2002). *Leishmania* promastigotes release a granulocyte chemotactic factor and induce interleukin-8 release but inhibit gamma interferon-inducible protein 10 production by neutrophil granulocytes. Infect. Immun. *70*, 4177–4184.

Inhibitors of DNA Topoisomerases as Potential Anti-leishmanial Agents

11

Sayan Chowdhury and Hemanta K. Majumder

Abstract

Leishmaniasis is a dynamic disease which is gradually spreading, with the high mortality rates and involved in the economic loss resulting from morbidity. The disease is mainly prevalent in the tropical and subtropical areas. The enormous development of molecular and cellular biology in recent times have provided opportunities for discovering newer molecular targets for drug designing, which now forms a rational basis for the development of improved anti-parasitic therapy. The diversity found in the life cycle of these organisms must be directed by genetic events, wherein topoisomerases play an important role in cellular processes affecting the topology and organization of intracellular DNA. Recently, emergence of the bi-subunit topoisomerase I in the kinetoplastid family has brought a new twist in topoisomerase research related to evolution, functional conservation and as a potential target that can be exploited in drug designing and development of new intervention strategies. Therefore, understanding the biology of kinetoplastid topoisomerases and the components and steps involved in this intricate process provide opportunities for target based drug designing against protozoan parasitic diseases. This review summarizes the biology of kinetoplastid topoisomerases, which are the key molecular targets in anti-leishmanial chemotherapy.

Introduction

Effective anti-parasitic chemotherapy requires proper identification of a suitable molecule or a cellular biochemical process that can be targeted by developing therapeutic agents in a selectively and exploitable approach. Discovery of key molecules in life processes like replication, transcription, translation, chromosome segregation etc. created new arena for targeted chemotherapy against several life-threatening diseases. In the early of 1970s, such a discovery was made when James C. Wang missed negative supercoiling in one of his *Escherichia coli* cell preparations. Though he stated the protein, he isolated for removing supercoiling as ω protein, later it was named as topoisomerase I. In its physiological state in the chromosome, the DNA helix is supercoiled and tightly packed into chromatin. Three decades ago, when the first DNA topoisomerase was discovered in extracts of *E. coli* cells (Wang, 1971), it was thought that DNA topoisomerase is required only for DNA replication. As the replication fork moves along, the upstream region should lead to overwinding of the DNA than in other regions. However, the bacterial enzyme as discovered, relaxes preferentially underwound or negatively supercoiled DNA. In 1976, the scientists discovered another topoisomerase with two special functions, i.e. one is like topoisomerase activity and other one is DNA-dependent ATPase. The enzymatic properties of DNA gyrase helped the researchers to solve concurrent problems associated with replication and segregation. First, unlike *E. coli* DNA topoisomerase I, gyrase is capable of removing positive supercoils in the presence of ATP (Gellert *et al.*, 1976) and thus appears to be more suitable in solving the problem of DNA overwinding during semiconservative replication. Although discovery of this 'cellular magicians' was made in 1971, the researches were not focused on the significance of these enzymes as potential therapeutic targets until 1980s. Over the past three decades, there has been an immense scientific research on topoisomerase mechanism of action with special interest on the development of compounds that can selectively inhibit this enzyme.

These DNA topoisomerases are ubiquitous enzymes that play a pivotal role in governing the dynamic nature of DNA secondary and higher order structures and thus solve topological hitch inside cells. These functions relate mainly to nucleic acid metabolism – namely replication, transcription, recombination and repair. DNA topoisomerases are DNA manipulators that either relieve the torsional strain in DNA or introduce torsional strain by introducing transient breaks in the nucleic acid backbone (Kikuchi and Asai, 1984; Stewart *et al.*, 1998). The mechanism by which these enzymes alter the DNA topology involves three major steps:

1. binding of an enzyme to the substrate DNA;
2. cleavage by trans-esterification reaction accompanied by the formation of a transient phosphodiester bond between a tyrosine residue in the protein and one of the ends of the broken strand and subsequent strand passage through the break leading to change in the linking number;
3. strand religation and release of the enzyme as the DNA is religated.

Based on their mechanism of action topoisomerases are broadly classified into two evolutionary independent families: Type I and Type II. Topoisomerases that cleave single strand DNA and reseal them are called as Type I topoisomerase. Type I topoisomerases can be categorized on the basis of their attachment to phosphodiester linkage. In the type IA the protein is linked to a 5′ phosphate in the DNA, and family members include *Escherichia coli* topoisomerase I and topoisomerase III, and *Saccharomyces cerevisiae* topoisomerase III and reverse gyrase. The prototype of type IB or I-3′ enzymes are found in all eukaryotes, in addition to vaccinia virus topoisomerase I where the protein is attached to a 3′ phosphate (Champoux, 2001). Though essentially similar in their action, these enzymes have a broader specificity than that of the *E. coli* enzymes. Despite the differences in the mechanism and specificity between the bacterial and eukaryotic enzymes, the yeast DNA topoisomerase I has been shown to functionally complement a bacteria mutant in DNA topoisomerase I (Bjornsti and Wang, 1987). A certain degree of divergence also exists in the substrate preference, cofactor requirement and subunit composition of different topoisomerase families. Type IA topoisomerases are able to relax only negatively supercoiled DNA, and require magnesium and single-stranded stretches of DNA for their function. However, topoisomerases IB are able to relax both positively and negatively supercoiled DNA with equal efficiency and do not require a single stranded region of DNA or metal ions for function (Stewart *et al.*, 1997).

The type II family includes *E. coli* DNA gyrase, *E. coli* topoisomerase IV (par E), all known eukaryotic type II topoisomerases and archaic topoisomerase VI. Type II enzymes are homodimeric (eukaryotic topoisomerase II) or tetrameric (gyrase), cleaving both strands of a duplex that changes in linking number in steps of two. The current mechanistic model for reactions catalysed by topoisomerase II involves the binding of two segments of DNA: a G (gate) segment, which is cleaved in both strands by the enzyme with the formation of an ester bond between active tyrosines and 5′-phosphates in the DNA; and a T (transport) segment, which is captured by an ATP-operated clamp and passed through the enzyme stabilized break in the G segment (Berger, 1998; Bakshi *et al.*, 2001).

The discovery of several new DNA topoisomerases has brought a deeper understanding of their important roles in living cells. The biological functions of DNA topoisomerases are deeply rooted in the double-helical structure of DNA, and the selection of double-stranded DNA as substrate has set the stage for their entrance (Wang, 1991). A broad classification of the different types of topoisomerases in different organisms is represented in Table 11.1.

Because DNA topoisomerases play key roles in cellular processes, affecting the topology and organization of intracellular DNA, it is

Table 11.1 Classification of type I and type II DNA topoisomerases from different species

Subfamily	Representative members
IA	Bacterial DNA topoisomerase I & II Yeast DNA topoisomerase III DNA topoisomerase IIIα and III Mammalian DNA topoisomerase IIIα and III
IB	Vaccinia and Pox virus monomeric topo I Kinetoplastida bi-subunit topoisomerase I Mammalian mitochondrial topoisomerase I Eukaryotic monomeric topoisomerase I
IIA	Bacterial gyrase, DNA topoisomerase IV Phage T4 DNA topoisomerase Yeast DNA topoisomerase II Drosophila DNA topoisomerase II Mammalian DNA topoisomerase IIα and IIβ
IIB	*Sulfolobus shibate* DNA topoisomerase VI (subunit A homologous to yeast SP011)

important to define the physiological functions and understand the molecular basis of their action. Moreover, beyond their normal cellular activities, these enzymes are proven molecular targets for clinically useful anti-tumour (Broxterman and Georgopapadakou, 2011; Nitiss, 2002; Denny and Baguley, 2003) and antimicrobial drugs (Shapiro, 1993; Bodley *et al.*, 1995; Bearden and Danziger, 2001). Under normal conditions, the covalent enzyme DNA cleavable complexes are fleeting catalytic intermediates and are present in low steady-state concentrations, which cells can tolerate. However, conditions that significantly decrease or increase the physiological concentrations of these breaks unleash a myriad of deleterious side effects, including mutations, insertions, deletions and chromosomal aberrations (Ferguson and Baguley, 1994). Thus, all topoisomerases are fundamentally dualistic in nature, catalysing essential cellular reactions and possessing an inherent dark side capable of inflicting great harm to the genome of an organism. For these reasons DNA topoisomerases have been recognized as potential chemotherapeutic targets for anti-tumours and antiparasitic agents (Liu, 1989; Burri *et al.*, 1996). In this context, work on parasite topoisomerases has been a growing focus of interest, particularly as toxic chemotherapy, increasing drug resistance of some parasite strains to classical drugs, and co-infection of *Leishmania* with HIV, have made parasites a severe threat to public health in developing countries. Development of vaccines is still under trial and improved therapy is desirable.

Kinetoplastid protozoans express two different kinds of topoisomerases

Topoisomerase I

Type I DNA topoisomerases were isolated from *Leishmania donovani* (Chakraborty and Majumder, 1988; Chakraborty *et al.*, 1993), *Trypanosoma cruzi* (Riou *et al.*, 1983) and *Crithidia fasciculata* (Melendy and Ray, 1987). The purified active enzymes (65–79 kDa) were ATP independent and found to be sensitive to the topoisomerase I-specific inhibitor camptothecin (Melendy and Ray, 1987). Although immunolocalization studies for C. *Fasciculata* topoisomerase I showed that it is situated in the nucleus rather than in the kinetoplast (Melendy and Ray, 1987), it has been demonstrated in trypanosomes that camptothecin treatment induces kDNA minicircle cleavage (Bodley and Shapiro, 1995). This observation suggests the possible existence of topoisomerase I in the kinetoplast of trypanosomes. Recently, it has been shown that *L. donovani* topoisomerase I also exists in both the nucleus and kinetoplast (Das *et al.*, 2004).

All eukaryotic type IB topoisomerase are monomeric and consist of four domains (Wang, 2002). The unconserved amino terminal domain contains putative signals for nuclear localization of the enzyme and is highly sensitive to proteolysis and dispensable for *in vitro* activity (Stewart *et al.*, 1996). The largest core domain is essential for enzyme activity and shows high phylogenic conservation, particularly in the amino acid residues interacting closely with DNA. The third domain is known as the linker, which is poorly understood and is variable in length. Finally, the carboxy-terminal domain is highly conserved and contains the SKINYL motif. Cleavage occurs by trans-esterification reaction involving nucleophilic attack by an active site tyrosine (Tyr 723 in human Topo I) on a DNA phosphodiester bond resulting in the formation of a covalent DNA 3′ phosphodiester linkage. In religation phase a similar trans-esterification reaction involves attack by the free DNA 5′ hydroxyl that releases the enzyme from the DNA (D'Arpa *et al.*, 1988; Liu and Miller, 1981).

The first DNA sequence of a topoisomerase I-like gene from the kinetoplastid *L. donovani* was reported by Broccoli *et al.* (1999). The deduced amino acid sequence of this gene showed an extensive degree of homology with the central core of other eukaryotic type IB topoisomerases, including several conserved motifs but having a variable C-terminus. Interestingly, this ORF exhibited considerable homology with the human topoisomerase I but was lacking the SKINYL motif that supplies the catalytic tyrosine. Villa and co-workers first reported the presence of a second subunit harbouring the SKINYL motif which when co-expressed with the larger subunit exhibits topoisomerase I activity *in vitro* (Villa *et al.*, 2003). The 636 amino acid long large subunit (LdTOP1L) is found on chromosome 34 and is closely homologous to the core domain of human topoisomerase I. LdTOPIS, in turn, the gene encoding for the small subunit, is found on *L. donovani* chromosome 4 and encodes for a 262-amino acid polypeptide with a predicted molecular mass of 28 kDa. The small subunit contains the phylogenetically conserved 'SKXXY' motif placed at the C-terminal domain of all type I DNA topoisomerases, which conserves a tyrosine residue playing a role in DNA cleavage (Fig. 11.1A). Bodley *et al.* (2003) have also identified a heterodimer topoisomerase I in *T. brucei* which is comprised of the 90 kDa large subunit and 36 kDa catalytic subunit. LdTOPIL shows about 54% identity with core subdomain of human topoisomerase I but less than 22% identity with the linker and the C-terminal domain. On the other hand, LdTOPIS shows 43.5% sequence identity with the C-terminal domain of human topoisomerase I, which includes alignment of conserved sequences surrounding the catalytic tyrosine residue. LdTOP1L also deviates significantly from human topoisomerase I at loop regions bounded by LdTOP1L residues Pro62-His63, Asp114-His118 and Pro314-Asp342 which do not share the conserved sequences. Overall, this similarity indicates that the structure and catalytic machinery of the two enzymes are highly conserved, despite the fact that one is monomer and the other one is heterodimer. Thus, a key question arises, what is the role of the small subunit (LdTOP1S) apart from supplying the catalytic tyrosine for trans-esterification. LdTOP1S contains a large non-conserved N-terminal extension (start-Met-Asn-210), enriched in serine residues which might be potential sites of phosphorylation. This may cause post-translational down-regulation of topoisomerase I, mediated by ubiquitin 26S proteasome pathway like in mammalian cells (Desai *et al.*, 2001).

Recent developments in the field of type I topoisomerases in kinetoplastids include the characterization of a heterodimeric type IB topoisomerase in *L. donovani* (Das, B.B. *et al.*, 2004), a new gene from *L. donovani* that contains the SKINYL motif containing the active site tyrosine implicated in DNA breakage and reunion. Das, B.B. *et al* (2004), described for the first time the *in vitro* reconstitution of the two recombinant proteins LdTOP1L and LdTOP1S corresponding to the large and small subunits The proteins were purified from bacterial extract and the activity was measured by plasmid DNA relaxation assay. LdTOP1L and LdTOP1S form a direct 1:1 heterodimer complex through protein–protein interaction. Under standard relaxation assay condition (50 mM KCl and 10 mM Mg^{2+}) reconstituted enzyme (LdTOP1LS) showed reduced processivity as well as 2-fold reduced affinity for DNA compared to eukaryotic monomeric rat liver topoisomerase I. Cleavage assay at various salt

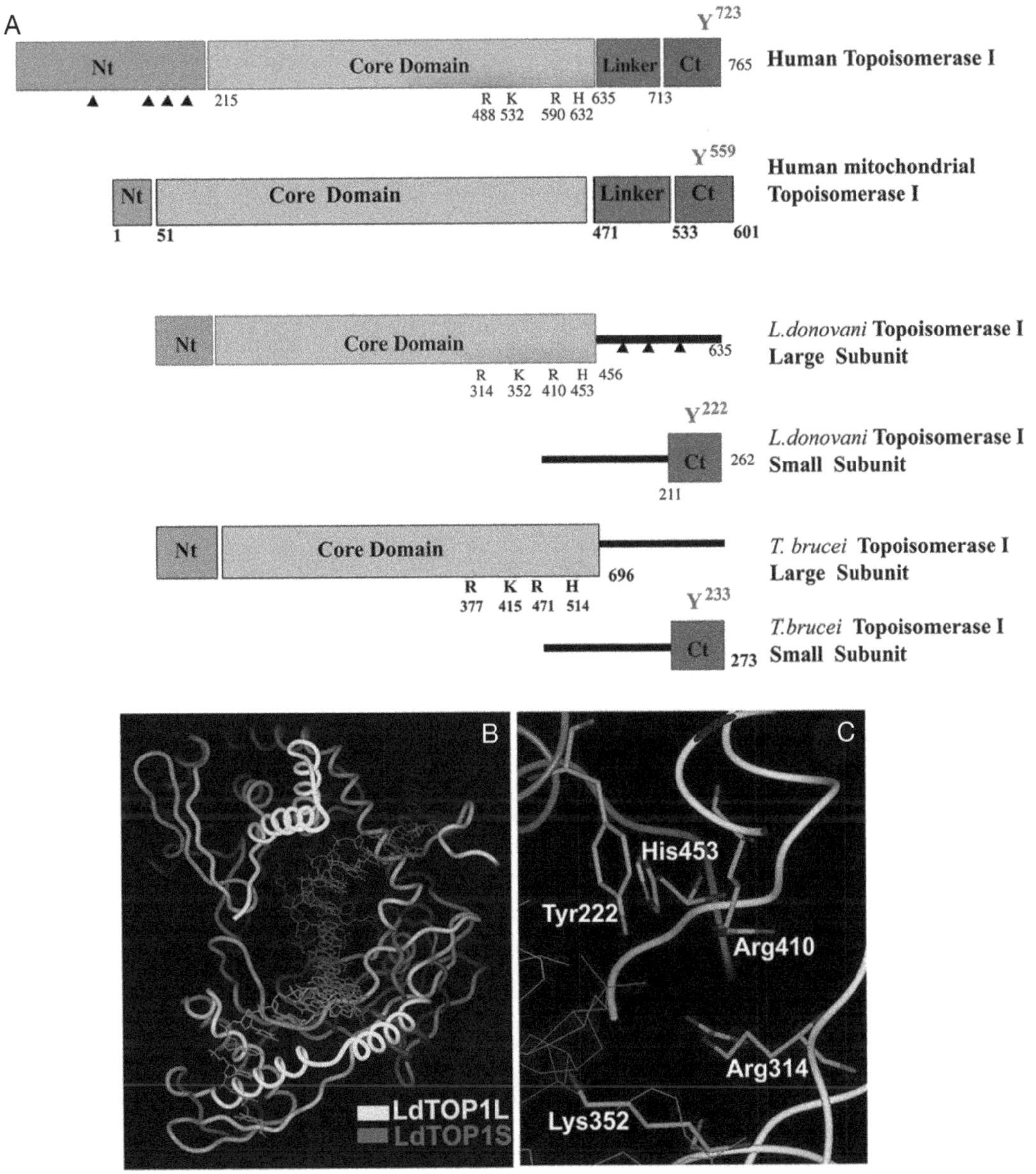

Figure 11.1 (A) Schematic representation of the domain organization of monomeric human topoisomerase I and *L. donovani* heterodimeric topoisomerase I. The alignments of the catalytic residues for both the enzyme are indicated in the figure. (B) Ribbon representation of truncated LdTOP1L/LdTOP1S protein with nicked DNA. Model generated on the coordinates of crystal structure of *L. donovani* topoisomerase I by Davies *et al*., (2006). (C) The conserved catalytic residues harbouring in LdTOP1L (light) and LdTOP1S (dark) are indicated in ball and stick representation and the nicked DNA substrate in blue. Figure is taken from Das *et al*. (2006) with permission.

concentrations reveal that camptothecin (CPT) enhanced the formation of 'cleavable complex' at low salt. Interaction between the two subunits leading to the formation of an active complex could be explored as an insight for development of new therapeutic agents with specific selectivity. This observation leads to the concept that, non covalent interaction of both subunits is necessary for the activity. This was further evidenced from the charge difference of the two subunits.

LdTOP1L has pI of 9.47 while that of LdTOP1S is 5.27. This charge difference clearly shows that these individual subunits are unstable until they interact with one another in the presence of salt. Recent findings by Das *et al.* (2005) reveal that deletion of 99 amino acids from the N-terminus of LdTOP1L results in a protein which failed to interact with the smaller subunit. This could be attributed to the presence of many polar residues in this region. Polar interactions are common between the subunits of heterocomplex proteins. The overall charge difference for the large and small subunits in conjunction with the unusual salt sensitivity of the parasite protein suggests that ionic interactions are important for holding the subunits together.

Moreover it was established that silencing of one subunit in *T. brucei* causes the coordinate loss of both subunits of DNA topoisomerase I as well as results in a rapid reduction in the synthesis of both DNA and RNA of kinetoplastid parasites (Bakshi and Shapiro, 2004). Das *et al.* (2005), also reveals that deletion of 39 amino acids from the N-terminus of LdTOP1L results in a protein with decreased cleavage activity and sensitivity to CPT. These data argued in favour of the interpretation that N-terminal amino acids of the large subunit regulates DNA dynamics during relaxation by controlling non-covalent DNA-binding or by coordinating DNA contacts by the other parts of the enzyme. Davies *et al.* (2006) have described a 2.27 Å crystal structure of an active truncated *L. donovani* TOPIL/TOPIS heterodimer bound to nicked double stranded DNA in the presence of vanadate (Fig. 11.1B). The structure predicts that the highly conserved constellation of the catalytic residues (Arg^{314}, Lys^{352}, Arg^{410} and His^{453} of LdTOP1L and the consensus catalytic residue Tyr^{222} in LdTOP1S) as shown in Fig. 11.1C, share a common module between *Leishmania* and human topoisomerase I. This study reveals that Arg^{410} residue of LdTOPIL (Arg^{590} in human topoisomerase I) which is positioned at 3.2 Å from the phenolic oxygen atom of Tyr^{222}, might play a pivotal role in activating Tyr^{222} residue of LdTOPIS (Tyr^{723} in human topoisomerase I) for attack on the scissile phosphate group. Moreover, it can also be predicted that Lys^{352} of LdTOPIL (Lys^{532} in human topoisomerase I) acts as the general acid in the cleavage reaction, while His^{453} of LdTOP1L close to the non-bridging oxygen atom of the vanadate, potentially mimics the transient state of enzyme–DNA covalent complex. Comparison of LdTOPILS to the structure of human topoisomerase I bound to DNA containing topotecan reveals that all of the amino acids that form the drug binding pocket are completely conserved between the two species.

Das, B.B. *et al.* (2004) showed that LdTOPILS was localized in both nucleus and kinetoplast of *L. donovani* (Fig. 11.2A). The existence of multiple localization signals have been mapped in the larger subunits of *Trypanosoma* and *Leishmania* topoisomerase I but no NLS has been found in smaller subunits of the enzyme (Das, B.B. *et al.*, 2004; Villa *et al.*, 2003). So it is likely that the subunits interact in the cytosol before nuclear and kinetoplast importation (Fig. 11.2B). But, whether the proteins perform separate functions in the cytoplasm is still unknown. A precise insight into the intracellular location of the expressed bi-subunit topoisomerase protein in *L. donovani* reveals that full-length LdTOPILS was localized in both nucleus and kinetoplast. In human, a type IB topoisomerase has been reported to act separately in the mitochondria (Zhang *et al.*, 2001). The enzyme is sensitive to CPT and exhibits 52% sequence identity with the nuclear topoisomerase I. The human mitochondrial topo I gene presumably arose by gene duplication and modification of the nuclear enzyme (Zhang *et al.*, 2001). Neither subunit contains a detectable mitochondrial targeting sequence but their localization in kinetoplast is an interesting finding. Direct evidence for the presence of a kinetoplast localized topoisomerase I in *T. brucei* was first provided by the loss of minicircles upon treatment with topoisomerase I poison CPT (Bodley and Shapiro, 1995). There are also reports of kDNA replication proteins like universal minicircle binding protein acting in the kinetoplasts that apparently lacks mitochondrial targeting sequences (Abu-Elneel *et al.*, 2001). In *Trypanosoma*, it has been shown that RNAi of any one of the subunits causes coordinated loss of both the subunit proteins (Bakshi and Shapiro, 2004). The loss of both the subunits could be due to progressive impairment of DNA metabolism because of the lack of topoisomerase

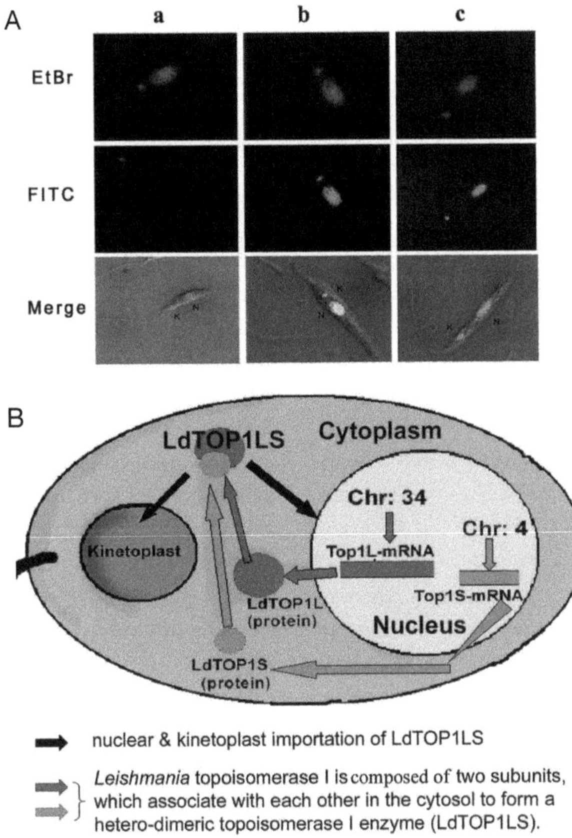

Figure 11.2 (A) Model representing the probable mechanism of synthesis and import of the bi- subunit *Leishmania* topoisomerase I. (B) Immunocytochemical localization of LdTOP1L and LdTOP1S. No fluorescence was observed when pre-immune serum was used as primary antibody (a). (b) Same as (a), but probed with anti-LdTOP1L. (c) Same as (a), but probed with anti-LdTOP1S primary antibody. Parasite cells were also stained with ethidium bromide to locate the nucleus and kinetoplast and the area of the overlapping FITC and ethidium bromide (EtBr) stain are shown in merged pictures. Cells were viewed at an original magnification of 100× under a Leica DM IRB inverted microscope. The nucleus (N) and kinetoplast (K) are indicated. Figure taken from Das *et al*. (2006).

activity. Alternatively, it could be hypothesized that individual proteins are stable only in association with each other and are highly unstable when separated. But the global loss of DNA biosynthesis precludes the estimation of any separate function for the topoisomerase subunits.

Topoisomerase II

Topoisomerase II activities have been purified from various kinetoplastid parasites (Das *et al.*, 2004). Topoisomerase II genes have also been cloned from *C. fasciculata, T. brucei, T. cruzi, L. donovani, L. infantum, L. chagasi* and *Bodo saltans* (Das *et al.*, 2004). The genes and proteins of the parasites were found to be smaller compared to higher eukaryotes. No gyrase-like activity (capable of introducing supercoils into DNA) has been found and the enzymatic activities and the genes are more like other eukaryotic counterparts. The topoisomerase activity isolated from *C. fasciculata* was shown to be immunolocalized in kinetoplast (but the overexpressed proteins from *L. donovani* and *B. saltans* were found to be localized both in the nucleus and kinetoplast (Das *et al.*, 2001; Gaziova *et al.*, 2003). Although it can be argued that the differences in cellular localization might be explained in terms of which epitopes were available for the recognition by different anti sera, none the less the existence of another topoisomerase II sequence (hypothesized to be a mitochondrial topoisomerase II) cannot be dismissed. It is likely that replication of catenated kinetoplast DNA requires another topoisomerase activity. Though all type II A topoisomerases are identical in one way that they change the linking number of DNA in an ATP-dependent manner, the eukaryotic type II enzymes are homodimers, while their bacterial counterparts like gyrase and topo IV are A_2B_2 tetramers, the B and A subunits being the N and C-terminal halves of their eukaryotic counterparts (Lynn *et al.*, 1996). Interest in parasite type II topoisomerases and their genes and proteins gains impetus from the fact that they are the key enzymes involved in replication of the massive kinetoplast DNA network and RNAi of topoisomerase II leads to the progressive degradation of mitochondria in *Trypanosoma*. Kinetoplastid parasites diverged early in the eukaryotic evolution at the base of the evolutionary tree well before the emergence of kingdom Metazoa. In spite of having a similarity and identity of 31% and 23% with yeast topoisomerase II, LdTOP2 was found to complement a temperature sensitive mutant yeast strain (Sengupta *et al.*, 2003). Just like other eukaryotic topoisomerase II, the parasite enzyme can also be divided into an N-terminal ATPase, a central DNA-binding and an unconserved C-terminal domain. In spite of being unconserved, the nuclear localization signal and the dimerization domain of this homodimeric enzyme have been mapped in the C-terminus (Fig. 11.3) (Sengupta *et al.*, 2005). The C-terminus also contains a stretch of 60 amino acids not present in the human host. Therefore this region can be exploited to develop anti-leishmanial targets. The parasite enzyme has a greater affinity for DNA and was also stable at a very high salt concentration as compared to its human host (Sengupta *et al.*, 2005). These findings were quite consistent with the greater susceptibility of the parasite protein to the anti-topoisomerase II agents. This is because of the fact that an enzyme with more affinity towards DNA would perform more DNA cleavage and thus a greater chance of being trapped in that state by an anti-topoisomerase II drug. The N-terminal 385 amino acids residues of LdTOP2 were found to possess the ATPase activity. Although the ATPase activity resides in the first 385 amino acid residues, only a larger protein was found to mimic the full-length enzyme kinetics in *in vitro* assay (Sengupta *et al.*, 2005b). The study identifies specific amino acids like Asn^{65}, Asn^{69}, Asn^{96} and Asp^{130} of the parasite protein that are involved in the interaction with ATP and etoposide. In contrast, the ATPase domain of human topoisomerase IIα (1–453 amino acids) displays similar catalytic properties, in terms of ATP turnover, to that of the full-length enzyme, except for the fact that the smaller fragment (1–420 amino acids) fails to be hyperstimulated by DNA (Campbell and Maxwell, 2002). Most interestingly the ATPase activity of the N-terminal 385 amino acids of the parasite protein was also found to be inhibited by etoposide. Thus, etoposide, in addition to being a poison for the parasite enzyme is also a catalytic inhibitor of the enzyme (Sengupta *et al.*, 2005b; Campbell and Maxwell, 2002). The active site tyrosine implicated in DNA breakage and rejoining for *L. donovani* topoisomerase II has been mapped to be Tyr775 (Fig. 11.3A) (Sengupta *et al.*, 2005b). This tyrosine is the only residue in the parasite protein, which is involved in the trans-esterification reaction and is also homologous to the Tyr^{804} of human (Tsai-Pflugfelder *et al.*, 1988). Surprisingly, the C-terminal truncation mutants of the parasite protein fail to be inhibited by etoposide (Sengupta *et al.*, 2005b) compared to the full-length enzyme. Like the human enzyme, the core domain of LdTOP2 contains all the elements essential for sequence preference in protein–DNA interaction, but unlike the human enzyme, the C-terminus of the parasite protein plays an important role in the *in vitro* topoisomerase II cleavage reaction. It was observed earlier that over expression of human N-terminal domain in yeast confers resistance to high concentrations of etoposide (Vilian *et al.*, 2003). The proposed model of topoisomerase II with etoposide is shown in Fig 11.3B. The observed phenotype was proposed to be due to the competition of the excess of the N-terminal domain with the full length enzyme for a limiting pool of inhibitor. So future challenge in the parasite topoisomerase II would be to develop drug resistant parasite strains and to see what causes this resistance and also to check what effect the individual domains of the enzyme have on the drug protein interaction in the context of the full-length enzyme.

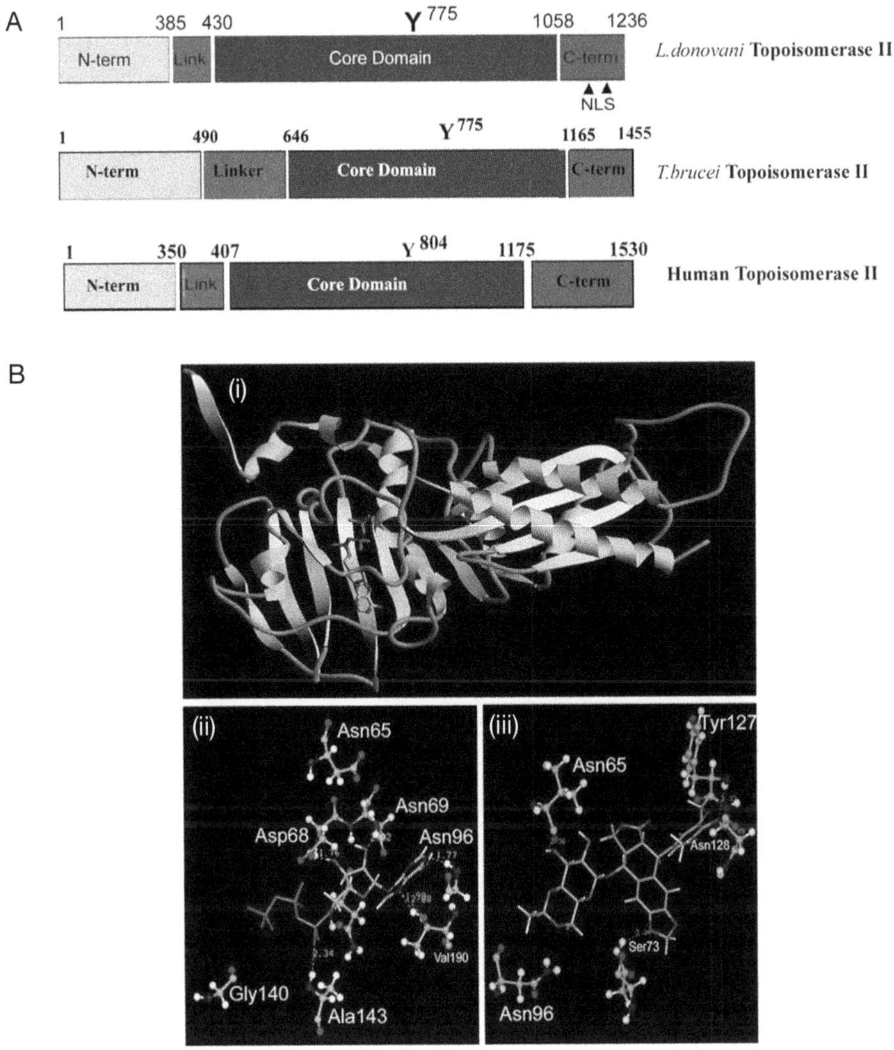

Figure 11.3 (A) Domain architecture of *L. donovani* topo II. (B) Ribbon representation of the 43 kDa ATPase domain of LdTOP2 with ATP depicting its secondary structures: (i) α-helices indicated in white, β-sheets in grey and random coils in dark grey. (ii) View of the model with etoposide. [iii] The amino acid side chains of the protein represented in ball and stick, ATP and etoposide in solid sticks. The H-bonds in dotted lines and bond distances in Å are shown. Figure taken from Das *et al*. (2006).

Evolutionary aspects of kinetoplastid topoisomerase I and topoisomerase II

Type IB topoisomerases including human topoisomerase I and vaccinia virus topoisomerase I share a common catalytic domain with the tyrosine recombinases that includes bacteriophage HP1 and certain phage integrases and the XerC/D, Cre and Flp recombinases (Corbett and Berger, 2004).

In prokaryotic tyrosine recombinases XerD, Cre, HP1 integrase and eukaryotic topoisomerases the active site residues RKHR (H/W)Y and RKRHY reside on a single monomer. In eukaryotic Flp recombinase and *Leishmania* heterodimeric topoisomerase I, the active site tyrosine is provided by one subunit and the other residue implicated in catalysis comes from another subunit (Chen and Rice, 2003). It may be possible that the active site in case of kinetoplastid topoisomerase I is created by the association of subunits similar to Flp recombinase. The structural diversity of the amino-terminal domain and the linker domain in various eukaryotic topoisomerase IB argues that they have probably evolved by independent gene fusion. Krogh and Shuman described that nuclear topoisomerase I enzymes evolved from the bacterial/poxvirus precursor (Krogh and Shuman, 2002). However, in contrast to vaccinia topoisomerase I the catalytic tyrosine of human topoisomerase I resides on a small 6 kDa domain and is connected to the central core domain by a 7 kDa coil–coil linker element (Champoux, 2001). It now seems simpler to posit that during evolution of eukaryotic IB topoisomerase, the kinetoplastid parasites (*Leishmania*, *Trypanosoma*) which arose earlier than multicellular eukaryotes gained some regulatory sequences (nuclear localization signals and sequences for CPT sensitivity) as split domain architecture.

Marcotte *et al.* (1999) demonstrated that the fusion of the subunits and insertion of additional domains allow for the rapid evolution of new signalling pathways by the incorporation of a novel interaction domain into a pre-existing peptide. Horizontal gene transfer has been shown to have an important role in the evolution of composite proteins (Simonson *et al.*, 2005). It is possible that horizontal gene transfer did play a key role in propelling the topoisomerase IB protein among the eukaryotes (Krogh and Shuman, 2002). Thus, we can suggest that eukaryotic monomeric type IB topoisomerase evolved from the common ancestral bi-subunit enzyme by fusion of the two subunits at genetic level (Fig. 11.4).

Eukaryotic type II topoisomerases are mainly involved in chromosome segregation and condensation via their decatenation activity. Topoisomerase II is available in two forms, topo IIA and topo IIB throughout the phylogeny which exhibits strikingly different distributions among the three domains of life (Gadelle *et al.*, 2003). The distribution of the enzyme among the kinetoplastid parasites is rather trivial. In the kinetoplastids, a mitochondrial topoisomerase II is well described, but little is known about a nuclear activity. To date, topoisomerase II activity has been purified from *T. cruzi*, *T. equiperdum* (Douc-Rasy *et al.*, 1986), *L. donovani* (Chakraborty and Majumder, 1987, 1991) and *C. fasciculata* (Shlomai *et al.*, 1984). The genes for the above enzymes exhibit high degree of homology among themselves. The enzyme from *L. donovani* is found to localize both in the nucleus and kinetoplast providing direct evidence for dual distribution of the same enzyme. Phylogenetic analysis reveals that the *L. donovani* topoisomerase II is closely related to the *C. fasciculata* (Das *et al.*, 2001) enzyme although the latter is predominantly localized in the kinetoplast (Melendy *et al.*, 1988). Recently, identification of distinct genes encoding nuclear and mitochondrial topoisomerase II in *T. brucei* (Kulikowicz and Shapiro, 2006) provides direct evidence for gene duplication in these lower eukaryotes.

Topoisomerases as therapeutic targets

Despite differences in catalytic mechanism and cellular functions, the critical feature of all topoisomerases is the DNA strand passage event. However, the ability to pass single or double-stranded segment of DNA freely through another comes with a heavy price; it requires enzymes that generate breaks in the genetic material. In an effort to maintain genomic integrity during this cleavage reaction, topoisomerases covalently attach to the newly generated DNA 3′ (eukaryotic topoisomerase I) or 5′ termini (all other topoisomerases) via phosphotyrosyl bonds. Under normal circumstances, these covalent enzyme–DNA cleavage complexes are transient catalytic intermediates and are present in low concentrations and consequently, they are tolerated by the cell. However, conditions that significantly increase the physiological concentrations cause deleterious side effects, including mutations, insertions, deletions and chromosomal aberrations. Thus, all topoisomerases are fundamentally dualistic in nature. Although they catalyse essential reactions in the cell, they possess an inherent dark side capable of inflicting great harm to the genome of an organism.

Classification of topoisomerase inhibitors

Topoisomerase-targeting therapeutics currently in use act by trapping the covalent enzyme–DNA

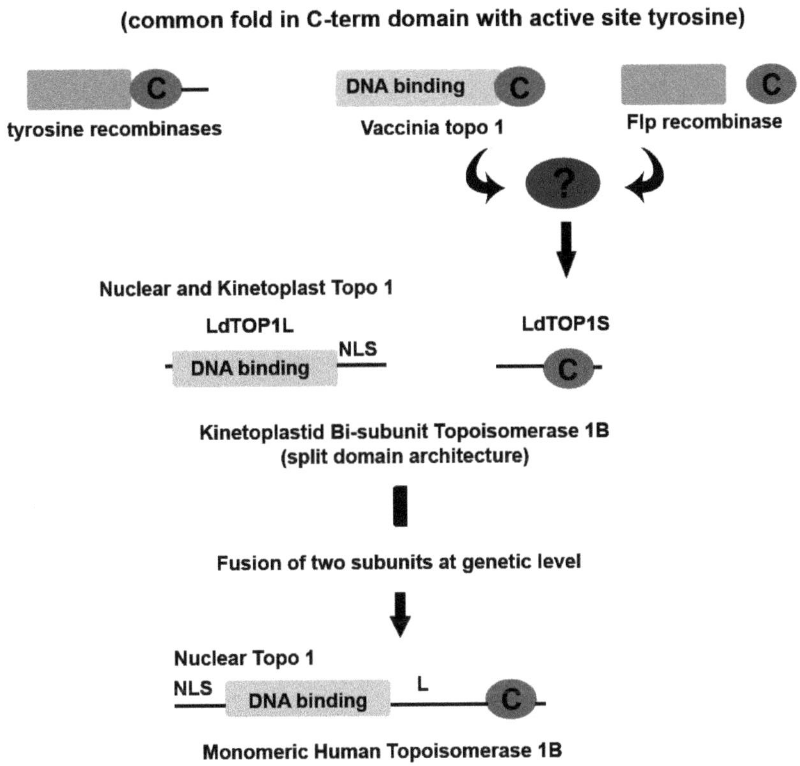

Figure 11.4 Model representing the probable route of evolution of eukaryotic type IB topoisomerase. Evolution from the common ancestral bi-subunit enzyme by fusion of the two subunits at genetic level. Figure modified from Das *et al.* (2006).

complexes of the first trans-esterification reaction. The known topoisomerase drugs can be divided into two classes, class I and class II (Liu, 1989; Wang, 1994). The class I drugs have been referred to as 'topoisomerase poison' where as the class two drugs are referred to as 'topoisomerase inhibitors'. The class I drug act by stabilizing the covalent topoisomerase-DNA covalent complexes. These include bacterial gyrase inhibitors quinolones, eukaryotic topoisomerase I inhibitor camptothecin and topoisomerase II inhibitors amsacrine, doxorubicin, etoposide and teniposide. The class II drugs interfere with catalytic function of DNA topoisomerase without trapping the covalent complexes (Fig. 11.5). These classes of drugs include the coumermycin family of antibiotics that act on bacterial gyrases, the eukaryotic DNA topoisomerase II inhibitor suramin, fostriecin, merbarone and bis-dioxopiperizines. Several inhibitors of eukaryotic topoisomerase I have also been reported. A major determinant of cytotoxicity for the class I drug is the conversion of a latent single- or double-stranded break in a drug–topoisomerase–DNA complex into an irreversible double-stranded break. Replication is the key cellular process that drives this conversion in case of the topoisomerase I drug camptothecin. However, for class II topoisomerase II drugs, processes other than replication might also be involved. Cell killing by class II topoisomerase II drugs may involve cell cycle progression through mitosis. Traversing of eukaryotic cells through mitosis in the absence of functional DNA topoisomerase II can lead to aneuploidy and chromosomal breakage. For class I drug, cytotoxicity increases with increasing cellular level of target enzyme where as for class II drugs opposite is true. Thus, increased levels of topoisomerases render cells hypersensitive to enzyme poisons but resistant to inhibitors. Conversely, decreased enzyme levels render cells resistant to poison but hypersensitive to inhibitors.

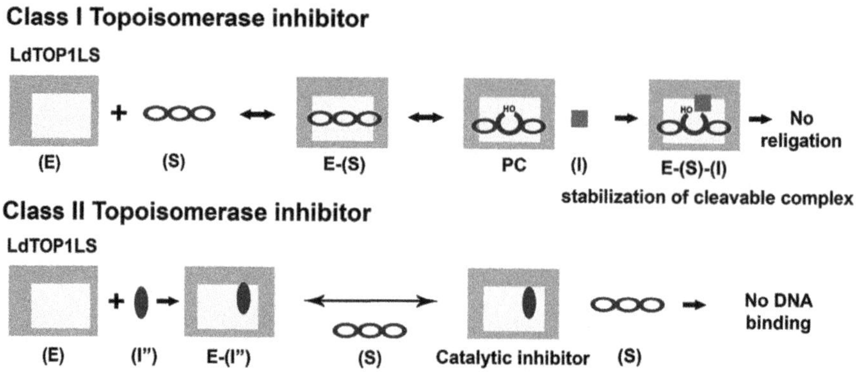

Figure 11.5 Schematic representation of mechanism of inhibition of bi-subunit topoisomerase I. Catalytic cycle of topoisomerase I is divided into DNA binding, cleavage and religation. (A) CPT, a Class1 inhibitor, binds to the enzyme-DNA post–cleavage complex (PC), and subsequently inhibits religation step, and thus stabilizes the catalytic intermediate, (B) DHBA a pentacyclic triterpenoid, a Class II inhibitor, binds in the catalytic site of the enzyme and prevents binding of the enzyme with the DNA. E- is reconstituted bi-subunit *L. donovani* topoisomerase I (LdTOP1LS), where the darker box represents the DNA binding large subunit (LdTOP1L) while the lighter small box is the catalytic subunit (LdTOP1S) harbouring SKXXY motif, S- substrate DNA (coiled), E-(S)- enzyme substrate complex, (I)- CPT, (I′)-DHBA. Figure modified from Das *et al.* (2007).

Topoisomerases as targets for antiparasitic agents

Sodium stibogluconate and ureastibamine, the two most potent and therapeutically used anti-leishmanial drugs have been reported by this laboratory to be specific inhibitors of *L. donovani* DNA topoisomerase I (Chakraborty and Majumder, 1988). Pentavalent antimonials, also used as anti-leishmanial drugs, have been found to stabilize cleavable complex with an ED_{50} of 16.7 μg/ml and 209.5 μg/ml for wild type and resistant strains respectively (Lucumi *et al.*, 1998). The structure of potential inhibitors of *Leishmania* topoisomerases are shown in Fig. 11.6.

Camptothecin

The first discovered and mostly studied topoisomerase poison is camptothecin (CPT). Camptothecin is a natural alkaloid isolated from the plant *Camptotheca acuminata*. As far as the chemical identity of camptothecin is concerned, it is a pentacyclic lactone. Under physiological condition, it equilibrates with opened-ring carboxylate configuration. The sodium salt of camptothecin carboxylate is highly soluble making it extensively studied for clinical purposes. CPT poisons topoisomerase IB in a non-competitive manner in micromolar and sub-micromolar range. It stabilizes the ternary complexes with topoisomerase IB and DNA, arresting DNA religation. This topoisomerase inhibiting property of CPT extends its utility from being anticancer drug to a potential agent for treating parasite disorders as well. It has been found that CPT generates covalent DNA–topoisomerase complex not only with human enzymes but with both nuclear and kinetoplast DNA from *Leishmania* as well.

Studies have rendered a thorough account on the mode of interaction of CPT with the *Leishmania* topoisomerase IB. Being structurally similar with a DNA base pair, CPT can successfully gets intercalated at the site of DNA cleavage. An aspartic acid residue located at the 353 position of the large subunit of *Leishmania* topoisomerase IB plays a crucial role in this interaction. This Asp^{353} forms a hydrogen bond with the 20(S)-hydroxyl moiety of the lactone form of CPT. This interaction is followed by another hydrogen bond establishment between nitrogen of CPT B-ring and an arginine residue located at 190th position. Apart from the roles of these two amino acid residues, asparagines located at 722nd position just adjacent to active site tyrosine, also assists in inhibition mediated by camptothecin. However, along with the individual amino–acids-CPT interactions, it has also been seen that a conserved FXGR motif present in the

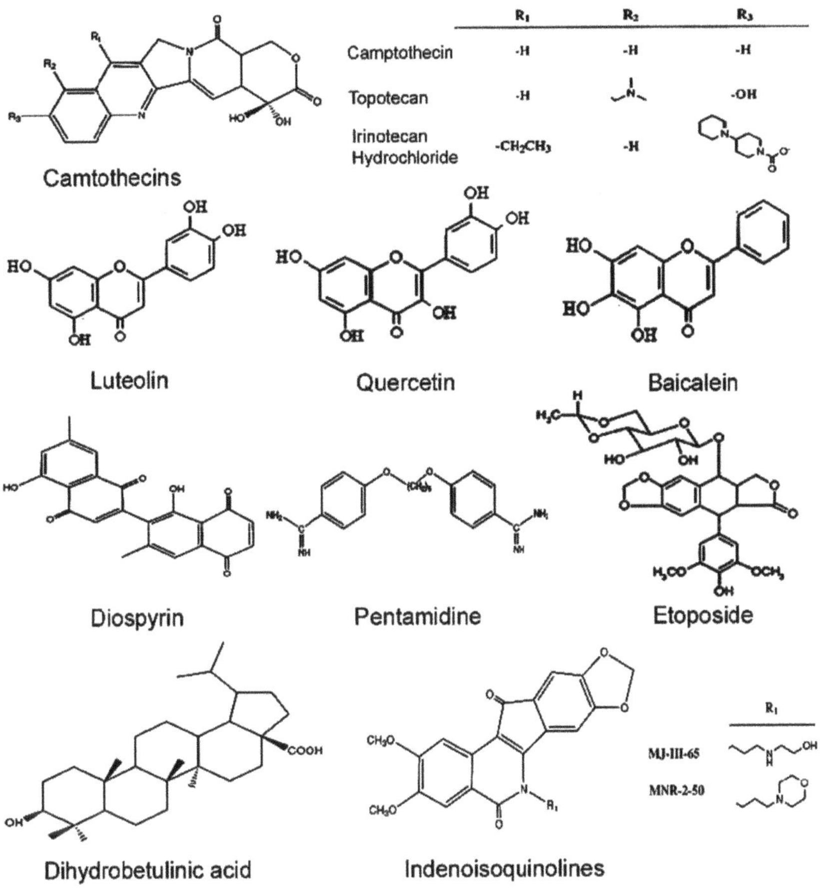

Figure 11.6 Structure of the potential *Leishmania* topoisomerases inhibitors.

core domain of *Leishmania* topoisomerase is also an important factor that mediates drug–enzyme interaction (Redinbo *et al.*, 1999).

CPT also hinders or blocks DNA rotation, which is evidenced by the crystal structure of the ternary complex between human topoisomerase I (topo 70) covalently linked to the DNA and the CPT derivative topotecan (Wang, 2002). Recent finding reveals that a highly CPT resistant *L. donovani* strain (LdRCPT.160) developed by stepwise exposure to CPT induces point mutations ($Gly^{185}Arg$ and $Asp^{325}Glu$) in the large subunit (LdTOP1L) of the bi-subunit topoisomerase I. The mutant enzyme shows reduced activity as well as reduced sensitivity towards CPT (Marquis *et al.*, 2005). The cytotoxicity of 9-substituted-10, 11-methylenedioxy analogues of camptothecin correlate well with cleavable complex formation in the nucleus and kinetoplast, and structural motifs have been identified that disproportionately increase toxicity to parasites, compared with mammalian cells. Sen *et al.* (2004a,b) have demonstrated that CPT induces programmed cell death (PCD) in both the amastigote and promastigote forms of *L. donovani*.

Drugs targeted to topoisomerase IB of kinetoplastid parasites

A huge array of compounds has been studied till date for their efficacy to inhibit topoisomerase IB of *Leishmania* by different group of researchers across the world. Derivatives of betulin and dihydrobetulin (Chowdhury *et al.*, 2011), 3,3′ di-indolylmethane (DIM) (Roy *et al.*, 2008), derivatives of DIM, camptothecin (CPT), baicalein, quercetin, etc. are some of such examples. However, the mechanisms of their action, level of toxicity towards the enzyme vary in case to case basis. We have shown that the flavonoids quercetin and luteolin, isolated from *Vitex nigundo*, have potent anti-leishmanial effect (Das *et al.*, 2006). The flavonoids inhibited the growth of *L. donovani* promastigotes and amastigotes *in vitro* and also promoted topoisomerase II-mediated linearization of kDNA minicircles. They arrest cell cycle progression in *L. donovani* promastigotes leading to apoptosis and reduced parasite burden in animal models (Mittra *et al.*, 2000). Recently, Das *et al.* (2006), described that naturally occurring flavones baicalein, luteolin and quercetin are potent inhibitors of the recombinant *L. donovani* topoisomerase IB. These compounds bind to the free enzyme and also intercalate into the DNA at a very high concentration (300 μM) without binding to the minor groove of DNA. The inhibition of topoisomerase I by these flavones is due to stabilization of topoisomerase I–DNA-cleavage complexes, which subsequently inhibit the religation step. Their ability to stabilize the covalent topoisomerase I–DNA complex *in vitro* and in living cells is similar to that of the known topoisomerase I inhibitor camptothecin (CPT). However, in contrast to CPT, baicalein and luteolin failed to inhibit the religation step when the drugs were added to preformed enzyme substrate binary complex. The most interesting part of the study reveals that baicalein and luteolin stabilize duplex oligonucleotide cleavage with CPT-resistant mutant enzyme LdTOP1Δ39LS lacking 1–39 amino acids of the large subunit (Das *et al.*, 2005). This observation was further supported by the stabilization of *in vivo* cleavable complex by baicalein and luteolin with highly CPT-resistant *L. donovani* strain. Thus, the interacting amino acid residues of *L. donovani* topoisomerase I may be partially overlapping or different for flavones and CPT. Drugs like CPT, DIM (Roy *et al.*, 2008), niranthin (Chowdhury *et al.*, 2012), lyoniside, saracoside (Saha *et al.*, 2013), etc. stabilize enzyme-DNA cleavable complex and subsequently arrest the religation step. On the contrary, betulin derivatives do not stabilize the cleavable complex and subsequently slow down the strand rotation step. Another point of concern is that, a particular drug would be of no use for anti-leishmanial therapy if it attacks both pathogen and host, i.e. human system. The human topoisomerase should either be completely resistant to the drug or the level of tolerance should be higher than that of *Leishmania* enzyme. It has been shown by Chowdhury *et al.* (2011) that novel derivatives of betulin and dihydrobetulin could successfully inhibit the activity of *L. donovani* topoisomerase IB in much lower concentration as compared to human topoisomerase I. Therefore, this particular class of compound could be further probed in for development of anti-leishmanial therapeutics. The basic take home message is that, all the drugs which possess topoisomerase IB inhibition property are either topoisomerase poisons which act by stabilizing the enzyme-DNA cleavable complex or they interfere with the catalytic activity of the enzyme without showing any cleavable complex stabilization.

It is also noteworthy that fatty acids and their derivatives have been found to possess antiprotozoal activity. A significant contribution in case of fatty acid mediated parasite killing has been made by Chaudhuri *et al.* (1986). They reported that antiprotozoal activity is proportional to the chain length of the fatty acid and level of toxicity increases with amount of unsaturation present. Keeping the chain length fixed in 18 carbon atoms oleic acid results in more toxicity as compared to octadecanoic acid. This is attributed to the presence of monounsaturation present in oleic acid. Again, C_{10} saturated fatty acid, i.e. decanoic acid, or capric acid is less potent than C_{18} saturated fatty acid, i.e. octadecanoic acid or stearic acid. Similar observations are found in case of acetylinic fatty acids. Icosynoic acid and icosenoic acid are more potent as compared to heptadecynoic acid and heptadecenoic acid. As far as inhibition of topoisomerase IB of *L. donovani* is concerned, saturated fatty acids do not show inhibitory effect as well. If the chain length is kept fixed, unsaturated fatty acids show more potency to inhibit the relaxation activity of *L. donovani* topoisomerase IB (Carballeira *et al.*, 2009). Again if nature of unsaturation is kept constant, then the enzyme inhibition efficiency increases with the increase in chain length. The phenomenon can be made clear through some examples. Icosynoic acid and icosenoic acid are more potent inhibitors of topoisomerase IB than heptadecynoic acid and heptadecenoic acid. This fact is again supported by 2-alkynoic acids. With respect to topoisomerase IB inhibiting effect, 2-octadecynoic acid shows significant action followed by 2-hexadecynoic acid and 2-tetradecynoic acid (Carballeira *et al.*, 2011). Therefore, both the carbon chain length and unsaturation present in the molecule are two important factors that govern the capability of fatty acids to inhibit the relaxation activity of topoisomerase IB of parasite. Again in the case of 2-methoxy-heptadecanoic acid, 2-methoxy-heptadecenoic acid, and 2-methoxy-heptadecynoic acid, the third compound, i.e. 2-methoxy-heptadecynoic acid shows the highest *L. donovani* topoisomerase IB inhibition followed by the second one, i.e. 2-methoxy-heptadecenoic acid. The first compound, i.e. 2-methoxy-heptadecanoic acid, however, does not have inhibition effect. Therefore, it is quite apparent that unsaturation is a major player for effective inhibition of the enzyme. It can be attributed to the weak intermolecular interactions between the active site of the enzyme and the presence of unsaturated bonds in the fatty acid. It has been found that C-6 triple bond coupled with α-methoxylation enhances the inhibition of C_{17} fatty acids towards topoisomerase IB of *L. donovani*. However, α-methoxylation decreases the toxicity of fatty acids towards *L. donovani* promastigotes but improves selectivity index (Carballeira *et al.*, 2012a). But as far as the geometry of the methoxylated analogues of the fatty acid mentioned here is concerned, 2-methoxy-heptadecanoic acid and 2-methoxy-heptadecynoic acid are not significantly different from each other. In this specific instance, intermolecular cation-π interaction may be the player near the binding sites of the fatty acid with the topoisomerase IB. Ammonium groups in the amino acids, particularly lysine and arginine, present in the enzyme active site might be interacting with double and triple bonds of these methoxylated fatty acids. Other factor like presence of a cis-cyclopropane group in a fatty acid molecule also reduces the enzyme inhibition potential of the fatty acid. Even topoisomerase IB not only from *L. donovani* but also from *L. infantum* has been found to be susceptible towards some compounds. Indenoisoquinolines compounds, viz. indotecan and AM13–55, are two such compounds which inhibit relaxation activity of *L. infantum* topoisomerase IB by stabilizing enzyme-DNA cleavable complex (Fouce *et al.*, 2012). Hence, they show mechanism similar to that of camptothecin. And more interestingly, these compounds show inhibition of human topoisomerase IB at much higher concentrations as compared to parasite enzyme.

Modification of core structure of known inhibitors has also been a strategy to develop novel drugs as well as drugs with improved efficacy. Benzimidazole and anthraquinone derivatives possess eukaryotic topoisomerase I inhibition property (Alper *et al.*, 2003; Zagotto *et al.*, 2000). Derivatives obtained through structural alterations made in mother compound result in more efficient compound with respect to enzyme inhibition. It has been reported that fusion of imidazole group with some anthraquinone derivatives results in improved *L. donovani* topoisomerase IB inhibitory activity (Chaudhuri *et al.*, 2007). Among such nine derivatives tested, tetrazole and morpholine derivative compounds showed remarkable potency. However, dialkylamine and piperazine derivative compounds were not significant enough. The investigation showed the correlation between the pK_a and electronegativity of the side-chain nitrogen with topoisomerase inhibiting properties. In order to achieve this side chains added with the aromatic core were varied. It is worth mentioning that p*K*a of the side chain nitrogen is an important factor to control enzyme inhibition potential of the compounds. This finding strongly supports the fact that presence of hydrogen bond acceptors on the side chain amine is favourable for enzyme inhibiting power (Nagarajan *et al.*, 2006). The underlying mechanism in this case, as hypothesized, is formation of hydrogen bond with positively charged amino acid residues in the ternary cleavable complex (Davis *et al.*, 2006).

However, finding out some novel compounds with topoisomerase IB inhibition property is not the ultimate success. Another issue which has elicited concern of the researchers is the emergence of drug resistant parasites. Therefore it has to be taken care of that novel compounds identified through extensive research should be equally potent against the topoisomerase IB enzyme of the drug resistant parasites. Betulin derivatives mentioned herein were found to be equally efficient in reducing the parasitic burden from macrophages infected with wild type as well as sodium–antimony gluconate-resistant parasite.

Topoisomerase II

Topoisomerase II, being a crucial enzyme in all organisms, is therefore a very important drug target as far as anticancer, antibacterial and even antiparasitic therapeutics are concerned. In this section, exploitation of topoisomerase II targeted drugs for anti-leishmanial therapeutics will be thoroughly discussed.

Since topoisomerase II plays an important role in many cellular events, targeting this particular enzyme has an important implication in development of antiproliferative therapeutics. Many of the anti-tumour drugs are targeted to topoisomerase II. There are several pathways by which these drugs interfere with the enzyme. A fundamental difference exists in the DNA intercalation property of such drugs (DNA intercalators: ellipticine, amiloside and DNA non-intercalators: etoposide, teniposide). Although these two categories of drugs have a basic distinction with respect to DNA intercalation property but they share a common characteristic of interfering with the cleavage/religation step of the catalytic cycle. The ultimate goal of their action is to stabilize the cleavable complex which subsequently leads to accumulation of double-strand breaks. However, it is attained either by enhancing cleavage reaction (e.g. ellipticine) (Froelich-Ammon *et al.*, 1995) or by inhibiting religation (e.g. etoposide) (Robinson *et al.*, 1991). These topoisomerase II targeted drugs not only hamper the cleavage step but also inhibit ATP hydrolysis (Robinson *et al.*, 1993) and strand passage (Chen *et al.*, 1984). All these drugs are ultimately targeted to generation of lethal double-stranded breaks, subsequently leading to apoptosis and therefore they are considered to be potential anti-tumour agents. Some other mechanisms adopted by other drugs are (i) inhibition of phosphodiester bond transfer from DNA to tyrosine residue, subsequently ceasing the enzyme catalytic activity, by staurosporine (Lassota *et al.*, 1996), (ii) arresting the enzyme in the 'closed clamp' by dioxopiperazine (Roca *et al.*, 1994), and (iii) inhibiting non-covalent DNA binding by aclarubicin (Sorensen *et al.*, 1992).

Topoisomerase II targeted drugs in case of eukaryotes possess a huge structural diversity. Chemical identity of these drugs comprises of an array of diverse compounds like anthracyclines, bisantrenes, ellipticins, fluoroquinolones, epipodophyllotoxins, triterpenoids, anthracenediones, flavonoids, aminocoumarin group antibiotics, isoquinolinedione, etc. Another critical issue for developing anti-parasitic drugs is the sensitivity of host's counterpart of the drug target towards the drug. A particular drug would be of no use if it simultaneously attacks pathogen as-well-as host system as mentioned earlier in the case of topoisomerase IB as well. As for example, anilinoacridine compounds, a class of topoisomerase II targeted drugs, exhibit cytotoxic effects towards human enzyme. These drugs also stabilize cleavable complex in *L. chagasi* nuclear as well as kinetoplast topoisomerase II. So this particular class of drug has less significance. But switching on to mitonafide analogues, they show selective targeting towards parasitic and human topoisomerase II whereas while considering kinetoplast and nuclear topoisomerase II, a differential targeting is observed (Slunt *et al.*, 1996). Hence, specificity of drugs towards host and pathogen enzyme is apparently a controlling factor as corroborated by these cases.

Another interesting aspect associated with topoisomerase II targeted drugs exploited for anti-leishmanial experimentation is the distinct existence of nuclear and kinetoplast topoisomerase II. It has been shown that some analogues of mitonafide stimulated kDNA linearization in *L. donovani* whereas another subset of analogues affected nuclear topoisomerase II activity. In contrast, anilinoacridines show similar response towards nuclear and kDNA topoisomerase II of *L. donovani* and human topoisomerase II as well. This difference in activity is attributed to the position and presence of nitro group. Switching on to topoisomerase II of *L. panamanensis*, second-generation fluoroquinolones, viz. enoxacin and ciprofloxacin; and flavonoid quercetin need a special mention due to their remarkably high potential to inhibit the catalytic activity of *L. panamensis* DNA topoisomerase II and at the same time substantially less toxicity towards human macrophage topoisomerase (Cortazar *et al.*, 2000). However, in some cases specificities get inversed. Flavonoid ellagic acid shows inhibition of macrophage enzyme but it cannot hamper protozoan enzyme function. Some drugs like protoberberine coralyne, diamidine diminazene are equally potent against both *L. panamensis* as well as macrophage enzyme. Luteolin, a plant-derived flavonoid, has been proved to be a potent anti-leishmanial agent. It induces topoisomerase II-mediated kinetoplast DNA cleavage leading to apoptosis (Mittra *et al.*, 2000). So, to summarize, compounds showing inhibition of protozoan enzyme without disturbing host enzyme are the ideal ones to be further investigated for developing anti-leishmanial drugs.

Recent advances in drugs targeting topoisomerases of *Leishmania* are presented in Table 11.2.

Conclusion

Recent years have witnessed tremendous progress in the understanding of parasitic protozoan topoisomerases. The many functions of the DNA topoisomerases within the kinetoplastid parasite cell, primarily catalysis of kDNA replication essential to parasitic survival, make it a challenge to establish links between their molecular roles and the physiological consequences of their inactivation. In addition, interactions between topoisomerases and other proteins that are related to mechanisms of parasite DNA replication and share common immunolocalizations, e.g. LACK antigen (analogue of receptor for activated protein kinase C), DNA polymerase β, and the mechanistic and functional consequence of these interactions, might be further explored to shed light on the many unanswered queries regarding the parasite biochemistry.

The search also continues for novel RNA topoisomerase activities in species of *Leishmania* and *Trypanosoma*. An extensive genetic recombination among the 10,000 minicircles, creating novel guide RNAs followed by RNA editing, might require distinct RNA topoisomerase. Topoisomerase genes and proteins characterized from these lower eukaryotes appear to share many characteristics associated with their human homologues, but certain striking differences, including different enzyme activity requirements, variable catalytic sites and different sensitivities to topoisomerase poisons, provide insight for the development of topoisomerase-directed antiparasitic therapeutics. It has been established by several studies that the inhibitors of topoisomerases convert these essential enzymes into intracellular proliferating cell toxins and thereby provide a good tool for preferentially killing of the highly replicative parasite cells within the host.

The ultimate challenge lies in over-expressing purified recombinant protozoan topoisomerases and elucidating their crystal structure. In this way, structural insights gained from molecular modelling can contribute to a better understanding of the molecular mechanisms of enzyme action *in vivo*, including their interaction with inhibitors and poisons screened from natural or synthetic sources. This will cause a paradigm shift in the quest to target the topoisomerase-based replication apparatus selectively as a means of therapeutic control of the parasites in the foreseeable future. Topoisomerase genes and proteins characterized from kinetoplastid parasite *Leishmania* appear to share many characteristics associated with their human homologues, but certain striking differences, including different enzyme activity requirements and different sensitivities to topoisomerase poisons provide insight for the development of topoisomerase-directed anti-parasitic therapeutics. It has been established by several studies that the inhibitors of topoisomerases convert these essential enzymes into intracellular proliferating cell toxins and thereby provide a good tool for preferentially killing of the highly replicative parasite cells within the host. The interaction of the enzyme with specific inhibitors and poisons screened from natural or synthetic sources will help in the quest to selectively target the topoisomerase-based replication apparatus as a means to therapeutically control the parasitic menace in the foreseeable future.

Acknowledgements

We thank Professor Sidhhartha Roy, present Director of this Institute, for his interest in this

Table 11.2 Different classes of DNA topoisomerase inhibitors

Drug	Chemical identity	Target	Mode of Inhibition	Reference
Mitonafide and mitonafide analogues	4-nitro-benzo-isoquinolinedione and derivatives	Topoisomerase II of *L. chagasi*	Linearization of kinetoplast minicircles via stabilization of DNA–enzyme complex	Slunt *et al.* (1996)
Dichloroacetamido bromobenzyl bromoindolyl bromoquinoline	Indolyl quinolines	Both topoisomerase IB and II of *L. donovani*	Dual inhibitor	Ray *et al.* (1997)
Diospyrin	Plant-derived bisnaphthoquinonoid	Topoisomerase IB of *L. donovani*	Stabilizes DNA-enzyme cleavable complex	Ray *et al.* (1998)
Luteolin	Plant-derived flavonoids	Topoisomerase II of *L. donovani*	Promotes topoisomerase II mediated linearization of kDNA minicircles and induces kDNA cleavage	Mitra *et al.* (2000)
Dihydrobetulinic acid	Pentacyclic triterpenoid	Both topoisomerase IB and II of *L. donovani*	Prevents enzyme-DNA binary complex formation	Chowdhury *et al.* (2003)
Novobiocin	Aminocoumarin antibiotic	Topoisomerase II of *L. donovani*	Catalytic inhibitor of topoisomerase II	Singh *et al.* (2005)
Enoxacin, ciprofloxacin	Fluoroquinolones	Topoisomerase II of *L. panamensis*	Inhibit the enzyme catalytic activity	Romero *et al.* (2005)
Anthra[1,2-d]imidazole-6,11-dione derivatives	Imidazole fused anthraquinone derivatives	Topoisomerase IB of *L. donovani*	Inhibit the enzyme catalytic activity	Chaudhuri *et al.* (2007)
Peganine hydrochloride dihydrate	Peganine hydrochloride dihydrate	Topoisomerase IB of *L. donovani*	Direct interaction with enzyme	Misra *et al.* (2008)
3,3′-di-indolylmethane	Acid condensation product of indole-3-carbinol	Topoisomerase IB of *L. donovani*	Non-competitive inhibitor, stabilizes DNA-enzyme cleavable complex	Roy *et al.* (2008)
16α–Hydroxycleroda-3,13(14) Z-dien-15,16-olide	Clerodane diterpenoids	Topoisomerase IB of *L. donovani*	Direct interaction with enzyme	Misra *et al.* (2010)
Disuccinyl betulin, diglutaryl dihydrobetulin, disuccinyl dihydrobetulin	Derivatives of betulin, a natural triterpene	Topoisomerase IB of *L. donovani*	Catalytic inhibitors that slow down the strand rotation event	Chowdhury *et al.* (2011)
Indotecan (LMP400), AM13–55	Indenoisoquinolines	Topoisomerase IB of *L. infantum*	Stabilize DNA-enzyme cleavable complex	Balaña-Fouce *et al.* (2012)
2-Octadecynoic acid, 2-hexadecynoic acid, 2-tetradecynoic acid	2-Alkylnoic fatty acid	Topoisomerase IB of *L. donovani*	Inhibition of enzyme activity	Carballeira *et al.* (2012a,b)
Niranthin, lyoniside, saracoside	Lignan compounds, i.e. diphenyl propanoids	Topoisomerase IB of *L. donovani*	Non-competitive inhibitors that stabilize DNA-enzyme cleavable complex	Chowdhury *et al.* (2012), Saha *et al.* (2013)
2-Methoxy-5,9-eicosadienoic acid, 2-methoxy-5,9-eicosadiynoic acid	α-Methoxylated fatty acid and acetylinic analogue	Topoisomerase IB of *L. donovani*	Inhibition of enzyme activity	Carballeira *et al.* (2013)

work. This work was supported by the grants from Department of Biotechnology (No. BT/PR4456/MED/29/355/2012), Government of India to H.K.M. Council of Scientific and Industrial Research (CSIR), Government of India supported S.C. with Senior Research Fellowships.

References

Abu-Elneel, K., Robinson, D.R., Drew, M.E., Englund, P.T., and Shlomai, J. (2001). Intramitochondrial localization of universal minicircle sequence-binding protein, a trypanosomatid protein that binds kinetoplast minicircle replication origins. J. Cell. Biol. *153*, 725–734.

Alper, S., Temiz Arpaci, O., Sener Aki, E., and Yalçin, I. (2003). Some new bi- and ter-benzimidazole derivatives as topoisomerase I inhibitors. Farmaco. *58*, 497–507.

Bakshi, R.P., and Shapiro, T.A. (2004). RNA interference of *Trypanosoma brucei* topoisomerase IB: Both subunits are essential. Mol. Biochem. Parasitol. *136*, 249–255.

Bakshi, R.P., Galande, S., and Muniyappa, K. (2001). Functional and regulatory characteristics of eukaryotic type II DNA topoisomerases. Crit. Rev. Biochem. Mol. Biol. *36*, 1–37.

Balaña-Fouce, R., Prada, C.F., Requena, J.M., Cushman, M., Pommier, Y., Álvarez-Velilla, R., Escudero-Martínez, J.M., Calvo-Álvarez, E., Pérez-Pertejo, Y., and Reguera, R.M. (2012). Indotecan (LMP400) and AM13-55: two novel indenoisoquinolines show potential for treating visceral leishmaniasis. Antimicrob. Agents Chemother. *56*, 5264–5270.

Bearden, D.T., and Danziger, L.H. (2001). Mechanism of action of and resistance to quinolones. Pharmacotherapy *21*, 224S-232S.

Berger, J.M. (1998). Structure of DNA topoisomerases. Biochim. Biophys. Acta. *1400*, 3–18.

Bjornsti, M.A., and Wang, J.C. (1987). Expression of yeast DNA topoisomerase I can complement a conditional-lethal DNA topoisomerase I mutation in *Escherichia coli*. Proc. Natl. Acad. Sci. U.S.A. *84*, 8971–8975.

Bodley, A.L., and Shapiro, T.A. (1995). Molecular and cytotoxic effects of camptothecin, a topoisomerase I inhibitor on Trypanosomes and *Leishmania*. Proc. Natl. Acad. Sci. U.S.A. *92*, 3726–3230.

Bodley, A.L., Wani, M.C., Wall, M.E., and Shapiro, T.A. (1995). Antitrypanosomal activity of camptothecin analogs. Structure–activity correlations. Biochem. Pharmacol. *50*, 937–942.

Bodley, A.L., Chakraborty, A.K., Xie, S., Burri, C., and Shapiro, T.A. (2003). An unusal type IB topoisomerase from African Trypanosomes. Proc. Natl. Acad. Sci. U.S.A. *100*, 7539–7544.

Broccoli, S., Marquis, J.F., Papadopoulou, B., Olivier, M., and Drolet, M. (1999). Characterization of a *Leishmania donovani* gene encoding a protein that closely resembles a type IB topoisomerase. Nucleic Acids Res. *27*, 2745–2752.

Broxterman, H.J., and Georgopapadakou, N. (2001). Cancer research 2001: drug resistance, new targets and drug combinations. Drug Resist. Updat. *4*, 197–209.

Burri, C., Bodley, A.L., and Shapiro, T.A. (1996). Topoisomerases in kinetoplastids. Parasitol. Today. *6*, 226–231.

Campbell, S., and Maxwell, A. (2002). The ATP-operated clamp of human DNA topoisomerase IIalpha: Hyperstimulation of ATPase by 'piggy-back' binding. J. Mol. Biol. *320*, 171–188.

Carballeira, N.M., Cartagena, M.M., Prada, C.F., Rubio, C.F., and Balaña-Fouce, R. (2009). Total synthesis and anti-leishmanial activity of the natural occurring acetylenic fatty acids 6-heptadecynoic acid and 6-icosynoic acid. Lipids, *44*:953–961.

Carballeira, N.M., Montano, N., Cintrón, G.A., Márquez, C., Rubio, C.F., Prada, C.F., and Balaña-Fouce, R. (2011). First total synthesis and anti-leishmanial activity of (Z)–16-methyl-11-heptadecenoic acid, a new marine fatty acidfrom the sponge Dragmaxia undata. Chem. Phys. Lipids. *164*, 113–117.

Carballeira, N.M., Cartagena, M., Li, F., Chen, Z., Prada, C.F., Calvo-Alvarez, E., Reguera, R.M., and Balaña-Fouce, R. (2012a). First total synthesis of the (±)–2-methoxy-6-heptadecynoic acid and related 2-methoxylated analogs as effectiveinhibitors of the *Leishmania* topoisomerase IB enzyme. Pure Appl. Chem. *84*, 1867–1875.

Carballeira, N.M., Cartagena, M., Sanabria, D., Tasdemir, D., Prada, C.F., Reguera, R.M., and Balaña-Fouce, R. (2012b). 2-alkynoic fatty acids inhibit topoisomerase IB from *Leishmania donovani*. *Bioorg*. Med. Chem. Lett. *22*, 6185–6189.

Carballeira, N.M., Montano, N., Alvarez-Velilla, R., Prada, C.F., Reguera, R.M., and Balaña-Fouce, R. (2013). Synthesis of marine α-methoxylated fatty acid analogs that effectively inhibit the topoisomerase IB from *Leishmania donovani* with a mechanism different from that of camptothecin. Mar. Drugs. *11*, 3661–3675.

Chakraborty, A.K., and Majumder, H.K. (1987). Decatenation of kinetoplast DNA by an ATP-dependent DNA topoisomerase from the kinetoplast hemoflagellate *Leishmania donovani*. Mol. Biochem. Parasitol. *26*, 215–224.

Chakraborty, A.K., and Majumder, H.K. (1988). Mode of action of pentavalent antimonials: specific inhibition of type I DNA topoisomerase of *Leishmania donovani*. Biochem. Biophys. Res. Commun. *152*, 605–611.

Chakraborty, A.K., and Majumder, H.K. (1991). An ATP-independent catenating enzyme from the kinetoplast hemoflagellate *Leishmania donovani*. Biochem. Biophys. Res. Commun *180*, 279–285.

Chakraborty, A.K., Gupta, A., and Majumder, H.K. (1993). A type I DNA topoisomerase from the kinetoplast hemoflagellate *Leishmania donovani*. Indian J. Biochem. Biophys. *30*, 257–263.

Champoux, J.J. (2001). DNA topoisomerases: structure, function and mechanism. Annu. Rev. Biochem. *70*, 369–413.

Chaudhuri, G., Ghoshal, K., and Banerjee, A.B. (1986). Toxic effects of fatty acids on *Leishmania donovani* promastigotes. Indian J. Med. Res. *84*, 361–365.

Chaudhuri, P., Majumder, H.K., and Bhattacharya, S. (2007). Synthesis, DNA Binding, and *Leishmania* Topoisomerase Inhibition Activities of a Novel Series of Anthra[1,2-d]imidazole-6,11-dione Derivatives. J. Med. Chem. *50*, 2536–2540.

Chen, G.L., Yang, L., Rowe, T.C., Halligan, B.D., Tewey, K.M., and Liu, L.F. (1984). Nonintercalative anti-tumor drugs interfere with the breakage-reunion reaction of mammalian DNA topoisomerase II. J. Biol. Chem. *259*, 13560–13566.

Chen, Y., and Rice, P.A. (2003). New insights into sight specific recombinations from Flp recombinase-DNA structures. Ann. Rev. Biophys. Biomol. Struc. *32*, 135–159.

Chowdhury, A.R., Mandal, S., Goswami, A., Ghosh, M., Mandal, L., Chakraborty, D., Ganguly, A., Tripathi, G., Mukhopadhyay, S., Bandyopadhyay, S., and Majumder, H.K. (2003). Dihydrobetulinic acid induces apoptosis in *Leishmania donovani* by targeting DNA topoisomerase I and II: implications in anti-leishmanial therapy. Mol. Med. *9*, 26–36.

Chowdhury, S., Mukherjee, T., Sengupta, S., Chowdhury, S.R., Mukhopadhyay, S., and Majumder, H.K. (2011). Novel betulin derivatives as anti-leishmanial agents with mode of action targeting type IB DNA topoisomerase. Mol. Pharmacol. *80*, 694–703.

Corbett, K.D., and Berger, J.M. (2004). Structure, mechanism and evolutionary relationship in DNA topoisomerases. Annu. Rev. Biophys. Biomol. Struc. *33*, 95–118.

Cortázar, T.M., Coombs, G.H., and Walker, J. (2007). *Leishmania panamensis*: comparative inhibition of nuclear DNA topoisomerase II enzymes from promastigotes and human macrophages reveals anti-parasite selectivity of fluoroquinolones, flavonoids and pentamidine. Exp. Parasitol. *116*, 475–482.

D'Arpa, P., Machlin, P.S., Ratrie, H. (3rd), Rothfield, N.F., Cleveland, D.W., and Earnshaw, W.C. (1988). cDNA cloning of human DNA Topoisomerase I: Catalytic activities of a 67.7kDa carboxy-terminal fragment. Proc. Natl. Acad. Sci. U.S.A. *85*, 2543–2547.

Das, A., Dasgupta, A., Sharma, S., Ghosh, M., Sengupta, T., Bandopadhyay, S., and Majumder, H.K. (2001). Characterisation of the gene encoding type II DNA topoisomerase from *Leishmania donovani*: a key molecular target in anti-leishmanial therapy. Nucl. Acids. Res. *29*, 1844–1851.

Das, A., Dasgupta, A., Sengupta, T., and Majumder, H.K. (2004). Topoisomerases of kinetoplastid parasites as potential chemotherapeutic targets. Trends Parasitol. *8*, 381–387.

Das, B.B., Sen, N., Ganguly, A., and Majumder, H.K. (2004). Reconstitution and functional characterization of bi-subunit topoisomerase I from *Leishmania donovani*. FEBS Lett. *565*, 81–88.

Das, B.B., Sen, N., Dasgupta, S.B., Ganguly, A., and Majumder, H.K. (2005). N-terminal region of the large subunit of *Leishmania donovani* bisubunit topoisomerase I is involved in DNA relaxation and interaction with the smaller subunit. J. Biol. Chem. *280*, 16335–16344.

Das, B.B., Sengupta, T., Ganguly, A., and Majumder, H.K. (2006). Topoisomerases of kinetoplastid parasites: why so fascinating? Mol Microbiol. *62*, 917–927.

Das, B.B., Ganguly, A., and Majumder, H.K. (2007). DNA topoisomerases of *Leishmania*: the potential targets for anti-leishmanial therapy. In Drug Targets in Kinetoplastid Parasites, Majumder, H.K., ed. (Springer, New York, USA), pp. 103–115.

Davies, D.R., Mushtaq, A., Interthal, H., Champoux, J.J., and Hol, W.G. (2006). The structure of the transition state of the heterodimeric topoisomerase I of *Leishmania donovani* as a vanadate complex with nicked DNA. J. Mol. Biol. *357*, 1202–1210.

Denny, W.A., and Baguley, B.C. (2003). Dual topoisomerase I/II inhibitors in cancer therapy. Curr. Top. Med. Chem. *3*, 339–353.

Desai, S.D., Li, T.K., Rodriguez-Bauman, A., Rubin, E.H., and Liu, L.F. (2001). Ubiquitin/26S proteasome-mediated degradation of topoisomerase I as a resistance mechanism to camptothecin in tumor cells. Cancer Res. *61*, 5926–5932.

Douc-Rasy, S., Kayser, A., Riou, J.F., and Riou, G. (1986). ATP-independent type II topoisomerase from trypanosomes. Proc. Natl. Acad. Sci. U.S.A. *83*, 7152–7156.

Ferguson, L.R., and Baguley, B.C. (1994). Topoisomerase II enzymes and mutagenicity. Environ. Mol. Mutagen. *24*, 245–261.

Froelich-Ammon, S.J., Patchan, M.W., Osheroff, N., and Thompson, R.B. (1995). Topoisomerase II binds to ellipticine in the absence or presence of DNA. Characterization of enzyme–drug interactions by fluorescence spectroscopy. J. Biol. Chem. *270*, 14998–15004.

Gadelle, D., Filee, J., Buhler, C., and Forterre, P. (2003). Phylogenomics of type II DNA topoisomerases. Bioessays *25*, 232–242.

Gaziova, I., and Lukes, J. (2003). Mitochondrial and nuclear localization of topoisomerase II in the flagellate *Bodo saltans* (Kinetoplastida), a species with noncatenated kinetoplast DNA. J. Biol. Chem. *278*, 10900–10907.

Gellert, M., Mizuuchi, K., O'Dea, M., and Nash, H. (1976). DNA gyrase: an enzyme that introduces superhelical turns into DNA. Proc. Natl. Acad. Sci. U. S. A. *73*, 3872–3876.

Kikuchi, A., and Asai, K. (1984). Reverse gyrase – a topoisomerase which introduces positive superhelical turns into DNA. Nature. *309*, 677–681.

Krogh, B.O., and Shuman, S. (2002). A poxvirus-like type IB topoisomerase family in bacteria. Proc. Natl. Acad. Sci. U.S.A. *99*, 1853–1858.

Kulikowicz, T., and Shapiro, T.A. (2006). Distinct genes encode type II Topoisomerases for the nucleus and mitochondrion in the protozoan parasite *Trypanosoma brucei*. J Biol Chem *281*, 3048–3056.

Lassota, P., Singh, G., and Kramer, R. (1996). Mechanism of topoisomerase II inhibition by staurosporine and other protein kinase inhibitors. J. Biol. Chem. *271*, 26418–26423.

Liu, L.F. (1989). DNA topoisomerase poisons as anti-tumor drugs. Annu. Rev. Biochem. *58*, 351–375.

Liu, L.F., and Miller, K.G. (1981). Eukaryotic DNA topoisomerases: Two forms of type I DNA topoisomerases from HeLa cell nuclei. Proc. Natl. Acad. Sci. U.S.A. *78*, 3487–3491.

Lucumi, A., Robledo, S., Gama, V., and Saravia, N.G. (1998). Sensitivity of *Leishmania viannia panamensis* to pentavalent antimony is correlated with the formation of cleavable DNA–protein complexes. Antimicrob. Agents Chemother. *42*, 1990–1995.

Lynn, R., Giaever, G., Swanberg, S.L., and Wang, J.C. (1986). Tandem regions of yeast DNA topoisomerase II share homology with different subunits of bacterial gyrase. Science. *233*, 647–649.

Marcotte, E.M., Pellegrini, M., Ng, H.L., Rice, D.W., Yeates, T.O., and Eisenberg, D. (1999). Detecting protein function and protein–protein interactions from genome sequences. Science. *285*, 751–753.

Marquis, J.F., Hardy, I., and Olivier, M. (2005). Topoisomerase I amino acid substitutions, Gly185Arg and Asp325Glu, confer camptothecin resistance in *Leishmania donovani. Antimicrob*. Agents. Chemother. *49*, 1441–1446.

Melendy, T., and Ray, D.S. (1987). Purification and nuclear localization of type I topoisomerase from *Crithidia fasciculata*. Mol. Biochem. Parasitol. *24*, 215–225.

Melendy, T., Shelina, C., and Ray, D.S. (1988). Localization of a type II DNA topoisomerase to two sites at the periphery of the kinetoplast DNA of *Crithidia fasciculata*. Cell *55*, 1083–1088.

Misra, P., Khaliq, T., Dixit, A., SenGupta, S., Samant, M., Kumari, S., Kumar, A., Kushawaha, P.K., Majumder, H.K., Saxena, A.K., *et al.* (2008). Anti-leishmanial activity mediated by apoptosis and structure-based target study of peganine hydrochloride dihydrate: an approach for rational drug design. J. Antimicrob. Chemother. *62*, 998–1002.

Misra, P., Sashidhara, K.V., Singh, S.P., Kumar, A., Gupta, R., Chaudhaery, S.S., Gupta, S.S., Majumder, H.K., Saxena, A.K., and Dube, A. (2010). 16alpha-Hydroxycleroda-3,13 (14)Z-dien-15,16-olide from Polyalthia longifolia: a safe and orally active anti-leishmanial agent. Br. J. Pharmacol. *159*, 1143–1150.

Mittra, B., Saha, A., Chowdhury, A.R., Pal, C., Mandal, S., Mukhopadhyay, S., Bandyopadhyay, S., and Majumder, H.K. (2000). Luteolin, an abundant dietary component is a potent anti-leishmanial agent that acts by inducing topoisomerase II-mediated kinetoplast DNA cleavage leading to apoptosis. Mol. Med. *6*, 527–541.

Nagarajan, M., Morrell, A., Ioanoviciu, A., Antony, S., Kohlhagen, G., Agama, K., Hollingshead, M., Pommier, Y., and Cushman, M. (2006). Synthesis and evaluation of indenoisoquinoline topoisomerase I inhibitors substituted with nitrogenheterocycles. J. Med. Chem. *49*, 6283–6289.

Nitiss, J.L. (2002). DNA topoisomerases in cancer chemotherapy: using enzymes to generate selective DNA damage. Curr. Opin. Investig. Drugs. *3*, 1512–1516.

Ray, S., Sadhukhan, P.K., Mandal, N.B., Mahato, S.B., and Majumder, H.K. (1997). Dual inhibition of DNA topoisomerases of *Leishmania donovani* by novel indolyl quinolines. Biochem. Biophys. Res. Commun. *230*, 171–175.

Ray, S., Hazra, B., Mittra, B., Das, A., and Majumder, H.K. (1998). Diospyrin, a bisnaphthoquinone: a novel inhibitor of type I DNA topoisomerase of *Leishmania donovani*. Mol. Pharmacol. *54*, 994–999.

Redinbo, M.R., Champoux, J.J., and Hol, W.G. (1999). Structural insights into the function of type IB topoisomerases. Curr. Opin. Struct. Biol. *9*, 29–36.

Riou, G.F., Gabillot, M., Douc-Rasy, S., Kayser, A., and Barrois, M. (1983). A type I DNA topoisomerase from Trypanosoma cruzi. Eur. J. Biochem. *134*, 479–484.

Robinson, M.J., Martin, B.A., Gootz, T.D., McGuirk, P.R., Moynihan, M., Sutcliffe, J.A., and Osheroff, N. (1991). Effects of quinolone derivatives on eukaryotic topoisomerase II. A novel mechanism for enhancement of enzyme-mediated DNA cleavage. J. Biol. Chem. *266*, 14585–14592.

Robinson, M.J., Corbett, A.H., and Osheroff, N. (1993). Effects of topoisomerase II-targeted drugs on enzyme-mediated DNA cleavage and ATP hydrolysis: evidence for distinct drug interaction domains on topoisomerase II. Biochemistry *32*, 3638–3643.

Roca, J., Ishida, R., Berger, J.M., Andoh, T., and Wang, J.C. (1994). Anti-tumor bisdioxopiperazines inhibit yeast DNA topoisomerase II by trapping the enzyme in the form of a closed protein clamp. Proc. Natl. Acad. Sci. U.S.A. *91*, 1781–1785.

Romero, I.C., Saravia, N.G., and Walker, J. (2005). Selective action of fluoroquinolones against intracellular amastigotes of *Leishmania (Viannia) panamensis in vitro*. J. Parasitol. *91*, 1474–1479.

Roy, A., Das, B.B., Ganguly, A., Bose Dasgupta, S., Khalkho, N.V., Pal, C., Dey, S., Giri, V.S., Jaisankar, P., Dey, S., and Majumder, H.K. (2008). An insight into the mechanism of inhibition of unusual bi-subunit topoisomerase I from *Leishmania donovani* by 3,3'-di-indolylmethane, a novel DNA topoisomerase I poison with a strong binding affinity to the enzyme. Biochem. J. *409*, 611–622.

Roy, A., Chowdhury, S., Sengupta, S., Mandal, M., Jaisankar, P., D'Annessa, I., Desideri, A., and Majumder, H.K. (2011). Development of derivatives of 3, 3'-diindolylmethane as potent *Leishmania donovani* bi-subunit topoisomerase IB poisons. PLoS One. *6*, e28493.

Saha, S., Mukherjee, T., Chowdhury, S., Mishra, A., Chowdhury, S.R., Jaisankar, P., Mukhopadhyay, S., and Majumder, H.K. (2013). The lignan glycosides lyoniside and saracoside poison the unusual type IB topoisomerase of *Leishmania donovani* and kill the parasite both *in vitro* and *in vivo*. Biochem. Pharmacol. *86*, 1673–1687.

Sen, N., Das, B.B., Ganguly, A., Mukherjee, T., Bandyopadhyay, S., and Majumder, H.K. (2004a). Camptothecin-induced imbalance in intracellular cation homeostasis regulates programmed cell death in unicellular hemoflagellate *Leishmania donovani*. J. Biol. Chem. *279*, 52366–52375.

Sen, N., Das, B.B., Ganguly, A., Mukherjee, T., Tripathi, G., Bandyopadhyay, S., Rakshit, S., Sen, T., and Majumder, H.K. (2004b). Camptothecin induced mitochondrial dysfunction leading to programmed cell death in unicellular hemoflagellate *Leishmania donovani*. Cell Death Differ. *11*, 924–936.

Sengupta, T., Mukherjee, M., Mandal, C., Das, A., and Majumder, H.K. (2003). Functional dissection of the C-terminal domain of type II DNA topoisomerase of kinetoplastid hemoflagellate *Leishmania donovani*. Nucleic Acids Res. *31*, 5305–5316.

Sengupta, T., Mukherjee, M., Das, A., Mandal, C., Das, R., Mukherjee, T., and Majumder, H.K. (2005a). Characterization of the ATPase activity of topoisomerase II from *L. donovani* and identification of the residues conferring resistance to etoposide. Biochem. J. *390*, 419–426.

Sengupta, T., Mukherjee, M., Das, R., Das, A., and Majumder, H.K. (2005b). Characterization of the DNA-binding domain and identification of the active site residue in the gyr A half of *Leishmania donovani* topoisomerase II. Nucleic Acids Res. *33*, 2364–2373.

Shapiro, T.A. (1993). Inhibition of topoisomerases in African trypanosomes. Acta Trop. *54*, 251–260.

Shlomai, J., and Zadok, A. (1983). Reversible decatenation of kinetoplast DNA by a DNA topoisomerase from trypanosomatids. Nucleic Acids Res. *11*, 4019–4034.

Shlomai, J., Zadok, A., and Frank, D. (1984). A unique ATP-dependent DNA topoisomerase from trypanosomatids. Adv. Exp. Med. Biol. *179*, 409–422.

Simonson, A.B., Servin, J.A., Skophammer, R.G., Herbold, C.W., Rivera, M.C., and Lake, J.A. (2005). Decoding the genomic tree of life. Proc. Natl. Acad. Sci. U.S.A. *102*, 6608–6613.

Singh, G., Jayanarayan, K.G., and Dey, C.S. (2005). Novobiocin induces apoptosis-like cell death in topoisomerase II over-expressing arsenite resistant *Leishmania donovani*. Mol. Biochem. Parasitol. *141*, 57–69.

Slunt, K.M., Grace, J.M., Macdonald, T.L., and Pearson, R.D. (1996). Effect of mitonafide analogs on topoisomerase II of *Leishmania chagasi*. Antimicrob. Agents Chemother. *40*, 706–709.

Sørensen, B.S., Sinding, J., Andersen, A.H., Alsner, J., Jensen, P.B., and Westergaard, O. (1992). Mode of action of topoisomerase II-targeting agents at a specific DNA sequence. Uncoupling the DNA binding, cleavage and religation events. J. Mol. Biol. *228*, 778–786.

Stewart, L., Ireton, G.C., and Champoux, J.J. (1996). The domain organization of human Topoisomerase I. J. Biol. Chem. *271*, 7602–7608.

Stewart, L., Ireton, G.C., and Champoux, J.J. (1997). Reconstitution of human topoisomerase I by fragment complementation. J. Mol. Biol. *269*, 355–372.

Stewart, L., Redinbo, M.R., Qiu, X., Hol, W.G., and Champoux, J.J. (1998). A model for the mechanism of human topoisomerase I. Science. *279*, 1534–1541.

Tsai-Pflugfelder, M., Liu, L.F., Liu, A.A., Tewey, K.M., Whang-Peng, J., Knutsen, T., Huebner, K., Croce, C.M., and Wang, J.C. (1988). Cloning and sequencing of cDNA encoding human DNA topoisomerase II and localization of the gene to chromosome region 17q21–22. Proc. Natl. Acad. Sci. U.S.A. *85*, 7177–7181.

Vilain, N., Tsai-Pflugfelder, M., Benoit, A., Gasser, S.M., and Leroy, D. (2003). Modulation of drug sensitivity in yeast cells by the ATP-binding domain of human DNA topoisomerase II alpha. Nucleic Acids Res. *31*, 5714–5722.

Villa, H., Otero Marcos, A.R., Reguera, R.M., Balaña-Fouce, R., García-Estrada, C., Pérez-Pertejo, Y., Tekwani, B.L., Myler, P.J., Stuart, K.D., Bjornsti, M.A., and Ordóñez, D. (2003). A novel active DNA topoisomerase I in *Leishmania donovani*. J. Biol. Chem. *278*, 3521–3526.

Wang, J.C. (1971). Interaction between DNA and an *Escherichia coli* protein ω. J. Mol. Biol. *55*, 523–526.

Wang, J.C. (1991). DNA topoisomerases: why so many? J. Biol. Chem. *266*, 6659–6662.

Wang, J.C. (1994). DNA topoisomerase as targets of therapeutics: an overview. Adv. Pharmacol. *29A*, 1–9.

Wang, J.C. (2002). Cellular roles of DNA topoisomerases: a molecular perspective. Nat. Rev. Mol. Cell. Biol. *6*, 430–440.

Zagotto, G., Supino, R., Favini, E., Moro, S., and Palumbo, M. (2000). New 1,4-anthracene-9,10-dione derivatives as potential anticancer agents. Farmaco. *55*, 1–5.

Zhang, H., Barceio, J.M., Kohlhagen, G., Zimonjic, D.B., Popescu, N.C., and Pommier, Y. (2001). Human mitochondrial topoisomerase I. Proc. Natl. Acad. Sci. U.S.A. *98*, 10608–10613.

Mechanism of Drug Resistance in Visceral Leishmaniasis

12

Shyam Sundar and Jaya Chakravarty

Abstract

The treatment of visceral leishmaniasis is far from satisfactory. It is further complicated by the growing resistance of the parasites to anti-leishmanials. The exact mechanisms of resistance to various anti-leishmanials are not known. However, increased intracellular thiol levels, reduced uptake of the drug, sequestration and rapid drug efflux are some of the mechanism known for antimony resistance. Reduced uptake and increased efflux of pentamidine can lead to resistance. *L. donovani* miltefosine transporter (LdMT) and the protein LdRos3 has shown to play an important role in resistance to miltefosine. Paromomycin, being an aminoglycoside, is at an increased risk of developing resistance; however, clinical resistance to the drug has not been reported as yet. Unresponsiveness and relapses after amphotericin B (AmB) therapy is rare however with the increasing use of AmB in lipid formulations that have longer half-lives the risk for developing resistance can become a reality. Thus, there is an urgent need to know the markers of drug resistance and to develop tools to determine the emergence of drug resistance in the field.

Introduction

Leishmaniasis, a vector-borne disease, is caused by an obligate intracellular protozoan of the genus *Leishmania*. It broadly manifests as visceral leishmaniasis (VL; also known as kala-azar), cutaneous leishmaniasis (CL) and mucocutaneous leishmaniasis (MCL). VL is caused by the *Leishmania donovani* complex: *L. donovani*, the causative organism of VL in the Indian subcontinent and Africa; *L. infantum* (*L. chagasi*) which causes VL in the Mediterranean basin, Central and South America. CL is caused by various *Leishmania* species.

In South Asia and the Horn of Africa, the predominant mode of transmission of VL is anthroponotic, and humans with kala-azar or post – kala-azar dermal leishmaniasis (PKDL) provide the major reservoir for transmission (Magill, 1995; Pearson RD, 1999). In the Mediterranean, the Middle East and Brazil, VL is zoonotic, with dog as the most important reservoir host sustaining transmission (Pearson, 1999). Most CL have zoonotic transmission except those caused by *L. tropica*, which is anthroponotic. The only proven vectors of human disease are sand fly of species *Phlebotomus* in the Old World and *Lutzomyia* in the New World (WHO, 2010).

Approximately 0.2 to 0.4 million VL cases and 0.7 to 1.2 million CL cases occur each year. More than 90% of global VL cases occur in just six countries: India, Bangladesh, Sudan, South Sudan, Brazil and Ethiopia. CL is more widely distributed, with about one-third of cases occurring in each of three regions, the Americas, the Mediterranean basin and western Asia from the Middle East to Central Asia. The 10 countries with the highest estimated case counts, Afghanistan, Algeria, Colombia, Brazil, Iran, Syria, Ethiopia, North Sudan, Costa Rica and Peru, together account for 70–75% of globally estimated incidence of CL (Alvar *et al.*, 2012).

HIV–VL co-infection is reported from more than 35 countries. Initially, most of these cases were from south-western Europe, but the number of cases is increasing in sub-Saharan Africa especially Ethiopia, Brazil and South Asia (Alvar *et al.*, 1997, 2008; Desjeux and Alvar, 2003).

For several decades, pentavalent antimonials (Sb^v) have been the standard first- line medicines for visceral leishmaniasis. With the increasing antimonial resistance in the Indian subcontinent amphotericin B deoxycholate has replaced Sb^v as the first line drug. In the recent years, drugs like oral miltefosine, and paramomycin and newer therapies like single dose liposomal amphotericin B (L-AmB) and combination therapies have been added in the treatment of VL. At the same time as these new therapies are becoming available drugs like Sb^v and miltefosine are being threatened by the development of resistance.

In this chapter we have reviewed the mechanism of resistance of various drugs used in the treatment of visceral leishmaniasis. A literature search on the mechanism of resistance was done on PubMed and through Google.

Mechanism of drug resistance

The efficacy of anti-leishmanial agents depends on factors such as the immune status of the host and sensitivity of the *Leishmania* species to the drugs. This is evident from experience with patients with HIV/VL co-infection, in whom anti-leishmanial therapies are less effective and more toxic, and who tend to relapse frequently.

Owing to the variation in the sensitivity of *Leishmania* species to different drugs, the efficacy and required dosage of the anti-leishmanial agents vary in different endemic regions and thus the treatment recommendation by WHO also have regional differences (WHO, 2010).

Antimonials

Mechanism of action

Although Sb^v antimonials have been use for more than eight decades, their molecular and cellular mechanisms of action are not yet well understood. It is believed that it behaves as a prodrug, which undergoes biological reduction to much more active trivalent form of antimony (Sb^{III}) that exhibits anti-leishmanial activity. However, the site of (amastigote or macrophage) and mechanism of reduction (enzymatic or non-enzymatic) remain unclear (Haldar *et al.*, 2011). Moreover, amastigotes are more susceptible to Sb^v as they can reduce Sb^v to Sb^{III} while promastigotes cannot (Callahan *et al.*, 1994; Ephros *et al.*, 1997, 1999; Goyard *et al.*, 2003). The reduction of Sb^v to Sb^{III} requires an active participation of thiol compounds of both mammalian host (glutathione (GSH), cysteine and cysteinyl-glycine) as well as of parasite origin (trypanothione (T(SH)2) (Ferreira Cdos *et al.*, 2003; Frezard *et al.*, 2001; Yan *et al.*, 2003). Acidic pH and slightly elevated temperature also favours the reduction of Sb^v to Sb^{III}. However, it was observed that promastigotes contain higher intracellular concentrations of T(SH)2 and GSH than amastigotes (Ariyanayagam and Fairlamb, 2001; Wyllie *et al.*, 2004), and both stages maintain an intracellular pH value close to neutral (Glaser *et al.*, 1988). Thus, non-enzymatic reduction of Sb^v to Sb^{III} fails to account for the insensitivity of promastigotes to Sb^v and suggests that there could be enzymatic reduction of Sb^v. Studies have shown that a parasite-specific enzyme, thiol-dependent reductase (TDR1), abundant in amastigotes play a role in reduction of Sb^v to Sb^{III} (Denton *et al.*, 2004). An arsenate reductase homologue in *Leishmania major* (LmACR2) has also been shown to catalyse the reduction of Sb^v in the presence of GSH (Zhou *et al.*, 2004). The mechanism of killing of parasite by Sb^{III} could be through trypanothione/trypanothione reductase (TR) system which keeps T(SH)2 in the reduced state and protects the parasites from oxidative damage (Krauth-Siegel and Comini, 2008). Trivalent antimonials interfere with T(SH)2 metabolism by inhibiting TR and inducing rapid efflux of intracellular T(SH)2 and GSH into intact *Leishmania* cells (Cunningham and Fairlamb, 1995; Wyllie *et al.*, 2004).

Some studies have shown that Sb^v may have intrinsic anti-leishmanial activity. Sb^v inhibits macromolecular biosynthesis in amastigotes (Berman *et al.*, 1985), possibly via perturbation of energy metabolism due to inhibition of glycolysis and fatty acid betaoxidation (Berman *et al.*, 1987). However, the specific targets in these pathways have not been identified. It is also hypothesized that antimonials clear intracellular *Leishmania* parasites via activation of host immune system.

History of antimonial resistance

Although the selection of resistant *Leishmania* has long been a part of laboratory studies, it is only in the past 20 years that acquired resistance has become a clinical threat. The first indication of drug resistance came from North Bihar, in the early 1980s, of about 30% patients not responding to the prevailing regimen of Sb^v, which was a small daily dose (10 mg/kg; 600 mg maximum) for short duration (6–10 days) (Peters, 1981). Then two 10-day courses with a 10-day interval therapy with sodium antimony gluconate was recommended by an expert committee leading to a marked improvement in the cure rates up to 99% (Aikat *et al.*, 1979). However, in 1984, it was seen that with 20 mg/kg (maximum 600 mg) for 20 days, 86% of patients were cured and cure rate with 10 mg/kg was quite low (Thakur *et al.*, 1984). In the same year, the WHO expert committee recommended that pentavalent antimony be used in doses of 20 mg/kg up to a maximum of 850 mg for 20 days, and a repetition of similar regimen for 20 days in cases of treatment failures. The WHO recommendations was evaluated a few years later by Thakur *et al.* (1988) and it was reported that only 81% of patients were cured by this regimen, although with an extension of the treatment for 40 days, 97% of patients could be cured. Three years later, the same group noted a further decline in cure rate to 71% after 20 days of treatment, and recommended extended duration of treatment in non-responders (Thakur *et al.*, 1991). However, by early 1990s, extending the therapy to 30 days could cure only 64% of patients in a hyperendemic district of Bihar (Jha TK, 1992). In two studies carried out under strictly supervised treatment schedules, it was observed that only about one-third of the patients could be cured with the currently prevailing regimen (Sundar *et al.*, 1997, 2000). The incidence of primary unresponsiveness was 52%, whereas 8% of the patients relapsed. Incidentally, only 2% of the patients from the neighbouring state of (Eastern) Uttar Pradesh, India, failed treatment (Sundar *et al.*, 2000). There were reports of antimony resistance spreading to the Terai regions of Nepal, especially from the district adjoining the hyperendemic areas of Bihar, where up to 30% of patients seem to be unresponsive, though in Eastern Nepal a 90% cure rate had been reported (Rijal *et al.*, 2003). These studies confirmed that a high level of antimony resistance existed in Bihar, whereas it was still effective in surrounding areas.

There had been speculations whether Indian *L. donovani* had become truly refractory to Sb^v or resistance occurred because of the inadequate doses being used in Bihar. In a study to determine whether acquired drug resistance was present in Bihar, *L. donovani* isolates were taken from responders and non-responders. *In vitro* amastigote-macrophage assay showed that isolates from patients who did respond to sodium stibogluconate treatment were threefold more sensitive, with 50% effective doses (ED_{50}) around 2.5 μg Sb/ml compared with isolates from patients who did not respond (ED_{50} around 7.5 μg Sb/ml) (Lira *et al.*, 1999). The significant differences in amastigote sensitivity supported the concept of acquired resistance in Bihar.

The reasons for the emergence of resistance were the widespread misuse of the drug which included free availability of drugs, use of drug by qualified medical practitioners and unqualified quacks and inappropriate and inadequate dose (Sundar *et al.*, 1994). Another reason for the growing resistance to Sb^v in India while it still remained sensitive all over the world could be due to the fact that in an anthroponotic cycle once Sb^v resistance gets established, it spreads exponentially and organisms sensitive to the drug get eliminated quickly, whereas the drug-resistant parasites continue to circulate in the community.

Mechanism of resistance

The mechanisms of resistance to antimony in *Leishmania* have largely emerged from studies conducted on laboratory generated drug-resistant cell lines generated through step wise exposure to either antimony or related metal arsenic (Ouellette *et al.*, 2004). Over the last few years, the focus shifted towards exploring the mechanisms of antimony resistance in clinical isolates. Interestingly, a common mechanism of drug resistance has not been observed operating in the isolates of either same species from different countries and in different species from same country. Mechanisms of *in vitro* antimonial resistance may differ from that in field isolates (Decuypere *et al.*, 2005).

Similarly, *in vitro* unresponsiveness does not necessarily translate to clinical resistance (Sharief *et al.*, 2006). Reduction of drug concentration within the parasite, either by decreasing drug uptake or by increasing efflux/sequestration of the drug, constitutes the primary mechanism of antimonial resistance; other potential resistance mechanisms include inhibition of drug activation, inactivation of active drug, and gene amplification (Choudhury *et al.*, 2008; Cortes-Selva *et al.*, 2005; Decuypere *et al.*, 2005; Haldar *et al.*, 2011; Kothari *et al.*, 2007; Maltezou, 2010; Mukherjee *et al.*, 2007; Singh, 2006).

Role of aquaporin

In *Leishmania*, aquaglyceroporin1 (AQP1), member of the aquaporin superfamily has been shown to facilitate uptake of Sb^{III} (Marquis *et al.*, 2005; Ouellette *et al.*, 2004). Overexpression of AQP1 in *L. major* (LmAQP1) produces hyper susceptibility to Sb^{III}, whereas gene deletion renders the parasite resistant (Gourbal *et al.*, 2004; Marquis *et al.*, 2005). Studies on clinical isolates from Nepal indicated down-regulation of AQP1 (Decuypere *et al.*, 2005, 2008) whereas AQP1 RNA levels remain unaltered in resistant isolates of *L. braziliensis* and *L. guyanensis* (Adaui *et al.*, 2011; Torres *et al.*, 2010). Studies from India did not show down-regulation of AQP1 RNA levels consistently in all the antimony-resistant isolates (Maharjan *et al.*, 2008) whereas in another study down-regulation of AQP1 correlated well with clinical drug resistance in a majority of Indian VL and PKDL isolates (Mandal *et al.*, 2010). A recent study from India reported a significant down-regulation of RNA transcripts of AQP1 in all the antimony resistant clinical isolates as well as the laboratory resistant mutant. The down-regulation was more significant in isolates with higher resistance indices (Rai *et al.*, 2013b). These studies suggest that the down-regulation of AQP1 may not be a universal mechanism for resistance to all *Leishmania* species but could be one of the resistance mechanisms in *L. donovani* isolates from the Indian subcontinent.

Thiol metabolism

Thiol metabolism has a central role in the maintenance of an intracellular reducing environment so that the cell can defend itself against the damage caused by oxidative stress. Thus, higher levels of thiols in resistant isolates protect the parasites from Sb-mediated oxidative stress (Haldar *et al.*, 2011). They also promote resistance by forming conjugates with Sb^{III} for efflux and/or sequestration. Studies have indicated that the parasite can limit drug exposure by forming metal–thiol conjugates with trypanothione, a parasite-specific thiol (Cunningham and Fairlamb, 1995; Mukhopadhyay *et al.*, 1996). These conjugates are then sequestered into an intracellular vesicle before extrusion by specific transporters (Haimeur and Ouellette, 1998; Legare *et al.*, 1997, 2001a).

Arsenite- or antimony-resistant laboratory mutants of all *Leishmania* species exhibit significantly increased levels of intracellular thiols, namely cysteine, GSH, and trypanothione (TSH), suggesting a role for thiols in resistance (Ashutosh *et al.*, 2007; Decuypere *et al.*, 2005). The synthesis of two precursors GSH and spermidine determines the level of TSH. The γ-GCS gene encoding γ-glutamyl cysteine synthetase, which catalyses the rate-limiting step in GSH biosynthesis, has been found to be amplified in arsenite-resistant *L. tarentolae* (Grondin *et al.*, 1997), while the gene ODC which encodes ornithine decarboxylase, an enzyme involved in the regulation of spermidine biosynthesis, was also overexpressed (Guimond *et al.*, 2003; Haimeur *et al.*, 1999). The inhibition of the γ-GCS and ODC genes by their specific inhibitors, L-buthionine-(SR)-sulphoximine (BSO) and D La-difluoromethylornithine (DFMO), respectively, resulted in the reversal of arsenite or antimony resistance in laboratory mutants (Arana *et al.*, 1998; Grondin *et al.*, 1997) as well as in clinical isolates (Carter *et al.*, 2003, 2005).

However, in clinical isolates results were variable. γ-GCS was neither amplified and nor up-regulated in *L. donovani* Indian resistant isolates (Mittal *et al.*, 2007; Mukherjee *et al.*, 2007) but down-regulation of γ-GCS was observed in Nepalese isolates (Decuypere *et al.*, 2005, 2008). In *L. guyanensis*, γ-GCS was over expressed in therapeutic failure isolates (Carter *et al.*, 2006). Similarly, the precursor protein of spermidine biosynthesis, ODC was amplified at the genetic and protein levels in Indian *L. donovani* resistant isolates (Mukherjee *et al.*, 2007; Rai *et al.*, 2013b) and in *L. braziliensis* (Adaui *et al.*, 2011), but the gene was down-regulated in isolates from Nepal (Decuypere *et al.*, 2005). Studies have shown that antimony-resistant isolates down-regulate the expression of γ-GCS of macrophages, probably by down-regulating host NF-κB, which is known to regulate γ-GCS expression (Carter *et al.*, 2006). This would result in the reduction of intramacrophage GSH levels and promote an intracellular oxidative environment, thereby minimizing the intramacrophage reduction of Sb^{V} to its toxic form Sb^{III} (Wyllie and Fairlamb, 2006). This indicates that SAG resistance in *L. donovani* is associated with manipulation of both host and parasite thiol levels.

Another pivotal enzyme of the thiol metabolism responsible for maintaining the intracellular reducing environment through trypanothione is trypanothione reductase (TR). Studies have shown increased RNA levels as well as enzyme activity of TR in Sb^{V} resistant isolates of *L. donovani* (Rai *et al.*, 2013b). In this study thiol up-regulation in *L. donovani* is mediated by the increased expression of ODC and TR. Expression rate of TR was also increased in Sb^{V} resistant clinical isolates of *L. braziliensis* (Adaui *et al.*, 2011).

Drug efflux

Another resistance mechanism, responsible for lower concentration of drug in the cell is rapid efflux. The ATP binding cassette superfamily of proteins has been widely reported to export xenobiotics (Gottesman *et al.*, 2002; Homolya *et al.*, 2003) outside the cell. These include the P-glycoprotein (P-gp) and multi-drug resistance related protein (MRP). P-gp type efflux pumps play role mostly in resistance to hydrophobic compounds while MRP type pumps are known for efflux of anionic compounds in conjugation with thiols (Gottesman *et al.*, 2002). In *Leishmania*, several ABC transporters have been reported and characterized in relation to drug resistance.

The first ABC transporter identified and characterized was MRPA. However, this transporter was not responsible for the drug efflux across the plasma membrane. It was shown to confer antimony resistance by sequestering thiol–metal conjugates in an intracellular vesicle (Legare *et al.*, 2001b). Overexpression of MRPA has been reported to decrease influx of antimony rather than increase efflux (Callahan *et al.*, 1994). Gene amplification accompanied with up-regulation of MRPA gene was observed in *L. donovani* isolates only from India (Mittal *et al.*, 2007; Mukherjee *et al.*, 2007) but neither from Nepal nor in *L. braziliensis* and *L. guyanensis* (Adaui *et al.*, 2011; Torres *et al.*, 2010). A recent study of resistant field isolates of *L. donovani* from India exhibited up-regulation of MRPA which was related to antimony resistance indices, i.e. higher the resistance index, higher was the up-regulation of MRPA thereby suggesting that MRPA may play a role in antimony resistance (Rai *et al.*, 2013b).

Vanadate-sensitive P-type ATPases included in the ABC superfamily of transporters such as P- glycoprotein involved in multidrug resistance in mammalian cells was observed to be significantly increased in the membrane of methotrexate resistant *L. tropica* (Sanchez *et al.*, 1994) and arsenite-resistant *L. donovani* (Prasad *et al.*, 2000) laboratory mutants. Recently functionality of verapamil sensitive P-gp type efflux pumps was demonstrated in *L. donovani* antimony resistant field isolates., There was increased activity of vanadate sensitive P-type ATPases in antimony-resistant isolates. These findings were further corroborated by the functional assays of efflux pumps. These resistant isolates also exhibited decreased accumulation and increased efflux of one of P-gp substrates, rhodamine 123, which was significantly inhibited in presence of verapamil, the P-gp pump inhibitor (Rai *et al.*, 2013b). This inhibition of efflux was partial and reversible whereas there was complete inhibition of Rho 123 efflux by trifluoperazine (TFP, another P-gp blocker), in both sensitive and resistant isolates further confirming the presence of P-gp-type efflux pumps in *L. donovani* promastigotes. The transmembrane efflux pump P-gp 170, was also found to play a key role in the phenomenon of drug resistance in a study including *L. infantum* and *L. donovani* in Cyprus (Messaritakis *et al.*, 2013).

Thus, the overall phenomenon of antimonial resistance is multifactorial. Increased intracellular thiol levels, reduced uptake of the drug, sequestration and rapid drug efflux are some of the mechanism known for antimony resistance. A recent study from Nepal showed that differences in genetic background were a major driving force in the development of heterogeneous SSG-R phenotypes. The parasite's genetic background determines the parasite's intrinsic capacity to cope with antimonial stress, and this intrinsic capacity is likely to vary to some extent between distinct genetic populations (Decuypere *et al.*, 2012).

Miltefosine (MIL)

It is an alkyl phospholipid (hexadecylphosphocholine) and the first oral anti-leishmanial agent registered for use in India from March 2002 following a Phase III trial in which 50–100 mg/day dose for 28 days resulted in a long-term cure rate of 94% (Sundar *et al.*, 2002).The efficacy of the drug varies with region: a study from Ethiopia showed that the final cure among non-HIV-infected patients 6 months after treatment in the miltefosine group was only 75.6% (Ritmeijer *et al.*, 2006). The drug was chosen for the elimination programme in India, Nepal and Bangladesh for its ease of use and applicability in the control program (Sundar *et al.*, 2008a). Frequent gastro-intestinal adverse events, quick recovery (within 10 days most patients feel better) and absence of a directly observed treatment program are likely to drive the patients to prematurely discontinue the treatment and suboptimal compliance will ultimately lead to the development of parasite resistance (Sundar and Murray, 2005). After a decade of use of the drug in the Indian subcontinent, its efficacy declined to 90.3% and the relapse rate doubled (Sundar *et al.*, 2012); 10.8% and 20.0% of patients from Nepal also relapsed within 6 or 12 months after MIL treatment, respectively (Rijal *et al.*, 2013). In another phase IV study enrolling 977 patients a final cure rate of 85% could only be achieved (Rahman *et al.*, 2011). An *in vitro* susceptibility assay based on intracellular amastigotes on host macrophages was done to assess the susceptibility of *L. donovani* clinical isolates from different geographical regions of India to miltefosine, paromomycin, and amphotericin B. Effective dose for 90% killing (ED90) values were significantly higher for miltefosine ($P = 0.005$) and paromomycin ($P = 0.02$) in isolates from the high endemic region as compared to non-endemic region (Prajapati *et al.*, 2012).

There is species-specific variation in sensitivity to miltefosine (Croft *et al.*, 2006). Among *L. donovani, L. major, L. tropica, L. aethiopica, L. mexicana, and L. panamensis, L donovani* was the most sensitive, and *L. major* the least sensitive species in *in vitro* assays (Escobar *et al.*, 2002). While clinical isolates using a murine macrophage–amastigote model showed significant lack of sensitivity of *L. braziliensis* and *L. guyanensis* isolates from patients in Peru (Yardley *et al.*, 2005).

Mechanism of action

The exact mechanism of action of miltefosine remains largely unknown. The intracellular accumulation of miltefosine includes the following steps: binding to plasma membrane, internalization in the parasite cell (two proteins, the miltefosine transporter LdMT and its beta subunit LdRos3, are the most significant), and intracellular targeting and metabolism. It has been found that miltefosine induces an apoptosis-like cell death in *L. donovani* by producing numerous defects (Perez-Victoria, F.J. *et al.*, 2006b). It has effects on phospholipid metabolism and the parasite's membrane composition: decreases phosphatidylcholine, but increases phosphatidylethanolamine (PE) (Rakotomanga *et al.*, 2007). This observation might be related to the miltefosine-mediated inhibition of the inward transport of exogenous choline into the parasite (Zufferey and Mamoun, 2002). Miltefosine's immunomodulatory properties have been proposed as an additional contributory factor to its anti-leishmanial activity (Wadhone *et al.*, 2009; Zeisig *et al.*, 1995).

Mechanism of resistance

Miltefosine resistance, or rather drug non-susceptibility, could relatively easily be induced *in vitro*, although it has not been characterized *in vivo*. *L. donovani* promastigote selected against high MIL concentration (40 μM) were observed to be 15-fold less susceptible to miltefosine with resistant parasites accumulating no more than 3% of the drug (Perez-Victoria *et al.*, 2003a; Seifert *et al.*, 2003). The transport of miltefosine over the parasite cell membrane is thought to be facilitated by a putative *L. donovani* miltefosine transporter (LdMT) and the protein LdRos3. It was shown that decreased miltefosine accumulation and defective inward translocation was the major determinant of decreased susceptibility, (Perez-Victoria *et al.*, 2003a) which was demonstrated to be mediated through inactivation of LdMT and LdRos3 (Perez-Victoria, F.J. *et al.*, 2003b, 2006a; Sanchez-Canete *et al.*, 2009). LdMT is a large plasma membrane protein that belongs to the P4 subfamily of P-type ATPases and LdRos3 is a non-catalytic subunit of this membrane protein which belongs to the CDC 50/Lem3 protein family, which together play an important role in maintaining the phospholipid asymmetry of the parasite membrane (Weingartner *et al.*, 2010). Wild-type parasites overexpressing LdMT became hypersensitive to the drug due to their ability to take up higher amounts of MIL (Perez-Victoria *et al.*, 2003b). LdMT associated miltefosine resistance could be transferred to the amastigote stage, with no apparent loss of infectivity, and even persists *in vivo* (Seifert *et al.*, 2007). In clinical isolates, low expression of the LdMT–LdRos3 complex was correlated to the natural non-susceptibility to miltefosine of *L. braziliensis* strains (Sanchez et al., 2009). Miltefosine resistance has been reported to be caused by single point mutations in these transporter proteins (Perez-Victoria *et al.*, 2003a; Seifert *et al.*, 2007).

Increased efflux of miltefosine has also been implicated in miltefosine resistance, mediated through the overexpression of an ABC transporter: the *Leishmania* P-glycoprotein-like transporter (*Leishmania* ABCB1 or LtrMDR1) (Perez-Victoria, J.M. *et al.*, 2001, 2006). An *L tropica* line overexpressing P-glycoprotein MDR1 showed cross-resistance to MIL and accumulated 5–10 fold less MIL than its parental line. The overexpression of two *Leishmania*-specific ABC subfamily G-like transporters (LiABCG6 and LiABCG4 half-transporters) has shown to be involved in MIL efflux in *Leishmania*.(Castanys-Munoz *et al.*, 2007, 2008).

Studies have suggested that miltefosine elicit its effects by mitochondrial dependent-programmed cell death associated with generation of reactive oxygen species (ROS). The ability to tolerate MIL-induced ROS and alterations at the level of apoptosis control could be one of the alternative mechanisms of drug resistance. A recent study has shown MIL-unresponsive *Leishmania* accumulates a lower number of reactive oxygen species (ROS) under conditions that are sufficient to cause ROS-induced mitochondrial membrane potential damage and apoptotic death of miltefosine responsive *Leishmania*. Additionally, there is higher superoxide dismutase and ascorbate peroxidise (APX) activity in miltefosine-unresponsive *Leishmania*. SOD is a vital enzyme that detoxifies superoxide to H_2O_2 (less toxic than superoxide) and oxygen, and APX scavenges H2O2and converts it in to water. SOD and APX protect the parasite from oxidative stress and subsequent damage (Das *et al.*, 2013). It has been reported that *Leishmania donovani* mitochondrial iron superoxide dismutase-A (LdFeSODA) overexpression protects parasites from miltefosine by protecting the mitochondria of *Leishmania* from oxidative stress, thereby inhibiting programmed cell death (Getachew and Gedamu, 2012). Gene expression levels for LdFeSODA was 5.3-fold higher in MIL-resistant phenotypes upon drug treatment than the sensitive strains (Mishra and Singh, 2013).

Whole genome sequencing recently revealed that miltefosine resistance in *L. major* mutants can be both genetically and phenotypically highly heterogeneous. Two of the three identified markers of miltefosine resistance in this study were implicated in drug susceptibility: the previously described P-type ATPase and pyridoxal kinase (Coelho *et al.*, 2012).

Although the relapse rate of patients on miltefosine has increased in the Indian subcontinent, decrease in the susceptibility of parasites to miltefosine *in vivo*, a precursor of the emergence of drug resistance, has not yet been observed (Dorlo *et al.*, 2012; Prajapati *et al.*, 2013; Rijal *et al.*, 2013). The plasma MIL concentrations in cured and relapsed Nepalese patients were similar, indicating a similar exposure to MIL (Rijal *et al.*, 2013). Another recent study from Nepal revealed that increased infectivity of the parasite is associated with miltefosine relapse (Rai *et al.*, 2013a). However, *L. infantum* isolates from a HIV/VL co infected patient on miltefosine maintenance showed a gradual decrease of the miltefosine susceptibility with the IC_{50} values of miltefosine increasing over time, from 5.00 μmol/l to 50.10 μmol/l and on sequencing the entire *LdMT* gene a new SNP,

L832F, was found in the miltefosine-resistant strain which reverted back to the wild-type allele 3 years after withdrawal from miltefosine (Cojean *et al.*, 2012).

The high miltefosine relapse rate especially in a anthroponotic focus with a history of rampant misuse of drugs poses a major challenge and therefore miltefosine monotherapy should be strongly discouraged.

Amphotericin B and its lipid formulations

Amphotericin B (AmB) is a polyene antibiotic has been used as a second line treatment for VL since the 1960s. This drug exhibits an excellent anti-leishmanial activity with > 90–95% cure rates in Indian VL cases.

Mechanism of action

AmB has high affinity for ergosterol, the predominant sterol in these microbes, over cholesterol, the predominant sterol in the mammalian host cells. It results in disorganization of the membrane (Bolard, 1986), possibly by formation of specific pores composed of small aggregates of AmB and sterol (Urbina *et al.*, 1987). These defects cause depolarization of the membrane and an increase in membrane permeability for protons and monovalent cations (Brajtburg *et al.*, 1990; Lamy-Freund *et al.*, 1985).

Mechanism of resistance

Unresponsiveness and relapses after AmB occurs rarely, except among HIV-infected patients, who tend to relapse frequently (Lachaud *et al.*, 2009; Minodier *et al.*, 1998). In HIV-VL co-infected patient with multiple relapses treated with AmB, no resistance to AmB was seen in clinical isolates analysed *in vitro* (Lachaud *et al.*, 2009). A rare case of AmB -unresponsive *L. donovani* infection was reported from India. *In vivo* studies showed more than three times greater inhibition of AmB-sensitive parasites compared to parasites with AmB relapse at a normal AmB dose in mice of 1.5 mg/kg body weight. Four single nucleotide polymorphisms (SNPs) were detected in this patient in the cysteine proteinase B gene affecting alignments in deduced amino acids (Srivastava *et al.*, 2011).

In another study from India an amphotericin B-resistant clinical isolate which demonstrated 8-fold higher 50% lethal doses (LD_{50}) than an amphotericin B-sensitive strain was used to explore the mechanism of amphotericin B resistance (Purkait *et al.*, 2012). Fluorimetric analysis demonstrated lower anisotropy in the motion of the diphenylhexatriene fluorescent probe in the resistant strain, which indicated a higher fluidity of the membrane for the resistant strain than for the sensitive strain. This change in membrane fluidity was probably a consequence of the resistance-induced modification of membranous lipid metabolism. SCMT (S-adenosyl-L-methionine:C-24-Δ-sterol methyltransferase) is an important enzyme in the sterol biosynthetic pathway, as it performs C-24 transmethylation, which is a key step in production of ergosterol. The expression patterns of the two transcripts of S-adenosyl-L-methionine: C-24-sterol methyltransferase and the absence of ergosterol, replaced by cholesta-5,7,24-trien-3β-ol in the membrane of the resistant parasite, indicate a decreased AmB affinity, which is evidenced by decreased amphotericin B uptake (Purkait *et al.*, 2012). Similar findings were observed in a laboratory-derived AmB-resistant *Leishmania* promastigote, where ergosterol is replaced by a precursor, cholesta-5,7,24-trien-3-ol (Mbongo *et al.*, 1998).

Expression levels of the ABC transporters reported to be involved in drug efflux/sequestration in *Leishmania* was explored in this study to understand the involvement of ABC transporters in conferring AmB resistance. The expression level of MDR1 was found to be higher in the resistant strain, suggesting a higher rate of efflux of AmB. The resistant parasite also up-regulated tryparedoxin cascade and a more-reduced intracellular thiol level, which helps in better scavenging of reactive oxygen species produced by AmB (Purkait *et al.*, 2012).

The use of amphotericin B and its lipid formulations have dramatically increased especially in the Indian subcontinent. A single dose of 10 mg/kg of liposomal amphotericin B has recently been recommended as the preferred treatment for VL in the Indian subcontinent (Sundar *et al.*, 2010; WHO, 2010). With the increasing use of AmB in lipid formulations that have longer half-lives, the possibility of resistance cannot be ignored.

Paromomycin (aminosidine)

Paromomycin (PM) is a broad-spectrum aminoglycosidic aminocyclitol belonging to the neomycin family. A Phase III trial in which a dose of 15 mg/kg PM sulfate (11 mg base) for 21 days gave a cure rate of 95%, and was approved by the Indian government in August 2006 for the treatment of patients with VL (Sundar *et al.*, 2007). The results were confirmed in a large Phase IV study in India (Sinha *et al.*, 2011). Encouraged by this, a large Phase III study was done in Sudan, Ethiopia and Kenya comparing the efficacy of PM alone at the dose shown to be efficacious in India against SSG alone (20 mg/kg/day for 30 days) and against a combination treatment of SSG and PM for 17 days. The overall efficacy of PM alone was significantly lower than SSG and it had to be discontinued (Hailu *et al.*, 2010). An *in vitro* analysis on the sensitivity of amastigotes in a murine macrophage model showed that *L. major* and *L. tropica* isolates were more sensitive than *L. braziliensis* and *L. Mexicana* isolates and *L. donovani* showed intermediate sensitivity (Neal *et al.*, 1995).

Mechanism of action

PM inhibits bacterial protein synthesis. It binds to the 30S ribosomal subunit and interferes with initiation of protein synthesis (Sundar and Chakravarty, 2008). The mechanism of action in *Leishmania* is not known. It has been proposed that it might alter membrane fluidity, interact with ribosomes, interfere with the mitochondrial membrane potential and inhibit respiration (Maarouf *et al.*, 1997). *In vitro* studies in *L. donovani* has shown that cationic PM binds to the negatively charged leishmanial glycocalyx followed by internalization by endocytosis. After 72 hours of exposure to the drug the mitochondrial membrane potential was significantly decreased, indicating that this organelle might be the ultimate target of the drug. Both cytoplasmic and mitochondrial protein synthesis were inhibited following paromomycin exposure (Jhingran *et al.*, 2009).

In the same study a line selected for resistance to the drug showed reduced paromomycin accumulation associated with a significant reduction in the initial binding to the cell surface. The drug induced reduction in membrane potential and inhibition of protein synthesis were less pronounced in the resistant strain in comparison with the wild-type (Jhingran *et al.*, 2009). In another study comparative proteomic analysis of the wild type and the paromomycin resistant *L. donovani* strains showed up-regulation of the ribosomal proteins in the PM-resistant strain (PRr) which could be one of the mechanisms utilized by the resistant parasites as a defence against PM. Up-regulation of proteins that may have a role in intracellular survival and vesicular trafficking in the PRr strain was observed. Ultrastructural analysis by electron microscopy demonstrated increased number of vesicular vacuoles in PRr strain when compared to the wild-type strain. vesicular sequestration of PM in to the vacuoles by these resistant parasites might be involved in conferring the resistance phenotype (Chawla *et al.*, 2011).

Clinical resistance to PM has not been reported most probably due to its limited use. In a study of cloned antimony-resistant *L. donovani* field isolates from India and Nepal exposed to stepwise increasing concentrations of PM (up to 500 μM), either as promastigotes or intracellular amastigotes. Resistance selection in promastigotes took about 25 weeks to reach the maximal 97 μM inclusion level that did not affect normal growth however, applying PM pressure to intracellular amastigotes produced resistance after just two selection cycles ($IC_{50}=199$ μM) compared with the parent strain ($IC_{50}=45$ μM) (Hendrickx *et al.*, 2012). Whether these *in vitro* phenotype translates to *in vivo* treatment failure remains an important area of research. Another, study of clinical isolates from endemic and non-endemic regions in the J774A.1 macrophage cell line found increased tolerance of miltefosine and paromomycin in isolates from a patient from a high endemic region. Effective dose for 90% killing (ED_{90}) values were significantly higher for miltefosine ($P=0.005$) and paromomycin ($P=0.02$) in isolates from the high endemic region, might be a prelude to increasing tolerance of these drugs and could result in increasing unresponsiveness in the future (Prajapati *et al.*, 2012). Being an aminoglycoside PM is at an increased risk of developing resistance thus it is imperative to monitor treatment and avoid its use as monotherapy.

Pentamidine

Pentamidine is another anti-leishmanial agent which suffered the same fate as Sb^V in North Bihar. It was the first drug to be used in patients refractory to Sb^V and cured 99% of these patients initially however in the next two decades its efficacy dwindled to approximately 70% of patients (Jha *et al.*, 1991; Jha, 1983). Its use in VL was ultimately abandoned due to its decreased efficacy and serious toxicities. However, it has been used to good effect in treatment of New World CL.

The anti-leishmanial mechanism of action of pentamidine, are still not clearly known, however possible mechanism include inhibition of polyamine biosynthesis, DNA minor groove binding, and effect on mitochondrial inner membrane potential (Bray *et al.*, 2003). Pentamidine-resistant promastigote clones of *L. donovani* and *L. amazonensis* were shown to have 18- and 75-fold reduced uptakes, respectively, and increased efflux (Basselin *et al.*, 2002). Specific transporters for pentamidine uptake have been characterized and might have a role in resistance (Bray *et al.*, 2003; Coelho *et al.*, 2003). Wild-type promastigotes accumulate more pentamidine in the mitochondrion in comparison to resistant cells. It is suggested that less organelle accumulation makes far more drug available for efflux (Basselin *et al.*, 2002).

Combination therapy

The growing resistance of the parasite to anti-leishmanial drugs suggests that the currently used monotherapy needs to be reviewed. The rationale behind multidrug therapy are increased activity through use of compounds with synergistic or additive activity acting at different sites, shorter duration of therapy and lower dose requirement, thereby reducing chances of toxic side effects and cost, and preventing the emergence of drug resistance (Sundar and Chakravarty, 2013). In an experimental study Seifert and Croft demonstrated activity enhancement index (AEI) of different drugs *in vivo*, where the highest potentiation of miltefosine activity was achieved with amphotericin B (AEI of up to 11.3). No significant interaction was observed when miltefosine was combined with SSG (AEI of up to 2.38). The potentiation of miltefosine *in vivo* was also achieved with the combination of miltefosine and PM (AEI of up to 7.22) (Seifert and Croft, 2006). The combination of Sb^V and PM has been extensively used in Southern Sudan by Médecins Sans Frontières (MSF) initially in patients who relapsed after conventional Sb^V and since 2002 as first-line therapy for VL (Sundar and Chakravarty, 2013). A phase III study in the Indian subcontinent, three drug combinations (single injection of 5 mg/kg/L AmB and 7-day 50 mg oral miltefosine or single 10-day 11 mg/kg intramuscular PM; or 10 days each of miltefosine and PM were tested for the treatment of VL. All the combinations showed an excellent cure rate (> 97%) (Sundar *et al.*, 2011). Based on these studies WHO recommended combination therapy to be one of the preferred therapy for the Indian subcontinent (Sundar *et al.*, 2008b, 2011; WHO, 2010) and the combination of SSG with PM for 17 days is treatment of choice in East Africa and Yemen (WHO, 2010).

Recently a study was designed to elucidate the ability of *L. donovani* to develop experimental resistance to anti-leishmanial drug combinations. The induction of resistance to amphotericin B/miltefosine, amphotericin B/paromomycin, amphotericin B/Sb^{III}, miltefosine/paromomycin, and Sb^{III}/paromomycin was determined using a step-wise adaptation process to increasing drug concentrations. Intracellular amastigotes resistant to these drug combinations were obtained from resistant *L. donovani* promastigotes, and the thiol and ATP levels and the mitochondrial membrane potential of the resistant lines were analysed. Resistance to drug combinations was obtained after 10 weeks and remained in the intracellular amastigotes. It was observed that promastigotes/amastigotes resistant to one drug combination showed a marked cross-resistant profile to other anti-leishmanial drugs. Additionally, the thiol levels were increased in resistant lines that remained protected against the drug-induced loss of ATP and mitochondrial membrane potential. Resistance to the combinations miltefosine/paromomycin and Sb^{III}/paromomycin was easily obtained experimentally (Garcia-Hernandez *et al.*, 2012).

This study reinforces the need for proper monitoring of treatment when combination therapy are given as resistance could be induced resulting in a rapid loss of efficacy of not one but two therapeutic options.

Conclusion and future trends

Inventory of anti-leishmanial drugs is very small, and emergence of drug resistance is further complicating the control of leishmaniasis. A better understanding of mechanism of action of the drugs as well as mechanism of drug resistance, with easy to use markers of resistance may pave the way for more rational use of drugs. Combination chemotherapy is rapidly emerging as the norm for treating several infective disorders like malaria, tuberculosis, HIV, etc., and its application is strongly advocated for VL. Directly observed therapy given free, in treatment centres manned by trained personnel, will go along way in controlling the disease as well as drug resistance.

References

Adaui, V., Castillo, D., Zimic, M., Gutierrez, A., Decuypere, S., Vanaerschot, M., De Doncker, S., Schnorbusch, K., Maes, I., Van der Auwera, G., *et al.* (2011). Comparative gene expression analysis throughout the life cycle of *Leishmania braziliensis*: diversity of expression profiles among clinical isolates. PLoS Negl. Trop. Dis. *5*, e1021.

Aikat, B.K., Sahaya, S., Pathania, A.G., Bhattacharya, P.K., Desai, N., Prasad, L.S., Mishra, S., and Jain, S. (1979). Clinical profile of cases of kala-azar in Bihar. Indian J. Med. Res. *70*, 563–570.

Alvar, J., Canavate, C., Gutierrez-Solar, B., Jimenez, M., Laguna, F., Lopez-Velez, R., Molina, R., and Moreno, J. (1997). *Leishmania* and human immunodeficiency virus coinfection: the first 10 years. Clin. Microbiol. Rev. *10*, 298–319.

Alvar, J., Aparicio, P., Aseffa, A., Den Boer, M., Canavate, C., Dedet, J.P., Gradoni, L., Ter Horst, R., Lopez-Velez, R., and Moreno, J. (2008). The relationship between leishmaniasis and AIDS: the second 10 years. Clin. Microbiol. Rev. *21*, 334–359.

Alvar, J., Velez, I.D., Bern, C., Herrero, M., Desjeux, P., Cano, J., Jannin, J., and den Boer, M. (2012). Leishmaniasis worldwide and global estimates of its incidence. PLoS One 7, e35671.

Arana, F.E., Perez-Victoria, J.M., Repetto, Y., Morello, A., Castanys, S., and Gamarro, F. (1998). Involvement of thiol metabolism in resistance to glucantime in *Leishmania tropica*. Biochem. Pharmacol. *56*, 1201–1208.

Ariyanayagam, M.R., and Fairlamb, A.H. (2001). Ovothiol and trypanothione as antioxidants in trypanosomatids. Mol. Biochem. Parasitol. *115*, 189–198.

Ashutosh, Sundar, S., and Goyal, N. (2007). Molecular mechanisms of antimony resistance in *Leishmania*. J. Med. Microbiol. *56*, 143–153.

Basselin, M., Denise, H., Coombs, G.H., and Barrett, M.P. (2002). Resistance to pentamidine in *Leishmania mexicana* involves exclusion of the drug from the mitochondrion. Antimicrob. Agents Chemother. *46*, 3731–3738.

Berman, J.D., Waddell, D., and Hanson, B.D. (1985). Biochemical mechanisms of the anti-leishmanial activity of sodium stibogluconate. Antimicrob. Agents Chemother. *27*, 916–920.

Berman, J.D., Gallalee, J.V., and Best, J.M. (1987). Sodium stibogluconate (Pentostam) inhibition of glucose catabolism via the glycolytic pathway, and fatty acid beta-oxidation in *Leishmania mexicana* amastigotes. Biochem. Pharmacol. *36*, 197–201.

Bolard, J. (1986). How do the polyene macrolide antibiotics affect the cellular membrane properties? Biochim. Biophys. Acta *864*, 257–304.

Brajtburg, J., Powderly, W.G., Kobayashi, G.S., and Medoff, G. (1990). Amphotericin B: current understanding of mechanisms of action. Antimicrob. Agents Chemother. *34*, 183–188.

Bray, P.G., Barrett, M.P., Ward, S.A., and de Koning, H.P. (2003). Pentamidine uptake and resistance in pathogenic protozoa: past, present and future. Trends Parasitol. *19*, 232–239.

Callahan, H.L., Roberts, W.L., Rainey, P.M., and Beverley, S.M. (1994). The PGPA gene of *Leishmania major* mediates antimony (SbIII) resistance by decreasing influx and not by increasing efflux. Mol. Biochem. Parasitol. *68*, 145–149.

Carter, K.C., Sundar, S., Spickett, C., Pereira, O.C., and Mullen, A.B. (2003). The *in vivo* susceptibility of *Leishmania donovani* to sodium stibogluconate is drug specific and can be reversed by inhibiting glutathione biosynthesis. Antimicrob. Agents Chemother. *47*, 1529–1535.

Carter, K.C., Hutchison, S., Boitelle, A., Murray, H.W., Sundar, S., and Mullen, A.B. (2005). Sodium stibogluconate resistance in *Leishmania donovani* correlates with greater tolerance to macrophage anti-leishmanial responses and trivalent antimony therapy. Parasitology *131*, 747–757.

Carter, K.C., Hutchison, S., Henriquez, F.L., Legare, D., Ouellette, M., Roberts, C.W., and Mullen, A.B. (2006). Resistance of *Leishmania donovani* to sodium stibogluconate is related to the expression of host and parasite gamma-glutamylcysteine synthetase. Antimicrob. Agents Chemother. *50*, 88–95.

Castanys-Munoz, E., Alder-Baerens, N., Pomorski, T., Gamarro, F., and Castanys, S. (2007). A novel ATP-binding cassette transporter from *Leishmania* is involved in transport of phosphatidylcholine analogues and resistance to alkyl-phospholipids. Mol. Microbiol. *64*, 1141–1153.

Castanys-Munoz, E., Perez-Victoria, J.M., Gamarro, F., and Castanys, S. (2008). Characterization of an ABCG-like transporter from the protozoan parasite *Leishmania* with a role in drug resistance and transbilayer lipid movement. Antimicrob. Agents Chemother. *52*, 3573–3579.

Chawla, B., Jhingran, A., Panigrahi, A., Stuart, K.D., and Madhubala, R. (2011). Paromomycin affects translation and vesicle-mediated trafficking as revealed by proteomics of paromomycin -susceptible -resistant *Leishmania donovani*. PLoS One *6*, e26660.

Choudhury, K., Zander, D., Kube, M., Reinhardt, R., and Clos, J. (2008). Identification of a *Leishmania infantum* gene mediating resistance to miltefosine and SbIII. Int. J. Parasitol. *38*, 1411–1423.

Coelho, A.C., Beverley, S.M., and Cotrim, P.C. (2003). Functional genetic identification of PRP1, an ABC transporter superfamily member conferring pentamidine resistance in *Leishmania major*. Mol. Biochem. Parasitol. *130*, 83–90.

Coelho, A.C., Boisvert, S., Mukherjee, A., Leprohon, P., Corbeil, J., and Ouellette, M. (2012). Multiple mutations in heterogeneous miltefosine-resistant *Leishmania major* population as determined by whole genome sequencing. PLoS Negl. Trop. Dis. *6*, e1512.

Cojean, S., Houze, S., Haouchine, D., Huteau, F., Lariven, S., Hubert, V., Michard, F., Bories, C., Pratlong, F., Le Bras, J., *et al.* (2012). *Leishmania* resistance to miltefosine associated with genetic marker. Emerg. Infect. Dis. *18*, 704–706.

Cortes-Selva, F., Jimenez, I.A., Munoz-Martinez, F., Campillo, M., Bazzocchi, I.L., Pardo, L., Ravelo, A.G., Castanys, S., and Gamarro, F. (2005). Dihydro-beta-agarofuran sesquiterpenes: a new class of reversal agents of the multidrug resistance phenotype mediated by P-glycoprotein in the protozoan parasite *Leishmania*. Curr. Pharm. Des. *11*, 3125–3139.

Croft, S.L., Sundar, S., and Fairlamb, A.H. (2006). Drug resistance in leishmaniasis. Clin. Microbiol. Rev. *19*, 111–126.

Cunningham, M.L., and Fairlamb, A.H. (1995). Trypanothione reductase from *Leishmania donovani*. Purification, characterisation and inhibition by trivalent antimonials. Eur. J. Biochem. *230*, 460–468.

Das, M., Saudagar, P., Sundar, S., and Dubey, V.K. (2013). Miltefosine-unresponsive *Leishmania donovani* has a greater ability than miltefosine-responsive *L. donovani* to resist reactive oxygen species. FEBS J. *280*, 4807–4815.

Decuypere, S., Rijal, S., Yardley, V., De Doncker, S., Laurent, T., Khanal, B., Chappuis, F., and Dujardin, J.C. (2005). Gene expression analysis of the mechanism of natural Sb(V) resistance in *Leishmania donovani* isolates from Nepal. Antimicrob. Agents Chemother. *49*, 4616–4621.

Decuypere, S., Vanaerschot, M., Rijal, S., Yardley, V., Maes, L., de Doncker, S., Chappuis, F., and Dujardin, J.C. (2008). Gene expression profiling of *Leishmania* (*Leishmania*) *donovani*: overcoming technical variation and exploiting biological variation. Parasitology *135*, 183–194.

Decuypere, S., Vanaerschot, M., Brunker, K., Imamura, H., Muller, S., Khanal, B., Rijal, S., Dujardin, J.C., and Coombs, G.H. (2012). Molecular mechanisms of drug resistance in natural *Leishmania* populations vary with genetic background. PLoS Negl. Trop. Dis. *6*, e1514.

Denton, H., McGregor, J.C., and Coombs, G.H. (2004). Reduction of anti-leishmanial pentavalent antimonial drugs by a parasite-specific thiol-dependent reductase, TDR1. Biochem. J. *381*, 405–412.

Desjeux, P., and Alvar, J. (2003). *Leishmania*/HIV co-infections: epidemiology in Europe. Ann. Trop. Med. Parasitol. *97 Suppl 1*, 3–15.

Dorlo, T.P., Balasegaram, M., Beijnen, J.H., and de Vries, P.J. (2012). Miltefosine: a review of its pharmacology and therapeutic efficacy in the treatment of leishmaniasis. J. Antimicrob. Chemother. *67*, 2576–2597.

Ephros, M., Waldman, E., and Zilberstein, D. (1997). Pentostam induces resistance to antimony and the preservative chlorocresol in *Leishmania donovani* promastigotes and axenically grown amastigotes. Antimicrob. Agents Chemother. *41*, 1064–1068.

Ephros, M., Bitnun, A., Shaked, P., Waldman, E., and Zilberstein, D. (1999). Stage-specific activity of pentavalent antimony against *Leishmania donovani* axenic amastigotes. Antimicrob. Agents Chemother. *43*, 278–282.

Escobar, P., Matu, S., Marques, C., and Croft, S.L. (2002). Sensitivities of *Leishmania* species to hexadecylphosphocholine (miltefosine), ET-18-OCH(3) (edelfosine) and amphotericin B. Acta Trop. *81*, 151–157.

Ferreira Cdos, S., Martins, P.S., Demicheli, C., Brochu, C., Ouellette, M., and Frezard, F. (2003). Thiol-induced reduction of antimony(V) into antimony(III): a comparative study with trypanothione, cysteinyl-glycine, cysteine and glutathione. Biometals *16*, 441–446.

Frezard, F., Demicheli, C., Ferreira, C.S., and Costa, M.A. (2001). Glutathione-induced conversion of pentavalent antimony to trivalent antimony in meglumine antimoniate. Antimicrob. Agents Chemother. *45*, 913–916.

Garcia-Hernandez, R., Manzano, J.I., Castanys, S., and Gamarro, F. (2012). *Leishmania donovani* develops resistance to drug combinations. PLoS Negl. Trop. Dis. *6*, e1974.

Getachew, F., and Gedamu, L. (2012). *Leishmania donovani* mitochondrial iron superoxide dismutase A is released into the cytosol during miltefosine induced programmed cell death. Mol. Biochem. Parasitol. *183*, 42–51.

Glaser, T.A., Baatz, J.E., Kreishman, G.P., and Mukkada, A.J. (1988). pH homeostasis in *Leishmania donovani* amastigotes and promastigotes. Proc. Natl. Acad. Sci. U.S.A. *85*, 7602–7606.

Gottesman, M.M., Fojo, T., and Bates, S.E. (2002). Multidrug resistance in cancer: role of ATP-dependent transporters. Nat. Rev. Cancer *2*, 48–58.

Gourbal, B., Sonuc, N., Bhattacharjee, H., Legare, D., Sundar, S., Ouellette, M., Rosen, B.P., and Mukhopadhyay, R. (2004). Drug uptake and modulation of drug resistance in *Leishmania* by an aquaglyceroporin. J. Biol. Chem. *279*, 31010–31017.

Goyard, S., Segawa, H., Gordon, J., Showalter, M., Duncan, R., Turco, S.J., and Beverley, S.M. (2003). An *in vitro* system for developmental and genetic studies of

Leishmania donovani phosphoglycans. Mol. Biochem. Parasitol. *130*, 31–42.

Grondin, K., Haimeur, A., Mukhopadhyay, R., Rosen, B.P., and Ouellette, M. (1997). Co-amplification of the gamma-glutamylcysteine synthetase gene gsh1 and of the ABC transporter gene pgpA in arsenite-resistant *Leishmania tarentolae*. EMBO J. *16*, 3057–3065.

Guimond, C., Trudel, N., Brochu, C., Marquis, N., El Fadili, A., Peytavi, R., Briand, G., Richard, D., Messier, N., Papadopoulou, B., *et al.* (2003). Modulation of gene expression in *Leishmania* drug resistant mutants as determined by targeted DNA microarrays. Nucleic Acids Res. *31*, 5886–5896.

Hailu, A., Musa, A., Wasunna, M., Balasegaram, M., Yifru, S., Mengistu, G., Hurissa, Z., Hailu, W., Weldegebreal, T., Tesfaye, S., *et al.* (2010). Geographical variation in the response of visceral leishmaniasis to paromomycin in East Africa: a multicentre, open-label, randomized trial. PLoS Negl. Trop. Dis. *4*, e709.

Haimeur, A., and Ouellette, M. (1998). Gene amplification in *Leishmania tarentolae* selected for resistance to sodium stibogluconate. Antimicrob. Agents Chemother. *42*, 1689–1694.

Haimeur, A., Guimond, C., Pilote, S., Mukhopadhyay, R., Rosen, B.P., Poulin, R., and Ouellette, M. (1999). Elevated levels of polyamines and trypanothione resulting from overexpression of the ornithine decarboxylase gene in arsenite-resistant *Leishmania*. Mol. Microbiol. *34*, 726–735.

Haldar, A.K., Sen, P., and Roy, S. (2011). Use of antimony in the treatment of leishmaniasis: current status and future directions. Mol. Biol. Int. *2011*, 1–23.

Hendrickx, S., Inocencio da Luz, R.A., Bhandari, V., Kuypers, K., Shaw, C.D., Lonchamp, J., Salotra, P., Carter, K., Sundar, S., Rijal, S., *et al.* (2012). Experimental induction of paromomycin resistance in antimony-resistant strains of *L. donovani*: outcome dependent on *in vitro* selection protocol. PLoS Negl. Trop. Dis. *6*, e1664.

Homolya, L., Varadi, A., and Sarkadi, B. (2003). Multidrug resistance-associated proteins: Export pumps for conjugates with glutathione, glucuronate or sulfate. Biofactors *17*, 103–114.

Jha, S.N., Singh, N.K., and Jha, T.K. (1991). Changing response to diamidine compounds in cases of kala-azar unresponsive to antimonial. J. Assoc. Physicians India *39*, 314–316.

Jha, T.K. (1983). Evaluation of diamidine compound (pentamidine isethionate) in the treatment resistant cases of kala-azar occurring in North Bihar, India. Trans. R. Soc. Trop. Med. Hyg. *77*, 167–170.

Jha, T.K., Singh, N.K., and Jha, S.N. (1992). Therapeutic use of sodium stibogluconate in kala-azar from some hyperendemic districts of N. Bihar, India. J. Assoc. Physicians India *40*, 868.

Jhingran, A., Chawla, B., Saxena, S., Barrett, M.P., and Madhubala, R. (2009). Paromomycin: uptake and resistance in *Leishmania donovani*. Mol. Biochem. Parasitol. *164*, 111–117.

Kothari, H., Kumar, P., Sundar, S., and Singh, N. (2007). Possibility of membrane modification as a mechanism of antimony resistance in *Leishmania donovani*. Parasitol. Int. *56*, 77–80.

Krauth-Siegel, R.L., and Comini, M.A. (2008). Redox control in trypanosomatids, parasitic protozoa with trypanothione-based thiol metabolism. Biochim. Biophys. Acta *1780*, 1236–1248.

Lachaud, L., Bourgeois, N., Plourde, M., Leprohon, P., Bastien, P., and Ouellette, M. (2009). Parasite susceptibility to amphotericin B in failures of treatment for visceral leishmaniasis in patients coinfected with HIV type 1 and *Leishmania infantum*. Clin. Infect. Dis. *48*, e16–22.

Lamy-Freund, M.T., Ferreira, V.F., and Schreier, S. (1985). Mechanism of inactivation of the polyene antibiotic amphotericin B. Evidence for radical formation in the process of autooxidation. J. Antibiot. (Tokyo) *38*, 753–757.

Legare, D., Papadopoulou, B., Roy, G., Mukhopadhyay, R., Haimeur, A., Dey, S., Grondin, K., Brochu, C., Rosen, B.P., and Ouellette, M. (1997). Efflux systems and increased trypanothione levels in arsenite-resistant *Leishmania*. Exp. Parasitol. *87*, 275–282.

Legare, D., Cayer, S., Singh, A.K., Richard, D., Papadopoulou, B., and Ouellette, M. (2001a). ABC proteins of *Leishmania*. J. Bioenerg. Biomembr. *33*, 469–474.

Legare, D., Richard, D., Mukhopadhyay, R., Stierhof, Y.D., Rosen, B.P., Haimeur, A., Papadopoulou, B., and Ouellette, M. (2001b). The *Leishmania* ATP-binding cassette protein PGPA is an intracellular metal-thiol transporter ATPase. J. Biol. Chem. *276*, 26301–26307.

Lira, R., Sundar, S., Makharia, A., Kenney, R., Gam, A., Saraiva, E., and Sacks, D. (1999). Evidence that the high incidence of treatment failures in Indian kala-azar is due to the emergence of antimony-resistant strains of *Leishmania donovani*. J. Infect. Dis. *180*, 564–567.

Maarouf, M., de Kouchkovsky, Y., Brown, S., Petit, P.X., and Robert-Gero, M. (1997). *In vivo* interference of paromomycin with mitochondrial activity of *Leishmania*. Exp. Cell Res. *232*, 339–348.

Magill, A.J. (1995). Epidemiology of the leishmaniases. Dermatol. Clin. *13*, 505–523.

Maharjan, M., Singh, S., Chatterjee, M., and Madhubala, R. (2008). Role of aquaglyceroporin (AQP1) gene and drug uptake in antimony-resistant clinical isolates of *Leishmania donovani*. Am. J. Trop. Med. Hyg. *79*, 69–75.

Maltezou, H.C. (2010). Drug resistance in visceral leishmaniasis. J. Biomed. Biotechnol. *2010*, 617521.

Mandal, S., Maharjan, M., Singh, S., Chatterjee, M., and Madhubala, R. (2010). Assessing aquaglyceroporin gene status and expression profile in antimony-susceptible and -resistant clinical isolates of *Leishmania donovani* from India. J. Antimicrob. Chemother. *65*, 496–507.

Marquis, N., Gourbal, B., Rosen, B.P., Mukhopadhyay, R., and Ouellette, M. (2005). Modulation in aquaglyceroporin AQP1 gene transcript levels in drug-resistant *Leishmania*. Mol. Microbiol. *57*, 1690–1699.

Mbongo, N., Loiseau, P.M., Billion, M.A., and Robert-Gero, M. (1998). Mechanism of amphotericin B resistance in *Leishmania donovani* promastigotes. Antimicrob. Agents Chemother. *42*, 352–357.

Messaritakis, I., Christodoulou, V., Mazeris, A., Koutala, E., Vlahou, A., Papadogiorgaki, S., and Antoniou, M. (2013). Drug resistance in natural isolates of *Leishmania donovani* s.l. promastigotes is dependent of Pgp170 expression. PLoS One *8*, e65467.

Minodier, P., Piarroux, R., Garnier, J.M., Unal, D., Perrimond, H., and Dumon, H. (1998). Pediatric visceral leishmaniasis in southern France. Pediatr. Infect. Dis. J *17*, 701–704.

Mishra, J., and Singh, S. (2013). Miltefosine resistance in *Leishmania donovani* involves suppression of oxidative stress-induced programmed cell death. Exp. Parasitol. *135*, 397–406.

Mittal, M.K., Rai, S., Ashutosh, Ravinder, Gupta, S., Sundar, S., and Goyal, N. (2007). Characterization of natural antimony resistance in *Leishmania donovani* isolates. Am. J. Trop. Med. Hyg. *76*, 681–688.

Mukherjee, A., Padmanabhan, P.K., Singh, S., Roy, G., Girard, I., Chatterjee, M., Ouellette, M., and Madhubala, R. (2007). Role of ABC transporter MRPA, gamma-glutamylcysteine synthetase and ornithine decarboxylase in natural antimony-resistant isolates of *Leishmania donovani*. J. Antimicrob. Chemother. *59*, 204–211.

Mukhopadhyay, R., Dey, S., Xu, N., Gage, D., Lightbody, J., Ouellette, M., and Rosen, B.P. (1996). Trypanothione overproduction and resistance to antimonials and arsenicals in *Leishmania*. Proc. Natl. Acad. Sci. U.S.A. *93*, 10383–10387.

Neal, R.A., Allen, S., McCoy, N., Olliaro, P., and Croft, S.L. (1995). The sensitivity of *Leishmania* species to aminosidine. J. Antimicrob. Chemother *35*, 577–584.

Ouellette, M., Drummelsmith, J., and Papadopoulou, B. (2004). Leishmaniasis: drugs in the clinic, resistance and new developments. Drug Resist. Updat. *7*, 257–266.

Pearson R.D., J.S., and de Queiroz Sousa A. (1999). Tropical infectious diseases: principles, pathogens and practice (Churchill Livingstone, Philadelphia, PA).

Perez-Victoria, F.J., Castanys, S., and Gamarro, F. (2003a). *Leishmania donovani* resistance to miltefosine involves a defective inward translocation of the drug. Antimicrob. Agents Chemother. *47*, 2397–2403.

Perez-Victoria, F.J., Gamarro, F., Ouellette, M., and Castanys, S. (2003b). Functional cloning of the miltefosine transporter. A novel P-type phospholipid translocase from *Leishmania* involved in drug resistance. J. Biol. Chem. *278*, 49965–49971.

Perez-Victoria, F.J., Sanchez-Canete, M.P., Castanys, S., and Gamarro, F. (2006a). Phospholipid translocation and miltefosine potency require both *L. donovani* miltefosine transporter and the new protein LdRos3 in *Leishmania* parasites. J. Biol. Chem. *281*, 23766–23775.

Perez-Victoria, F.J., Sanchez-Canete, M.P., Seifert, K., Croft, S.L., Sundar, S., Castanys, S., and Gamarro, F. (2006b). Mechanisms of experimental resistance of *Leishmania* to miltefosine: Implications for clinical use. Drug Resist. Updat. *9*, 26–39.

Perez-Victoria, J.M., Perez-Victoria, F.J., Parodi-Talice, A., Jimenez, I.A., Ravelo, A.G., Castanys, S., and Gamarro, F. (2001). Alkyl-lysophospholipid resistance in multidrug-resistant *Leishmania tropica* and chemosensitization by a novel P-glycoprotein-like transporter modulator. Antimicrob. Agents Chemother. *45*, 2468–2474.

Perez-Victoria, J.M., Cortes-Selva, F., Parodi-Talice, A., Bavchvarov, B.I., Perez-Victoria, F.J., Munoz-Martinez, F., Maitrejean, M., Costi, M.P., Barron, D., Di Pietro, A., *et al.* (2006). Combination of suboptimal doses of inhibitors targeting different domains of LtrMDR1 efficiently overcomes resistance of *Leishmania* spp. to Miltefosine by inhibiting drug efflux. Antimicrob. Agents Chemother. *50*, 3102–3110.

Peters, W. (1981). The treatment of kala-azar – new approaches to an old problem. Indian J. Med. Res. *73 Suppl*, 1–18.

Prajapati, V.K., Mehrotra, S., Gautam, S., Rai, M., and Sundar, S. (2012). In vitro anti-leishmanial drug susceptibility of clinical isolates from patients with Indian visceral leishmaniasis – status of newly introduced drugs. Am. J. Trop. Med. Hyg. *87*, 655–657.

Prajapati, V.K., Sharma, S., Rai, M., Ostyn, B., Salotra, P., Vanaerschot, M., Dujardin, J.C., and Sundar, S. (2013). In vitro susceptibility of *Leishmania donovani* to miltefosine in Indian visceral leishmaniasis. Am. J. Trop. Med. Hyg. *89*, 750–754.

Prasad, V., Kaur, J., and Dey, C.S. (2000). Arsenite-resistant *Leishmania donovani* promastigotes express an enhanced membrane P-type adenosine triphosphatase activity that is sensitive to verapamil treatment. Parasitol. Res. *86*, 661–664.

Purkait, B., Kumar, A., Nandi, N., Sardar, A.H., Das, S., Kumar, S., Pandey, K., Ravidas, V., Kumar, M., De, T., *et al.* (2012). Mechanism of amphotericin B resistance in clinical isolates of *Leishmania donovani*. Antimicrob. Agents Chemother. *56*, 1031–1041.

Rahman, M., Ahmed, B.N., Faiz, M.A., Chowdhury, M.Z., Islam, Q.T., Sayeedur, R., Rahman, M.R., Hossain, M., Bangali, A.M., Ahmad, Z., *et al.* (2011). Phase IV trial of miltefosine in adults and children for treatment of visceral leishmaniasis (kala-azar) in Bangladesh. Am. J. Trop. Med. Hyg. *85*, 66–69.

Rai, K., Cuypers, B., Bhattarai, N.R., Uranw, S., Berg, M., Ostyn, B., Dujardin, J.C., Rijal, S., and Vanaerschot, M. (2013a). Relapse after treatment with miltefosine for visceral leishmaniasis is associated with increased infectivity of the infecting *Leishmania donovani* strain. MBio *4*, e00611–e00613.

Rai, S., Bhaskar, Goel, S.K., Nath Dwivedi, U., Sundar, S., and Goyal, N. (2013b). Role of efflux pumps and intracellular thiols in natural antimony resistant isolates of *Leishmania donovani*. PLoS One *8*, e74862.

Rakotomanga, M., Blanc, S., Gaudin, K., Chaminade, P., and Loiseau, P.M. (2007). Miltefosine affects lipid metabolism in *Leishmania donovani* promastigotes. Antimicrob. Agents Chemother. *51*, 1425–1430.

Rijal, S., Chappuis, F., Singh, R., Bovier, P.A., Acharya, P., Karki, B.M., Das, M.L., Desjeux, P., Loutan, L., and Koirala, S. (2003). Treatment of visceral leishmaniasis in south-eastern Nepal: decreasing efficacy of sodium stibogluconate and need for a policy to limit further decline. Trans. R. Soc. Trop. Med. Hyg. *97*, 350–354.

Rijal, S., Ostyn, B., Uranw, S., Rai, K., Bhattarai, N.R., Dorlo, T.P., Beijnen, J.H., Vanaerschot, M., Decuypere, S., Dhakal, S.S., *et al.* (2013). Increasing failure of miltefosine in the treatment of kala-azar in Nepal and the potential role of parasite drug resistance, reinfection, or noncompliance. Clin. Infect. Dis. *56*, 1530–1538.

Ritmeijer, K., Dejenie, A., Assefa, Y., Hundie, T.B., Mesure, J., Boots, G., den Boer, M., and Davidson, R.N. (2006). A comparison of miltefosine and sodium stibogluconate for treatment of visceral leishmaniasis in an Ethiopian population with high prevalence of HIV infection. Clin. Infect. Dis. *43*, 357–364.

Sanchez-Canete, M.P., Carvalho, L., Perez-Victoria, F.J., Gamarro, F., and Castanys, S. (2009). Low plasma membrane expression of the miltefosine transport complex renders *Leishmania braziliensis* refractory to the drug. Antimicrob. Agents Chemother. *53*, 1305–1313.

Sanchez, A., Castanys, S., and Gamarro, F. (1994). Increased P-type ATPase activity in *Leishmania tropica* resistant to methotrexate. Biochem. Biophys. Res. Commun. *199*, 855–861.

Seifert, K., and Croft, S.L. (2006). *In vitro* and *in vivo* interactions between miltefosine and other anti-leishmanial drugs. Antimicrob. Agents Chemother. *50*, 73–79.

Seifert, K., Matu, S., Javier Perez-Victoria, F., Castanys, S., Gamarro, F., and Croft, S.L. (2003). Characterisation of *Leishmania donovani* promastigotes resistant to hexadecylphosphocholine (miltefosine). Int. J. Antimicrob. Agents *22*, 380–387.

Seifert, K., Perez-Victoria, F.J., Stettler, M., Sanchez-Canete, M.P., Castanys, S., Gamarro, F., and Croft, S.L. (2007). Inactivation of the miltefosine transporter, LdMT, causes miltefosine resistance that is conferred to the amastigote stage of *Leishmania donovani* and persists *in vivo*. Int. J. Antimicrob. Agents *30*, 229–235.

Sharief, A.H., Gasim Khalil, E.A., Theander, T.G., Kharazmi, A., Omer, S.A., and Ibrahim, M.E. (2006). *Leishmania donovani*: an *in vitro* study of antimony-resistant amphotericin B-sensitive isolates. Exp. Parasitol. *114*, 247–252.

Singh, N. (2006). Drug resistance mechanisms in clinical isolates of *Leishmania donovani*. Indian J. Med. Res. *123*, 411–422.

Sinha, P.K., Jha, T.K., Thakur, C.P., Nath, D., Mukherjee, S., Aditya, A.K., and Sundar, S. (2011). Phase 4 pharmacovigilance trial of paromomycin injection for the treatment of visceral leishmaniasis in India. J. Trop. Med. *2011*, 645203.

Srivastava, P., Prajapati, V.K., Rai, M., and Sundar, S. (2011). Unusual case of resistance to amphotericin B in visceral leishmaniasis in a region in India where leishmaniasis is not endemic. J. Clin. Microbiol. *49*, 3088–3091.

Sundar, S., and Chakravarty, J. (2008). Paromomycin in the treatment of leishmaniasis. Expert Opin. Investig. Drugs. *17*, 787–794.

Sundar, S., and Chakravarty, J. (2013). Leishmaniasis: an update of current pharmacotherapy. Expert Opin. Pharmacother. *14*, 53–63.

Sundar, S., and Murray, H.W. (2005). Availability of miltefosine for the treatment of kala-azar in India. Bull. World Health Organ. *83*, 394–395.

Sundar, S., Singh, V.P., Sharma, S., Makharia, M.K., and Murray, H.W. (1997). Response to interferon-gamma plus pentavalent antimony in Indian visceral leishmaniasis. J. Infect. Dis. *176*, 1117–1119.

Sundar, S., More, D.K., Singh, M.K., Singh, V.P., Sharma, S., Makharia, A., Kumar, P.C., and Murray, H.W. (2000). Failure of pentavalent antimony in visceral leishmaniasis in India: report from the center of the Indian epidemic. Clin. Infect. Dis. *31*, 1104–1107.

Sundar, S., Jha, T.K., Thakur, C.P., Engel, J., Sindermann, H., Fischer, C., Junge, K., Bryceson, A., and Berman, J. (2002). Oral miltefosine for Indian visceral leishmaniasis. N. Engl. J. Med. *347*, 1739–1746.

Sundar, S., Jha, T.K., Thakur, C.P., Sinha, P.K., and Bhattacharya, S.K. (2007). Injectable paromomycin for Visceral leishmaniasis in India. N. Engl. J. Med. *356*, 2571–2581.

Sundar, S., Mondal, D., Rijal, S., Bhattacharya, S., Ghalib, H., Kroeger, A., Boelaert, M., Desjeux, P., Richter-Airijoki, H., and Harms, G. (2008a). Implementation research to support the initiative on the elimination of kala azar from Bangladesh, India and Nepal – the challenges for diagnosis and treatment. Trop. Med. Int. Health *13*, 2–5.

Sundar, S., Rai, M., Chakravarty, J., Agarwal, D., Agrawal, N., Vaillant, M., Olliaro, P., and Murray, H.W. (2008b). New treatment approach in Indian visceral leishmaniasis: single-dose liposomal amphotericin B followed by short-course oral miltefosine. Clin. Infect. Dis. *47*, 1000–1006.

Sundar, S., Chakravarty, J., Agarwal, D., Rai, M., and Murray, H.W. (2010). Single-dose liposomal amphotericin B for visceral leishmaniasis in India. N. Engl. J. Med. *362*, 504–512.

Sundar, S., Sinha, P.K., Rai, M., Verma, D.K., Nawin, K., Alam, S., Chakravarty, J., Vaillant, M., Verma, N., Pandey, K., *et al.* (2011). Comparison of short-course multidrug treatment with standard therapy for visceral leishmaniasis in India: an open-label, non-inferiority, randomised controlled trial. Lancet *377*, 477–486.

Sundar, S., Singh, A., Rai, M., Prajapati, V.K., Singh, A.K., Ostyn, B., Boelaert, M., Dujardin, J.C., and Chakravarty, J. (2012). Efficacy of miltefosine in the treatment of visceral leishmaniasis in India after a decade of use. Clin. Infect. Dis. *55*, 543–550.

Sundar, S., Thakur, B.B., Tandon, A.K., Agrawal, N.R., Mishra, C.P., Mahapatra, T.M., and Singh, V.P. (1994). Clinicoepidemiological study of drug resistance in Indian kala-azar. BMJ *308*, 307.

Thakur, C.P., Kumar, M., Singh, S.K., Sharma, D., Prasad, U.S., Singh, R.S., Dhawan, P.S., and Achari, V. (1984). Comparison of regimens of treatment with sodium stibogluconate in kala-azar. Br. Med. J. (Clin. Res. Ed.) *288*, 895–897.

Thakur, C.P., Kumar, M., Kumar, P., Mishra, B.N., and Pandey, A.K. (1988). Rationalisation of regimens of treatment of kala-azar with sodium stibogluconate in India: a randomised study. Br. Med. J. (Clin. Res. Ed.) *296*, 1557–1561.

Thakur, C.P., Kumar, M., and Pandey, A.K. (1991). Evaluation of efficacy of longer durations of therapy of fresh cases of kala-azar with sodium stibogluconate. Indian J. Med. Res. *93*, 103–110.

Torres, D.C., Adaui, V., Ribeiro-Alves, M., Romero, G.A., Arevalo, J., Cupolillo, E., and Dujardin, J.C. (2010). Targeted gene expression profiling in *Leishmania braziliensis* and *Leishmania guyanensis* parasites isolated from Brazilian patients with different antimonial treatment outcomes. Infect. Genet. Evol. *10*, 727–733.

Urbina, J.A., Cohen, B.E., Perozo, E., and Cornivelli, L. (1987). Spin-labeled amphotericin B: synthesis, characterization, biological and spectroscopic properties. Biochim. Biophys. Acta *897*, 467–473.

Wadhone, P., Maiti, M., Agarwal, R., Kamat, V., Martin, S., and Saha, B. (2009). Miltefosine promotes IFN-γamma-dominated anti-leishmanial immune response. J. Immunol. *182*, 7146–7154.

Weingartner, A., Drobot, B., Herrmann, A., Sanchez-Canete, M.P., Gamarro, F., Castanys, S., and Gunther Pomorski, T. (2010). Disruption of the lipid-transporting LdMT–LdRos3 complex in *Leishmania donovani* affects membrane lipid asymmetry but not host cell invasion. PLoS One *5*, e12443.

WHO (2010). Control of the Leishmaniasis. In Report of a meeting of the WHO Expert Committee on the Control of Leishmaniases (WHO, Geneva, Switzerland).

Wyllie, S., Cunningham, M.L., and Fairlamb, A.H. (2004). Dual action of antimonial drugs on thiol redox metabolism in the human pathogen *Leishmania donovani*. J. Biol. Chem. *279*, 39925–39932.

Wyllie, S., and Fairlamb, A.H. (2006). Differential toxicity of antimonial compounds and their effects on glutathione homeostasis in a human leukaemia monocyte cell line. Biochem. Pharmacol. *71*, 257–267.

Yan, S., Li, F., Ding, K., and Sun, H. (2003). Reduction of pentavalent antimony by trypanothione and formation of a binary and ternary complex of antimony(III) and trypanothione. J. Biol. Inorg. Chem. *8*, 689–697.

Yardley, V., Croft, S.L., De Doncker, S., Dujardin, J.C., Koirala, S., Rijal, S., Miranda, C., Llanos-Cuentas, A., and Chappuis, F. (2005). The sensitivity of clinical isolates of *Leishmania* from Peru and Nepal to miltefosine. Am. J. Trop. Med. Hyg. *73*, 272–275.

Zeisig, R., Rudolf, M., Eue, I., and Arndt, D. (1995). Influence of hexadecylphosphocholine on the release of tumor necrosis factor and nitroxide from peritoneal macrophages *in vitro*. J. Cancer Res. Clin. Oncol. *121*, 69–75.

Zhou, Y., Messier, N., Ouellette, M., Rosen, B.P., and Mukhopadhyay, R. (2004). *Leishmania major* LmACR2 is a pentavalent antimony reductase that confers sensitivity to the drug pentostam. J. Biol. Chem. *279*, 37445–37451.

Zufferey, R., and Mamoun, C.B. (2002). Choline transport in *Leishmania major* promastigotes and its inhibition by choline and phosphocholine analogs. Mol. Biochem. Parasitol. *125*, 127–134.

Index

D

E

N

O

U

V

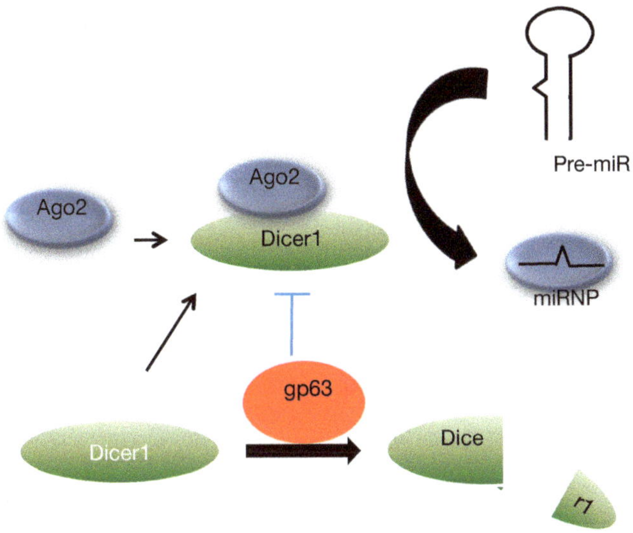

Figure 1.1 *Leishmania* derived exosomes contain GP63 metalloprotease that target Dicer1 in host cells and cleave it. Dicer1 depletion prevents pre-miRNA processing and production of mature miR-122. This subsequently prevents miR-122 activity and resulted lowering of cholesterol production in infected liver.

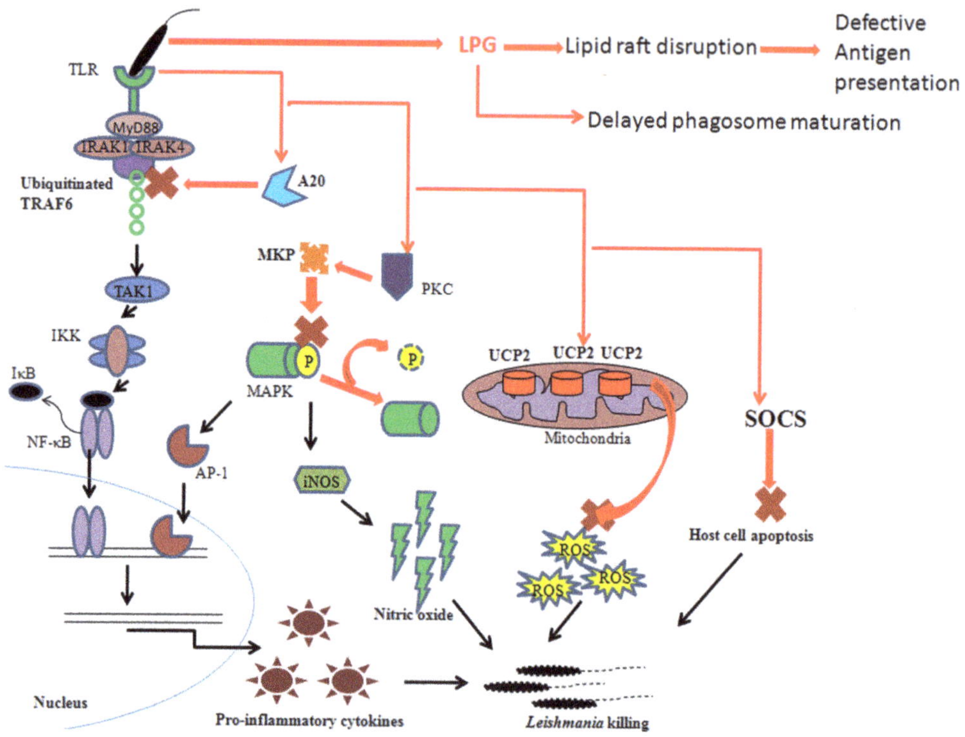

Figure 6.1 Modulation of macrophage immune-defensive signalling cascades by *Leishmania*: attachment of *Leishmania* parasites on macrophage surface results in interfering with TLR-mediated proinflammatory response through de-ubiquitinating enzyme A20 which disrupts the upstream TRAF6-TAK-TAB assembly thereby inhibiting NF-κB activation. This allows the parasite to silently enter macrophages without being recognized by pathogen recognition receptors. The *Leishmania* surface molecule LPG depletes membrane cholesterol hence disrupting lipid rafts and suppressing antigen presentation. This molecule also leads to delayed phagosome maturation thus allowing the promastigotes more time to convert to amastigotes which are better adapted to survive in the harsh acidic environment of the phagolysosome. Upon entering macrophages *Leishmania* activate a number of host molecules like PKC, MAPK and UCP2 which help the pathogen to inhibit harmful microbicidal molecules like nitric oxide and ROS. Further, activation of host SOCS proteins help the parasite to suppress macrophage apoptosis, thereby protecting its niche for survival. Normal macrophage signalling cascades are shown with black arrows. Red arrows indicate the modulations caused by *Leishmania* parasites or their surface molecule LPG.

www.ingramcontent.com/pod-product-compliance
Lightning Source LLC
LaVergne TN
LVHW071337210926
848486LV00003B/166
9781908230522